Contributions

of the

American Entomological Institute

Volume 19, 1982-1983

Contributions

of the

American Entomological Institute

Vol. 19, Parts 1-5, 1982

A REVISION OF THE GENUS *DELOMERISTA*
(HYMENOPTERA: ICHNEUMONIDAE)

Virendra Gupta

Center for Parasitic Hymenoptera
Department of Entomology and Nematology
University of Florida, Gainesville, FL 32611

Delomerista Foerster belongs to the Tribe Theroniini, subfamily Pimplinae (=Ephialtinae). Some authors tend to place it in a different tribe Delomeristini, but as Carlson (1979) has stated, there is really no evidence of its relationship either way, and therefore it is here left within the tribe Theroniini, as it is placed by Townes (1969).

Delomerista is a moderate-sized genus of Holarctic distribution. Species of it parasitise saw-fly cocoons, though in literature there are records of their being parasitic upon lepidopterous hosts as well as on a weevil. More recent studies on the genus are those of Walkley (1960) on the North American species and of Kasparyan (1977) on the European species. Additional species have been discovered from India, Europe, and North America in the collections of Townes, Gupta, Canadian National Collection, and of the U. S. National Museum. Some of the American species appeared to have previously been mixed up. Types and authentic determined specimens of several European, Japanese and American species were studied to elucidate their taxonomic relationships and identities. This resulted in the recognition of several new species which occur sympatrically with common species like *D. novita, borealis,* etc. and were previously mixed up with them, in the separation of *D. japonica* and *D. diprionis* as valid species, and the discovery of *D. indica*, n. sp., from the Himalayan mountains of India. *D. indica* extends the range of the genus to India.

A total of 17 species and 4 subspecies are now recognized, of which 6 taxa are new to science.

Genus DELOMERISTA Foerster

> *Delomerista* Foerster, 1869. Verh. Naturh. Ver. Rheinlande, 25: 164.
>   Type-species: *Pimpla mandibularis* Gravenhorst; designated by
>   Schmiedeknecht, 1888.
> Taxonomy: Walkley, 1960: 362-372. Townes, 1969: 127. Kasparyan,
>   1977: 69-74.
> Biology: Morris, et al., 1937: 360-361.

Body black. Face and clypeus of male yellowish-white (except in *laevis*). Moderately slender, medium-sized species. Face a little convex and minutely to moderately strongly punctate. Scape punctate. Clypeus usually lighter colored than face, flat (convex in *D. laevis*), its apical margin concave and without a median tooth. Mandibular teeth equal. Malar space 0.25 to 1.0 the basal width of mandible, yellow. In males malar space correspondingly shorter than in females. Flagellar segments with linear sensillae which are often absent on the basal one or two segments. Thorax usually subpolished and with minute setiferous punctures. Notauli indicated anteriorly. Epomia present. Prepectal carina present. Propodeum areolated, both longitudinal as well as transverse carinae present, but only the carinae bordering areola and petiolar areas prominent. Costula present or absent, or indistinct and its place indicated by the different sculpture of first and second lateral areas; second lateral area often somewhat depressed and rough. Apical transverse carina almost circular and more prominent. Tarsal claws simple, of moderate size, without an enlarged bristle having a spatulate tip (cf. *Theronia*), though a long slender bristle present on each claw. Nervellus intercepted variously, at, above or below the middle (even varying within the same species). Abdomen strongly granulose, granuloso-mat, or coriaceous with small scattered hairs, or closely punctate (as in two species *pfankuchi* and *kusuoi*). First tergite with a weak to strong dorsal curvature. Median carinae usually ending at this dorsal hump. Lateral carina passing just above the spiracle and often weak at this point. Ovipositor moderately compressed, moderately long (0.7 to 1.0 the length of abdomen), or very long (2.0 the abdomen in *longicauda*), its tip not sinuate but variously modified, either nodiform, or upper valve thick, or upper valve occupying a greater part of the depth of ovipositor tip (figs. 47-49, 51-53). The apical slope of the upper valve also variable and of diagnostic value for species.

HOST ASSOCIATIONS

Species of *Delomerista* seem to be chiefly ectoparasitic on pupating sawfly larvae within the cocoons. Host records of many species are lacking, while some species have been reported ectoparasitic upon a variety of lepidopterous larvae, and even a weevil larva (*Mononychus vulpeculus* for *D. novita*). Aubert (1969), quoting Kolomiets, 1962) even mentions *D. mandibularis* as being hyperparasitic upon larvae of *Rhogas dendrolimi* Matsumura, through *Dendrolimus sibiricus*. Most authors, however, believe that all hosts other than sawfly hosts are probably erroneous, as they are not confirmed by subsequent rearing records. For the same reason, Walkley (1960) stated that it is unwise to disregard the host records from Lepidoptera and Coleoptera until further rearings are done.

LARVAL MORPHOLOGY
(Figs. 63-65)

The Larvae of *Delomerista* have a large tooth-like projection at the junction of the blade and base of mandible. According to Short (1978) this character is found among other genera of the Theronini,in the Rhyssini, certain Ephialtini, and in the Orthopelmatinae. The epistoma is incomplete, hypostoma and hypostomal spur well sclerotized, lateral parts of labial sclerite slender and ventral part broad, about twice as deep as the width of lateral part. Maxillary and labial palps flattened and disc-shaped. Labral sclerite complete. Mandible with a broad base and its blade bifurcated—the longer part with dorsal and ventral teeth and smaller part toothless. Maxillary and labial palps with two sensillae each. Closing apparatus of spiracle not adjoining atrium. Antenna papiliform. Length of skin setae equal to the larger toothed part of mandibular blade.

Short does not separate out Delomeristini from Theroniini.

SPECIES

A list of species so far known, together with their host associations as far as known, is given below, including the new taxa described in the text. Species no longer valid, nor belonging to *Delomerista* are placed within brackets [ ]. Only valid species are numbered.

Four species groups are recognized, chiefly on the nature of the ovipositor tip. They are the Novita Group, the Japonica Group, the Frigida Group, and the Mandibularis Group. They are characterized and keyed in the key that follows the species list. The figures of the ovipositors should help identify the species-groups as well as the species.

[*Ephialtes albicinctus* Desvignes, 1862]. Type ♂, ?Great Britain (London).
　　Name preoccupied. A synonym of *Delomerista mandibularis* Graven-
　　horst cf. Fitton (1978).
　　?Great Britain. Host: Unknown.

[*Pimpla bilineata* Brullé, 1846]. Type ♀, Algeria (Paris).
　　Oehlke (1967) synonymized it under *Coccygomimus contemplator*
　　(Mueller) with a query, but Aubert (1969) placed it under *Delomerista*.
　　The latter situation followed by Constantineanu & Pisica (1977).
　　According to original description, apparently a species of *Coccygomi-
　　mus*.
　　Algeria. Host: Unknown.

1. *Delomerista borealis* Walkley, 1960. Type ♀, N.W.T. (Washington).
　　Also reported from Alaska and Quebec by Walkley.
　　Aubert (1969) placed it under *D. mandibularis* with a query. Kasparyan
　　(1977) reported it from the European USSR.
　　A valid species belonging to the Novita Group.
　　Occurring in Northwestern parts of North America and European USSR.
　　Host: Unknown.

[*Ephialtes desvignesii* Marshall, 1870]. New name for *E. albicinctus*
　　Desvignes, 1862.
　　Oehlke (1967) placed it under *Delomerista*. Fitton (1978) placed it as a
　　synonym of *D. mandibularis* (Gravenhorst).
　　England. Host: Unknown.

2. *Delomerista diprionis* Cushman, 1939. Type ♀, Ontario, Canada
　　(Washington).
　　Walkley (1960) synonymized it under *D. japonica* Cushman.
　　Hereby considered as a valid species, belonging to the Japonica Group.
　　Nearctic. Hosts: Several species of *Diprion*, *Neodriprion* and *Gilpina*,
　　including *D. similis*, *N. lecontei*, *N. nanulus nanulus*, *N. pratti
　　banksianae*, *N. sertifer*, *N. tsugae*, *Gilpina frutetorum*, and
　　G. hercyniae.

3. *Delomerista excavata* Ulbricht, 1913. Type ♀, Germany (?).
　　Aubert placed it under *mandibularis* with a query, while other authors
　　are not certain about its identity.

The type could not be located and therefore its taxonomic identity could not be ascertained.
Germany. Host: Unknown.

4. *Delomerista frigida* Kasparyan, 1977. Type ♀, European USSR (Leningrad).
Belongs to the Frigida Group.
Eastern Palaearctic. Host: Unknown.

[*Delomerista gelida* Walkley, 1960]. Type ♀, N.W.T. (Washington).
Hereby synonymized under *D. mandibularis* Gravenhorst. N. SYN.
Nearctic. Host: Unknown.

5. *Delomerista indica* Gupta, new species. Type ♀, Himalaya: Dalhousie (Gupta).
Belongs to the Japonica Group.
India (Himalayan mountains). Host: Unknown.

6. *Delomerista japonica* Cushman, 1937. Type ♀, Japan (Washington).
Walkley (1960) synonymized *D. diprionis* Cushman under it, but both species are here considered distinct but related, belonging to the Japonica Group.
Japan. Host: *Diprion nipponicus*.

7. *Delomerista kusuoi* Uchida & Momoi, 1957. Type ♀, Japan (Sapporo).
A valid species related to *D. pfankuchi* and belonging to the Novita Group.
Japan. Host: Unknown.

[*Pimpla laevifrons* Thomson, 1877]. Type ♂, Germany (Lund).
Townes (1944) synonymized it under *D. texana* (Cresson), while Oehlke (1966) synonymized *texana* under *laevis* (Gravenhorst).
Aubert (1969) considered it distinct and synonymized *strandi* Ulbricht under it. Constantineanu and Pisica (1977) treated it as a separate species. Kasparyan (1977) studied the lectotypes of *laevis* and *laevifrons* and confirmed the synonymy of this species. *D. strandi* is not a synonym of it, though a cotype of it is conspecific with it. The lectotype of *strandi* is different.
Europe. Host: Unknown.

8. *Pimpla laevis* Gravenhorst, 1829. Type ♀, Italy (Wroclaw).
= *laevifrons* Thomson.
= *texana* Cresson (cf. Oehlke, 1966).
A cotype of *D. strandi* belongs here, which is different from the lectotype of *strandi* designated by Oehlke, 1967.
A valid species, belonging to the Mandibularis Group.
Holarctic. Hosts: *Rhyacionia buoliana* (Lep.: Tortricidae) in Europe (cf. Aubert). Carlson (1979) does not mention any host in North America.

9. *Delomerista lepteces* Walkley, 1960. Type ♀, Alaska (Washington).
Belongs to the Novita Group.
Nearctic. Host: Unknown.

10. *Delomerista longicauda* Kasparyan, 1973. Type ♀, Russia (Leningrad).
A species belonging to the Mandibularis Group. A subspecies of it
(*longicauda americana*) is described from Northern Nearctic.
Holarctic. Host: Unknown.

11. *Pimpla mandibularis* Gravenhorst, 1829. Lectotype ♀, Poland (Wroclaw).
= *albicinctus* Desvignes (cf. Fitton, 1978)
= *desvignesii* Marshall (cf. Fitton, 1978)
= *gelida* Walkley (new synonym).
Aubert (1969) placed *excavata* Ulbricht as a synonym of it with a query,
but Kasparyan (1977) did not confirm this synonymy. He also placed
D. borealis Walkley as a doubtful synonym of it, apparently after the
suggestion of Walkley herself, which was because of misdetermined
specimens of *mandibularis* in Washington. In Washington there are
specimens from Europe determined as *mandibularis* by Roman and
Heinrich, which are specifically identical with the specimens of *borealis*
Walkley, but not with specimens determined as *mandibularis* by Perkins
and Kasparyan.
According to Oehlke (1967), Perkins selected a lectotype in 1936 and so
labelled the specimen, which was subsequently designated lectotype by
Oehlke (1967). Kasparyan (1977) followed this interpretation. Oehlke
mentions Breslau, Poland as the lectotype locality, while Kasparyan
mentions Warmbrunn, Poland!
A valid species belonging to the Mandibularis Group.
Hosts: As mentioned by Aubert from Europe: *Strongylogaster* sp.,
Euura amerinae (Hym.: Tenthredinidae).

12. *Delomerista masoni* Gupta, new species. Type ♀, Michigan (Townes).
Belongs to the Mandibularis Group.
Nearctic. Host: Unknown.

13. *Pimpla novita* Cresson, 1870. Type ♀, Mass. (Philadelphia).
Kasparyan (1977) reported it from European USSR.
A valid species belonging to the Novita Group. A new subspecies of it
(*novita europa*) is described from Poland and Kasparyan's specimens
may belong to the European subspecies.
Holarctic. Hosts: *Mononychus vulpeculus* (Curculionidae), *Acrobasis
rubrifasciella*, *Acrobasis* spp., *Exartema olivaceanum*, *Eublemma
minima* (= *Thalpochares carmelita*) (Lepidoptera). *Macremphytes* sp.,
Diprion similis (Hymenoptera).

14. *Delomerista pfankuchi* Brauns, 1905. Type ♀, Bremen (Berlin).
Troctocerus unicolor Hedwig, 1959. Syn. by Horstmann, 1981.
Europe. Russia. Walkley (1960) reported it as "Probably America",
but it does not occur in the Nearctic Region.
Valid species belonging to Novita Group.
Palaearctic. Hosts: *Chionodes tragicella*, *Psyche viciella*, *Talaeporia
tubolosa*, and *Diplodoma marginepunctella* (Lepidoptera). *Diprion pini*
(Hymenoptera).

15. *Pimpla strandi* Ulbricht, 1911. Lectotype ♀, Norvegia (Berlin).
Lectotype designated by Oehlke, 1967.
Another cotype conspecific with *laevifrons* and so labelled by Aubert,

led him to consider the two synonymous.
A valid species, also occurring in the Nearctic Region. Belongs to the
Mandibularis Group.
Holarctic. Host: Unknown.

[*Pimpla texana* Cresson, 1870]. Type ♀, Texas (Philadelphia).
P. *laevifrons* considered as a synonym of it by Townes (1944), while
Oehlke (1966) synonymized it under *laevis* (Gravenhorst).
A synonym of *Delomerista laevis*.
Nearctic. Host: Unknown.

16. *Delomerista townesorum* Gupta, new species. Type ♀, Michigan (Townes).
Belongs to the Frigida Group.
Nearctic. Host: Unknown.

[*Troctocerus unicolor* Hedwig, 1959]. Holotype ♀, ''Paitzkofen b. Straubing.
A junior synonym of *Delomerista pfankuchi*, vide Horstmann, 1981.

17. *Delomerista wakleyae* Gupta, new species. Type ♀, Alaska (Townes).
Belongs to the Frigida Group.
Nearctic. Host: Unknown.

MALES

The matching of the males with the females has posed a problem, as most
males available look alike and apparently males of several species are unknown.
Attempts have been made to identify the males in association with the females.
This resulted in certain broad groupings. These groupings are described below
and should aid in the identification of the males.

I. Males with face black, yellow only on sides. Malar space long, 0.8 to
1.0 the basal width of mandible. Clypeus convex. Abdominal tergites finely
granular. Only one species, *Delomerista laevis*, belongs here.
All other males so far as are known, have face yellowish-white, or white,
malar space 0.2 to 0.33 the basal width of mandible, and clypeus flat.

II. Areola small, broadly triangular due to the apical transverse carina
being close to the middle of propodeum. Propodeum short, sloping from base
to apex. Apical transverse carina often the strongest and very convexly arched.
This group includes the males of the Japonica Group, *D. japonica, diprionis*,
and *indica*, all of which have yellow malar space, tegula and ventral aspects of
scape and pedicel. *D. diprionis* has the strongest apical transverse carina,
which even encroaches the basal half of propodeum.

III. Propodeum smooth and shiny, with areola horse-shoe shaped, not
elongate. Costula distinct. Propodeal carinae rather strong and sharp.
Malar space, scape and pedicel ventrally, and tegula yellow. Flagellum
brownish. Propodeum with a short dorsal face. Hind femur usually wholly
brownish-yellow. Includes *D. masoni* of the Mandibularis Group.

IV. Areola elongate. Costula indistinct to absent. Propodeum long,
smoother, or granulose laterally. Apical transverse carina of propodeum

usually in apical 0.3 of propodeum. Dorsal face of propodeum long and not
sloping. Propodeum sloping only apicad of transverse carina. The rest of the
known males belong here.

A. Tegula black. Scape and pedicel black. *Delomerista borealis* belongs
here.

B. Tegula yellow. Malar space black. Three species, *D. novita, longicauda* and *lepteces* belong here. In *D. novita*, the scape and pedicel are yellow
ventrally, malar space short, 0.2 the basal width of mandible, nervellus intercepted at lower 0.4, face as long as wide, fore and middle coxae generally
white, and hind femur short, compressed and largely reddish. In *D. longicauda americana* (male of nominate subspecies unknown), the scape and pedicel
are black, malar space 0.33 the basal width of mandible, nervellus intercepted
at lower 0.3, face wider, about 1.5 as wide as long, shiny, fore coxa alone
white, and hind femur normal, largely to wholly blackish or blackish-brown.
D. lepteces is very much like *D. longicauda,* but malar space 0.3 as long as
basal width of mandible, face 1.3 as wide as long, dull, and hind femur is
largely reddish-brown or reddish, except in apical 0.3.

C. Tegula yellow. Malar space yellow. *Delomerista townesorum* and
D. mandibularis belong here. Both these species show variations in the coloration of scape, pedicel, and in the markings on the femora.

D. townesorum has a rugulose, rectangular face, hind femur with a black
apical ring, and nervellus intercepted at its lower 0.4 to 0.45 (appears usually
in the middle).

D. mandibularis generally has a smoother, squarish face, hind femur
usually wholly orange-brown, and nervellus intercepted in the lower 0.3 to
0.35. The scape and pedicel are usually yellow ventrally. There are variations, however. Some males have a rugulose face and scape appears partly to
largely black. A large number of males from British Columbia in Townes
Collection are tentatively placed here. They may be different.

V. Abdomen rugoso-punctate. The male of *D. pfankuchi* has a rugoso-
punctate abdomen, like that in female (all other males have coriaceous
abdomen), scape and pedicel yellow ventrally, flagellum brownish and hind
femur wholly orange colored.

In the key that follows, only the females have been taken into consideration.
D. excavata Ulbricht is unknown to me and therefore not included in the key,
nor in the text.

KEY TO THE SPECIES GROUPS AND SPECIES OF DELOMERISTA

(Females only. *Delomerista excavata* Ulbricht excluded)

1. Ovipositor with a distinct subapical node whence it tapers gradually to a
       point (figs. 3, 6, 9). Ventral margin of upper valve of ovipositor con-
       cave subapically where the lower valve widened and occupying the
       greater part of the depth of the ovipositor. Basal ridges on lower valve
       strongly inclined and rest of the ridges somewhat sinuate. Basal two
       flagellar segments with several rows of sensillae (cf. Frigida Group)
       (figs. 12, 15, 18, 19, 20). (In one species, *lepteces*, ovipositor finely
       tapered and needle-like, fig. 16). A. The Novita Group. 4

Ovipositor tip not nodiform, a little bulbous or more conspicuously widened (often a few low ripples seen in profile on the upper valve, figs. 23, 28). Upper valve heavier or wider preapically than the lower valve and occupying the greater part of the preapical depth of ovipositor. Ventral margin of upper valve convex. Ridges on lower valve (except the basal 3-4) almost vertical to reclining. 2

2. Upper valve of ovipositor not unusually heavy or widened, with an even gradual slope and with one or two ripple-like preapical formations (figs. 23, 28). Ridges on lower valve vertical (except the basal ones). First flagellar segment almost devoid of sensillae or only one or two present. II segment with fewer sensillae. (This group is sympatric with the Novita Group and the American species of it have been mixed up with the species of the latter. The ovipositor looks somewhat similar, but is definitely not nodiform and is a little bulbous in outline and sensillae are virtually absent). Propodeum with a flat dorsal face and then sloping only in apical half. B. The Frigida Group. 8

Upper valve of ovipositor heavier and widened preapically, bulbous or with a short abrupt apical slope (figs. 33, 38, 39, 44, 47-49, 51-53). Apical ridges on lower valve vertical to reclining. Sensillae on basal two flagellar segments numerous or sometimes fewer in number. Propodeum sloping from base to apex, though sometimes basal part convex (figs. 40, 50, 54). 3

3. Upper valve of ovipositor bulbous in outline, widened preapically and then abruptly sloping (figs. 33, 38, 39, 44), with faint ripples. Apical ridges on lower valve reclining while the basal ones inclining. The two valves meeting almost in a straight line: this line at mid height of ovipositor at base and usually faint. Basal two flagellar segments with numerous sensillae (figs. 31, 32, 36, 37, 45, 46). Propodeum in profile view with an even slope from base to apex (fig. 40). C. The Japonica Group. 10

Upper valve of ovipositor rather heavy, and occupying the greater part of the depth of ovipositor (figs. 47-49, 51-53), and without or with only faint ripples. Apical slope abrupt, short, except in *longicauda* (fig. 53), where ovipositor 2.0 as long as abdomen. Ovipositor usually arched. Ridges on lower valve vertical to slightly inclined (basal 3 ridges strongly inclined). The two valves meeting along an arched line, which is conspicuous throughout the length of ovipositor and is above the mid-height at base of ovipositor. Basal two flagellar segments with fewer sensillae (fig. 59-60). Propodeum in profile view (figs. 50, 54) strongly convex basally and then abruptly sloping to apex. D. The Mandibularis Group. 12

(The Novita Group)

4. Abdomen closely punctate (figs. 5, 11). Apical margins of tergites banded with yellow or red. Malar space 0.25 the basal width of mandible. Propodeum subpolished and with a few scattered punctures to punctate (figs. 4, 10). Costula faint to distinct.

Abdomen granulose to granuloso-coriaceous (figs. 2, 8, 14), without apical pale bands on tergites. Malar space 0.3 to 0.5 the basal width of mandible. Propodeum polished to subpolished. Costula absent. 6

5. Propodeum (fig. 4) subpolished, with a few scattered punctures laterally.
    Costula very faintly indicated. Abdominal tergites with yellow apical
    bands. Nervellus intercepted below the middle. Palaearctic.
    4. <u>pfankuchi</u> Brauns (p. 15)
    Propodeum (fig. 10) punctate, second lateral area rugoso-punctate.
    Costula distinct. Abdominal tergites apically banded with red and punc-
    tation coarser than in *pfankuchi* (fig. 11). Nervellus intercepted at its
    middle. Japan. . . . . 5. <u>kusuoi</u> Uchida & Momoi (p. 16)

6. Ovipositor finely tapered apically, needle-like (fig. 16), nodus rather
    indistinct. Malar space 0.45-0.5 the basal width of mandible.
    Propodeal areola (fig. 13) horse-shoe shaped. Nearctic.
    3. <u>lepteces</u> Walkley (p. 15)
    Ovipositor typical for the group with a distinct node and then long tapering
    (figs. 3,9). Malar space 0.3-0.45 the basal width of mandible.
    Propodeal areola usually elongate, more strongly defined and pentagon-
    al (figs. 1,7). 7

7. Tegula yellow. Hind tibia yellow ventrally or at base. Hind tarsus basally
    yellow. Abdomen granuloso-coriaceous (fig. 2). Malar space 0.3 the
    basal width of mandible. Nervellus intercepted at its lower 0.3-0.4.
    Holarctic. . . . . 1. <u>novita</u> (Cresson) (p. 11)
    Tegula black. Hind tibia and tarsus wholly black. Abdomen granulose
    (fig. 8). Malar space 0.4-0.45 the basal width of mandible. Nervellus
    intercepted at its lower 0.4-0.45. Holarctic.
    2. <u>borealis</u> Walkley (p. 14)

(The Frigida Group)

8. Face granuloso-punctate medially. Tegula black. Tergite I convex,
    granulose laterally, its dorsal and lateral carinae weak and confined only
    along the basal declivity. Fore leg wholly reddish-brown. Europe.
    6. <u>frigida</u> Kasparyan (p. 18)
    Face subpolished, with or without fine punctures. Tegula black or yellow.
    Tergite I flattened laterally and more rugulose rather than granulose,
    its median and lateral carinae distinct. Fore leg with yellow ventral
    marks. North America. 9

9. Tegula black. Propodeum rugulose in pleural area, second lateral area
    and in petiolar area. Legs dark reddish-brown with fore leg brownish
    (except rarely). Malar space 0.5 the basal width of mandible. Ovi-
    positor long, a little up-curved and longer than abdomen. Abdomen
    granulose. Nearctic. . . . . 7. <u>walkleyae</u>, n. sp. (p. 18)
    Tegula yellow. Propodeum smoother to subpolished, particularly in
    lateral and petiolar areas. Legs light yellowish-brown. Malar space
    0.3-0.4 the basal width of mandible. Ovipositor as long as abdomen,
    usually straight. Abdomen granuloso-mat (finer than in *walkleyae*).
    Nearctic. . . . . 8. <u>townesorum</u>, n. sp. (p. 19)

(The Japonica Group)

10. Ovipositor tip a little swollen preapically and then abruptly sloping (short
taper, fig. 33, 39).   First abdominal tergite thinner in profile view,
its dorsomedian carinae angled at 30⁰ and with a conspicuous flange at
its junction with lateral carina. Basal declivity smoother and shiny.
Postpetiole flat and rugulose. Malar space 0.33 to 0.45 the basal
width of mandible. Eastern Palaearctic: Japan, USSR.
                                   10. japonica Cushman (p. 22)
     Ovipositor tip in profile view more parallel-sided, with a small convex
slope (figs. 38, 44). First tergite convexly arched, its median carinae
moderately raised and making an angle of 45⁰ with the horizontal axis,
not conspicuously flanged at base. Malar space 0.5-0.7 the basal
width of mandible. 11

11. Malar space 0.6-0.7 the basal width of mandible. Nervellus intercepted
below the middle (lower 0.33). Junction of median and lateral carinae
of first tergite not flanged, the lateral carina angled at spiracle and
usually erased there or apically. Basal declivity of first tergite dull.
Ovipositor tip as in fig. 38. Nearctic. . 9. diprionis Cushman (p. 20)
     Malar space 0.5 the basal width of mandible. Nervellus usually inter-
cepted at the middle (some exhibit variations 0.5 ± 0.15). Junction of
median and lateral carinae of first tergites moderately flanged, the
lateral carina complete and strong, arched at spiracle. Basal declivity
of first tergite rough. Ovipositor as in fig. 44. India.
                                   11. indica, n. sp. (p. 23)

(The Mandibularis Group)

12. Malar space about as long as or longer than basal width of mandible (0.8 in
♂). Clypeus convex. Face smooth (face of male black). Propodeum
largely granulose. Abdominal tergites finely granulose (shagreened in
♂). Ovipositor tip as in fig. 52. Holarctic.
                                   15. laevis (Gravenhorst) (p. 28)
     Malar space 0.4-0.6 the basal width of mandible. Clypeus flat. Face with
minute punctures, subpolished. 13

13. Ovipositor very long, longer than the body and about 2.0 the length of
abdomen. Ovipositor tip upcurved, narrowed preapically (fig. 53).
Holarctic. . . . . 16. longicauda Kasparyan (p. 29)
     Ovipositor shorter, only about as long as abdomen or shorter, straight
preapically and not narrowed. 14

14. Ovipositor almost parallel-sided in profile view, uniformly slightly arched
upwards, particularly near tip, not conspicuously widened preapically
(fig. 48, 49). First flagellar segment almost devoid of sensillae
(fig. 59) (one or two in a row sometimes present). Ovipositor as long
as abdomen. Holarctic. . .   12. mandibularis (Gravenhorst) (p. 24)
     Ovipositor rather widened preapically (about 1: 1.5) (figs. 47, 51, 52) and
almost straight (or a little downcurved). First flagellar segment with
several rows of sensillae. Ovipositor shorter than abdomen. . . . 15

15.   Tegula black. Ovipositor very stout (fig. 47), slightly arched upwards.
        Malar space 0.5 to 0.6 the basal width of mandible. Areola pentagonal
        to a little elongate. Costula incomplete to indistinct. Second lateral
        area of propodeum rugose and depressed. Head as wide as high.
        Inner eye orbits shallowly indented. Holarctic.
                                            13. strandi (Ulbricht) (p. 26)
      Tegula yellow. Ovipositor moderately stout with a weak median bend down-
        wards. Malar space 0.4-0.5 the basal width of mandible. Areola
        semicircularly rounded, broader (sometimes elongate). Costula com-
        plete and distinct. Second lateral area of propodeum shiny (sometimes
        rugulose), generally not depressed. Head wider than high. Inner eye
        orbits moderately indented. Nearctic. . . 14. masoni, n. sp. (p. 27)

                          A. THE NOVITA GROUP

      Flagellum with sensillae, basal two flagellar segments with numerous
sensillae which are sometimes fewer in number.
      Propodeum long, with a longer dorsal face. Propodeum dorsally smoother.
Areola elongate. Costula absent or faintly indicated (distinct in *kusuoi*, where
propodeum is somewhat punctate). Ovipositor tip nodiform, seen in profile
upper valve with a slight protuberance subapically whence it evenly tapers to
an apical point. At nodus, depth of upper and lower valves about equal; beyond
this point lower valve slightly encroaching upon the upper valve and lower
margin of upper valve concave. Ridges on lower valve (except the basal ones)
vertical and slightly sinuate. Ovipositor 0.8 to 1.2 as long as abdomen.
      This group includes five species: *D. pfankuchi* from Europe, *D. kusuoi*
from Japan, *D. lepteces* from North America, and *D. borealis* and *D. novita*
from Europe as well as North America.

1. DELOMERISTA NOVITA (Cresson)

      Female: Face convex, punctate in the middle. Clypeus flat. Malar space
0.3-0.33 the basal width of mandible. Head a little wider than long. Eye
moderately strongly notched a little above antennal socket. Frons smooth.
Vertex with minute setiferous punctures. Interocellar distance 0.6-0.7 the
ocellocular distance. First flagellar segment with linear sensillae. Second
and third segments also with linear sensillae. Sensillae on basal two flagellar
segments in *novita europa* fewer in number. Mesoscutum subpolished and
more hairy than other parts. Pronotum, mesopleurum and metapleurum shiny,
polished and with scattered hairs. Propodeum smooth dorsally and a little
punctate laterally and in pleural area, a little rough in the petiolar area.
Areola fully formed, but basal area and costula absent (fig. 1). Apical transverse carina circular and strong. Dorsal face of propodeum appears flat and
about as long as its apical slope, which is gradual. Nervellus intercepted at
its lower 0.3-0.45. Abdomen granuloso-coriaceous (fig. 2). Ovipositor long,
straight, parallel-sided, about 0.5 as long as body and 0.8 as long as abdomen,
its tip with a subapical node and thence evenly and gradually tapered to a point.
Lower valve with vertical riges (fig. 3), with basal ridges inclined.
      Male: Similar to the female but face smooth. Pleural area of propodeum
often rugulose. Propodeum dorsally flatter and more elongate and often with
minute punctures.
      Two subspecies are recognized: *Delomerista novita novita* from North

America and *D. novita europa* from Europe. They differ as follows:

1. Sensillae on basal flagellar segments numerous. Areola generally hexagonal
    and a little longer than wide (though in reared specimens variations are
    seen from elongate ∩-shaped to somewhat broadly triangular). Legs
    reddish-brown with fuscous and pale marks. Fore and middle tarsi
    fuscous, their tibiae particularly the middle tibia, ventrally yellow and
    with apical fuscous marks. Hind tibia largely yellowish-white ventrally.
    Nearctic. . . . . 1a. novita novita (Cresson) (p. 12)
    Sensillae on basal two flagellar segments fewer in number. Areola usually
    elongate, narrowed basally, and somewhat pentagonal. Legs more
    uniformly brownish. Fore and middle tibiae and tarsi without fuscous
    or pale marks. Hind tibia wholly brownish-black and only with a basal
    white ring. Europe. . . . . . . 2b. novita europa, n. subsp. (p. 13)

1a. DELOMERISTA NOVITA NOVITA (Cresson) (figs. 1-3)

   Pimpla novita Cresson, 1870. Trans. Amer. Ent. Soc., 3: 146. ♀. des.
      Type ♀, Massachusetts (Philadelphia). Homotype examined in Townes
      Coll., 1980.
   Taxonomy: Walkley, 1960: 373. Kasparyan, 1977: 74. Carlson, 1979:
      349.
   Morphology: Finlayson, 1967: 1247 (larva). Short, 1978: 25,174 (larva).

   Female: Basal flagellar segments with numerous sensillae. Areola generally hexagonal and a little longer than wide, but areola variable, particularly in
reared specimens, from elongate ∩-shaped to somewhat broad-triangular.
   Black. Mandible, malar space, palpi, hind corner of pronotum, tegula,
and fore trochanter and trochantellus, yellowish-white. All coxae, femora,
and middle and hind trochanters and trochantellus   reddish-brown, with fore
coxa often dark brown. Fore and middle tibiae reddish-brown, with their ventral surface yellowish-white. Hind tibia with a yellowish-white annulus at
base, black dorsally, and yellowish-white ventrally, which mark variable, but
not extending to apex. Fore tarsus brown and middle and hind tarsi black.
Base of hind basitarsus white. Apical 0.22 of hind femur black. Sometimes
legs more uniformly reddish-brown, but hind tibia and tarsus exhibit the
characteristic pattern.
   Male: Malar space 0.2 to 0.25 the basal width of mandible and black.
Face and clypeus white. Face subpolished, with scattered punctures. Scape
and pedicel white ventrally. Tegula yellowish-white. Fore and middle coxae
and trochanters white. Middle femur ventrally and middle tibia often wholly
white. Hind femur short, reddish-brown with a blackish mark in apical 0.25
to 0.4 and merging with the reddish color of femur. Sometimes fore leg darker
with coxa brownish.
   One female labelled, *"Delomerista gracilis* Cushman, Type", "No. 19179-
USNM", from Santa Cruz, California, is a variant specimen of *novita novita*
with legs more reddish-brown, areola triangular, dorsal face of propodeum
shorter, and apical transverse carina of propodeum more strongly arched.
Some reared specimens of *novita novita* from Wisconsin also exhibit this sort
of propodeum and a flatter first tergite.
   Length: ♀, 9-13 mm. Fore wing 7-9.5 mm. Ovipositor 5-7 mm. ♂,
7-11 mm. Fore wing 4.5-8 mm.
   Specimens: 29♂ males and 59♀, from the Nearctic Region: *British*

Columbia (Hudson Hope, Robson, and Stone Mt. Park); *California* (Santa Cruz Mts.); *Michigan* (Ann Arbor, Brevort, Emmet Co., Huron Mts., Houghton Co., Iron River, Marquette Co., Schoolcraft Co., Roscommon Co.); *Minnesota* (Stacy and Chisago Co.); *New Hampshire* (Mt. Madison, Pinkham Notch, and White Mts.); *New Jersey* (High Point State Park); *New York* (Farmingdale, Hancock, L. Sebago, Moss Lake, Oneota); *North Carolina* (Linville Falls); *Ontario* (Bells Corners, Chaffeys Locks, Constance Bay, Cumberland, and Ottawa); *Oregon* (Mt. Hood and Selma); *Pennsylvania* (Bald Eagle State Park, Hamilton, Spring Brook), *Quebec* (Gracefield, Knowlton, Lacoste, Lac Mondor, Cap Rouge, Gracefield, Ste Flore, and Sweetsburg); *South Carolina* (Cleveland); *Vermont* (Laurel Lake); *Virginia* (Mountain Lake); *West Virginia* (Cranberry Gls.); *Wisconsin* (Gibson Lake, Madison, and Polk Co.); and *Yukon Territory* (Rampart House).

Specimens reported by Walkley (1960) from Saskatoona, Sask., Robson, BC, and Golden Lake, Ontario, are now referred to *Delomerista townesorum*, n. sp. Specimens reported by her from Maine, Maryland, Massachusetts, Connecticut and Washington could not be examined.

Hosts: The host records on specimens seen are from *Diprion similis* at Gibson Lake, Wisconsin, Stacy, Minnesota, and at Linville Falls, N.C., the adults occur mostly during August. Walkley reported other host records (quoting earlier authors), like *Mononyhus vulpeculus*, *Acrobasis rubrifasciella*, *Exartema oliva* and *Eublemma minima*, stating that the last two records are doubtful and the collector and identifier are not given. Finlayson (1967) records this species from *Acrobasis* (Phycitidae).

Distribution: This species is widely distributed in Eastern North America. It ranges northwestwards to British Columbia and Yukon Territory. In these areas it is partially replaced by a closely related species, *D. borealis* Walkley.

1b. DELOMERISTA NOVITA EUROPA, n. subsp.

This subspecies is extremely close to the nominate subspecies from the Nearctic Region, but the hind tibia and tarsus are black except for whitish basal rings, in which character it somewhat approaches *Delomerista borealis*.

The distinguishing characters are: Sensillae on basal two flagellar segments fewer in number. Propodeum smoother dorsally. Areola elongate, narrowed basally, almost pentagonal in outline. Nervellus intercepted in lower 0.3-0.45. Hind corner of pronotum yellow. Tegula yellow, with a brownish spot near wing base. All coxae, trochanters and femora brownish, shiny. Fore and middle tibiae and tarsi brown. Hind femur only faintly infuscate apically. Hind tibia and tarsus wholly black except for narrow basal yellowish-white rings.

Male has the usual characters like face, scape and pedicel ventrally, fore and middle leg wholly, and hind tibia ventrally, yellowish-white. Hind femur black only apically. Malar space 0.2-0.25 the basal width of mandible, black. Nervellus intercepted at lower 0.25-0.3.

This subspecies has been confused with *D. mandibularis* Gravenhorst, according to misdeterminations of Roman and others. Specimens reported as *novita* by Kasparyan (1977) probably belong to this subspecies. His specimens were not checked.

Length: ♀, 11-13 mm. Fore wing 9-11 mm. Ovipositor 6-7 mm. ♂ 9-11 mm. Fore wing 6-8 mm.

Holotype: ♀, *Germany*: Bayr. Wald Waldmunchen, 800 m, VII-1948,

G. Heinrich (labelled *D. mandibularis*) (TOWNES).

Paratypes (2♂, 9♀): *Germany:* Hahnheide b. Trittau, near Hamburg, 1♀, VIII-1945, G. Heinrich (Townes). Fürstenberg i.M., Fr. W. Konow, 4♀, Baker Collection (one det. as *mandibularis*) (Washington). Austria: Osterich-Tal. Allgau, 1100 m, 1♀, 31-VII-1949, G. Heinrich (det. *mandibularis*) (Townes) Poland: 1♀, *ex Diprion similis* (abnormal, propodeum and abdomen banded with red) (Washington). Sweden: Skåne, Trap 2, 1♀, VII-1969, B. Svensson (Townes).Messaure, 1♂, 27-VI-1971, K. Muller; Loderupstrandbad, 1♂, 9-VI-1961, M. Townes & C. West (Townes). *England:* Oxford, Bagley Woods ♀, 25-IX-1960, H. K. Townes (Townes).

Host: Diprion similis.

Distribution: Germany, Austria, Poland, Sweden, England, and probably U.S.S.R.

2. DELOMERISTA BOREALIS Walkley (figs. 7-9)

Delomerista borealis Walkley, 1960. *In* Townes and Townes: Bull.
    U. S. Natl. Mus., 216(2): 370. ♀. Type ♀, NWT: Long 141° W,
    Lat 69° 20´N (Border of Alaska and Yukon Territory) (Washington).
    Examined in 1980. Paratypes from Alaska, Yukon Territory, North-
    western Territory, Quebec, and Colorado.
Taxonomy: Kasparyan, 1977: 71.

This species is extremely similar to *D. novita* and could well be considered as a subspecies of the latter, but for distributional overlaps and a few structural differences, as follows:

Female: First flagellar segment with fewer rows of sensillae. Face smoother and with sparser punctures. Inner margin of eye only slightly indented opposite antennal socket. Frons more deeply excavated. Malar space 0.4 to 0.45 the basal width of mandible. Nervellus intercepted at its lower 0.4 to 0.45. Propodeum subpolished. Areola more rounded basally and elongate than in *novita*. Abdomen granulose. Ovipositor longer, about 0.75 to 0.8 as long as the body, and often longer than the abdomen.
    Black. Malar space and mandible white. Hind corner of pronotum black, or white at its junction with tegula. Tegula black. Legs more uniformly reddish-brown with fore coxa and trochanters often blackish-brown. Middle tibia and tarsus fuscous. Hind tibia and tarsus black, the tibia often with a small white spot dorsally near base. Apex of hind femur sometimes blackish.
    Male: Face and clypeus white, with clypeus black margined. Malar space yellow, 0.3 the basal width of mandible. Tegula black. Scape and pedicel black. Flagellum blackish. Hind tibia often yellow underneath. Hind femur black on apical 0.2, this black mark rather sharply contrasting with the reddish-brown color of femur.
    Length: ♀, 7-13 mm. Fore wing 5.5-9 mm. Ovipositor 6-9 mm. ♂, 7-11 mm. Fore wing 5-8 mm.
    Specimens examined: 12♂ and 40 ♀ from the Nearctic Region, from the following localities: *Alaska* (Anchorage, Deering, Isabella Pass, Mt. McKinley, Valdez); *British Columbia* (Racing River, Stone Mt. Park); *Colorado* (Poudre Lake, Fall River Pass in Rocky Mt. Nat. Park and Gould); *Northwest Territories* (North Shore of Lac Mounoir, Mackenzie Riber Delta, Norman Wells, and Tuktoyaktuk); and *Yukon Territory* (Dawson, Dempster Highway, mile 51, Herschel Island).

Walkley (1960) also reported this species from Quebec, Canada. Kasparyan (1977) reported it from European USSR. These specimens have not been seen.

Three paratypes of this species from Rampart House (YT), Raindeer Depot, Mackenzie Delta, (NWT), are now assigned to another species, *D. walkleyae*, n. sp.

Host: Unknown.

Distribution: Northwestern parts of North America and also in U.S.S.R.

3. DELOMERISTA LEPTECES Walkley (figs. 13, 14, 15, 16, 19)

Delomerista lepteces Walkley, 1960. *In* Townes and Townes: Bull.
U. S. Natl. Mus., 216(2): 368. ♀. Type: ♀, Alaska: Mt. McKinley,
1600 ft (Washington). Examined in 1980. Paratypes from Quebec and
Colorado.

This species is characterized by having a slender ovipositor with a pointed and needle-like tip, both valves tapering apically and lower valve with weak teeth. Other characteristic features of the species are:

Female: First flagellar segment (fig. 15) with one or two rows of sensillae, about 1.4 the second segment. Face with indistinct punctures, leathery in appearance. Malar space 0.45-0.5 the basal width of mandible. Propodeum a little rough and shiny dorsally, sparsely and shallowly punctate laterally. Areola horse-shoe shaped. Petiolar area rugulose. Costula absent. Second lateral area somewhat depressed. Nervellus intercepted at its lower 0.33. Discoidella bent down in the middle (in other species it is straight or only weakly curved). First tergite granulose, its lateral carina incomplete, partly erased apicad of spiracle. Ovipositor fine, needle-like, about 0.6 as long as the body length and only a little shorter than the abdomen (0.87).

Black. Mandible, malar space and palpi yellowish-white. Clypeus brown. Hind corner of pronotum elongately yellow. Tegula yellow to brownish. Fore coxa black. Fore femur and tibia black dorsally and yellow ventrally. Fore tarsus blackish-brown. Middle and hind coxae reddish-brown. Middle tibia and femur yellow in front, with their apices and tarsus brownish-black. Hind femur reddish-brown with apical third of it and hind tibia and tarsus black except for one or two yellow spots on tibia ventrally.

Male: Face and clypeus white. Face wider. Malar space black, 0.3 the basal width of mandible. Scape and pedicel black. Flagellum black. Tegula yellow. Hind femur reddish or reddish-brown with a blackish mark on apical 0.3. Propodeal areola weaker than in the female.

Length: ♀, 8-10.5 mm. Fore wing 7-8 mm. Ovipositor 4-5 mm. ♂, 8-9 mm.

Specimens: *Alaska:* Mt. McKinley, 1♀ (type) (Washington). *British Columbia:* Stone Mt. Park, 3800±ft., 1♀, 20-VII-1973, ♂, 12-VII-1973, H. & M. Townes (Townes). *Colorado:* Poudre Lake, Rocky Mountain National Park, 1100 ft., 1♀ (paratype), 12-VIII-1948, H. G. & D. Townes (Townes).

Host: Unknown.

Distribution: Alaska, Quebec, British Columbia, and Colorado. Specimen from Quebec not seen.

4. DELOMERISTA PFANKUCHI (Brauns) (figs. 4, 5, 6, 20)

[*Pimpla*] (*Delomerista*) *pfankuchi* Brauns, 1905. Ztschr. Syst. Hym. Dipt.,
5: 131. ♀. des. Type ♀, Bremen, Germany (Berlin). Examined in
1980.

Troctocerus unicolor Hedwig, 1959. Nachr. Naturw. Mus. Aschaffenburg,
    62: 96. ♀. Syn. by Horstmann,1981. South Italy and Turkey.
Taxonomy: Walkley, 1960: 371. Kasparyan, 1977: 74.

The ovipositor of this species is also long tapered but the lower valve is
more tapered apically than the upper valve and has more distinct teeth along its
tapered part. This species is distinctive in having the abdominal tergites
closely irregularly punctate, rather than granulose or coriaceous.
    Female: Antenna 32 segmented. Subapical segments a little longer than
wide. First flagellar segment with two rows of two or three sensillae each.
Malar space 0.25 the basal width of mandible. Interocellar distance equal to
ocellocular distance. Face minutely punctate. Mesopleurum and metapleurum
shiny and with scattered minute punctures. Propodeum subpolished, with a few
scattered punctures laterally. Areola elongate, more pentagonal. Propodeal
carina not strong. Costula faintly indicated, rather weak. Abdomen closely
irregularly punctate. Nervellus intercepted in lower 0.35. Ovipositor as
long as abdomen, its tip with a weak node, long tapered apically.
    Blackish-brown. Two small marks on face below antennal sockets,
mandible, hind corner of pronotum, tegula, and apical margins of second and
the following tergites, yellow; these bands wider on succeeding tergites.
Malar space yellowish-brown. Legs yellowish-brown. Hind tibia and tarsus
blackish-brown, with small yellow spots at base of tibia, underneath in middle
of tibia, and at base of first and second trochanteral segments.
    Male: Face, clypeus, mandible, and malar space, yellow. Face more
convex and punctate. Scape and pedicel yellow ventrally. Second lateral area
of propodeum a little rough. Costula more prominent in female. Tegula
yellow. Fore and middle legs pale-yellow. Hind tibia yellowish-white and only
apically black. Hind femur orange colored. Abdomen rugoso-punctate, with
narrow apical brownish bands.
    Length: 8 mm. Fore wing 7.5 mm. Ovipositor 5 mm.
    Specimens: Germany: Bremen, 1♀ (type), *ex Psyche viciella, ex* Coll.
Pfankuch (Berlin). *Poland,* 1♂, *ex Diprion similis,* 28-VIII-1938 (Washington).
Specimen reported by Walkley not examined.
    Distribution and Hosts: According to Oehlke (1966) this species is Holarc-
tic and has been reported as parasitic on diprionid cocoons (*Diprion pini*) by
Sitowski (1925) and de Fluiter (1932) in Poland and Netherlands, respectively.
Aubert (1969) reported this species from Germany, Poland and Netherlands,
mentioning the following hosts: *Chionodes tragicella* (Gelechiidae), and
Diprion pini L. (Diprionidae). Kasparyan (1977) reported it from European
USSR.
    A male, reared from *Diprion similis* in Poland (U.S.N.M.) has punctate
banded abdomen and is placed here. A female with the same data, however,
has granulose abdomen and is more like *D. novita europa,* although it has
banded abdomen, which is never the case in *novita.*
    This species does not occur in the Nearctic Region. The specimen reported
by Walkley as from "W.Va." must have come from Europe. This specimen
has not been examined.

5. DELOMERISTA KUSUOI Uchida & Momoi (figs. 10, 11, 12, 17, 18)

Delomerista kusuoi Uchida & Momoi, 1957. Insecta Matsumurana, 21(1-2):
    10. des., fig. Type ♀, Japan: Kyoto (Sapporo). Examined in 1980.

This species was described from two females from Kyoto. It is related to
D. pfankuchi Brauns in having a punctate abdomen and a tapered ovipositor with
a subapical nodus. It also has the two yellow subantennal marks on face as in
pfankuchi. *D. kusuoi*, however, is larger and stouter with body black rather
than brownish, with narrowly red apical bands on abdominal tergites, and propodeum densely punctate and strongly carinate. Other important diagnostic
characters are:

Female: Antenna longer, 41-segmented, subapical segments short, and
thick, as long as wide (cf. *pfankuchi*). Basal flagellar segments with numerous
sensillae. Interocellar distance about 1.3 the ocellocular distance. Face
punctate. Middle lobe of mesoscutum a little elongate anteriorly. Propodeum
densely punctate. Second lateral area rugose. Areola apically tending to be
rugose. Areola well defined, rounded basally and broadly horse-shoe shaped.
Costula distinct and complete, though low as compared to other carinae.
Apical transverse carina bounding the petiolar area very strong. Abdomen
punctate, more strongly so than in *pfankuchi*.

Black. Antenna ventrally, two triangular spots below antennal sockets,
base of mandible, malar space, tegula, and fore and middle legs, yellowish-
brown. Antenna a little darker. Hind coxa and femur reddish-brown, its
trochanters yellowish-brown, its tibia black dorsally and apically and yellowish
ventrally, its tarsus blackish-brown with base of first segment yellowish-brown.
Hind femur blackish at extreme apex. Abdomen black with narrow, faintly
reddish apical bands, particularly on second to fifth segments. Intersegmental
membranes, where visible (fifth segment onwards), yellow.

Male: Unknown.

Length: ♀, 13 mm. Fore wing 10.5 mm. Ovipositor 6 mm.

Specimen: *Japan:* Kyoto (Honshu), 1♀ (type), 29-X-1955, Kusuo Iwata
(Sapporo).

In the original description the month of collection mentioned is May. There
was also a paratype with the same data, which has not been seen.

Host: Unknown.

Distribution: Japan.

II. THE FRIGIDA GROUP

First flagellar segment without sensillae or only one or two present.
Second segment with few sensillae. Malar space 0.4 to 0.5 the basal width of
mandible. Notauli deeper than in Novita Group. Propodeum rough. Petiolar
area rugulose. Areola as long as wide, sometimes longer. Second lateral
area depressed and differently sculptured. Costula indicated by a wrinkled
line. Propodeum with a dorsal flat face, not evenly sloping to apex. Oblique
grooves on second tergite shallower than in Novita Group. Ovipositor tip not
compressed, a little bulbous. Upper valve not nodiform, slightly thicker than
the lower valve and with one or two indentations or ripple-like formations
preapically whence it evenly tapers apically. Lower margin of upper valve
slightly convex to almost straight (in Novita Group concave and less in depth
than lower valve). Ovipositor as long as abdomen or a little longer.

The American species of this group are similar to and sympatric with the
species of the Novita Group, and have been mixed up with them in the past.
Besides the ovipositor tip, other useful characters are the yellow ventral
aspects of fore femur and tibia and the absence or near absence of sensillae on
the first flagellar segment, and the second segment also with no or fewer sensillae. The European species *frigida* Kasparyan has wholly reddish fore legs

and is closely related to the American species, *townesorum* and *walkleyae*, described below as new species.

6. DELOMERISTA FRIGIDA Kasparyan

Delomerista frigida Kasparyan, 1977. New and Little-known Insects of
    European U.S.S.R. Acad. Sci. USSR, 1977: 71. ♀. key, des., fig.
    Type: ♀, Russia (Leningrad). Paratype examined, 1980.

Female: Face somewhat granuloso-punctate medially. Malar space 0.5 the basal width of mandible. Eye only slightly indented opposite antennal socket. Thorax as usual, with mesoscutum somewhat hairy and pleural areas smooth. Metapleurum with minute punctures. Propodeum rugulose in pleural area, second lateral area and petiolar area; smooth basodorsally. Areola elongate horse-shoe shaped. Costula represented by depressions and a wavy line. Abdomen granulose. First tergite more convex and granulose laterally, its median and lateral carinae somewhat indistinct and confined around its basal declivity. Ovipositor straight, its tip a little bulbous and evenly sloping. Ovipositor about 0.75 the length of body and a little longer than the abdomen.
    Black. Malar space and mandible yellow. Clypeus blackish-brown. Tegula black. Hind coxa blackish-brown. Legs reddish-brown. Middle tibia and tarsus fuscous. Hind femur and tarsus black. Hind tibia dirty white sub-basally.
    Male: Unknown.
    Length: ♀, 7 mm. Fore wing 6 mm. Ovipositor 5 mm.
    Specimens: Russia: 1♀, (paratype) (Townes). Germany: Haag Amper, Ober Bayern, 1♀, May 1948 (det. as *D. laevis*) (Washington).
    Host: Unknown.
    Distribution: Palaearctic Region: U.S.S.R. and Germany.

7. DELOMERISTA WALKLEYAE, n. sp. (figs. 21, 22, 23, 24, 25)

Female: Face subpolished, with distinct minute scattered punctures. Malar space about 0.45 to 0.5 the basal width of mandible. Eye moderately indented opposite antennal socket. Thorax as usual for the genus, but mesopleurum with scattered minute punctures, which are less evident in smaller specimens. Propodeum rugulose in pleural area, petiolar area, and in second lateral area (as in *frigida*) and smoother basodorsally. Areola horse-shoe shaped. Costula faintly indicated (fig. 21). Lateral aspect of postpetiole flat to a little concave, more rugulose rather than granulose. Lateral carina more prominent than in *frigida*. Abdominal tergites finely granulose. Oblique grooves on second tergite weaker. Ovipositor long, about 0.7 as long as body and longer than abdomen, a little arched in larger specimens.
    Black. Malar space and mandible yellowish-white. Clypeus brown. Hind corner of pronotum black to narrowly yellow. Tegula black. Fore leg largely blackish (type) with femur and tibia yellow ventrally. Mandible and hind coxae, trochanters, and femora reddish-brown, and their tibiae and tarsi black. Middle tibia yellow ventrally. Hind tibia dirty yellow to brownish yellow basoventrally (more often brownish rather than yellow). Hind femur with a narrow apical fuscous dorsal mark. Abdomen black.
    Male: Unknown.
    Length: ♀, 8-10 mm. Fore wing 6-7 mm. Ovipositor 5-6 mm.
    Holotype ♀, *Alaska:* Tsaina River, 17.VIII.1973, H. & M. Townes (Townes).

Paratypes: 16♀. *Alaska:* Anchorage, ♀, 6 to 12-VII-1976, Peter A. Rush. Thomson Pass, ♀, 15-VIII-1973, H. & M. Townes (Townes). King Salmon, Naknek River, 1♀, 19-VII-1952, J. B. Hartley (Ottawa). *Northwest Territories:* Mackenzie River Delta, 68° 43´ N, 134° 15´ W, 2♀, 18 to 19-VII-1979, L. Humble (det. *D. borealis*) (Ottawa). Mackenzie Delta, Raindeer Depot, 1♀, 10-VII-1948, J. P. Vockeroth (paratype of *D. borealis* Walkley) (Ottawa). North Shore of Lac Maunoir, 2♀, 15-VII-1969, G. E. Shewell (Ottawa). Kovaluk R., 69° 11´ N, 131° W, 2♀, 19 to 24-VI-1971, W. R. M. Mason (Ottawa). Norman Wells, ♀, 27-VI-1969, G. E. Shewell (Ottawa). *Yukon Territory:* Rampart House, ♀, 20-VII-1951 and ♀, 11-VII-1951 (paratypes of *D. borealis* Walkley), J. E. H. Martin (Ottawa). Dempster Highway, mile 51, ♀, 7 to 12-VII-1973, G. & D. M. Wood (Ottawa). *British Columbia:* Stone Mt. Park, 3800±ft., 2♀, 12 and 19-VII-1973, H. & M. Townes (Townes).

Host: Unknown.

Distribution: Northwestern Nearctic Region. Sympatric with *D. borealis.*

8. DELOMERISTA TOWNESORUM, n. sp. (figs. 26, 27, 28)

Female: Face subpolished, somewhat leathery, without distinct punctures or laterally with a few scattered punctures. Malar space 0.3 to 0.4 the basal width of mandible. Eye shallowly indented opposite antennal socket. Mesopleurum polished, with only minute setiferous punctures. Propodeum subpolished to a little wrinkled laterally. Areola as is *walkleyae,* but costula faint and second lateral area smoother. Nervellus intercepted in upper 0.4. Abdomen including postpetiole granuloso-mat. Ovipositor straight or a little upcurved, about as long as the abdomen, and 0.55 to 0.6 as long as the body.

Black. Malar space and mandible yellow. Clypeus brownish-black (sometimes face also like clypeus). Hind corner of pronotum and tegula yellow. All legs pale yellowish-brown with fore tibia dorsally fuscous and ventrally yellowish. Middle and hind tibiae yellow with dorsal and apical blackish marks. Middle and hind tarsi black with base of basitarsus yellow.

Male: Face and clypeus yellowish-white. Malar space 0.25 to 0.3 the basal width of mandible, yellow. Scape and pedicel yellow ventrally, scape often only narrowly so. Tegula yellow. Nervellus intercepted almost at its center or in upper 0.45. Fore and middle coxae yellowish-white. Hind femur reddish-brown with only its apex black marked.

Length: ♀, 7-10 mm. Fore wing 6-8 mm. Ovipositor 5-6 mm. ♂, 7-10 mm. Fore wing 6-8 mm.

Holotype: ♀, *Michigan:* Midland County, 21-31-V-1961, R. R. Dreisbach (Townes).

Paratypes (25♂, 22♀): *Michigan:* Ann Arbor, 4♀, 15 to 17-V-1960, 28-V-1962, 30-V-62, 7-VI-1962, H. & M. Townes (Townes). Schoolcraft Co., ♂, 8-VI-1960, R. R. Dreisbach. Emmet Co., 5♀, ♂, 27-V-1960, R. R. Dreisbach. Charlevois Co., ♀, 31-V-1960, R. R. Dreisbach (all Townes). *Oregon:* Mt. Hood, 5400´, ♀, 24-VII-1978, 2♂, 20-VII-1978, 1♀, 26 to 28-VII-78, H. & M. Townes (Townes). *Washington:* Mt. Rainier, 2700 ft., 2♂, 3♀, 11-VII to 16-VIII-1940, H. & M. Townes (Townes). *Arizona:* Nr. Alpine, 5♂, 24 to 29-V-1947, H. & M. Townes (Townes). *British Columbia:* Robson, ♀, 10-VI-1945, H. R. Foxlee (Townes), ♀, 24-V-1950, H. R. Foxlee (Ottawa). Stone Mt. Park, 3800±ft., 14♂, 12 to 22-VII-1978, H. & M. Townes (Townes). *Saskatchewan:* Regina, ♀, 14-VI-1940, T. B. Rempel (Townes). Saskatoon, ♀, 22-VI-1923, N. J. Atkinson (labelled *D. novita*) (Ottawa). *Ontario:* Golden Lake, ♀, 9-VI-1952, G. S. Walley (Ottawa). *Iowa:* ♀, 11-VI-1938, B. Berger (det. *D. novita*)

(Washington). *Virginia:* ?Blacksburg, 1♀ (labelled *D. novita*) (Washington).
   Host: Unknown.
   Distribution: This species is sympatric with *D. novita* and has been con-
fused with the same in the past. Apparently a widespread species in North
America, as is *D. novita*.
   A large female from Washington: Mt. Rainier, approaches *D. walkleyae*
in facial punctures, pleural punctures, but otherwise agrees with *townesorum*.
Similarly a male from Emmet Co., Michigan, has black scape and pedicel,
but otherwise resemble the males of *townesorum*.

C. THE JAPONICA GROUP

   Basal flagellar segments with sensillae. Propodeum short and slanting,
with a short dorsal face (figs. 29, 34, 41). In profile view evenly sloping from
base to apex (fig. 40). Areola broadly triangular, or semicircularly arched.
Region of costula linearly depressed and wrinkled, with an appearance of a
carina, though costula often faint to indistinct. Apical transverse carina
strongly arched and petiolar area occupying the apical half of propodeum.
Petiolar area subpolished and with a few rugosities. First tergite widened
basally, almost parallel-sided, or a little flanged at the junction of dorsal
and lateral carinae towards base; this segment comparatively thinner and more
compressed than in species of groups A and B. Median dorsal carinae not very
distinct on postpetiole. Upper valve of ovipositor tip a little bulbous and
widened, its apical slope abrupt, the sloping portion demarcated by two little
humps on the dorsal valve, and about as long as the width of ovipositor at this
point (which is the maximum width of the ovipositor). Ovipositor tip demarcated from the rest of ovipositor by a weak constriction, which appears characteristic of the group. Line of junction of upper and lower valves straight.
Upper and lower valves of equal depth. Apical ridges on lower valve reclining,
while the basal ones inclining. Ovipositor about 0.75 the length of abdomen.
   This group includes three closely related species: *D. japonica* Cushman
from Japan and USSR, *D. diprionis* Cushman from North America, and
D. indica, n. sp. from the Himalayan mountains of India. The former two
species have often been synonymized or considered subspecies of each other.
A key to these species is not very satisfactory because of certain overlapping
characters. Reference should therefore be made to the diagrams and descriptions of the ovipositor tips (figs. 33, 38, 39 and 44), to the shape of the propodeum (figs. 29, 34, 40 and 41), and also to the nature of the first tergite, malar
space and nervellus, to separate the species.

9. DELOMERISTA DIPRIONIS Cushman (figs. 34-38)

   Delomerista diprionis Cushman, 1939. J. Washington Acad. Sci., 29: 398.
       ♂, ♀. Type ♀, Canada: Oakville, Ontario (Washington). Examined in
       1980. Hosts: *Diprion* spp., *Neodriprion* spp.
   Delomerista japonica Cushman: Walkley, 1960. *In* Townes and Townes:
       Bull. U. S. Natl. Mus., 216(2): 367. Syn. in part.
   Delomerista japonica diprionis Cushman: Carlson, 1979. *In* Krombein
       et al. : Catalog of Hymenoptera north of Mexico, 1: 349.
   Biology: Furniss & Dowden, 1941: 49-51. Griffiths, 1960: 656. Torger-
       sen, 1969: 60.
   Morphology: Finlayson, 1960: 25 (larva). Short, 1978: 25, 174 (larva).

This species was synonymized with *D. japonica* Cushman by Walkley
(1960). The types have been examined as well as other material of the two
taxa. In my opinion, the two species are related but distinct. Although
diprionis shows a great degree of variability, it can be separated from *japonica*
by the combination of characters mentioned above.

This species has been reared in North America on a number of hosts,
belonging to the genera *Diprion, Gilpina* and *Neodiprion.* The variations could
not be correlated either with the host associations or with the distribution of
the species, and therefore *D. diprionis* is considered to be a polymorphic and
widely distributed species in the Nearctic Region. The larval head (fig. 64) is
somewhat different from that of *japonica* (fig. 65).

Female: Head wider than long. Face a little convex and punctate. In
smaller-sized specimens face with a few longitudinal striations medially.
Orbital areas smooth. Clypeus smooth. Malar space 0.6 to 0.7 the basal
width of mandible. Frons and vertex smooth. Mesoscutum evenly convex,
with minute setiferous punctures. Scutellum also with minute setiferous punctures. Pronotum smooth. Mesopleurum with a few minute and scattered punctures but more shiny than mesoscutum. Metapleurum shiny but with minute and
denser punctures, more so than on mesopleurum or mesoscutum. (In specimens from Alaska and British Columbia, the metapleurum is variable from
smooth to a little rugulose). Propodeum variable. Areola varying from triangular to wider and more semicircularly arched or crescentic in outline.
Costula faint to distinct (even paratypes from same host and locality vary).
Propodeum smooth baso-dorsally, somewhat punctate laterally, and petiolar
area usually rugulose or with a few rugosities. Second lateral area often
depressed and punctate. Propodeum in profile appearing short and abruptly
sloping, with dorsal face short, less than half the length of propodeum, and
apical slope nearly vertical. Nervellus intercepted at lower 0.3. First tergite
rather stocky and thicker medially, with its dorsomedian carinae making an
angle of 45° with the horizontal axis. Its lateral carina either interrupted in
the region of spiracle, or faded out apically. Area between lateral carina and
ventral carinae wider and rough, tending to be rugose. Juction of lateral and
median carinae not conspicuously flanged out. Basal declivity of first tergite
dull. Central raised area of postpetiole granulose and laterally rugulose. All
abdominal tergites coarsely granulose. Ovipositor tip (fig. 38) thicker and with
a convex even apical slope. Upper valve a little widened preapically. In profile view ovipositor appears more parallel-sided and without a preapical constriction as seen in *japonica* (fig. 39).

Black. Mandible, malar space, hind corner of pronotum and tegula,
yellowish-white. Clypeus brown to black. Legs reddish-brown to yellowish-
brown, with hind femur just apically, hind tibia except for an elongate white
baso-ventral mark and hind tarsus (except for a white basal mark), black.
Fore coxae sometimes brownish. Fore and middle tibia whitish ventrally and
faintly blackish dorsally, though often more uniformly brownish. Hind tibia
sometimes largely black but always with a yellowish-white basal annulus.
Color of tegula variable from yellow to black.

Male: Face and propodeum a little more dull rugulose. Scape and pedicel
yellow. Face yellowish white. Malar space yellow and about 0.3 the basal
width of mandible. Fore and middle legs yellow with dorsal fuscous marks on
femur, tibia and tarsus. Hind trochanters yellow. Hind femur reddish-brown.
Rest of coloration and sculpture as in female. Tegula usually yellow.

Variations: The majority of specimens reared from *Diprion,* as well as
from *Neodiprion,* show pale yellow tegula and hind tibia white or pale yellow on

the underside, rather than this color confined to base. Usually specimens reared from *Neodiprion* are smaller than those reared from *Diprion*. Specimens from Quebec generally have hind tibia brownish-black except at base.

Specimens from British Columbia, Northwest Territory and Yukon Territory show a tendency of having brownish-black tegula and specimens from Alaska, as well as from British Columbia mountains have a wholly black tegula. Usually all coxae are reddish-brown, but often specimens from Alaska as well as from British Columbia show black or blackish fore coxae and hind tibia and tarsus wholly black or only their bases white.

In smaller-sized specimens reared from *Neodiprion* in Ontario, the costula is absent and propodeum is more shiny, with only a faint depression in the second lateral area. In larger sized specimens from Ontario, reared from *Diprion similis*, the costula is marked by a distinct crease with second lateral area depressed and punctate.

Length: ♀, 4.5-10 mm. Fore wing 4-7 mm. Ovipositor 3-4 mm. ♂, 4-8 mm. Fore wing 3.5-6 mm.

Specimens: Several males and females from the Nearctic Region as follows: *Alaska* (Anchorage, Richardson Highway, at Mi 249, Taku Harbor, Union Bay); *Alberta* (Miette Valley, Jasper Park); *British Columbia* (Awun River, Allard Lake, Clinton, Cottonwood, Goderich, Lac la Hache, Maynard Lake, Revelstoke, Sayward, Victoria, Robson); *California* (Fish Camp, Sherwood, Strawberry); *Connecticut* (New Haven); *Idaho* (Fairfield); *Manitoba* (Aweme, Sand Hills); *New Brunswick* (Fredricton); *New Hampshire* (Mt. Madison); *New Jersey* (Moorestown); *New York* (Watson); *North Carolina* (Clingmans Dome, Mt. Mitchell); *Northwest Territories* (North Shore of Lac Maunoir, Norman Wells, McConnell, Kovaluk River, Raindeer Depot, Tuktoyaktuk); Nova Scotia (Liscombe River, Lum Co.); Ontario (Biscotasing, Cobden, Constance Bay, Grand Bend, Hawk Lake, Merivale, Oakville, Renfrew, Southampton, Sydney); *Quebec* (Church Cr., Cumshewa Inlet, Brome, Great Whale River, Georgeville, Knob Lake, Laniel, Montreal, Norway Bay); *Washington* (Mt. Rainier, 4700 ft., Snoqualmie Pass); *Wisconsin* (Gibson Lake, Gordon, Polk Co.); *Yukon Territory* (Dempster Highway at mile 37, Herschel Island, White Horse).

Specimens reported by Walkley from Maine, Michigan, Minnesota, Oregon, and Vermont were not seen by me.

Hosts: *Diprion similis*, *Diprion polytomum*, *Gilpina frutetorum*, *G. hercyniae*, *Neodiprion lecontei*, *N. nanulus nanulus*, *N. pratti banksianae*, *N. sertifer*, *N. tsugae*, *N. abietes*.

The larval head is figured in fig. 64.

Distribution: Widespread in northern North America.

10. DELOMERISTA JAPONICA Cushman (figs. 29-33, 39)

Delomerista japonica Cushman, 1937. Insecta Matsumurana, 12: 35.
     ♂, ♀. des. Type ♀, Japan: Nagawa-Mura, Nagano-ken (Washington).
     Examined in 1980. Host: *Diprion nipponicus*.
Taxonomy: Walkley, 1960: 367 (in part). Kasparyan, 1977: 73 (in part).
Morphology: Short, 1978: 25, 175 (larva).

This species is extremely similar to *D. diprionis* and the two were considered synonymous by Walkley, and as subspecies by Carlson (1979). The nature of the ovipositor tip (figs. 38, 39) is different in the two: in *japonica* the tip is with an abrupt apical slope and with a preapical constriction. The differences in the two subspecies are rather subtle, but the combination of

characters mentioned in above should distinguish the two, as well as the
related *D. indica*. Larval head as in figure 65.

Male and *Female:* Malar space 0.33 to 0.45 the basal width of mandible.
Face a little more strongly arched and punctate than *diprionis* specimens of
the same size. Mesoscutum somewhat subpolished, leathery. Mesopleurum
and metapleurum smoother and more polished. Propodeal areola more hexagonal in outline. In profile view, propodeum rather evenly convex and with a
convex slope. Dorsal face of propodeum about half the length of propodeum,
apical slope more inclined rather than vertical. First tergite thinner in depth,
with dorsomedian carinae making an angle of 30⁰ with the horizontal axis.
Lateral carina complete to apex and area between it and ventral carina narrow
and granulose to rugose. Junction of lateral and median carinae conspicuously
flanged out. Basal declivity of first tergite smoother and shiny. Central
raised area of postpetiole flat and rugulose. Abdomen more rugulose rather
than granulose, more often ruguloso-granulose. Ovipositor tip small, tapered,
in profile view a little widened preapically (fig. 33), somewhat bulbous and then
abruptly sloping to a point. Upper valve with a straight and short slope.

Color essentially similar to that of *D. diprionis*. Fore leg usually
yellowish-brown and a little lighter in color than middle and hind legs. Hind
femur with only a faint infuscate apical mark. Sometimes tegula and apex of
hind femur brownish.

Length: ♀, 8-9 mm. Fore wing 6.5-7 mm. Ovipositor 3.5-4 mm.
♂, 8-9 mm. Fore wing 6.5-7 mm.

Specimens: Japan: Nagawa-Mura, Nagano-ken, ♀ (type) Jan.-March 1937
(Washington). Same locality, ♀, ♂, (Washington). Kamikochi,Japan, ♂, 3♀,
July 22 to 23, 1954, Townes family (Townes).

Host: Diprion nipponicus.

Distribution: Eastern Palaearctic Region: Japan. Kasparyan (1977)
reported it also from USSR.

11. DELOMERISTA INDICA, n. sp. (figs. 40-46)

Male and *Female:* Face a little protuberant medially, punctate. Clypeus
flat, subpolished. Malar space 0.5 the basal width of mandible. Frons, vertex, occiput and temple subpolished, shiny, with vertex posteriorly and temple
sparsely hairy. Interocellar distance equal to the ocellocular distance. Mesoscutum and scutellum mat, evenly convex, with minute setiferous punctures.
Side of thorax shiny. Metapleurum with a few distinct scattered punctures
and area near hind coxa rugose. Propodeum subpolished, with irregular punctures laterally on pleural areas and scattered rugosities on lateral area and
petiolar area. Areola more squarish or triangular, with lateral carinae semicircularly arched. Costula interrupted, represented by a linear wrinkled line.
Propodeum depressed in this area. Nervellus intercepted at the middle, with
some specimens showing variations: intercepted a little above or below the
middle. Abdomen finely granulose to granuloso-coriaceous; apical margins
and terminal tergites mat. First tergite evenly arched (fig. 42), depressed
medially, its dorsomedian carinae not very distinct and bordering a shallow
basal declivity, which is dull and granuloso-mat. Lateral carina just above
spiracle complete, moderately flanged out basally, and area below it rugulose.
Central depression on postpetiole granulose, bordered by rugosities (granuloso-
rugose). Ovipositor tip evenly tapering and not very bulbous (fig. 44), the
upper valve a little heavier near tip and with an even convex slope, as in
D. diprionis.

Female color: Black. Malar space, mandible except teeth, hind corner of pronotum, tegula (except for a black apical spot), and wing bases, yellow. Legs reddish-brown with trochanters and stripes on anteroventral aspects of fore and middle femora and ventral region of basal 0.75 of hind tibia, yellow. Apex of hind femur, hind tibia otherwise, and hind tarsus black. Apical tarsal segment of fore and middle legs darker.

Male color: Face, clypeus, mandible except teeth, yellow. Scape and pedicel yellow ventrally. Hind corner of pronotum and tegula yellow. Fore and middle coxae and their trochanters yellow, their tibiae and femora marked with brownish, apical tarsal segments darker. Hind coxa and femur reddish-brown, hind trochanters yellow, hind tibia and tarsi black with tibia ventrally (except at apex) and base of first tarsal segment, yellow.

Length: ♀, 8.5 to 9.5 mm. Fore wing 7.5-8 mm. Ovipositor 3.5-4 mm. ♂, 8-9 mm. Fore wing 7-8 mm.

Holotype: ♀, *India:* Himachal Pradesh: Dhenkund (near Dalhousie), 2743 m, 25-IX-1971, Coll. Gulati. (Gupta).

Allotype: ♂, same data, Coll. Tulsi, DJD-168 (Gupta).

Paratypes: Several ♂, ♀. *India:* Himachal Pradesh (Dalhousie, 2132 m, Dhenkund 2743 m, Ahla 2286 m, Kalatop 2488 m, Khajiar, 1828 m, all localities in Dalhousie Hills, collected from May to October 1971 and July 1965.) (Gupta).

Host: Unknown.

Distribution: India: Mountains of Himachal Pradesh.

D. THE MANDIBULARIS GROUP

Basal flagellar segments with fewer to several sensillae (fig. 55-62). Malar space 0.4 to 0.6 the basal width of mandible (except in *laevis*, where it is 1.0 to 1.2). In males malar space 0.33 to 0.4 the basal width of mandible (in *laevis* 0.8). Propodeum convex; in profile view (fig. 50, 54) with basal part convex and apical part abruptly sloping, almost vertically so. Costula faint to indistinct. Propodeum generally granulose to rugulose in pleural area, second lateral area and petiolar area. Often petiolar area with a few wrinkles. First tergite convex, with lateral carina distinct and usually complete. Median carinae usually distinct to the dorsal hump and then weak. Ovipositor stout. Upper valve heavy (figs. 47-49, 51-53) and occupying the greater part of the depth of ovipositor, except in *D. longicauda*, where the tip is narrowed preapically and ovipositor is twice as long as the abdomen. Ridges on lower valve vertical to a little inclined. Basal three ridges strongly inclined. Upper and lower valves of ovipositor meeting in an arched line and in a slant. Ovipositor 0.8 to 1.0 as long as abdomen (2.0 in *longicauda*), usually slightly arched upwards or bent downwards.

This group includes five species: *Delomerista longicauda* (with ovipositor 2.0 the length of abdomen), *D. laevis* (malar space long, 0.8 to 1.2 the mandible width), *D. strandi* (ovipositor very stout and straight), *D. mandibularis* (= *gelida*, ovipositor slightly curved upwards), and *D. masoni* (ovipositor bent downwards and costula strong).

12. DELOMERISTA MANDIBULARIS (Gravenhorst) (figs. 48, 49, 50, 59, 60)

Pimpla mandibularis Gravenhorst, 1829. Ichneumonologia europaea, 3: 180. ♀. Lectotype (labelled by Perkins, 1936, designated by Oehlke, 1967). ♀, Poland: Breslau (Wroclaw). Specimens det. Perkins and Kasparyan examined, 1980.

Ephialtes albicinctus Desvignes, 1862. Trans. Ent. Soc. London, 1: 226.
♂. des. Type ♂, ?Great Britain (London). Name preoccupied. Syn.
after Fitton (1978). Examined, 1981.

Ephialtes desvignesii Marshall, 1870. Ichneumonidum Britannicorum
Catalogus, p. 20. New name for *Ephialtes albicinctus* Desvignes.

Delomerista gelida Walkley, 1960. *In* Townes and Townes: U. S. Natl.
Mus., Bull. 216(2): 366. ♀. des. fig. Type ♀, N. W. T.: Cameron Bay,
Great Bear Lake (Washington). Examined, 1980. New synonym.
Paratypes from Sask: Waskesiu, Quebec: Great Whale River.

Taxonomy: Oehlke, 1967: 34. Aubert, 1969: 99. Fitton, 1976: 326. Fitton,
1978: 14. Kasparyan, 1977: 74.

The lectotype locality mentioned by Oehlke (1967), who designated the
lectotype, is Breslau (=Wroclaw), while Kasparyan mentions Warmbrunn as
the lectotype locality. Gravenhorst had specimens from both the localities.
The type of *Ephialtes albicinctus* Desv. has a label by Perkins (1933) mentioning its synonymy with *D. mandibularis* Grav. This synonymy was never
published by him. Fitton (1978) first published this synonymy.

Female: First and second flagellar segments with few sensillae (fig. 59, 60).
Face convex medially and with scattered punctures on a shiny surface. Sometimes
punctures a little denser. Clypeus smooth and flat. Malar space 0. 5 the basal width
of mandible. Mesoscutum subpolished with setiferous punctures. Side of thorax
polished and with minute scattered setiferous punctures. Propodeum smooth dorsally and rugulose laterally and in the petiolar area. Second lateral area depressed and
rough. Areola smooth, horse-shoe shaped, sometimes a little squarish or pentagonal. Costula faintly indicated by a broken carina. Seen in profile, propodeum with a
convex slope (fig. 50). Nervellus intercepted below the middle. Abdomen including
median area of postpetiole granulose. First tergite as in fig. 50, with dorsal and lateral carinae distinct and ruguloso-coriaceous on side. Ovipositor stout, of uniform
width, and preapically slightly curved upwards, as long as abdomen. Ovipositor tip
a little widened (fig. 48, 49), with dorsal valve occupying nearly 2/3 of the depth of ovipositor near apex, while the lower valve a occupying 2/3 of the depth of ovipositor
near base (the two valves meeting in a slant).

Black. Malar space, mandible, and hind corner of pronotum, yellowish-white.
Tegula black, though often narrowly to partly yellowish-white. Legs dark reddish-
brown to orange-brown, with fore leg partly to largely brownish in American populations. Hind tibia and tarsus black. Apex of hind femur often black. Middle tibia and
tarsus reddish-brown to blackish-brown. Fore coxa brownish-black in Swedish specimens also.

Male: Face yellowish-white, smoother or a little rugulose in larger specimens. Malar space yellow, 0. 3 to 0. 4 the basal width of mandible. Scape and
pedicel yellow ventrally, or black. Propodeal carinae strong. Costula faint
to obsolete. Areola elongate. Tegula yellow. Fore and middle legs largely
pale yellow, with their femora and tibiae and tarsi brownish dorsally. Hind
coxa, trochanter and femur reddish-brown. Trochanters often yellow ventrally.
Hind femur sometimes apically black marked. Hind tibia and tarsus blackish-
brown dorsally, with tibia ventrally largely yellowish-white in the middle.
Postpetiole rugose. Abdominal tergites granuloso-mat.

Length: ♀, 8. 5 to 9. 5 mm. Fore wing 7-8 mm. Ovipositor 6. 5-7. 5 mm.
♂, 6-9 mm. Fore wing 5-8 mm.

Specimens examined: 1♂, ?*England*, type of *E. albicinctus* (London).
England: Newton Abbot, ♀, 8-VI-1941, as *mandibularis* by Perkins), J. F.
Perkins (Washington). *Russia*, ♂, ♀, from *Betula*, 18-V-1890 (det. as *mandi-
bularis* by Kasparyan) (Townes). *Germany:* Siegmundung, ♀, 7-VI-1919 (det.

as *mandibularis* by Habermehl) (Ottawa). *Northwest Territories:* Cameron Bay, Great Bear Lake, ♀ (type of *D. gelida* Walkley), 1-VII-1957, T. N. Freeman (Washington). *Quebec:* Great Whale River, ♀, 20-VII-1949 (paratype of *D. gelida*) (Ottawa).*Saskatchewan:* Waskesiu, ♂, ♀, 20-VI-1938, J. G. Rempel (♀ paratype of *gelida*) (Townes). In addition 28♂ and 22♀ from the following localities: *British Columbia* (Hixon, Ft. Nelson, Racing River, Stone Mt. Park); *Alaska* (Anchorage, Delta Junction, Tsaina River); *Quebec* (Great Whale River),*Saskatchewan*(Conquest); *Northwest Territories* (Normal Well, Kovaluk River, Tuktoyaktuk), *Yukon Territory* (Dawson), and *Sweden* (Messaure, Skåne).

Only males from the following states: Arizona (near Alpine), Alberta (Banff), California (Leevining), where they occur sympatrically with *D. masoni* or *D. townesorum*. The males show variations and their identities are not very clear.

In addition, Walkley also reported this species from Ontario, Colorado, New York and Pennsylvania. Those specimens were not examined.

Hosts: Euura amerinae and *Strongylogaster* sp. (Tenthredinidae) in Europe.

Distribution: Holarctic. In the Nearctic Region apparently restricted to Northern and Northwestern parts.

13. DELOMERISTA STRANDI (Ulbricht) (figs. 47, 54, 57, 58)

Pimpla strandi Ulbricht, 1911. Arch. Naturgsch., 77: 149. ♀.
  Lectotype (designated by Oehlke, 1967), ♀, "Norvegia, E., Coll.
  Strand, Rosvana" (Berlin). Examined in 1980.
Delomerista strandi: Oehlke, 1967. Hymenopterorum Catalogus (nova
  editio),2: 34. Kasparyan, 1977: 74. Russia.

Female: Face centrally a little rugulose and punctate. Malar space 0.5 to 0.6 the basal width of mandible. Basal flagellar segments with fewer hairs and with several sensillae (figs. 57, 58). Frons, vertex and temple smoother. Thorax dull, hairy. Mesoscutum with close setiferous punctures, particularly on the middle lobe. Mesopleurum with scattered punctures, smoother in smaller-sized specimens. Metapleurum partly to largely rugulose. Pleural area of propodeum rugulose. Second lateral area rugulose and depressed. Rest of propodeum smoother but with a few wrinkles around carinae. Areola elongate, somewhat pentagonal in outline. Costula incomplete. In profile, propodeum as in fig. 54. First tergite strongly convex (fig. 54), rugulose to rugose laterally, its median carinae weak beyond the dorsal hump. Its lateral carina strong throughout. Abdomen finely granulose. Ovipositor stout, straight and compressed, its upper valve very wide (fig. 47), and occupying almost 2/3 the apical depth of ovipositor. Ovipositor narrow basally and widened apically (1:1.5), the line of junction of the upper and lower valves running obliquely from base to apex of ovipositor. Teeth on lower valve vertical, with basal 3-4 teeth inclined. Ovipositor about 0.8 as long as abdomen.

Black. Malar space and mandible yellow. Tegula black. Legs reddish-brown with hind tibia and tarsus black. Fore and middle tarsi brownish-black. Middle tibia fuscous dorsally (legs in type-specimen from Europe a little paler). Fore coxa and trochanter brownish in the lectotype and in a few American specimens). Clypeus brown. Sometimes mandible brown rather than yellow. Hind tibia varying from pale brown to blackish brown. Wings with a purple iridescence in specimens from Mackenzie River Delta, which are also larger and stouter specimens.

Male: Unknown.

Length: ♀, 7-11 mm. Fore wing 5-8 mm. Ovipositor 4-6 mm.

The lectotype, which is a smaller-sized specimen, has the face smoother, dull and not distinctly punctate. Otherwise similar to other specimens.

Specimens: "Norvegia, E. Ros Vania", Coll. Strand. ♀ (lectotype) (Berlin) (A cotype of this species is actually a specimen of *D. laevis* Gravenhorst). *Alaska:* mile 28, Richardson Highway, ♀, (det. *D. gelida* by Walkley), 27-VII-1951, W. R. M. Mason (Ottawa). *British Columbia:* Stone Mt. Park, 3800±ft., ♀, 13-VII-1973, H. & M. Townes (Townes). *Alberta:* Jasper, ♀, 26-VII-1949, C. P. Alexander (Townes). This specimen has longer (0.8) malar space). *Northwest Territories:* Mackenzie River Delta, 68° 43´ N., 134° 15´ W., 3♀, 18 to 29-VII-1979, L. Humble (Ottawa) (det. as *D. borealis* in 1980). Norman Wells, ♀, 29-VI-1969, S. E. Shewell (Ottawa). *Yukon Territory:* mile 87 on Dempster Highway, ♀, 18 to 28-VII-1971, G. & D. M. Wood (Ottawa).

Host: Unknown.

Distribution: Holarctic Region. In the Nearctic Region, apparently confined to the Northwestern parts.

14. DELOMERISTA MASONI, n. sp.

This species has apparently been confused with *Delomerista gelida* = *mandibularis* in the past, as some specimens from Alaska and British Columbia, in Canadian National Collection, Ottawa bear determination label of *D. gelida* Walkley. It can be readily separated from *gelida* = *mandibularis* by the nature of the ovipositor, tegula, propodeum, etc.

Female: Head wide, punctate. Eye moderately deeply notched. Malar space 0.4 to 0.5 the basal width of mandible. Basal flagellar segments with sensillae. Mesoscutum subpolished, with numerous setiferous punctures. Side of thorax polished and with minute setiferous punctures on mesopleurum and metapleurum; those on metapleurum a little coarser than on mesopleurum. Pleural area of propodeum irregularly punctate. Propodeum dorsally smoother and shiny. Areola semicircularly rounded. Costula strong and distinct. All propodeal carinae strong. Petiolar area subpolished with weak rugulosities around carinae. Nervellus intercepted in lower 0.3. First tergite short, humped dorsally, where median carinae angled and getting weaker. Lateral carina distinct throughout. Abdomen granulose, with postpetiole laterally and basomedian area of second tergite rugose. Ovipositor 0.8 the length of abdomen, widened apically from base, with a weak median bend.

Black. Malar space, mandible, hind corner of pronotum, and tegula yellowish-white. Antenna and clypeus brownish. Legs reddish-brown, with hind tibia and tarsus largely black. Hind tibia with a basal and ventral yellow mark. Hind tarsus with a basal yellow ring. Fore femur and tibia and middle tibia yellowish ventrally.

Male: Scape and pedicel ventrally yellowish-white. Flagellum brown. Face, clypeus, mandible and malar space yellowish-white. Fore and middle legs and hind trochanters, tegula, and hind corner of pronotum yellowish-white. Hind coxa and femur yellowish-brown. Apex of hind femur narrowly black, or hind femur wholly yellowish-brown. Hind tibia and tarsus with black marks and yellow ventrally. Malar space 0.3 the basal width of mandible. Nervellus intercepted at lower 0.4 to upper 0.4. Dorsal carina of first tergite usually strong to apex of tergite. Costula present. Propodeal carinae like those in female.

Variations: This species is rather close to *D. strandi* from the northern latitudes and some specimens of it approach that species. The chief differences being its yellow tegula, hind tibia not wholly black, costula of propodeum strong

and complete and propodeum dorsally smoother. However, there are some
specimens where the costula appears a little weak, but complete, second lateral
area rugulose, hind leg more extensively black, and tegula may have a black
spot. In these specimens, generally, the apex of hind femur is also black or
blackish, which is not the case in *strandi*, and usually there is a basal yellow
ring on hind tarsus and tibia. The shape of areola is also variable, being a
little elongate and not semicircular in some specimens.

Length: ♀, 9-11 mm. Fore wing 5.5-8.5 mm. Ovipositor 4 -5.5 mm.
♂, 9-10 mm. Fore wing 6-8 mm.

Holotype: ♀, *Michigan*: Ann Arbor, 30-V-1962, H. & M. Townes (Townes).
Paratypes: New York: Ithaca, 1♀, 3-V-1936, H. K. Townes. *Pennsylvania:*
Spring Brook, ♀, 8-VI-1945, H. & K. Townes. *Nebraska:* Valentine Refuge,
2♂, 2♀, 7-VI-1972, H. & M. Townes (Townes). *Ontario:* Chaffeys Locks, ♀
(face narrower), 21-VI-1975, J. Belwood (Townes). Ottawa, ♀, (det. *gelida* by
Walkley), 29-V-1941, G. S. Walley. *Oregon:* Pinehurst, ♀, 2-VII-1978. H. &
M. Townes (Townes). 7♂ from Ochoco Creek, Hyatt Reservoir, Mt. Hood,
June-July 1978, H. & M. Townes (Townes). *Colorado:* Gould, 1♂, 6-VIII-
1974, H. & M. Townes. *Arizona:* 15♂, near Alpine, Oak Cr. Canyon.
(Townes).

Distribution: Nearctic Region, New York to Rocky Mountains.

15. DELOMERISTA LAEVIS (Gravenhorst) (figs. 52, 61, 62)

Pimpla laevis Gravenhorst, 1829. Ichneumonologia europeaea, 3: 180.
   ♀. Type ♀, Piemont, Italy (Wroclaw).
Pimpla texana Cresson, 1870. Trans. Amer. Ent. Soc., 3: 145. ♀.
   Type ♀, Texas (Philadelphia). Homotype examined in Townes
   Collection, 1980. Synonymized by Oehlke, 1966.
Pimpla laevifrons Thomson, 1877. Opusc. Ent., 8: 750. ♂, ♀.
   Lectotype (labelled by Aubert, 1968) ♀, "Norl" (Lund). Synonymized
   with *texana* by Townes, 1944.
Taxonomy: Walkley, 1960: 365. Oehlke, 1966: 816. Aubert, 1979: 73.

This species is rather characteristic in having a smooth face, convex
clypeus, convex temple and malar space about as long or longer than the basal
width of mandible. The ovipositor tip (fig. 52) assigns it to the Mandibularis
Group.

Female: Basal two flagellar segments hairy and with very few sensillae.
Head a little narrowed ventrally. Malar space 1.0 to 1.2 as long as the basal
width of mandible, longest among the species of *Delomerista*. Clypeus basally
convex, smooth. Face smooth, subpolished, with a few minute scattered punctures. Frons, vertex and temple smooth. Temple slightly more convex and
wider than in other species. Thorax subpolished. Mesopleurum smoother
than in other species, with only minute scattered punctures. Metapleurum
finely granular, at least on apical half. Propodeum granulose in pleural area,
second lateral area, and in petiolar area. Areola elongate, roughly triangular,
narrowed and rounded basally. Costula faintly indicated or partly indistinct.
Nervellus intercepted at lower 0.33 to 0.5. First tergite angled dorsally,
where median carinae almost end. Lateral carina more or less complete.
Postpetiole laterally granuloso-rugose. Abdomen finely granular with apical
tergites becoming shagreened or coriaceous. Ovipositor stout, short, 0.7 to
0.8 the length of abdomen, tip heavy, thick, (fig. 52).

Black. Malar space and mandible pale yellow. Clypeus brownish. Tegula
black, sometimes partly yellow. Legs reddish-brown with apices of hind femur

and tibia and all tarsi fuscous to black. Color of hind tibia varying from wholly black to wholly reddish-brown (particularly in European specimens). Wings lightly clouded (European specimens) or darker fuscous (American specimens). Apical fuscous marks on hind femur often light to absent.

Male: Face black with whitish marks on sides (face of males of other species wholly yellowish-white). Clypeus yellowish-white. Coxae black. Trochanters fuscous. Propodeum wholly granulose. Costula usually distinct. Abdomen more shagreened and coriaceous than granulose.

Length: ♀, 7.5 to 10.5 mm. Fore wing 6.0-9.0 mm. Ovipositor 5.0-6.0 mm. ♂, 5.5-9.5 mm. Fore wing 4.5-8.0 mm.

Specimens: Europe: "Norvegia E., Hemnesberget Ranen", 1♀ (cotype of D. strandi), 13-VII-1903, Coll. Strand (Berlin). Several males and females from Alaska, British Columbia, Manitoba, Maine, New Hampshire, Quebec, Northwest Territory, and Yukon Territory.

Host: Aubert (1969) mentions *Rhyacionia buoliana* as a host.
Distribution: Europe.

16. DELOMERISTA LONGICAUDA Kasparyan

This species is characterized by having a very long ovipositor, which is about 2.0 as long as abdomen and longer than the body length. The ovipositor tip is also characteristic (fig. 53). The ovipositor tip places it under the Mandibularis Group.

Female: First flagellar segment with few sensillae (fig. 55). Face wider than long, smooth, subpolished, with a few indistinct scattered punctures. Malar space 0.5± the basal width of mandible. Clypeus flat, smooth. Frons, vertex and temple smooth. Thorax subpolished. Mesoscutum more hairy than mesopleurum and metapleurum, which are beset with scattered setiferous punctures. Propodeum subpolished, finely rugulose apicolaterally, or largely smooth and shiny. Pleural area of propodeum rugulose. Areola horse-shoe shaped, often open basally or apically, smooth. Costula indistinct. Second lateral area of propodeum a little depressed, smooth to somewhat wrinkled. Petiolar area smooth or with a few wrinkles. Nervellus intercepted at lower 0.3. First tergite 1.5 to 1.75 as long as wide, evenly convex dorsally, finely granuloso-rugulose to shagreened, its lateral carina weak, but distinct beyond the spiracle, its median carinae weak. Second and third tergites shagreened to granulose. Apical tergites shagreened to coriaceous. Ovipositor long, about 2.0 as long as abdomen, and longer than the body, curved upwards, more so in apical 0.2. Its tip demarcated by a shallow constriction (fig. 53), swollen preapically, with upper valve wider and with a long apical slope.

Two subspecies are recognized: *Delomerista longicauda longicauda* Kasparyan from USSR and *D. longicauda americana* from North America. The two differ mainly in coloration of legs and tegula, and also show minor differences in the sculpture of propodeum and abdomen.

Key to the subspecies of *Delomerista longicauda*

1. Tegula black. Legs including hind femur, yellowish-brown. Hind tibia and tarsus fuscous. Propodeum subpolished and finely rugulose in petiolar and second lateral area. Basal three abdominal tergites granulose. USSR. . . . 16a. longicauda longicauda Kasparyan (p. 30)

Tegula yellow. Legs yellowish-brown but fore leg brownish-black.
Middle and hind legs beyond trochanters blackish-brown (blackish marks
on middle and hind femur varying in extent). Propodeum shiny with
second lateral area and petiolar area largely smooth. Basal three
abdominal tergites finely ruguloso-granulose, leaning toward shagreened.
North America. . . . . 16b. longicauda americana, n. subsp. (p. 30)

16a. DELOMERISTA LONGICAUDA LONGICAUDA Kasparyan

Delomerista longicauda Kasparyan, 1973. Zool. J., 52: 1877. ♀. des.
     fig. Type ♀, Russia (Leningrad). Paratype examined, 1980.
Taxonomy: Kasparyan, 1977: 72.

Female: A paratype from Russia has been examined and the chief diagnostic features are given in the key to subspecies. The first tergite is stocky
and its lateral carina is indistinct in the specimen before me.
Distribution: Russia.

16b. DELOMERISTA LONGICAUDA AMERICANA, n. subsp. (figs. 53, 55, 56)

Female: Characterised as in the key. Black. Malar space, mandible,
hind corner of pronotum, and tegula, whitish-yellow. Fore leg brownish-black
with yellow marks on underside of femur, tibia, and trochanters. Middle coxa
and trochanters yellowish-brown. Middle femur, tibia and tarsus fuscous.
Tibia often yellowish marked. Hind coxa, trochanters, and base of femur
yellowish-brown. Femur otherwise, tibia and tarsus blackish-brown. Base
of tibia and tarsus yellowish-white.
Male: Face white, about 1.5 as wide as long, shiny, with minute punctures. Malar space black, 0.33 the basal width of mandible. Scape and
pedicel black. Propodeum subpolished, finely rugulose around carinae.
Abdomen mat to shagreened. Postpetiole with a median depression.
Median carinae on first tergite indistinct. Fore coxa white. Middle and
hind coxae brownish. Fore and middle femora, tibiae and tarsi fuscous
marked, with a white base color. Hind femur usually largely black or
blackish, though the extent variable. Hind tibia black with a ventral white
mark. Hind tarsus black.
     Length: ♀, 8-10 mm. Fore wing 6.5-9 mm. Ovipositor 8-11 m.
♂, 7-8 mm. Fore wing 6-7 mm.
     Holotype ♀, Alaska: Anchorage, 6 to 12-VII-1976, Peter A. Rush
(Townes).
     Paratypes: Alaska: Anchorage, ♀, 25 to 30-VI-1976; ♂, 11 to 18-
VI-1976, Peter A. Rush (Townes). Tsaina River, ♀, 18-VIII-1973, H. &
M. Townes (Townes). *British Columbia:* Stone Mt. Park at 3800± and
5500 ft., 3♀, 10♂, 13 to 20-VII-1973, H. & M. Townes (Townes).
     Host: Unknown.
     Distribution: Alaska and British Columbia in Northwestern Nearctic
Region.

ACKNOWLEDGMENT

I am thankful to Dr. Henry Townes for providing me with the necessary facilities to undertake this work and for his advice.

REFERENCES

1. Aubert, Jacques-F. 1969. Les Ichneumonides Ouest-Palearctiques et leurs hôtes. 1. Pimplinae, Xoridinae, Acaenitinae. Ouvrage Publee avec le Concours du CNRS, p. 1-304.

2. Aubert, Jacques-F. 1972. Etude Commentee de Nouveaux Lectotypes Choisis dans les Collections Holmgren et Thomson (Hym. Ichneumonidae). Ent. Scand. 3: 145-152.

3. Carlson, R. W. 1979. Family Ichneumonidae. *In* Krombein *et al.*: Catalog of Hymenoptera in America North of Mexico 1: 315-740.

4. Constantineanu, M. I. & Pisica, Constantin, 1977. Fauna Republicii Socialiste Romania. Insecta, Hymenoptera, Ichneumonidae 9(7): 1-310.

5. Finlayson, Thelma 1960. Taxonomy of Cocoons and Puparia, and their Contents, of Canadian Parasites of *Neodiprion sertifer* (Geoff.). Canad. Ent. 92: 20-47.

6. Finlayson, Thelma 1967. Taxonomy of Final Instar Larvae of the Hymenopterous and Dipterous Parasites of *Acrobasis* spp. (Lepidoptera: Phycitidae) in the Ottawa Region. Canad. Ent. 99: 1233-1271.

7. Fitton, M. G. 1976. The Western Palaearctic Ichneumonidae (Hymenoptera) of British Authors. Bull. Brit. Mus. (Nat. Hist.) Ent. 32(8): 303-373.

8. Fitton, M. G. 1978. Family Ichneumonidae. *In* Kloet and Hincks: A Check List of British Insects, 2nd Edition. Handbooks for Identification of British Insects, Vol. 11 (4, Hymenoptera): 12-45. Royal Ent. Soc. London.

9. Fluiter, H. J. de, 1932. Bijdrage tot de Kennis der Biologie en Epidemiologie van de gewone Dennenbladwasp, *Pteronus (Lophyrus) pini* L. in Nederland. Tijdschr. Plantz. 38: 125-196.

10. Furniss, R. L. & Dowden, P. B. 1941. The Western Hemlock Sawfly *Neodiprion tsugae* Middleton, and its parasites in Oregon. J. Econ. Ent. 34: 46-52.

11. Griffiths, K. J. 1960. Parasites of *Neodiprion pratti banksianae* Rohwer in Northern Ontario. Canad. Ent. 92: 653-658.

12. Horstmann, K. 1981. Typenrevision der von Karl Hedwig Beschriebenen Arten und Formen der Familie Ichneumonidae (Hymenoptera). Ent. Mitt. Zool. Mus. Hamburg 7(112): 65-82.

13. Kasparyan, D. R. 1973. A New Species of the Genus *Delomerista* (Hymenoptera, Ichneumonidae) (in Russian). Jool. Z. 52: 1877-1878.

14. Kasparyan, D. R. 1977. Review of the European Species of Ichneu-
    monids of the Genus *Delomerista* Foerster (Hymenoptera, Ichneu-
    monidae). (in Russian). New and Little-Known Species of Insects
    of the European part of USSR. Acad. Sci. USSR. 1979: 69-75.

15. Kolomiets, N. G. 1962. Akad. Nauk SSSR Sibirskoe Otd. Parazishi
    i Khishchniki Sibirskogo Shelkopryada, p. 95.

16. Morris, K. R. S., Cameron, E. & Jepson, W. F. 1937. The Insect
    Parasites of the Spruce Sawfly (*Diprion polytomum* Thg.) in
    Europe. Bull. Ent. Res. 28: 341-393.

17. Oehlke, J. 1966. Die in Europäischen Kiefernbuschhornblattwespen
    (Diprionidae) Parasitierenden Ichneumonidae. Beitr. z. Ent.
    15(7-8): 791-879 (1965).

18. Oehlke, J. 1967. Westpaläartische Ichneumonidae I: Ephialtinae. *In*
    Junk: Hymenopterorum Catalogus *(nova editio)* Part 2: 1-48.

19. Short, J. R. T. 1978. The Final Larval Instars of the Ichneumonidae.
    Mem. Amer. Ent. Inst. 25: 1-508 (*Delomerista*, p. 25).

20. Sitowski, L. 1925. Do Biologji Pasorzytow borecznixa (*Lophyrus* Latr.),
    sur la Biologie des Parasites de *Lophyrus* Latr. Poln. m. Dtsch.
    Zuzammenf. 14: 1-25.

21. Torgersen, T. R. 1969. Hymenopterous Parasites of the Hemlock Sawfly,
    Neodiprion tsugae Middleton, in Southern Alaska with a key to Larval
    Remains. J. Ent. Soc. Brit. Columbia 66: 53-62.

22. Townes, H. K. 1944. A Catalog and Reclassification of the Nearctic
    Ichneumonidae. Part I. Mem. Amer. Ent. Soc. 11(1): 1-477.
    (*Delomerista*, pp. 48-49).

23. Townes, H. and M. 1960. Ichneumon-flies of America North of Mexico:
    2. Subfamilies Ephialtinae, Xoridinae, Acaenitinae. U. S. Natl.
    Mus. Bull. 216(2): 1-676.

24. Townes, H., Momoi, S. & Townes, M. 1965. A Catalog and Reclassifica-
    tion of Eastern Palearctic Ichneumonidae. Mem. Amer. Ent. Inst.
    5: 1-661.

25. Townes, H. 1969. The Genera of Ichneumonidae, Part I. Mem. Amer.
    Ent. Inst. 11: 1-300.

26. Ulbricht, A. 1911. Ichneumonidenstudien. Arch. Naturg. 77(1 Bd. 2
    Heft): 144-152.

27. Ulbricht, A. 1913. Ichneumoniden der Umgegend Krefelds II. Nachtrag.
    Mitt. Naturw. Mus. Stadt. Crefeld. 1913: 1-17.

28. Walkley, Luella M. 1960. The genus *Delomerista*. In Townes & Townes:
    Ichneumon-flies of America North of Mexico: 2. U. S. Natl. Mus. Bull.
    216(2): 362-373.

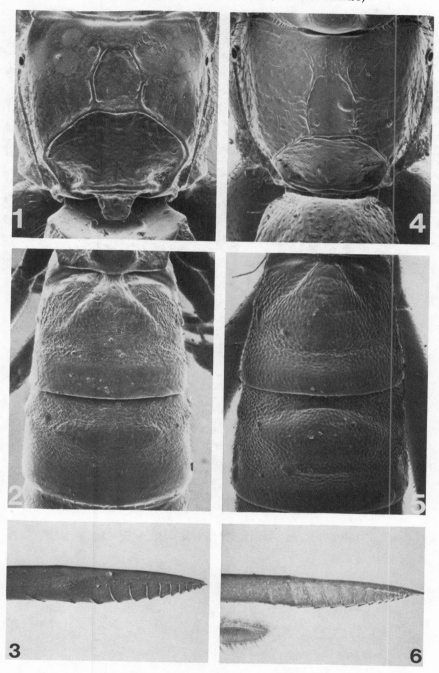

Figs. 1-3. *Delomerista novita:* 1, Propodeum. 2, Tergites II,III. 3, Ovipositor tip. Figs. 4-6. *D. pfankuchi:* 4, Propodeum. 5, Tergites II,III. 6, Ovipositor tip.

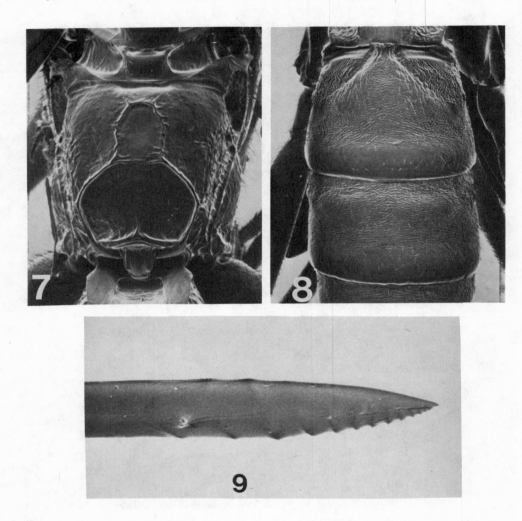

Figs. 7-9. *Delomerista borealis:* 7, Propodeum.   8, Tergites II, III.
9, Ovipositor tip.

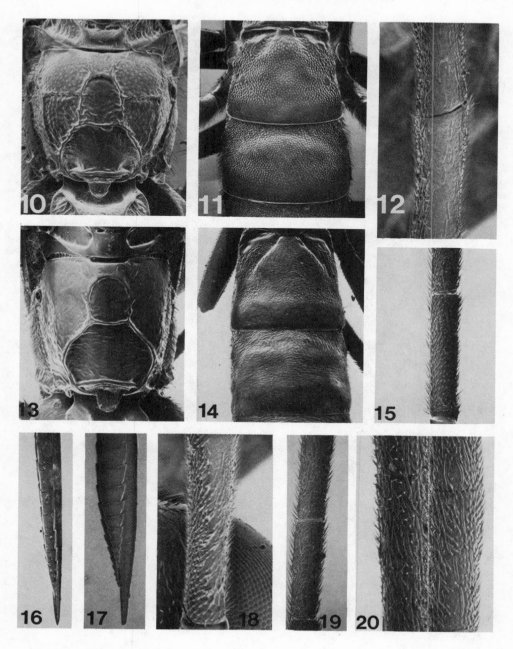

Figs. 10-20. *Delomerista kusuoi:* 10, Propodeum. 11, Tergites II, III. 12, Flagellar segments II, III. 17, Ovipositor tip. 18, Flagellar segment I. *D. lepteces:* 13, Propodeum. 14, Tergites II, III. 15, Flagellar segment I. 16, Ovipositor tip. 19, Flagellar segments II, III. *D. pfankuchi:* 20, Flagellar segment I.

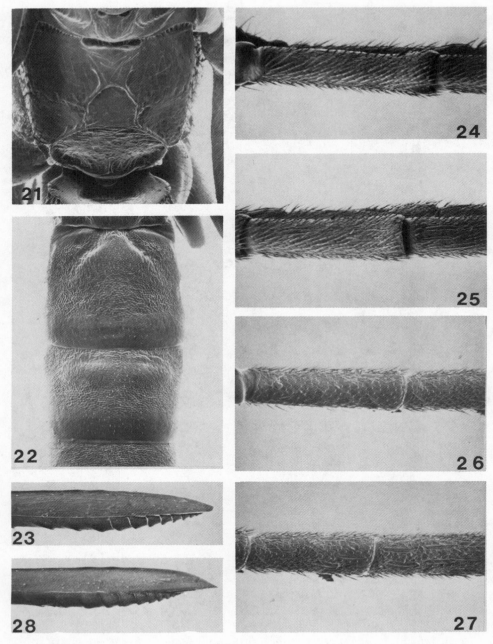

Figs. 21-25. *Delomerista walkleyae:* 21, Propodeum. 22, Tergites II, III. 23, Ovipositor tip. 24, Flagellar segment I. 25, Flagellar segments II, III. Figs. 26-28. *D. townesorum:* 26, Flagellar segment I. 27, Flagellar segments II, III. 28. Ovipositor tip.

Figs. 29-33. *Delomerista japonica:* 29, Propodeum.　30, Tergites II, III.
31, Flagellar segment I.　32, Flagellar segments II, III.　33, Ovipositor tip.

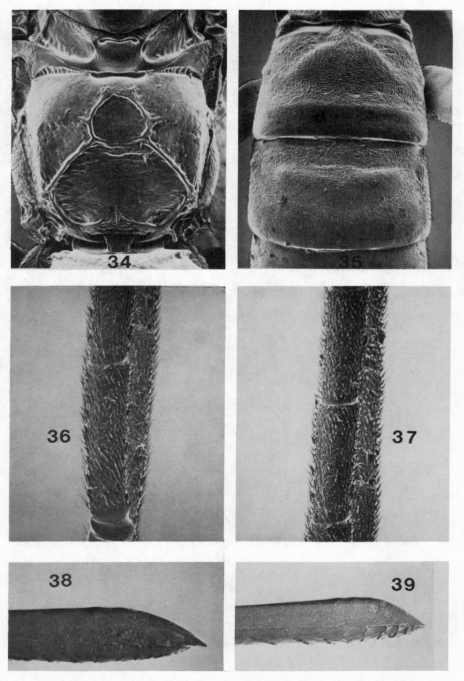

Figs. 34-38. *Delomerista diprionis:* 34, Propodeum. 35, Tergites II, III. 36, Flagellar segment I. 37, Flagellar segments II, III. 38, Ovipositor tip. Fig. 39. *D. japonica,* ovipositor tip.

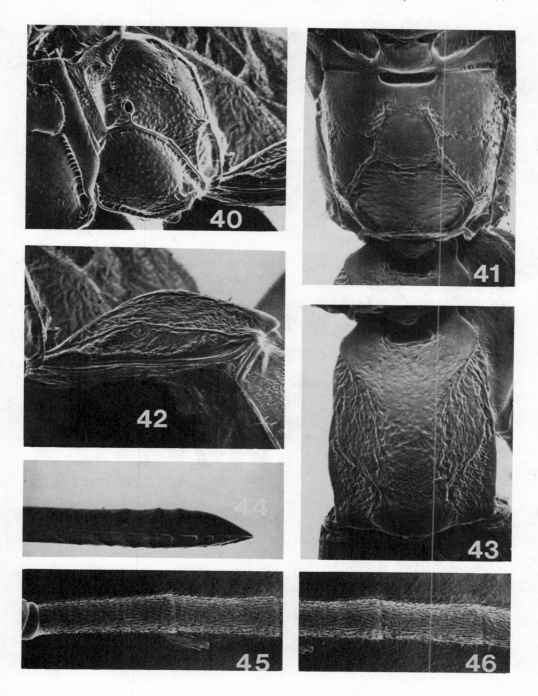

Figs. 40-46. *Delomerista indica:* 40, 41, Propodeum. 42, 43, Tergite I. 44, Ovipositor tip. 45, Flagellar segments I, II. 46, Flagellar segments II, III.

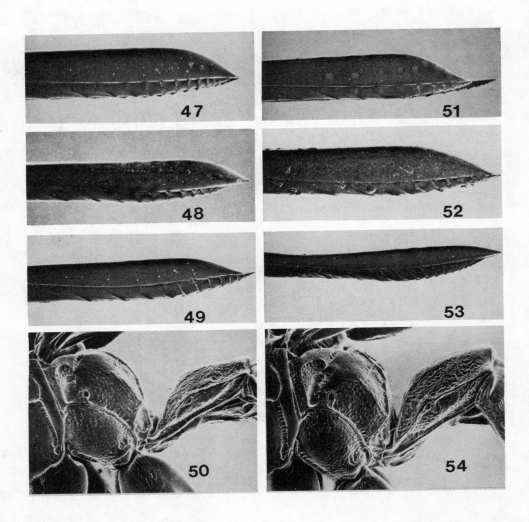

Figs. 47-54. *Delomerista strandi:* 47, 51, Ovipositor tip. 54, Propodeum and first tergite. *D. mandibularis:* 48, 49, Ovipositor tip. 50, Propodeum and first tergite. *D. laevis:* 52, Ovipositor tip. *D. longicauda americana:* 53, Ovipositor tip.

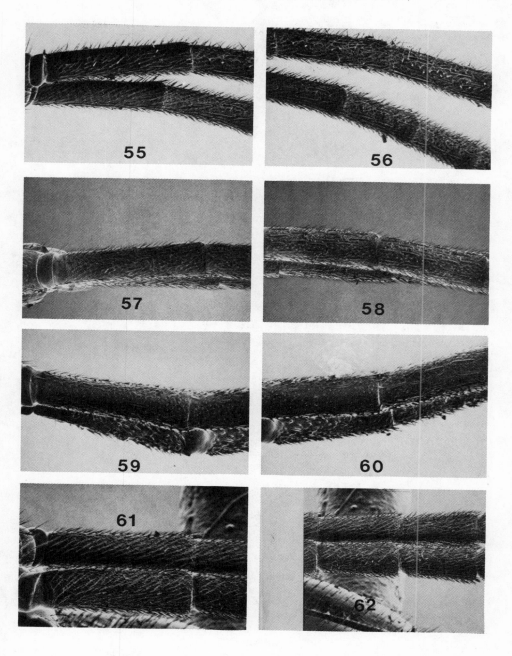

Figs. 55-62. Basal flagellar segments. *Delomerista longicauda americana:* 55, Segments I, II. 56, Segments II, III. *D. strandi:* 57, Segment I. 58, Segments II, III. *D. mandibularis:* 59, Segment I. 60, Segments II, III. *D. laevis:* 61, Segment I. 62, Segments II, III.

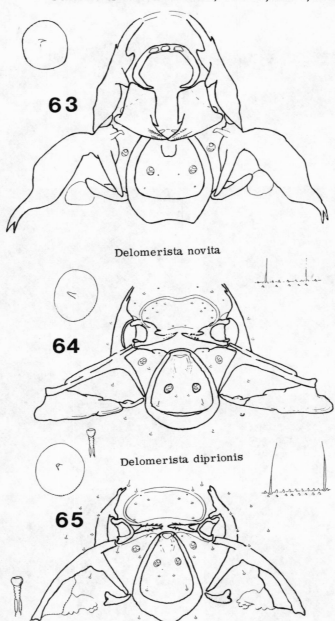

Delomerista novita

Delomerista diprionis

Delomerista japonica

Figs. 63-65. Larval heads of: 63, *Delomerista novita* 64, *D. diprionis*. 64, *D. japonica*

A REVISION OF THE GENUS *AGONOCRYPTUS*

(HYMENOPTERA: ICHNEUMONIDAE)

by
Santosh Gupta *

American Entomological Institute
Ann Arbor, Michigan, 48105

Agonocryptus Cushman, 1929, belongs to the tribe Gabuniini of the sub-family Mesosteninae. It was originally proposed for *Mesostenus discoidaloides* Viereck, 1905, from Kansas, U.S.A., which happens to be the only Nearctic species to date under the genus. The genus, however, is widespread in the Neotropical Region. In addition to the type-species mentioned above, Cushman (1929) also placed *Mesostenus (Mesostenus) chichimecus* Cresson, 1873 from Mexico and *Cryptus heathi* Brues, 1912 from Brazil under this genus. Subsequently five additional species were transferred to it: *Mesostenus (Mesostenus) admirandus* Cresson, 1873 from Mexico by Townes (1946), *M. varus* Brullé, 1846 from Guiana, and *M. physocnemis* Brullé, 1846, *M. luxuriosus* Taschenberg, 1876, and *M. violascens* Taschenberg, 1876 from Brazil, by Townes and Townes (1966). They also synonymized *M. luxuriosus* with *physocnemis*. Thus a total of seven species are so far known from the Nearctic and Neotropical regions.

In the present paper 18 new species and 6 new subspecies are described from the Neotropical Region. The types of all but those of Taschenberg and one of Brullé (*physocnemis*, supposedly lost) have been studied. *A. violascens* (Taschenberg) is not included as I could not find diagnostic characters in the original description for inclusion in the key. *A. discoidaloides* is considered a subspecies of *chichimecus*. Thus a total of 24 species and 7 subspecies are treated in the present paper.

Genus *AGONOCRYPTUS* Cushman   (fig. 1)

> *Agonocryptus* Cushman, 1929. Proc. U. S. Natl. Mus., 74(16): 6.
>> Type: (*Mesostenus discoidaloides* Viereck) = *chichimecus discoidaloides;* original designation.
>> Taxonomy: Townes and Townes, 1962: 502. Townes and Townes, 1966: 128.

Body 8-18 mm. long. Clypeus small, its basal half convex and apical portion flattened, its apical margin truncate, with or without a median tooth. Mandible short, narrow apically, its lower tooth always longer than the upper. Face usually trans-striate. Occipital carina may or may not reach hypostomal carina. Mesoscutum smooth, punctate or rugoso-punctate. Epomia present or absent. Sternaulus weak, reaching middle coxa. Prepectal carina usually 0.8 to 0.9 the height of mesopleurum (except in *A. admirandus* Cresson, where it is only 0.5 the height of mesopleurum). Pleural carina of propodeum absent. Apical transverse carina complete or interrupted medially. Basal transverse carina always

---

* Present address: Center for Parasitic Hymenoptera, Dept. of Entomology & Nematology, Univ. of Florida, Gainesville, FL 32611.

complete and evenly arched. Propodeal spiracle small, roundish. Propodeum between the two transverse carinae variously sculptured. Areolet small, pentagonal to rectangular (wider than high), sometimes a little trapezoidal. Nervulus always basad of basal vein. Second recurrent vein meeting areolet at its middle to apical 0.33. Nervellus intercepted at its middle to upper 0.33. Petiole stout to slender, in female 2.0 to 3.0 as long as its apical width, its sternite 0.4 to 0.75 the length of first tergite. First tergite without sub-basal lateral triangular tooth. In *bicolor* with a ventrolateral tooth on either side. Dorsolateral carina of first tergite usually absent, sometimes in traces near its base. First tergite smooth and shiny, impunctate to strongly punctate. Second tergite impunctate to punctate. Ovipositor tip usually with six teeth. Tip of lower valve of ovipositor with a dorsal lobe enclosing the upper valve.

Hosts: Host records are known only for *Agonocryptus chichimecus*. *Podosesia syringae*, *Eupogonius vestitus* and *Psyrassa unicolor*, are the hosts of *A. chichimecus discoidaloides*, and *Aerenicopsis championi* of *A. chichimecus chichimecus*.

Six species groups are recognized, based on the sculpture of frons, vertex, and mesoscutum, length of petiole, and position of first abdominal sternite relative to the tergite.

List of species of *Agonocryptus* reported in the paper:

I. The Chichimecus Group

1a.	*chichimecus chichimecus* (Cresson)	Mexico
1b.	*chichimecus discoidaloides* (Viereck)	U.S.A.
2.	*bicolor*, new species	Mexico
3.	*ruficrus*, new species	Mexico
4.	*rufithorax*, new species	Brazil
5.	*russulus* new species	Brazil

II. The Physocnemis Group

6a.	*physocnemis physocnemis* (Brullé)	Brazil, Argentina
6b.	*physocnemis nigristernum*, new subspecies	Brazil, Argentina?
7a.	*lioneli lioneli*, new subspecies	Argentina, Brazil
7b.	*lioneli coxinota*, new subspecies	Ecuador
8a.	*argentinus argentinus*, new subspecies	Central Argentina
8b.	*argentinus tucumanus*, new subspecies	Northern Argentina

III. The Heathi Group

9.	*heathi* (Brues)	Brazil

IV. The Varus Group

10a.	*A. varus varus* (Brullé)	French Guinea, Panama
10b.	*A. varus nigrifemur*, new species	Argentina
11.	*A. rufigaster*, new species	Surinam
12a.	*A. adustus adustus*, new subspecies	Peru, Brazil, Ecuador
12b.	*A. adustus paulus*, new subspecies	Colombia, Ecuador
13.	*A. gossypii*, new species	Brazil
14a.	*A. leurosus leurosus*, new species	Brazil
14b.	*A. leurosus flavosternum*, new subspecies	Paraguay, Argentina, Bolivia

15. *A. erugatus*, new species   Panama, Colombia

V. The Amoenus Group

16. *A. admirandus* (Cresson)   Mexico
17. *A. amoenus*, new species   Brazil
18. *A. bispotus*, new species   Mexico
19. *A. tricolor*, new species   Brazil

VI. The Rugifrons Group

20. *A. rugifrons*, new species   Argentina
21. *A. mulleus*, new species   Argentina
22. *A. infuscatus*, new species   Brazil
23. *A. fumosus*, new species   Brazil, Argentina

Group?

24. *A. violascens* (Taschenberg)   Brazil

KEY TO THE SPECIES GROUPS AND SPECIES OF *AGONOCRYPTUS*

1. Frons smooth and shiny, often with a few scattered punctures or dull
    impunctate (*argentinus*). First abdominal tergite impunctate, or with
    a few punctures subapically or apicolaterally. Vertex usually smooth
    and shiny, sometimes with a few minute punctures (in *heathi* vertex
    granulose and in *chichimecus* area behind ocelli punctate). 2
    Frons strongly punctate to rugoso-punctate. First tergite wholly closely
    punctate. Vertex punctate. (Face always rugoso-striate, epomia
    always absent). Rugifrons Group. 27

2. Prepectal carina extending only half the height of mesopleurum. Vertex
    and frons shiny, impunctate. Mesopleurum and metapleurum shiny
    with separated minute punctures. Propodeum striate. Mesoscutum
    with a few scattered punctures. First tergite shiny, impunctate.
    Petiole long, slender, in female nearly 3.0 as long as its apical width.
    First sternite 0.7 as long as petiole. Amoenus Group. Mexico.
                    16. admirandus (Cresson) (p. 31)
    Prepectal carina extending up to 0.8 the height of mesopleurum or up to
    the base of subtegular ridge. Vertex, mesopleurum, propodeum and
    abdominal tergites variously sculptured. 3

3. Mesoscutum punctate or rugoso-punctate. Vertex granulose or shiny with
    scattered minute punctures. Propodeum basad of basal carina punctate,
    sometimes punctures running into striations. 4
    Mesoscutum smooth and shiny, impunctate, sometimes only basally with
    a few punctures. Vertex shiny. Propodeum basad of basal carina shiny
    or with few punctures or fine striations. Body generally shiny. . . 16

4. Vertex dull granulose (sometimes appearing mat). Mesoscutum moderately
       punctate. Basal area of propodeum with yellow elongated mark.
       (Occipital carina joining hypostomal carina at a distance almost equal
       to the basal width of mandible. Lower portion of occipital carina wavy.
       Fore wing with dark band at apex and just before stigma. All abdominal
       tergites with yellow apical bands). Heathi Group. Brazil.
                                    9. heathi (Brues) (p. 20)
    Vertex (including ocellar area) polished and impunctate or area behind
       ocelli punctate. Mesoscutum punctate to rugoso-punctate. Basal area
       of propodeum without any yellow mark. 5

5. Mesoscutum (at least the middle lobe) strongly and densely punctate,
       punctures coalescing and tending to be rugose, or sometimes mat
       areas seen on lateral lobes. Tooth on apical margin of clypeus absent
       or minute. Occipital carina never touching hypostomal carina or in
       chichimecus very close or weakly touching. Basal flagellar segments
       rounded, not flattened. Apical transverse carina of propodeum usually
       complete and distinct or only medially interrupted. Male flagellum
       devoid of hairs and bristles (except for the regular minute pilosity).
       Nervellus intercepted at the middle, except in *rufithorax*. Chichimecus
       Group. 6
    Mesoscutum punctate, punctures well separated, often with smooth areas
       in the middle and on lateral lobes. Tooth on apical margin of clypeus
       always present and distinct, though small. Occipital carina either
       meeting hypostomal carina or strongly bent inwards and coming very
       close to hypostomal carina but not touching it (*physocnemis*). Basal
       flagellar segments moderately flattened. Apical transverse carina of
       propodeum distinct only laterally and broadly interrupted medially.
       Male flagellum with small hairs and with stout small bristles in between.
       Nervellus intercepted above the middle. Physocnemis Group. . . . 11

(I. Chichimecus Group)

6. Thorax wholly reddish-brown, without any marks. Abdomen red or black
       with reddish marks. Face trans-striate. Clypeus laterally striate.
       Epomia absent or faintly indicated in middle of pronotal sulcus. . . . 7
    Thorax black with extensive yellow marks. Abdomen black with yellow
       bands, or reddish with yellow bands. Face dull or smooth in the middle,
       obliquely striate laterally. Clypeus shiny, without striations. Epomia
       present. 8

7. Areolet rectangular, closed, wider than high. Wings hyaline, without any
       brown marks. Abdomen dorsally black, with reddish brown marks.
       Nervulus intercepted above its middle. Brazil.
                                    4. rufithorax, n. sp. (p. 13)
    Areolet more squarish, as wide as high, second intercubitus unpigmented
       (areolet open), intercubiti convergent. Fore wing with fuscous marks
       medially and apically. Abdomen reddish-brown, without distinct black
       marks. Nervellus intercepted at its middle. Brazil.
                                    5. russulus, n. sp. (p. 13)

8. Abdomen reddish-brown, with yellow apical bands. All coxae reddish-brown, rest of the legs yellowish-brown. Scape red. First abdominal tergite punctate centrally. Mesopleurum closely and deeply punctate. Mexico. . . . . 3. ruficrus, n. sp. (p. 12)

    Abdomen black, with yellow bands. Hind coxa black with yellow marks. Femora usually black marked, except in *A. chichimecus discoidaloides*. Scape black, with or without a white mark. First tergite smooth and shiny, or punctate apicolaterally. Mesopleurum with punctures which are quite separate (*A. bicolor*), or punctato-striate (*A. chichimecus*). 9

9. Frons smooth or with scattered punctures. First tergite stouter, bent medially and punctate apicolaterally. Mesopleurum punctate to rugoso-punctate. Metapleurum rugose to rugoso-punctate (punctate in smaller specimens). Apical propodeal carina broadly arched, complete or incomplete medially (in type of *chichimecus* complete and double bent medially). Hind femur reddish-brown or black dorsally and partly on the inner side. U. S. A. and Mexico.
$$\text{1. chichimecus (Cresson) . . . . 10}$$

    Frons rugulose. First tergite narrower apically, smooth, its base slender and with a pointed tooth on either side subbasally. Mesopleurum largely shiny, with minute punctures. Metapleurum punctate, punctures well separated. Apical transverse carina of propodeum broadly arched and very narrowly interrupted medially. Hind femur black dorsally. Mexico. . . . . 2. bicolor, n. sp. (p. 11)

10. Hind femur black-marked dorsally. Hind trochanters black-marked. Hind femur comparatively thicker in the middle. Mesopleural punctures often coalescing. Metapleurum more rugoso-punctate than punctate (sculpture similar to that of propodeum). Male with all abdominal tergites black-banded. Mexico.
$$\text{1a. chichimecus chichimecus (Cresson) (p. 10)}$$

    Hind femur and trochanters uniformly orange-brown (similar to tibia). Femur slender. Mesopleural punctures separated from each other. Metapleurum more punctate than rugose (sculpture weaker than that of propodeum). Male with apical abdominal tergites reddish. U. S. A.
$$\text{1b. chichimecus discoidaloides (Vier.) (p. 11)}$$

(II. Physocnemis Group)

11. Occipital carina interrupted near hypostomal carina, bent inwards. Frons minutely punctate. Subapical flagellar segments not compressed, about 2.0 as long as wide. Rather stout species with yellow marks on thorax and second tergite. (Mark on pronotal collar in female interrupted medially). Brazil and Argentina.
$$\text{6. physocnemis (Brullé) . . . . 12}$$

    Occipital carina meeting hypostomal carina. Frons smooth. Subapical flagellar segments compressed, about as long as wide. Moderately stout species with thorax yellow marked, the pronotal mark complete, or slender species with reddish thorax. 13

12. Mesosternum white marked. First tergite basally reddish to yellow.
    Clypeal groove usually not black marked. Fore and middle coxae white
    or white-marked. . .   6a. physocnemis physocnemis (Brullé) (p. 16)
    Mesosternum black, without white marks. First tergite basally black.
    Clypeal groove black marked. Fore and middle coxae black to largely
    black. . . . . . .   6b. physocnemis nigristernum, n. subsp. (p. 17)

13. Wings clear hyaline, only the apex of fore wing a little fuscous. Thorax
    black, marked with yellow. Mesopleurum minutely punctate and shiny.
    Ecuador, Brazil, and Argentina. . . .   7. lioneli, n. sp. 14
    Wings wholly purple-brownish or fore wing with brownish bands. Thorax
    largely red to largely black, without yellow marks. Mesopleurum punc-
    tate to punctato-striate. Argentina. . 8. argentinus, n. sp. 15

14. Hind coxa and middle coxa reddish-orange. Pronotal collar wholly yellow.
    Second and third abdominal tergites without yellow and black marks.
    Pronotum mostly punctate. Argentina and Brazil.
                              7a. lioneli lioneli, n. subsp. (p. 17)
    Hind coxa and middle coxa yellow, with black marks. Yellow mark on
    pronotal collar interrupted. Second and third tergites without yellow or
    black marks. Pronotum mostly punctato-striate. Ecuador.
                              7b. lioneli coxinota, n. subsp. (p. 18)

15. Wings wholly smoky brownish-black with a purple tinge. Mesopleurum
    largely punctate or in larger and darker specimens somewhat striate.
    Hind tibia black on apical 0.15 to 0.2. Scape and pedicel reddish-brown.
    Central Argentina. . . .   8a. argentinus argentinus, n. subsp. (p. 19)
    Wings hyaline. Fore wing with two brownish bands separated by a hyaline
    area. Mesopleurum striato-punctate. Hind tibia black on apical 0.35.
    Scape and pedicel black. Northern Argentina.
                              8b. argentinus tucumanus, n. subsp. (p. 20)

16. First abdominal segment of female 2.0 to 2.4 x as long as its apical width,
    quadrangular in cross-section at subbasal region, its sternite extending
    up to 0.4 to 0.6 the length of tergite, usually to the level of spiracle.
    First sternite separated from its tergite by a ridge. First tergite with
    ventrolateral triangular projections at base (sometimes weak). Post-
    genal area evenly arched and not separated by a crease from the lower
    part of gena. Occipital carina not strongly deflected inwards and point
    ing towards base of mandible, away from hypostomal carina and touch-
    ing it only in gossypii. Varus Group.   17
    First abdominal segment long and slender, in female 3.0 x as long as its
    apical width, rounded in cross-section at subbasal region, its sternite
    extending up to 0.75 the length of tergite. First sternite and tergite
    fused together, without any trace of ventrolateral carina except in
    tricolor, where ventrolateral projections are seen at base. Postgenal
    area somewhat flattened and widened and separated from the lower por-
    tion of gena by a crease (in profile view). Occipital carina deflected
    inwards either strongly or weakly, coming close to joining hypostomal
    carina or erased for a distance equal to the basal width of mandible.
    Amoenus Group. 25

(IV. Varus Group)

17. Abdomen impunctate and polished. First abdominal sternite short of the level of spiracle, 0.4 the length of tergite. Propodeum basad of basal carina smooth and shiny. Abdominal tergites sparsely hairy dorsally, except tips of seventh and eighth. Body rather stout. Panama and Colombia. . . . .   15. erugatus, n. sp. (p. 29)
  Abdomen with punctures, at least on second and third tergites laterally. First abdominal sternite extending up to the level of spiracle or slightly beyond it. Basal area of propodeum trans-striate (except in *adustus*). Fourth and following tergites rather uniformly hairy dorsally, with their margins fringed with hairs, especially seventh and eighth. Body slender. 18

18. Pronotum striate, particularly along the sulcus and along posterior margin. Pronotal collar and upper margin with scattered punctures. First sternite ending at the level of spiracle and 0.5 the length of tergite. Areolet 1.5 to 1.75 as wide as high. Brazil, Paraguay, Bolivia, and Argentina. . . . .   14. leurosus, n. sp. 19
  Pronotum punctate to mat. Pronotal collar sometimes smooth and polished. First sternite either ending at the level of spiracle or extending slightly beyond it, 0.5 to 0.6 the length of tergite. Areolet 1.25 to 1.5 as wide as high. 20

19. Mesosternum and middle coxa dorsally black. First tergite wholly reddish. Brazil. . . . .   14a. leurosus leurosus, n. subsp. (p. 28)
  Mesosternum with yellow marks. Middle coxa dorsally red. First tergite with apical and basal yellow marks and medially usually blackish, sometimes second tergite with blackish marks. Paraguay, Bolivia, and Argentina. . . . . . .   14b. leurosus flavosternum, n. subsp. (p. 28)

20. Thorax wholly yellowish-brown and abdomen from second segment onwards brownish-black. Wings tinged with fuscous brown. Male abdomen yellowish-brown, as is thorax. Flagellar segments not flattened and without long hairs. Small sized species. Brazil.
                              13. gossypii, n. sp. (p. 26)
  Thorax black with yellow marks. Abdomen reddish-brown. Wings clear hyaline, with or without fuscous patches. Basal flagellar segments in female usually flattened. Male flagellum usually with long hairs. . 21

21. Propodeum largely smooth, with weak to indistinct striations, particularly on the basal area. Apical transverse carina indistinct, or if faintly indicated, more semicircular, like carinae bordering petiolar area. First sternite 0.6 the length of tergite, ending slightly beyond the level of spiracle. Fore wing with fuscous bands at middle and at its apex. Peru, Brazil, Ecuador, and Colombia.
                              12. adustus, n. sp. 22
  Propodeum largely strongly striate to rugoso-striate. Apical transverse carina distinct, strongly arched medially. First sternite 0.5 to 0.6 the length of its tergite. Fore wing with or without fuscous bands. . 23

22. Face finely striate. First tergite, hind coxa, trochanters, and femur reddish-brown, without any marks. Hind tibia and tarsus yellow with fourth tarsal segment apically and fifth either partly or wholly brownish. Pronotal collar wholly yellow. Apex of fore wing very lightly fuscous. Basal flagellar segments flattened (as usual). Size 8-14 mm. Peru, Brazil and Ecuador. . . . . 12a. adustus adustus, n. subsp. (p. 25)

  Face smooth and shiny, without striations. First tergite apically and basally yellow. Hind coxa, femur and trochanters reddish-brown with fuscous marks. Apex of hind femur blackish. Hind coxa with a dorsal yellow line. Hind tibia and tarsus yellow, but fourth and fifth tarsal segments black. Apical fuscous mark on fore wing more conspicuous. Yellow mark on pronotal collar interrupted medially. Basal flagellar segments not conspicuously flattened. Size 10 mm. Colombia and Ecuador. . . . . 12b. adustus paulus, n. subsp. (p. 26)

23. Propodeum strongly rugose medially. Petiolar area with vertical striations. First sternite extending beyond the level of the spiracle, 0.6 the length of its tergite. Pronotum irregularly punctate, the groove along its hind margin wrinkled. Areolet rectangular, about 1.5 as wide as high. Mesoscutum with yellow crescentic mark on side of lateral lobe. Yellow mark on pronotal collar interrupted medially. Surinam.
                                      11. rufigaster, n. sp. (p. 23)

  Propodeum striate. Petiolar area of propodeum smooth to dull mat. First sternite ending at the level of spiracle. Pronotum with minute punctures, shiny, the groove along its hind margin smoother. Areolet more squarish, about 1.2 as wide as high. Lateral lobes of mesoscutum without any yellow marks. Yellow mark on pronotal collar not interrupted medially. . . . . 10. varus (Brullé) 24

24. Femora yellowish-brown. Yellow mark on pronotal collar interrupted medially. Abdomen yellowish-brown. French Guiana and Panama.
                              10a. varus varus (Brullé) (p. 22)

  Fore and middle femora ventrally and hind femur wholly black. Pronotal collar wholly yellow. Abdomen reddish-brown. Argentina.
                              10b. varus nigrifemur, n. subsp. (p. 23)

(V. Amoenus Group)

25. Occipital carina making a right angle with hypostomal carina, strongly deflected towards base of hypostomal carina, away from the base of mandible by 1.5 the basal width of mandible. Abdomen smooth and shiny, impunctate. Tergites without hairs. Basal area of propodeum with conspicuous semicircular striations. Brazil.
                              17. amoenus, n. sp. (p. 32)

  Occipital carina normal, making an angle of less than 45⁰ with the hyposto-mal carina and joining it. Occipital carina not deflected towards base of hypostomal carina, running towards base of mandible, but often erased (except in tricolor, in which propodeum smoother). Abdomen with setiferous punctures and apical tergites hairy. Basal area of propodeum without striations, often punctate laterally. 26

26. Epomia present. Propodeum apicad of basal carina with strong striations and punctures in between. Abdomen banded with black and yellow. Mexico. . . . . 18. bispotus, n. sp. (p. 33)
Epomia absent. Propodeum with broken to indistinct striations apicad of basal transverse carina. Abdomen reddish, without stripes. Brazil.
19. tricolor, n. sp. (p. 34)

(VI. Rugifrons Group)

27. Wings hyaline. Face strongly rugose in the middle. Clypeus granulose with a few punctures. Thorax black, with yellow marks. Abdomen reddish-brown or body wholly reddish-brown. 28
Wings strongly infuscate. Face rugoso-striate. Clypeus finely striate. Thorax reddish-brown, without any marks. Abdomen black with reddish marks. 29

28. Thorax black with yellow marks. First tergite white apically. Tarsal segments of middle leg reddish-brown to brownish-black. First tergite closely punctate. Occipital carina meeting hypostomal carina. Argentina. . . . . 20. rugifrons, n. sp. (p. 35)
Body wholly reddish-brown. First tergite apically not white. Tarsal segments of middle leg white. First tergite only centrally with sparse punctures. Occipital carina not meeting hypostomal carina. Argentina. . . . . 21. mulleus, n. sp. (p. 36)

29. Occipital carina not meeting hypostomal carina. Face reddish-brown, without yellow orbital rings. Apical 0.6 of first tergite and hind femur black. Second and third tarsal segments of hind leg yellow. Brazil.
22. infuscatus, n. sp. (p. 36)
Occipital carina meeting hypostomal carina. Face black, with broken yellow orbital rings. First tergite, hind coxa and hind femur reddish-brown. In addition to second and third segments, the apical 0.6 of first tersal segment of hind leg yellow. Argentina and Brazil.
23. fumosus, n. sp. (p. 37)

I. THE CHICHIMECUS GROUP

This group is characterized by having the frons smooth; vertex generally smooth with minute scattered punctures or punctures closer behind ocelli; occipital carina almost touching to widely separated from hypostomal carina, not deflected inwards; tooth on apical margin of clypeus absent or minute; basal flagellar segments of female antenna not compressed; flagellar segments in male devoid of long hairs (males of only *chichimecus* and *russulus* are known); pronotum striate to striato-punctate; epomia absent or present; mesoscutum strongly to densely punctate; punctures coalescing and tending to be rugose; propodeum punctate basally, rugoso-striate medially, its apical transverse carina present, complete or broken medially; abdomen strongly punctate. The nervellus is intercepted at the middle, except in *rufithorax*.

This group includes five species: *A. chichimecus* (Cresson) from the U.S. A. and Mexico, and four new species, *bicolor* and *ruficrus* from Mexico and *rufithorax* and *russulus* from Brazil   (maps 1, 2).

A. chichimecus and *bicolor* are distinctive by the absence of tooth on clypeal margin and by the presence of epomia. The body is marked with yellow. They can be differentiated among themselves by the nature of frons, apical transverse carina of propodeum, shape and punctation of first tergite, and by the color of hind leg. *A. rufithorax* and *A. russulus* are distinctive in having reddish thorax and abdomen, without yellow marks, meso- and metapleurum punctate, clypeal margin with a minute tooth, and by the absence of epomia. They can be differentiated among themselves by the nature of areolet, markings on wings, and by the position of nervellus. *A. russulus* is the only species with fuscous marks on the fore wing in the female. *A. ruficrus* has many characters in common with *A. chichimecus* and *bicolor*, but abdomen is reddish-brown, coxae reddish-brown, and clypeal margin with a minute tooth.

1. AGONOCRYPTUS CHICHIMECUS (Cresson)

Female: Face shiny with a few trans-striations below antennal sockets to granulose with submedian oblique striations. Clypeus shiny, without any punctures or striations, its apical margin truncate, without a median tooth. Malar space granulose, nearly equal to the basal width of mandible. Temple shiny, impunctate. Frons shiny, with a few fine superficial punctures in ocellar area. Vertex shiny, closely punctate behind ocelli. Occipital carina close to hypostomal carina and separated from it by a distance equal to 0.25 to 0.5 the basal width of mandible. Pronotum punctate to striate, its hind margin and pronotal sulcus striate. Epomia present, strong or weak. Mesoscutum closely and deeply punctate. Prepectal carina reaching subtegular ridge. Scutellum sparsely punctate. Mesopleurum punctate to rugoso-punctate, punctures close (punctate in smaller specimens). Metapleurum rugose to rugoso-punctate. Apical transverse carina of propodeum broadly arched, complete or incomplete medially. Propodeum strongly punctate to rugoso-punctate, centrally more rugose, its petiolar area with irregular striations or longitudinal ridges. Areolet trapezoidal to pentagonal (first intercubitus shorter than the second). Second intercubitus pigmented or unpigmented. Second recurrent vein meeting areolet at its apical 0.4 to 0.3. Nervellus intercepted at its middle. Petiole stout, its sternite reaching the spiracle. First tergite shiny, punctate dorsolaterally. Second and third tergites wholly closely and finely punctate. The following tergites punctate. Ovipositor 0.6 to 0.7 the length of abdomen.

Two subspecies, *A. chichimecus chichimecus* (Cresson) and *A. c. discoidaloides* (Viereck) *(New Status)* are recognized.

1a. AGONOCRYPTUS CHICHIMECUS CHICHIMECUS (Cresson)

Mesostenus (Mesostenus) chichimecus Cresson, 1873. Proc. Acad. Nat. Sci. Philadelphia, 1873: 155. ♀. key, des. Lectotype: ♀, Mexico: Orizaba (Philadelphia). Mexico: Cordoba. Type examined.
Mesostenus chichimecus: Cresson, 1916. Mem. Amer. Ent. Soc., 1: 23. Lectotype designation.
Agonocryptus chichimecus: Cushman, 1929. Proc. U. S. Natl. Mus., 74(16): 7. syn. Townes, 1946. Bol. Ent. Soc. Venezolana, 5: 34. Townes, 1966. Mem. Amer. Ent. Inst., 8: 128.

Male and *Female*: Punctures on mesopleurum often coalescing. Metapleurum rugoso-punctate. Hind femur thicker than in *discoidaloides*. Black, with yellow marks and bands on abdomen. Face, clypeus, labrum, a mark on base of mandible, orbital ring, temple broadly, a mark on mesoscutum, scutellum, metascutellum, propleurum, upper margin of pronotum, pronotal collar completely or interrupted medially, tegula, subtegular ridge, speculum, a large mark covering most of mesopleurum, mesosternum large, a mark below wings, metapleurum, and a dagger-shaped mark on petiolar area of pro-podeum, yellow. Wings hyaline. Legs yellow. Fore femur ventrally and fore tarsus brown. Hind coxa basally and with an oblong mark dorsally, trochanters partly, and femur dorsally, black. First tergite apically and basally, and the following tergites apically and laterally yellow.

Length: 11-16 mm. Fore wing 8-12 mm.

Specimens examined: 22♀, 4♂. *Mexico:* Orizaba, 1♀ (lectotype). *Mexico:* Cordoba, 1♀ (labelled paratype). [Specimens bear label only as Mexico. Locality data are in the original descriptions]. *Mexico:* 1♀ (paratype, Washington). *Mexico:* San Rafael, Jocoltepec, 1♀. Ticul Yucatan, 3♀, June 1969, CIBC, *ex* Cerambycid on *Eupatorium odoratum.* Vera Cruz, 3♀, 2♂, *ex Aerenicopsis championi,* Krauss, No. 5776. Vera Cruz, Ver., 6♀, May 1959, N. L. H. Krauss: 1♀, June 1955, *ex Aerenicopsis championi.* Ma Cambo, Vera Cruz, Ver. 1♀, June 1965, N. L. Krauss *ex Aerenicopsis championi* in branch of *Lantana camara.* Vera Cruz., Ver., 4♀, 2♂, May 1955, *ex Aerenicopsis championi* burrows in *Lantana camara* (all Washington).

Host: Aerenicopsis championi.

Distribution: Mexico.

1b. AGONOCRYPTUS CHICHIMECUS DISCOIDALOIDES (Viereck), new status

Mesostenus discoidaloides Viereck, 1905. Trans. Kansas Acad. Sci., 19: 319. ♀. Type: ♀, Rock Creek, 900 ft., Douglas Co., Kansas (Lawrence).

Agonocryptus discoidaloides: Cushman, 1929. Proc. U. S. Natl. Mus., 74(16): 6. Townes and Townes, 1962. Bull. U. S. Natl. Mus., 216(3): 502. des., fig.

Female: Punctures on mesopleurum a little separated from each other. Metapleurum more punctate than rugose. Clypeal groove and base of clypeus usually black. Second intercubitus pigmented. Hind femur and trochanters uniformly orange brown.

. *Male:* Slender. Structure similar to that of female except that pronotum is shiny, without striations or punctures and punctation of mesopleurum and meta-pleurum a little sparse and superficial. Apical abdominal tergites usually reddish. First abdominal tergite basally black. Hind tibia apically darker. (In *chichimecus chichimecus* all abdominal tergites are black, banded with yellow.)

Length: 10-18 mm. Fore wing 6-14 mm.

Specimens examined: U. S. A: *Alabama* (Gulf Shores). *Florida* (Key Largo, Larkins, Paradise Key, Tarpon Springs). *New Jersey* (Moorestown). *New York* (Farmingdale). *North Carolina* (Wake Co.). *South Carolina* (McClellanville). *Wisconsin* (Milwaukee).

Hosts (From Carlson's Catalog): *Podosesia syringae, Eupogonius vestitus, Psyrassa unicolor.*

Distribution: U. S. A.: Southeastern United States, Texas, Kansas, and some localities in Midwestern States. Distributional Map 288 in Townes and Townes, 1962.

2. AGONOCRYPTUS BICOLOR, n. sp.

Female: Face centrally smooth and shiny, laterally with a few oblique striations. Clypeus smooth and shiny, its apical margin truncate, without a median tooth. Malar space granulose, 1.0 the basal width of mandible. Temple shiny, impunctate. Frons rough or rugulose, a little densely rugulose in ocellar area. Vertex medially punctate. Occipital carina close to hypostomal carina and separate from the latter by a distance about half the base of mandible. Pronotum punctate above, striate in the sulcus, striate along margins. Epomia present. Mesoscutum deeply and closely punctate. Prepectal carina reaching close to subtegular ridge. Mesopleurum minutely punctate, with shiny areas between punctures. Area below subtegular ridge with a few striations. Metapleurum punctate, punctures well separated. Propodeum with narrowly interrupted apical transverse carina which is strongly curved submedially. Basal area of propodeum striato-punctate. Area between the two transverse carinae punctato-striate. Petiolar area punctate, with a few longitudinal ridges. Areolet pentagonal. Second recurrent vein meeting areolet in the middle. Nervellus intercepted at its middle. Petiole moderately stout, its base slender and with a ventro-lateral tooth on either side. First sternite reaching apical 0.4 of petiole. First tergite smooth and shiny, with a few scattered punctures. Second and the third tergites basally and medially closely and deeply punctate. The following tergites finely mat.

Black, with yellow marks and bands. Face, broad orbital rings, clypeus except margins, temple, base of mandible, and labrum, yellow. A mark on upper margin of pronotum, pronotal collar, mark on propleurum, central mark on mesoscutum, scutellum, metascutellum, tegula, subtegular ridge, speculum, a large mark covering most of mesopleurum, mesosternum, a mark below wings, metapleurum, and an inverted T-shaped mark on petiolar area of propodeum, yellow. Legs yellow with black marks. Fore femur dorsally and tarsus brownish. Middle coxa ventrally with a basal black mark. Middle femur and tibia ventrally, and tarsus, brown. Hind coxa basally and dorsally black marked. Trochanters and femur dorsally black. Wings hyaline. First tergite apically and the following tergites apically and laterally, yellow.

Male: Unknown.

Length: 15 mm. Fore wing 12 mm.

Holotype: ♀, *Mexico:* Oaxaca, Metate, 85 kilometers south of Tuxtepec, 9000 m., Oct. 20, 1962, H. & M. Townes (Townes).

Distribution: Mexico (map 1).

3. AGONOCRYPTUS RUFICRUS, n. sp.

Female: Face dull, with a few punctures and oblique striations submedially. Clypeus dull, its apical margin with a minute tooth. Malar space granulose, 0.9 the basal width of mandible. Temple shiny, impunctate. Frons with a few minute scattered punctures. Vertex with fine punctures in ocellar area. Occipital carina close to hypostomal carina and separated from it by a distance of 0.25 the basal width of mandible. Pronotum deeply punctate, its hind margin and sulcus with striations. Epomia present, though not complete. Mesoscutum closely and deeply punctate. Scutellum deeply but sparsely punctate. Mesopleurum closely and deeply punctate. Metapleurum with irregular punctures, punctures close and deep. Propodeum rugoso-punctate to rugose, its apical transverse carina broadly interrupted medially. Areolet slightly trapezoidal, second recurrent vein meeting areolet at its apical 0.4. Nervellus intercepted at its middle. Petiole stout, its length less than 2.0 the apical width of first tergite, its sternite reaching the spiracle. First tergite punctate centrally. Second and third tergites closely punctate. The following tergites finely mat. Ovipositor 0.9 the length of abdomen.

Black with yellow marks. Abdomen reddish with yellow bands. Scape reddish-brown. Face, clypeus except its apical margin, labrum, base of mandible, broad orbital rings, and temple, yellow. Upper margin of pronotum, pronotal collar, propleurum, a mark on mesoscutum, scutellum, metascutellum, tegula, subtegular ridge, speculum, mesopleurum broadly, mesosternum, area below wings, metapleurum wholly, and a large dagger-shaped mark on propodeum extending up to basal transverse carina, yellow. Wings slightly clouded apically. Legs reddish-yellow. First tergite apically and basally, and the following tergites apically, with yellow bands.

Male: Unknown.

Length: 16 mm. Fore wing 12 mm.

Holotype: ♀, [Baja California, *Mexico*]: San Jose del Cabo (Washington).

Distribution: Mexico (map 1).

4. AGONOCRYPTUS RUFITHORAX, n. sp.

Female: Face granulose with striations submedially and a few scattered punctures laterally. Clypeus granulose, laterally finely striate, its apical margin with a minute median tooth. Malar space obliquely aciculate, 0.8 the basal width of mandible. Temple smooth and shiny. Frons shiny, with fine punctures close to ocelli. Vertex shiny, with a few fine superficial punctures. Occipital carina directed towards base of mandible and separated from the latter by a distance about equal to the basal width of mandible; this area finely striate. Pronotum striato-punctate, its hind margin striate. Epomia absent. Mesoscutum closely and deeply punctate. Scutellum with fine, deep punctures. Prepectal carina reaching subtegular ridge. Mesopleurum and metapleurum closely and deeply punctate. Propodeum deeply punctate, punctures running into transverse striations. Petiolar area sparsely punctate, with a few lateral longitudinal ridges. Apical transverse carina present, but a little irregular submedially. Areolet rectangular, closed, wider than high. Second recurrent vein meeting areolet at its middle. Nervellus intercepted above the middle. Petiole moderately long, not very stout, 2.0 as long as its apical width, smooth and shiny and with a few fine superficial punctures. Second and third tergites closely and finely punctate. The following tergites finely mat. Ovipositor 0.7 the length of abdomen.

Brownish-red. Abdomen brownish-black. Interrupted orbital rings, basal half of clypeus, and labrum, yellow. Wings hyaline. Legs reddish-brown. Middle tibia lighter in color. Hind femur and tibia black. Hind tarsus yellow. Second and following abdominal tergites blackish with reddish-brown marks submedially.

Male: Unknown.

Length: 12-15 mm. Fore wing 8-10 mm.

Holotype: ♀, *Brazil:* Nova Teutonia, Santa Catarina, April 13, 1948, Fritz Plaumann (Townes).

Paratypes: 3♀, *Brazil:* Nova Teutonia, Santa Catarina, Feb. 1946 and Dec. 1970 (Townes). Nova Teutonia, 27º 11' B. 52º 23' L., 300-500 m., 1♀, Nov. 1966, Fritz Plaumann (Ottawa).

Distribution: Brazil (map 2).

5. AGONOCRYPTUS RUSSULUS, n. sp.

Female: Face and clypeus finely striate. Apical margin of clypeus truncate and with a minute median tooth. Malar space granulose, equal to the basal width of mandible. Temple shiny. Occipital carina directed towards base of mandible and separated from the latter by a distance equal to the basal width of mandible; this area finely striate. Frons shiny and with a few superficial punctures. Pronotum punctate with a few oblique striations, its hind margin striate. Epomia absent. Mesoscutum, mesopleurum, and metapleurum punctate. Scutellum sparsely punctate. Propodeum basally rugoso-punctate, medially rugoso-striate. Petiolar area somewhat irregularly striate. Areolet squarish, as wide as high, intercubiti convergent, but second intercubitus largely unpigmented (areolet open). Second recurrent vein meeting areolet at its middle. Nervellus intercepted at its middle. First tergite slender, punctate apicolaterally, its sternite reaching the spiracle. Second and third tergites closely and finely punctate. The following tergites finely mat. Ovipositor 0.8 as long as the abdomen.

Reddish-brown. Median flagellar segments, a mark on labrum (and sometimes on clypeus), interrupted orbital rings, a mark on pronotal collar, upper edge of pronotum, tegula, and subtegular ridge, white. Ventral aspect of first trochanter of fore leg, and hind tarsus, white. Middle coxa and femur, hind coxa, trochanters, femur and tibia, and abdominal tergites apically, brownish. Wings hyaline and with brownish marks apically and medially.

Male: Antenna spinose, with small fine hairs inbetween. Face and clypeus finely striate, punctate inbetween. Frons granulose with fine transverse aciculations. Vertex shiny, sparsely superficially punctate. Pronotal collar sparsely superficially punctate, its groove shiny and without striations or punctures. Mesoscutum closely superficially punctate. Mesopleurum shiny. Metapleurum with a few superficial punctures. Apical and basal transverse propodeal carinae complete and strong. Basal area of propodeum close to spiracle irregularly punctate, its lateral area inbetween the two transverse carinae rugose and medially with a few trans-striations. Petiolar area shiny, with two longitudinal striate, one on each side of middle. Second recurrent vein meeting areolet at its apical 0.3.

Yellowish-brown. Tenth to fifteenth flagellar segments yellow. Face, clypeus, labrum, base of mandible, temple broadly, interrupted orbital rings, propleurum, pronotal collar, tegula, subtegular ridge, a mark on scutellum, fore coxa and trochanter, and hind tarsus yellow. Second and following abdominal tergites basally, hind tibia apically, and first hind tarsal segment darker. Wings hyaline.

Length: 10-15 mm. Fore wing 8-12 mm.

Holotype: ♀, *Brazil:* Santa Barbara, Minas Gerais, Serra do Caraca, Jan. 1970, Oliveira (Townes).

Allotype: ♂, Same locality and data as the type (Townes).

Paratypes: 3♀, 3♂. *Brazil:* Same locality as the type, 1600 m., 1♂, 1♀, Jan. 1970 and Feb. 1969, F. M. Oliveira. Guanabara, Represa Rio Grande, 1♀, March 1972, M. Alvarenga. Murique, Rio de Janeiro, Mangaratiba, 1♀, July, 1969, M. Alvarenga. S. J. Barreiro, Sao Paulo, Serra da Bocaína, 1650 m., 2♂, Nov. 1969, M. Alvarenga and Seabra (Townes).

Variations: Two males from Bocaína are a little darker than the others, especially the mesoscutum and pronotum.

Distribution: Brazil (map 2).

II. THE PHYSOCNEMIS GROUP

This group is characterized by having the frons smooth and shiny; vertex with a few scattered minute punctures behind ocelli; occipital carina meeting hypostomal carina or bent inwards and close to it but not actually touching it (*physocnemis*); clypeus with a median tooth along its apical margin; basal flagellar segments somewhat compressed; pronotum striate in middle and punctate along upper and hind margins; epomia present but not complete; mesopleurum punctate but punctures well separated; metapleurum punctate; propodeum basally strongly punctate; apical transverse carina distinct only laterally, broadly interrupted medially; and abdomen punctate. The male flagellum is beset with small hair with stout bristles in between. The nervellus is intercepted above the middle.

The Physocnemis Group includes three species: *A. physocnemis* (Brullé) from Brazil and Argentina; *A. lioneli* from Argentina, Brazil and Ecuador; and *A. argentinus* from *Argentina*. Each species has two subspecies (map 3).

A. physocnemis has its occipital carina bent inwards, coming close to the middle of hypostomal carina, but not meeting it, frons minutely punctate and subapical flagellar segments not compressed. In the other two species the occipital carina meets the hypostomal carina and the subapical flagellar segments are compressed. In *A. argentinus* the wings are either wholly purple-brownish or with blackish-brown bands on the fore wing. In *A. lioneli* the wings are hyaline or a little brown-tinged apically. The mesopleurum of *A. lioneli* is minutely punctate and shiny, while *A. argentinus* exhibits variations in having punctate to rugoso-punctate mesopleurum. These two species can be further differentiated by the color of abdomen and thorax.

6. AGONOCRYPTUS PHYSOCNEMIS (Brullé)

Female: Subapical flagellar segments not compressed, about 2.0 as long as wide. Face granulose, with a few punctures laterally; medially and submedially with fine trans-striations. Clypeus granulose, with a few impressions of punctures or a few trans-striations laterally, its apical margin with a median tooth. Malar space granulose. Temple shiny, finely granulose close to malar space. Frons shiny, with minute punctures. Vertex shiny, with a few scattered punctures behind ocelli. Occipital carina angularly bent and coming very close to hypostomal carina, sometimes bent portion erased and replaced by one or two striations. Pronotum deeply punctate. Pronotal collar finely and closely punctate. Pronotal sulcus with striations. Epomia absent. Mesoscutum punctate, with shiny areas in-between. Scutellum sparsely punctate, punctures superficial. Mesopleurum punctate, but punctures in yellow area not prominent. Area below subtegular ridge with a few trans-striations. Metapleurum punctate. Basal area of propodeum strongly punctate to punctato-striate. Propodeum medially rugoso-striate or punctato-striate. Apical transverse carina broadly interrupted medially. Petiolar area irregularly longitudinally striate. Areolet wider than high. Second recurrent vein meeting areolet at its middle. Nervellus intercepted at its upper 0.4. Petiole stout to very stout, its sternite 0.4 to 0.5 the length of petiole. First tergite shiny and with a few punctures. Second and third tergites with close and fine punctures. The following tergites finer and mat. Ovipositor 0.6 as long as abdomen.

Black, with yellow marks. Abdomen brown. Face, clypeus except its apical margin, labrum, orbital rings, temple, propleurum apically, upper part of pronotum, interrupted mark on pronotal collar, a mark on mesoscutum, scutellum, metascutellum, tegula, subtegular ridge, speculum, a broad oblong mark on mesopleurum, a mark on mesosternum (present or absent), mark on metapleurum (except basal black band), and an inverted T-shaped mark on apical slope of propodeum, yellow. Wings hyaline. Fore and middle legs yellow with their coxae and femora ventrally black and tarsi brownish-black. Hind coxa, trochanters, and femur reddish-brown, tibia and tarsus yellow. First tergite black, with yellow marks at base and apex. (In subspecies *nigristernum* yellow only apically). The following tergites brown with the second tergite basally black and apically yellow. Sometimes third and fourth tergites also narrowly brownish-black basally and yellow apically.

   Male: Antennal flagellum with small hairs and spines in between. Pronotum shiny, without punctures. Otherwise sculpture similar to that of the female.

   Black with yellow marks. Resembles female with the following differences: Pronotal collar wholly yellow. Mesopleural mark larger. Mesosternum wholly yellow or wholly black. Wings hyaline. Fore and middle coxae and trochanters wholly white. Hind coxa black with yellow marks. Hind femur black, yellowish-brown basally. Hind tibia basally yellow and apically black. First tarsal segment black. Abdomen black with apical yellow bands.

   Two subspecies are recognized: *Agonocryptus physocnemis physocnemis* (Brullé) and *A. physocnemis nigristernum,* n. subsp. Their chief difference is in the coloration of mesosternum and first tergite, as given in the key.

6a. AGONOCRYPTUS PHYSOCNEMIS PHYSOCNEMIS (Brullé)

Mesostenus physocnemis Brullé, 1846. *In* Lepeletier: Histoire Naturelle
   des Insectes. Hyménoptères, 4: 236. ♀. des. Type: ♀, Brazil (lost).
Mesostenus luxuriosus Taschenberg, 1876. Ztschr. f. die Gesam. Naturw.
   Halle, 48: 94. ♀. des. Type: ♀, Brazil: [Nova Friburgo] (Halle).
   Synonymized by Townes, 1966.
Agonocryptus physocnemis: Townes, 1966. Mem. Amer. Ent. Inst.,
   8: 128.

   Characterized as under the species and in the key. This subspecies is more reddish with fore and middle coxae white or white marked and mesosternum with white marks. The base of first tergite is reddish or yellow.

   Length: ♀: 20-12 mm. Forewing 15-10 mm. ♂: 15-12 mm. Forewing 6-8 mm.

   Specimens examined: 21♀, 8♂. *Brazil:* Linhares, Espirito Santo, 5♀, 6♂, Sept. 1972. Encruzilhada, Bahia, 980 m., 3♀, 1♂, Nov. 1974. Represa Rio Grande, Guanabara, 1♀, June 1967; 1♀, Oct. 1967; 1♀, Jan. 1968; 1♂, March 1972; 4♀, May 1972. Espirito Santo, Castelo, 1♀, Nov. 1976; all above collected by M. Alvarenga.Pedra Azul, Minas Gerais, 800 m., 1♀, Nov. 1970, F. M. Oliveira; 2♀, Nov. 1971, Nov. 1972, Seabra & Olivera (Townes). *Argentina:* Corrientes, Las Marias, Ca. Virasoro, ♀, Nov. 10-15, 1969, C. C. Porter (Tucuman). One specimen shows some characters of *physoenemis nigristernum.* This one has not been designated paratype.

   Distribution: Argentina, Northern Brazil (map 3).

6b. AGONOCRYPTUS PHYSOCNEMIS NIGRISTERNUM, n. subsp.

This subspecies is differentiated from the typical subspecies in having the first tergite basally black, clypeal groove black marked, mesosternum black, without yellow marks and fore and middle coxae black. Otherwise agrees with the typical subspecies. The two subspecies intergrade in Argentina and further study may reveal that the two segregats may not have yet assumed subspecific status.

Length ♀: 20-12 mm. Fore wing 15-10 mm. ♂: 10-15 mm. Fore wing 5-8 mm.

Holotype: ♀, *Brazil:* Campina Grande, (near Curitiba), Feb. 15, 1966, H. & M. Townes (Townes).

Allotype: ♂, same locality and data as the holotype (Townes).

Paratypes: 10♀, 6♂. *Brazil:* Same locality as the holotype, 1♂, Feb. 17, 1966, H. & M. Townes (Townes). Curitiba, 1♀, 1♂, Jan. 20-31, 1969, L. J. Stange. Nova Teutonia, Santa Catarina, 3♂, 9♀, collected on various dates from March to December by Fritz Plaumann. Floresta da Tijuca, Guanabara, 1♂, Apr. 1966, Alvarenga & Seabra (Townes). Nova Teutonia, 300-500 m., 1♀, Oct. 28, 1957; 1♀, Oct. 30, 1958, Fritz Plaumann (Ottawa).

Distribution: Southern Brazil, ?Argentina. (map. 3).

7. AGONOCRYPTUS LIONELI, n. sp.

Female: Subapical flagellar segments compressed, about as long as wide. Face and clypeus mat, with fine sparse punctures. Face with a few trans-striations in-between. Apical margin of clypeus with a median tooth. Malar space aciculate, equal to the basal width of mandible. Temple shiny. Frons shiny, impunctate. Vertex shiny, impunctate. Occipital carina touching hypostomal carina. Pronotum punctate to punctato-striate. Pronotal collar and upper part of pronotum with fine punctures. Pronotal sulcus with a few or more striations. Epomia faintly indicated. Mesoscutum shiny, its middle lobe sparsely punctate and lateral lobes closely punctate with shiny central areas. Scutellum with a few superficial punctures. Mesopleurum shiny, punctate to minutely punctate. Metapleurum with close and deep punctures. Basal area of propodeum punctate, its central area rugoso-striate. Propodeum between the two trans-carinae rugose and in the yellow area trans-striate to rugose. Petiolar area longitudinally striate. Areolet a little wider than high. Second recurrent vein meeting areolet at its middle or in its apical 0.4. Nervellus intercepted at its upper 0.4. Petiole stout. First sternite reaching up to the spiracle of first tergite. First tergite shiny, with or without a few fine punctures apicolaterally. Second and third tergites closely and finely punctate. The following tergites mat.

Two subspecies, *A. lioneli lioneli,* from Argentina and Brazil, and *A. lioneli coxinota,* from Ecuador, are recognized by the characters mentioned in the key.

7a. AGONOCRYPTUS LIONELI LIONELI, n. subsp.

Female: Pronotum largely punctate, only with a few striations in the pronotal sulcus. Punctures on mesopleurum fine and a little apart. Propodeum centrally rugose, but its yellow area mostly trans-striate. First tergite with a few fine punctures apicolaterally.

Black. Head and thorax black with yellow marks. Abdomen yellowish-brown to reddish-brown, with black and yellow marks. Face, clypeus except apical margin, labrum, temple, orbital rings, pronotal collar, upper part of pronotum, propleurum apically, a mark on mesoscutum, scutellum, meta-scutellum, tegula, subtegular ridge, speculum, a large oblong mark on meso-pleurum, a mark on mesosternum, a mark below wings, a mark on metapleur-um, and a dagger-shaped mark on the apical slope of propodeum, yellow. Wings hyaline, with brownish marks medially and apically. Legs yellowish. Fore coxa, trochanters, and femur ventrally black-marked. Middle coxa reddish-orange, and femur ventrally black. Hind coxa, trochanters, and femur reddish-orange, its tibia and tarsus yellowish-brown. Abdomen reddish-brown to brown. First tergite apically and basally yellow, black in the middle, laterally brownish-red. Second and third tergites without yellow and black marks.

Male: Flagellar segments hairy and with spines. Male more shiny than the female, with pronotum with a few punctures on the lower side only. Apical transverse carina of propodeum strong. Second abdominal tergite medially, and third only basally, punctate.

Black with yellow marks. Color of head and thorax similar to that of female except that yellow mark on pronotal collar is interrupted. Hind coxa black with white marks, its trochanters, femur and tibia reddish-brown, and tibia apically and first tarsus basally brown. Rest of hind tarsus white. Fore and middle coxae white with blackish marks baso-dorsally, trochanters yellow, darker dorsally, femora and tibiae yellowish-brown, and tarsi brownish. Abdominal tergites 1-3 basally black. Tergites 1-2 apically yellow. Rest of abdomen reddish-brown.

Length: 10-18 mm. Fore wing 7-13 mm.

Holotype: ♀, *Argentina:* Tucuman, 11 kilometers west of Las Cejas, March 9-April 11, 1968, Lionel Stange (Townes).

Allotype: ♂, Same locality and data as the holotype, (Townes).

Paratypes: 13♀, 25♂. *Argentina:* Same data as the holotype, but collected on different dates from February to April, 5♀, 21♂ (Townes). Positos, Salta, 1♀, 2♂, Jan. 1971, M. A. Fritz (Townes). *Brazil:* Colatina, Espirito Santo, 1♂, Oct. 1969, F. M. Oliveira. Bahbalha, Ceara, 400 m., 1♂, May 1969, M. Alvarenga (Townes). Santo Grande, 2♀, 1968, M. Fritz (Townes). *Argen-tina:* Corrientes, Las Marias, Ca Virasoro, 1♀, Nov. 10-15, 1969, C. Porter. Salta. Rte 34, 12 km. NE Urundel-Arroyo Riacho Seco, 2♀, July 24-29, 1978, Porter and Fidalgo (Porter). Salta, Rio Pescado Ca., 1♀, Feb. 19, 1959, Atmat Bennagar (Porter). Isla, Martin Garcia, 1♂, Jan. 1971, H. Zimmer-mann (Washington).

Distribution: Argentina and Brazil (map 3).

7b. AGONOCRYPTUS LIONELI COXINOTA, n. subsp.

Female: Pronotum mostly punctato-striate. Punctures on mesopleurum closer and deeper. Propodeum centrally rugose, including the yellow area. First tergite mostly impunctate.

Color similar to that of *lioneli lioneli,* except that the yellow mark on pronotum interrupted, middle and hind coxae yellow and black marked ventrally, second and third tergites with black subbasal and yellow apical marks.

Male: Unknown.

Length: 18 mm. Fore wing 12-13 mm.

Holotype: ♀, *Ecuador:* San Rafael, 200 m., 1930, Campos R. (Washington).

Distribution: Ecuador (map 3).

8. AGONOCRYPTUS ARGENTINUS, n. sp.

Female: Subapical flagellar segments compressed, about as wide as long. Face dull, weakly to moderately striate. Clypeus dull, mat with a few lateral striations. Malar space aciculate mat, 0.9 times the basal width of mandible. Frons dull, slightly protuberant in the middle. Temple shiny, impunctate. Vertex shiny, with a few minute scattered punctures. Occipital carina meeting hypostomal carina. Pronotum punctate. Pronotal sulcus striate. Epomia indistinct, sometimes faintly visible in the sulcus. Mesoscutum with scattered punctures on the middle lobe, its lateral lobes sparsely punctate with smooth central areas. Scutellum with scattered punctures, which are close in the apical half. Mesopleurum punctate to punctato-striate in larger and darker specimens. Metapleurum rugose to reticulato-rugose. Petiolar area of propodeum longitudinally striate. Apical transverse carina broadly interrupted medially. Areolet rectangular. Second recurrent vein meeting areolet at its middle or in apical 0.45. Nervellus intercepted at its upper 0.4. Petiole stout, its sternite reaching spiracle. Postpetiole with scattered punctures. Second and third tergites closely punctate. The following tergites finely punctate to mat. Ovipositor 0.6 to 0.7 as long as the abdomen.

Reddish-brown to black. Scape and flagellum reddish-brown to black. Flagellar segments 6-10 white. Head reddish-brown. Apex of mandible, malar space, temple medially, and area just below antennal sockets, black. Inner orbital rings, incomplete outer orbital rings, labrum, and a mark on clypeus, yellow. Thorax reddish-brown. Pronotal collar, middle lobe of mesoscutum medially, tegula, subtegular ridge, lower portion of mesopleurum, mesosternum, area behind wings, metapleurum, and propodeum dark brown. Sometimes pronotal collar, mesoscutum, mesopleurum, and mesosternum, reddish-brown. A specimen from Potrerillos, Mendoza, with thorax wholly black, while in some other specimens from Cordoba, thorax black with reddish marks on various parts. Wings smoky brownish-black with purple tinge or hyaline. Fore wing with broad apical and medial dark brown bands. Fore, middle and hind coxae, and trochanters reddish-brown, with a few fuscous marks. All femora and tibiae yellow with blackish marks on tibia apically. Fore and middle tarsi yellowish-brown. Hind tarsus black. Abdomen brownish-black to black, with first tergite lighter in color.

The specimens exhibit considerable variations in sculpture and color. However, two subspecies are recognized: *A. argentinus argentinus* from Central Argentina, and *A. argentinus tucumanus* from Northern Argentina.

8a. AGONOCRYPTUS ARGENTINUS ARGENTINUS, n. subsp.

Female: Mesopleurum punctate to striato-punctate in larger and darker specimens. Color exhibiting considerable variation as described under species description. Scape and pedicel reddish-brown. Femur and tibia yellow, sometimes darker but not reddish brown. Tibia black in apical 0.14 to 0.12. Wings smoky brownish-black and with a purple tinge.

Male: Unknown.

Length: 14-18 mm. Fore wing 10-13 mm.

Holotype: ♀, *Argentina:* Cordoba, Davis (Cambridge).

Paratypes: 4♀. *Argentina:* Cordoba, 3♀, W. M. Davis. Mendoza, Potrerillos, 1♀, Feb. 20, 1966, C. C. Porter (Cambridge).

Distribution: Central Argentina (map 3).

8b. AGONOCRYPTUS ARGENTINUS TUCUMANUS, n. subsp.

Female: Mesopleurum striato-punctate. Scape and pedicel black. Face black, with two small roundish marks just below antennal sockets. Inner orbits, outer orbits (interrupted), and clypeus except its apical margin, yellow. Thorax reddish-brown. Hind margin of pronotum, apical corner of scutellum, metascutellum, tegula, subtegular ridge, mesopleural area just below speculum, metapleurum, and central area of propodeum, brownish-black. Petiolar area yellowish-brown. Wings hyaline, dark brown medially and apically. Fore leg yellowish-brown, its coxa and first trochanter dorsally blackish. Middle coxa, trochanters and femur, reddish-brown, its tibia yellowish-brown and tarsus blackish. Middle coxa and trochanters with brownish marks. Hind coxa, trochanters, and femur reddish-brown, with brownish marks. Hind tibia yellow with its apical 0.35 black. Hind tarsus black. Abdomen black with basal 0.5 of first tergite yellowish-brown.
    Male: Unknown.
    Length: ♀, 17 mm. Fore wing 13 mm.
    Holotype: ♀, *Argentina:* Tucuman, 11 kilometers west of Las Cejas, March 7-26, 1967, Lionel Stange (Townes).
    Distribution: Northern Argentina (map 3).

III. THE HEATHI GROUP

    This group is characterized by having a granulose vertex; closely punctate mesoscutum, with smooth areas in between; and strongly punctate metapleurum, with punctures stronger than on mesopleurum. The lower portion of occipital carina is wavy and it meets hypostomal carina at a distance equal to the basal width of mandible. It includes only one species, *Agonocryptus heathi* (Brues), which has an elongated yellow mark on basal area of propodeum and all abdominal tergites are with yellow apical bands.

9. AGONOCRYPTUS HEATHI (Brues)

    Cryptus heathi Brues , 1912. Ann. Ent. Soc. America, 5: 196. ♀. des.,
        fig. Type: ♀, Brazil: Guarabira (= "Independencia") in Paraiba
        (Cambridge).
    Agonocryptus heathi Cushman, 1929. Proc. U. S. Natl. Mus., 74(16): 7.
        Townes, 1966. Mem. Amer. Ent. Inst., 8: 128.

Female: Face with semicircular striations and fine scattered punctures. Clypeus with a few fine punctures. Malar space mat, equal to the basal width of mandible. Frons shiny, with a few fine punctures. Vertex granulose. Occipital carina wavy in the lower portion and meeting hypostomal carina at a distance equal to the basal width of mandible. Pronotum punctate. Epomia absent. Mesoscutum punctate, with shiny areas in between. Mesopleurum finely punctate. Metapleurum more strongly punctate than mesopleurum. Scutellum finely and sparsely punctate. Propodeum basally closely and deeply punctate, centrally rugoso-striate. Apical transverse carina complete. Areolet pentagonal. Second intercubitus not as strong as the first. Second recurrent vein meeting areolet at its apical 0.4. Nervellus intercepted at its upper 0.35. Petiolar area irregularly striate. Petiole stout, its sternite extending up to spiracle. Petiole and postpetiole shiny, with a few impressions of punctures laterally. Second and third tergites closely and finely punctate, the following tergites finely mat.

Black. Thorax yellow marked. Abdomen yellowish-brown with yellow bands. Face yellow with a central triangular black mark. Malar space, apex of mandible, frons and vertex medially, black. Pronotal collar, upper edge of pronotum, a mark on mesoscutum, tegula, subtegular ridge, scutellum, meta-scutellum, an oblong mark on mesopleurum, a mark below wings, a mark on mesosternum, a broad mark on metapleurum, a small mark at the center of basal area of propodeum, and a dagger-shaped mark at the apical slope of propodeum, yellow. Hind coxa, trochanters, femur and tibia yellowish-brown, with tibia basally lighter and apically a little darker. Hind tarsus and a small mark on coxa dorsally, white. Fore and middle legs largely yellowish, with darker patches. Wings hyaline. Fore wing with dark band at apex and just before stigma. Abdomen yellowish-brown with apical yellow bands on tergites.

Length: 12 mm. Fore wing 8 mm.

Specimens examined: 2♀. *Brazil:* Independencia (Guarabira), Paraiba, 1♀ (type), Mann & Heath (Cambridge). *Brazil:* Caruaru, 900 m., 1♀, April, 1972, M. Alvarenga (Townes). This specimen exhibits slight variations from the type, with face more strongly striate, a small yellow mark on side of pro-podeum, and middle coxa with a small yellow mark.

Distribution. Brazil (map 2).

IV. THE VARUS GROUP

Related and rather similar to the Amoenus Group in having smooth frons, mesoscutum and vertex, but the occipital carina is not strongly deflected inwards, rather pointing towards base of mandible or towards apex of hyposto-mal carina, always away from it by varying distances (except in *gossypii*). Postgenal area not conspicuously flattened or widened, more uniformly arched with the lower part of gena. Basal flagellar segments in female flattened (less so in *adustus paulus*, and not so in *gossypii*). First abdominal segment not very slender, in female about 2.0 to 2.4 its apical width, and quadrangular in cross-section in subbasal region, its sternite extending up to the middle (level of spiracle), or slightly beyond it (0.4-0.6 length of tergite). Epomia absent (or faintly seen in *adustus paulus*).

The males have long hairs on flagellar segments (except in *gossypii*), each segment also usually with long spine-like stout seta amongst the hairs at the apex of segments, particularly from fourth flagellar segment onwards. Males of most species are similar looking and not very diagnostic. They are slender in build. They even resemble the males of the Amoenus Group.

A. gossypii is exceptional in many characters, particularly in having the occipital carina touching the hypostomal carina, and in other characters mentioned in parenthesis above. However, it fits better in this group than any where else.

This group includes *A. varus* (Brullé), and five new species: *gossypii, adustus, erugatus, leurosus* and *rufigaster*, all from the Neotropical Region. Of the species placed in this group, *A. erugatus* is rather stout with abdomen smooth and polished, hairs on apical tergites rather sparse dorsally, and fourth to sixth abdominal tergites without apical fringe of hairs. The pro-podeum basad of basal carina is smooth and shiny. *A. leurosus* has the pro-notum striate and the areolet is rather wide, about 1.5-1.75 x as wide as high. *A. adustus* and *A. rufigaster* have the first abdominal sternite extending up to 0.6 the length of tergite, but in *adustus* the propodeum basad of basal carina is polished with only a few punctures or weak striations and apical transverse carina of propodeum is indistinct, while in *rufigaster* the propodeum is striate basally and rugose in the central yellow area and the apical transverse carina is distinct. The petiolar area is with vertical striations. *A. varus* has the first sternite ending at the level of spiracle, propodeum striate medially rather than rugose, petiolar area mat, wings without fuscous marks, pronotum mat, and areolet more squarish. Otherwise it comes close to *rufigaster* in the sculpture of basal area of propodeum and in the shape of apical transverse carina of propodeum. *A. gossypii* is rather distinctive. It is a small sized species, with occipital carina meeting hypostomal carina, and flagellar segments not flattened and without long hairs.

10. AGONOCRYPTUS VARUS (Brullé)

Female: Face with fine trans-striations. Clypeus with a few fine striations laterally. Malar space finely aciculate, equal to the basal width of mandible. Temple close to the malar space with a few fine aciculations. Frons dull, without punctures. Temple and vertex shiny, impunctate. Occipital carina away from hypostomal carina by a distance equal to 0.5 the basal width of mandible. Pronotum smooth, with scattered punctures, groove along its hind margin smooth. Mesoscutum, scutellum, mesopleurum and metapleurum shiny, with minute setiferous punctures. Basal area of propodeum finely striate. Central area of propodeum between the basal and apical transverse carinae trans-striate. Apical transverse carina strongly arched medially. Petiolar area smooth to dull, mat. Areolet more squarish, about 1.2 as wide as high; second recurrent vein meeting in its middle. First abdominal segment 2.2 as long as its apical width; its sternite ending at the level of spiracle, about 0.5 the length of its tergite. All abdominal tergites dull, with hairs, which become denser on apical tergites. Second abdominal tergite punctate medially and third at base. Ovipositor about 0.72 the length of abdomen.

Two subspecies are recognized: *A. varus varus* (Brullé) and *A. varus nigrifemur*, n. subsp., based upon the color of legs, pronotal collar and abdomen, as given in the key.

A specimen from Jamaica (from Washington Museum) is damaged and has a reddish thorax and black abdomen, and probably represents a distinct subspecies. It is not named because of its damaged condition.

10a. AGONOCRYPTUS VARUS VARUS (Brullé)

Mesostenus varus Brullé, 1846, *In* Lepeletier: Histoire naturelle des
   Insectes, Hyménoptères, 4: 235. ♀. des. Type: ♀, "Guyana"
   (Paris).
Agonocryptus varus: Townes, 1966, Mem. Amer. Ent. Inst., 8: 128.

Female: Black with yellow marks; abdomen yellowish-brown. Face, clypeus (except apical margin), labrum, orbital rings, a mark on mesoscutum, scutellum, metascutellum, pronotal collar, upper part of pronotum, propleurum, tegula, subtegular ridge, speculum, a mark below wings, a large oblong mark on mesopleurum, a mark on metapleurum, and a dagger-shaped mark on apical slope of propodeum, yellow. Legs yellow, with hind coxa, trochanters and femur brownish. Fore coxa with a black mark at base ventrally. Wings hyaline, without fuscous marks.
Male: Face transversely striated. Clypeus shiny. Malar space granu-

lose, slightly less than the basal width of mandible. Temple close to the malar space with a few fine aciculations. Frons dull without punctures. Temple and vertex shiny, impunctate. Occipital carina reaching the hypostomal carina. Pronotum smooth, its upper part with close superficial punctures, its hind margin punctate with a few striations. Mesoscutum, scutellum, mesopleurum and metapleurum shiny, with minute setiferous punctures. Basal area of propodeum striate; central area of propodeum between basal transverse and apical transverse carinae trans-striate. Apical transverse carina strongly arched medially. Petiolar area longitudinally striate. Areolet slightly wider than high; second recurrent vein meeting at its middle. First abdominal tergite 3.5 x its tergite. Abdominal tergites shiny. Second abdominal tergite medially and third basally punctate.

Black with yellow marks. Abdomen yellowish-brown, with black bands. Face, clypeus, labrum, base on mandible, temple, malar space, orbital rings, a mark on mesoscutum, scutellum, matascutellum, pronotal collar, pronotum, (except upper sulcus) propleurum, tegula, subtegular ridge, speculum, a mark below wing, lower part of mesopleurum, mesosternum, metapleurum and a dagger-shaped mark on apical slope of propodeum, yellow. Legs yellow. Fore femur and tibia dorsally with an elongate brown mark, third and fourth tarsal segments and claws black. Middle femur and tibia dorsally with an elongate brown mark, tarsus and claws black. Hind coxa and trochanters dorsally black marked, femur apically and apical 0.6 of tibia and apical 0.6 of basitarsus black. First abdominal tergite apically and basally and second to fifth abdominal tergites apically with yellow bands.

Length ♀: 16-20 mm. Fore wing 12-15 mm. ♂: 10-15 mm. Fore wing 6-12 mm.

Specimen examined: 3♀♀, 2♂♂. 1♀ (type), (Paris). The type has no locality label; only a green circular label, and an identification label. *Panama:* Barro Colo, Is., CZ, 1♀, June 1940, Zetek 4669 (Washington). Tobago Is., 1♀, Feb. 14, 1912, A. Busck Coll. (Washington). Tobago Is., 1♂, Feb. 24, 1912, 1♂, Feb. 23, 1912, A. Busck Coll. (Washington).

Distribution: Panama and "Guiana" (map 4).

10b. AGONOCRYPTUS VARUS NIGRIFEMUR, n. subsp.

Female: Pronotal collar wholly yellow. Mesosternum with yellow marks. Fore leg yellow, its coxa and femur black ventrally, and tarsus brownish. Middle coxa black with dorsal yellow mark, trochanters and femur black, tibia yellow, tarsus brownish. Hind coxa reddish-brown, darker ventrally, trochanters brown, femur black, tibia and tarsus yellow. Abdomen brown with blackish areas in middle of first, second, third, and fourth tergites.

Male: Unknown.

Length: 15 mm. Fore wing 10 mm.

Holotype: ♀, *Argentina*: Salta, Route 34, 12 kilometers northeast of Urundel-Arroyo Riacho Seco, July 24-29, 1978, Porter and Fidalgo (Porter).

Paratype: 1♀, *Argentina*: Salta, Rio Pescado, Ca Oran, May 23, 1970, C. Porter (Porter).

Distribution: Argentina (map 4).

11. AGONOCRYPTUS RUFIGASTER, n. sp.

Female: Face with fine trans-striations. Clypeus with a few fine striations laterally and centrally with a few punctures in between. Malar space finely aciculate, equal to the basal width of mandible. Temple close to malar space with a few fine aciculations. Frons, vertex and temple shiny, impunctate. Occipital carina very close to hypostomal carina but not meeting it. Pronotum shiny, punctate, its hind margin smooth above, striate below. Middle lobe of mesoscutum only basally with a few punctures. Mesoscutum otherwise smooth and shiny. Scutellum, mesopleurum and metapleurum shiny with minute setiferous punctures. Basal area of propodeum finely striate; area above spiracle striato-punctate; central area of propodeum strongly rugose; petiolar area with vertical striations; apical transverse carina distinct, complete and strongly arched medially. Areolet 1.5 as wide as high. Second recurrent vein meeting at its apical 0.40. Nervellus intercepted at its upper 0.40. First abdominal segment 2.2 as long as its apical width, its sternite extending up to 0.6 its length. All abdominal tergites shiny, with hairs. Fourth tergite onwards with an apical fringe of hairs. Second and third tergites minutely and finely punctate. Ovipositor about 0.84 the length of abdomen.

Black, with yellow marks; abdomen yellowish-brown. General coloration similar to that of *A. varus*. Differences are: yellow mark on pronotal collar interrupted medially. Lateral lobes of mesoscutum with a crescentic mark. Fore wing infuscate centrally and apically.

Male: Similar to that of *A. adustus paulus* in structure and color and differs only in the lighter color of hind coxa and hind femur.

Length: 11-16 mm. Fore wing 8-13 mm.

Holotype: ♀, *Surinam:* Paramaibo, malaise trap, March 2-4, 1964, D. C. Geijskes (Townes).

Allotype: ♂, *Surinam:* 45 kilometers south of Paramairbo, Oct. 3-8, 1963, D. C. Geijskes (Townes).

Distribution: Surinam (map 4).

12. AGONOCRYPTUS ADUSTUS, sp. nov.

Female: Face granulose with submedian trans-striations or with a few punctures. Clypeus granulose with a few superficial punctures medially. Malar space granulose. Frons, temple and vertex shiny, impunctate. Occipital carina away from hypostomal carina by a distance equal to the basal width of mandible. Pronotum with scattered punctures, its scrobe long, hind margin wrinkled. Middle lobe of mesoscutum basally punctate. Mesoscutum otherwise smooth and shiny. Scutellum, mesopleurum, mesosternum and metapleurum mostly shiny and with a few minute setiferous punctures. Propodeum largely smooth, with weak to indistinct striations, particularly on the basal area; apical transverse carina indistinct, if faintly indicated, more semicircular, like carinae bordering petiolar area. Areolet 1.3 to 1.4 as wide as high. Second recurrent vein meeting areolet at middle or in its apical 0.3. Nervellus intercepted at its upper 0.4. First abdominal segment 2.2 as long as its apical width, shiny; its sternite extending up to 0.6 the length of tergite. Second and third tergites with a few fine punctures basolaterally. The following tergites smooth and shiny. Fourth and following tergites rather uniformly hairy dorsally with their margins fringed with hairs, especially on seventh and eighth. Ovipositor 0.7 the length of abdomen.

Black, the head and thorax marked with yellow and abdomen reddish-brown. Head yellow except clypeal margin, malar space, mandible, frons and vertex medially, and whole of occiput, black. Pronotal collar wholly yellow or interrupted in the middle. Upper margin of pronotum, propleurum, a mark at apex of middle lobe of mesoscutum, scutellum, metascutellum, tegula, subtegular ridge, speculum, a mark behind hind wing, a broad and oblong mark on mesopleurum, a large mark on metapleurum, and an inverted T-shaped mark on apical slope of propodeum, yellow. Leg color as described under the subspecies. Wings hyaline with fore wing marked brown medially and apically. Abdomen wholly reddish-brown or first tergite with basal and apical yellow marks and second tergite apically yellow, bordered with black. Tip of abdomen fuscous.

Male: Rather thin and slender as compared to the female, with antennal segments with long hairs and a stiff seta at apex of each flagellar segment among the hairs. Body sculpture generally similar to that of female, but males show considerable variation. Face shallowly punctate to trans-striate. Frons smoother to punctate. Basal area of propodeum smooth to rough. Propodeum centrally striate to rugose. Apical transverse carina distinct.

Black. Antenna wholly black. Head yellow with mandible, frons and vertex medially and upper half of occiput, black. Thorax black with pronotal collar, upper margin of pronotum, propleurum, a mark at apex of middle lobe of mesoscutum, scutellum, metascutellum, tegula, subtegular ridge, speculum, area behind hind wing broadly, and oblong mark on meso and metapleurum, and petiolar area, yellow. Wings hyaline. Fore wing only apically lightly to moderately fuscous; fore and middle coxae and trochanters yellow, their femora, tibiae and tarsi yellowish-brown with middle tarsus blackish. Hind leg either wholly black with coxa encircled with white, or coxa, first trochanteral segment and femur reddish-brown. Abdomen either largely black with apices of first to third segments and sides of fourth to sixth segments yellow, or apical half of all tergites yellowish-brown; color of hind leg and abdominal tergites highly variable in the two subspecies described below.

Two subspecies are recognized, which have been keyed out along with the key to the species. *A. adustus adustus* occurs in Peru and at lower elevations in Brazil and Ecuador bordering Peru southeast of the high Andes running across Colombia and Ecuador. *A. adustus paulus* occurs on the Andes in Colombia and Ecuador and also west of the same at somewhat lower elevations.

12a. AGONOCRYPTUS ADUSTUS ADUSTUS, n. subsp.

Female: Face finely shagreened with trans-striations. Basal area of propodeum smooth and shiny, with a few fine punctures. Central area finely to moderately striate. Pronotum with minute punctures. Basal flagellar segments somewhat flattened.

General color as described under the species. Differences and variations are: Face with small lateral black marks just above clypeus in the paratypes. Yellow mark along pronotal collar complete. Mesosternum black. Speculum only with a small yellow mark. Sometimes apex of first tergite yellow. Fore and middle legs largely yellow with coxae blackish on their posterior sides, their second trochanteral segments and femora blackish-brown dorsally and apical tarsal segments brownish. (In one paratype the brownish mark on middle femur appears more ventral in position.) Hind coxa, trochanters and femur reddish-brown, tibia and tarsus yellow with fifth and sometimes fourth also, blackish-brown.

Male: Face shallowly striate; in one paratype smoother and with shallow punctures. Frons punctate to smooth. Propodeum basally smooth and shiny and centrally striate.

Mesosternum black, sometimes mesosternum yellow near middle coxa.

Hind leg wholly black or hind femur brownish-black. Fore wing of allotype darkly fuscous apically; other paratypes with lightly to darkly fuscous fore wing tip. Abdomen largely black with only apices of tergites brownish. Sometimes yellow marks on sides of thorax smaller.

Length: ♀, 8-14 mm. Fore wing 6-10 mm.

Holotype: ♀, *Peru:* Avispas, 30 m. nr. Marcapata, Sept. 1962, Luis Peña (Townes).

Allotype ♂, same data as the holotype, except collected on Oct. 1-15, 1962 (Townes).

Paratypes: 2♀, 5♂. *Peru:* Quincemil, 750 m, near Marcapata, 1♀, Nov. 1962; 3 ♂, Oct. 20-30, 1962. Avispas, 30 m. near Marcapata, 1♂, Sept. 1962, all collected by Luis Peña (Townes). *Brazil:* Jacareacanga, Para, 1♀, Dec. 1968, Moacir Alvarenga (Townes). *Ecuador:* Napo and Coca Rivers, 1♂, May 2-10, 1965, Luis Peña (Townes).

Distribution: Peru, and at lower elevations in Brazil and Ecuador (map 4).

12b. AGONOCRYPTUS ADUSTUS PAULUS, n. subsp.

Female: Basal flagellar segments not conspicuously flattened (exception, perhaps due to smaller sized specimens). Face shiny with a few scattered punctures, without striations; pronotum shallowly punctate and shiny. Propodeum shallowly striate in the middle. Otherwise smooth and shiny.

Black. Vertex with a Y-shaped yellow mark. Mesosternum with yellow marks. Yellow mark on pronotal collar interrupted medially. Fore and middle legs similarly colored. Coxae and first trochanteral segments whitish-yellow. Coxa ventrally black. Second trochanter, femur and tibia yellow, femur dorsally brownish-black; tarsus brownish. Hind coxa, trochanters and femur reddish-brown with fuscous marks, coxa with a dorsal yellow line, apex of femur blackish, tibia and tarsus yellow but fourth and fifth tarsal segments black. Apical fuscous mark on fore wing more conspicuous. First tergite apically and basally broadly yellow. Second tergite apically yellow, subapically black adjacent to the yellow mark.

Male: Face shallowly striate with a few punctures in between striations; frons and pronotum smooth and shiny; propodeum centrally rugoso-striate.

Black. Mesosternum largely yellow. Hind coxa and femur reddish-brown, trochanters and tibia black, tarsus either wholly black or third segment and sometimes second segment partly yellow. Sometimes hind femur brownish-black as is tibia and tarsus, but coxa always reddish-brown. Mesosternum at least partly yellow. Abdominal segments black basally and reddish-brown apically. First tergite yellow basally and apically and second tergite largely black. Color otherwise similar to that of adustus adustus.

Length: ♀, 10 mm. Fore wing 8 mm.

Holotype: ♀, *Colombia:* Cali, Oct. 1971, M. J. W. Eberhard (Townes).

Allotype ♂, same data as holotype (Townes).

Paratypes 8 ♂. *Colombia:* Cali, 3 ♂, same data as holotype. *Ecuador :* Santo Domingo, 680 m, 3 ♂, May 15-30, 1975, S. & J. Peck. Loja: Latoma, 1500 m, 2 ♂, Nov. 18-19, 1970, Luis Peña (Townes).

Distribution: Colombia and Ecuador (map 4).

13. AGONOCRYPTUS GOSSYPII, n. sp.

Female: Face granulose with a few scattered punctures. Clypeus shiny, impunctate, its apical margin without a median tooth. Malar space mat, 0.6 as long as the basal width of mandible. Temple, frons, and vertex shiny, impunctate. Occipital carina meeting hypostomal carina. Pronotum shiny, impunctate. Epomia absent. Middle lobe of mesoscutum with a few punctures basally. Scutellum impunctate. Mesopleurum with a few fine punctures. Metapleurum with a little coarser punctures as compared to mesopleurum, but punctures not deep. Propodeum basally with a few large scattered punctures. Propodeum medially transversely rugose. Petiolar area a little smoother. Areolet rectangular, 1.5 as wide as high. Second intercubitus absent. Second recurrent vein meeting areolet at its apical 0.33. First abdominal segment 2.0 as long as its apical width, its sternite ending at the level of its spiracle. First tergite shiny, impunctate. Second tergite with moderately sparse punctures, punctures large. Third and fourth tergites basally with close, fine punctures. The following tergites mat. Ovipositor 0.8 as long as the abdomen.

Yellowish-brown. Abdomen brownish-black. Head, scape, and antenna brown, except interrupted inner and outer orbital rings. Clypeus broadly, and labrum, yellow. Wings tinged with fuscous-brown. Legs yellowish-brown, with middle tarsus, hind trochanters, femur, tibia, basal 0.5 of hind basitarsus, and fourth tarsal segment, brown. Abdomen brownish-black. First segment brownish.

Male: Generally similar to the female, the flagellar segments without long hairs but with spines in between. Face wholly yellow. Petiolar area of propodeum yellow. Hind trochanters, femur and basal 0.5 of tibia yellowish-brown. First tergite of the same color as the rest of the abdomen.

Length: 8-10 mm. Fore wing 6 mm.

Holotype: ♀, *Brazil:* Sao Paulo, July 27, 1936, E. J. A. Hambleton, ex cotton stalks infested with *Gasterocercodes gossypii,* No. 27 (Washington).

Allotype: ♂, same locality and data as the holotype (Washington).

Paratypes: 1♀, 1♂, same data as the holotype (Washington).

Distribution: Brazil (map 4).

14. AGONOCRYPTUS LEUROSUS, n. sp.

Female: Face and clypeus finely, transversely striate with punctures in between. Malar space longitudinally aciculate and 0.9 the basal width of mandible. Temple close to malar space aciculate, otherwise smooth and shiny. Frons smooth and shiny, with a few minute superficial punctures. Vertex smooth and shiny, impunctate. Occipital carina away from hypostomal carina by basal width of mandible, pronotum striate, particularly along the sulcus and along posterior margin. Pronotal collar and upper margin with scattered punctures. Mesoscutum smooth and shiny, only the basal area of middle lobe with a few close punctures. Scutellum, mesopleurum, and mesosternum with moderately sparse fine punctures. Metapleurum shiny, with a few fine punctures to moderately deep punctures. Propodeum basally finely wholly striate, centrally striate; petiolar area semicircularly or irregularly striate. Areolet rather wide, about 1.5 to 1.75 as wide as high. Second recurrent vein meeting areolet at its apical 0.4. Nervellus intercepted at its upper 0.45. First abdominal segment shiny, about 2.2 as long as its apical width. First sternite ending at the level of spiracle, 0.5 the length of tergite. Second and third abdominal tergites closely, finely punctate. Abdominal tergites hairy, especially fourth tergite onwards. Ovipositor 0.85 the length of abdomen.

Black with yellow marks. Abdomen reddish. Face, clypeus except its apical margin, labrum, orbital ring, a triangular mark on mesoscutum, scu-

tellum, metascutellum, an interrupted mark on pronotal collar, upper margin
of pronotum, apical part of propleurum, tegula, subtegular ridge, a mark
below hind wing, a large mark on metapleurum and an inverted T-shaped mark
on apical slope of propodeum, yellow. Yellow mark on mesosternum present
or absent depending upon the subspecies. Wings hyaline, without fuscous
bands. Abdomen reddish-brown either wholly or with yellow and black marks
on first and second tergites.

14a. AGONOCRYPTUS LEUROSUS LEUROSUS, subsp. nov.

Female: Characterized by having the second abdominal tergite less
strongly punctate, more shiny, and abdomen wholly reddish-brown, without
any fuscous or yellow marks. Mesosternum wholly black, without yellow
marks. Fore leg yellow with coxa underneath and femur except dorsally,
black. Fourth and fifth tarsal segments blackish. Middle coxa yellow dorsally and black ventrally and on the inner side. Femur yellowish-brown with
black ventral and lateral lines. Trochanters and tibia yellow. Basal two tarsal segments yellowish-brown and rest black. Hind coxa, trochanters and
femur reddish-brown. Hind tibia and basal three tarsal segments yellow,
apical two tarsal segments black.
    Male: Similar to the female, flagellar segments with long hairs, each
segment also with a long spine-like stout seta among the hairs at the apex of
the segment, particularly segment four onwards. Frons punctate. Pronotum
shiny, with a few striations in the sulcus. Propodeum striato-rugose. Apical
transverse carina of propodeum present.
    Black with yellow marks. Abdomen a little brownish apically. Antenna
black, flagellar segments with yellow marks. Head yellow. Apex of mandible,
frons, vertex, and occiput medially black. Pronotal collar, upper part of
pronotum, propleurum, a mark on middle lobe of mesoscutum, scutellum,
metascutellum, tegula, subtegular ridge, a mark behind hind wing, mesopleurum broadly, mesosternum wholly, metapleurum except basal region,
and a small T-shaped mark at the apical slope of propodeum, yellow. Coxae
and first trochanters of fore and middle legs yellow. Femora and tibia of both
legs yellowish-brown. Fourth and fifth tarsal segments of fore leg and all tarsal segments of middle leg brownish-black. Hind leg black with femur lighter
in color at base. Wings hyaline, a little clouded apically. First tergite yellow,
medially black. Second tergite with an apical yellow band. Third and fourth
tergites basally black and apically yellow. The following tergites yellowish-
brown.
    Length: 14-18 mm. Fore wing 10-14 mm.
    Holotype: ♀, *Brazil:* Matto Grosso, Sinap, Oct. 1975, M. Alvarenga
(Townes).
        Allotype: ♂, *Brazil:* Goias, Jatai, Nov. 1972, F. M. Oliveira (Townes).
        Paratypes: 24♂, *Brazil:* Goias, Jatai, 14♂, Nov. 1972, F. M. Oliveira
(Townes). Vila Vera, 3♂, Oct. 1973, M. Alvarenga (Townes). Vilhena, Rond,
3♂, Nov. 1977, M. Alvarenga (Townes).
        Distribution: Brazil (map 4).

14b. AGONOCRYPTUS LEUROSUS FLAVOSTERNUM, subsp. nov.

Female: Characterized by having the abdomen reddish-brown as in the
typical subspecies, but the first tergite has a yellow mark at apex, and two
black marks laterally. Second tergite also with a triangular broad apical black

mark. Punctures on second tergite rather strong and more distinct and apical abdominal tergites more hairy, apical tergite blackish in the middle. Mesosternum with yellow marks. Fore leg yellow, coxa with a black ventral mark, trochanter and femur dorsally blackish-brown, apical tarsal segments brownish. Middle coxa reddish-brown with a yellow triangular spot, trochanters yellow, femur black dorsally, apical tarsal segments brownish, rest of middle leg yellowish. Hind coxa, trochanters and femur reddish-brown. Hind tibia and tarsus yellow with apical two tarsal segments brownish.

Male: Similar to that of *leurosus leurosus* except that pronotum shiny, without striations and propodeum striate.

Color similar to that of *leurosus leurosus* except that yellow mark on mesopleurum not extensive and yellow mark on apical slope of propodeum not T-shaped.

Length: 20 mm. Fore wing 16 mm.

Holotype: ♀, *Paraguay:* Villarica, Dec., F. Schade (Cambridge).

Allotype: ♂, *Paraguay:* Carumbe, Feb. 1, 1966, R. Golbach (Townes).

Paratypes: 4♀, 14♂. Same data as the holotype, 3♀ (Cambridge). Carumbe, 1♂, March 8, 1966, R. Golbach. Pirareta, 2♂, Dec. 26, 1971, Luis Peña. *Bolivia:* Altobeni, Palos Blancos, 600 m., 1♂, Luis Peña. *Argentina:* Salta Pocitos, 3♂, Jan. 1972, Manfredo Fritz. Jujuy, Aguas Calientes, 650 m, 2♂, Dec. 18-20, 1968, L. Peña. Salta, Tartagal, 1♂, Jan. 1972, Fritz (Townes). *Argentina:* Formosa, Mision-Aishi, 1♀, Dec. 15, 1948, R. Solbach. Salta, Campamento Jakulica, 40 kilometers east of Aguas Blancas, 3♂, C. Porter. Corrientes Paso de la Patria, 1♂, Nov. 5-7, 1969, C. Porter (Porter).

Distribution: Paraguay, Bolivia, and Argentina (map 4).

15. AGONOCRYPTUS ERUGATUS, n. sp.

Female: Rather stout species. Face finely striate. Clypeus with a few fine punctures. Malar space aciculate. Frons and vertex smooth and shiny. Occipital carina close to hypostomal carina, but not touching. Pronotum minutely punctate, shiny in between. Pronotal sulcus with wrinkles, which are more prominent in lower half, upper half of groove along hind margin smooth. Mesoscutum smooth and shiny, only the middle lobe at base with a few punctures. Scutellum, mesopleurum and metapleurum with a few fine scattered punctures. Propodeum basad of basal carina smooth and shiny, with a few scattered punctures, area basad of spiracle with crowded punctures. Central area of propodeum with rather weak striations with punctures in between. Striations faded laterally; petiolar area smooth, impunctate. Areolet about 1.5 as wide as high, with recurrent vein meeting areolet in the middle. Nervellus intercepted in its upper 0.45. First abdominal segment 2.0 its apical width. First sternite 0.4 the length of first tergite. All abdominal tergites shiny, impunctate, sparsely hairy dorsally. Tips of seventh and eighth with sparse hairs. Ovipositor 0.67 the length of abdomen.

Black with yellow marks; abdomen reddish-brown. Scape ventrally with two yellow marks. Head yellow. Face with two small marks, two dots along epistomal groove, margin of clypeus, malar space, mandible except basally, frons and vertex medially, and upper half of occiput, black. Thorax black with yellow spots on pronotum dorsally, pronotal collar, upper margin of pronotum, propleurum, a triangular mark at apex of middle lobe of mesoscutum, scutellum, metascutellum, tegula, subtegular ridge, speculum, a mark below hind wing, long oblong areas on meso- and metapleurum, an elongate mark on mesosternum and an inverted T-shaped mark on propodeum, yellow. Fore and middle legs yellow with their coxae basally black marked, their femora dorsally brownish-black and apical three tarsal segments brownish-black. Hind coxa yellow with a large brownish spot basoventrally and along inner side. Hind trochanters and femur reddish-brown, tibia and tarsus yellow with apical two tarsal segments dark brown; apex of femur fuscous dorsally. Wings hyaline. First abdominal segment yellow with a black subapical irregular mark. Second tergite with a triangular basal yellow mark, an apical yellow stripe, and two irregular yellow spots on sides, also with two triangular and linear black marks between the yellow and brownish color of the tergite. Third tergite with an elongate black basal mark and two yellow spots basolaterally. Apex of abdomen brownish-black. Intersegmental membranes yellow.

Male: Very slender as compared to the female. Flagellar segments with long hairs; each segment with a long spine-like seta at its apex, particularly from fourth flagellar segment onwards. Sculpture generally similar to female but frons with a few punctures. Pronotum smoother. Propodeum basally with sparse scattered punctures and a few striations, centrally striate between basal and apical transverse carinae. Petiolar area of propodeum longitudinally and irregularly striate (in some males smoother).

Black with yellow marks. Scape with a small yellow mark. Flagellar segments 10 to 13 dorsally yellow. Color of thorax similar to that of female except that black marks on face are absent and yellow mark on propodeum very small. Fore and middle legs yellow with their femora, tibia and tarsi reddish-brown. Middle tarsus blackish. Hind coxa yellowish-brown, dorsally with a dark brown line, femur dark brown, trochanters, tibia and tarsus (except second and third segments), blackish, second and third segments white, second often partly blackish. Wings hyaline. First abdominal segment yellow apically and basally, medially black. Second segment black in basal 0.75. Third and often the fourth basally black. Rest of tergites reddish-brown with their bases a little darker.

Length: 8-20 mm; fore wing 5-16 mm.

Holotype: ♀, *Panama:* Canal Zone: Margarita, Feb. 1960, S. Breeland (Townes).

Allotype: ♂, same locality and collector as the holotype, March 1960 (Townes).

Paratypes: 4♂. Same data as allotype, 2♂, *Colombia:* Rio Atrato Camp Sautata, 1♂, Nov. 11-Dec. 14, 1967. *Panama:* Barro Colorado Is., 1♂, March 11-31, 1963, C. & M. Rettenmeyer (Townes).

Distribution: Panama and Colombia (map 4).

V. THE AMOENUS GROUP

This group is characterized by having the frons smooth; face smooth to finely striate; vertex smooth; occipital carina deflected inwards either strongly (pointed towards base or middle of hypostomal carina) or weakly (in *tricolor*), coming very close to joining hypostomal carina, or erased for a distance equal to the basal width of mandible; postgenal area somewhat flattened and widened and separated from the lower portion of gena by a crease (seen in profile). Mesoscutum smooth. Thorax generally smooth. Propodeum striate (except in *tricolor*). Apical transverse carina of propodeum absent in females and present in males (of *tricolor* only known). Sometimes transverse carinae at the junction of yellow and black color prominent and appearing like transverse carina, though not

really present. Abdomen smooth or with scattered minute punctures, particularly on second and third tergites. Petiole long and slender, rounded in cross-
section near its base. Length of first abdominal segment about 3.0 its apical
width (a little shorter in *admirandus*, about 2.75 x), its sternite long and
extending to 0.70 to 0.8 the length of tergite. Ovipositor about half the length
of abdomen (0.5-0.65).

This group includes *A. admirandus* Cresson from Mexico and three new
species, *amoenus* and *tricolor* from Brazil and *bispotus* from Mexico (Map 5).
A. admirandus can be readily distinguished by having the prepectal carina
short, extending only in the lower 0.5 of mesopleurum; occipital carina
erased for a distance about 0.75 the basal width of mandible; pronotum punctate with area along its hind margin wrinkled, epomia absent, first abdominal
sternite extending to 0.65 the length of tergite, and hind femur largely black.
All other species have the prepectal carina extending up to 0.8 the height of
mesopleurum. *A. amoenus* is distinctive in having the occipital carina
strongly deflected inwards and coming very close to hypostomal carina near its
base rather than in the apical region; pronotum smooth, epomia absent, groove
along hind margin wrinkled first sternite long, extending to a distance about
0.8 the length of tergite, abdominal tergites without hairs. Hind femur with
extensive blackish marks. In all other species of this group the apical abdominal tergites are beset with conspicuous hairs.

A. bispotus is the only species of this group with epomia. It has the
occipital carina deflected inwards but is erased for a distance equal to the
basal width of mandible, pronotum punctate in the middle groove and striate
along hind margin, first sternite about 0.75 the length of its tergite, hind femur
largely yellow, and the speculum yellow (as is the case in *admirandus*, while in
other species it is black). *A. tricolor* is the only species in which the propodeum is smoother in the female, with only a few trans-striations centrally. In
male the propodeum is striate and the apical transverse carina is distinct. It
has the occipital carina erased as in *bispotus*, but not deflected inwards, pronotum with minute scattered punctures, groove along hind margin smooth,
epomia absent, first sternite extending up to 0.75 the length of its tergite,
and hind femur red.

The males have the occipital carina a little closer to the hypostomal carina
and the hind leg largely black. They also have large hairs on the antennal
flagellum. White marks on flagellar segments are present dorsally or absent,
but not forming a complete ring.

16. AGONOCRYPTUS ADMIRANDUS (Cresson)

Mesostenus (Mesostenus) admirandus Cresson, 1873, Proc. Acad. Nat.
    Sci. Philadelphia, 1873: 155. ♀. key, des. Type: ♀, Mexico: Orizaba
    (Philadelphia).
Agonocryptus admirandus: Townes, 1946, Bol. Ent. Venezolana, 5: 31.
    Townes, 1966. Mem. Amer. Ent. Inst. 8: 128.

Female: Face and clypeus shiny with a few fine striations. Frons shiny
with a few fine punctures close to ocelli. Vertex medially with a few superficial punctures. Malar space granulose, equal to the basal width of mandible.
Temple smooth and shiny, a little granulose close to malar space. Occipital
carina erased for a distance about 0.75 the basal width of mandible. Mesoscutum shiny, with scattered punctures. Pronotum with scattered punctures.
Pronotal sulcus with a few striations. Scutellum punctate. Mesopleurum and

metapleurum shiny and with scattered superficial fine punctures. Mesopleurum just below subtegular ridge striate. Prepectal carina reaching 0.5 the height of mesopleurum. Propodeum basally between the two yellow marks semicircularly striate, punctate basolaterally just above the spiracle; centrally punctatostriate. Petiolar area of propodeum laterally punctate, in the middle shiny and with a few trans-striations. Areolet pentagonal, wider than high. Second recurrent vein meeting areolet at its middle. Nervellus intercepted at its upper 0.33. First abdominal segment slender, nearly 2.75 x its apical width, impunctate. First sternite extending up to 0.7 its length. Second abdominal tergite with fine scattered superficial punctures medially and submedially. Third tergite basally finely and closely punctate. The following tergites finely mat. Ovipositor about 0.65 the length of abdomen.

Black with yellow marks and bands. Head (including face) yellow except frons, vertex and occiput medially, malar space, clypeal margin and apex of mandible, black. The following parts are yellow: A mark on pronotal collar, upper part of pronotum, propleurum apically, a central mark on mesoscutum, scutellum and metascutellum wholly, tegula, subtegular ridge, speculum, a mark below wings, mesopleurum, metapleurum and mesosternum broadly, two squarish marks at base of propodeum, and a dagger-shaped mark at the apical area of propodeum (this mark reaching up to basal transverse carina). Fore and middle legs yellow, except coxa, trochanters and femur dorsally with black line, their tarsi brownish. Hind coxa yellow with two lateral longitudinal black lines. Hind trochanters and femur black with a yellow longitudinal ventral line, third tibia and tarsus yellow. Wings hyaline, first abdominal tergite yellow with a subapical black mark. Second tergite with a triangular yellow basal mark, its sides with yellow marks connected to the apical transverse yellow band. All other abdominal tergites with yellow apical bands, their sides also yellow. Seventh and eighth tergites broadly yellow laterally, with the eighth tergite black up to the apex.

Male: Unknown.

Length: 18mm; fore wing 12 mm.

Specimen examined: Mexico: 1♀ (type) (Philadelphia).

The locality label on specimen does not state precise type-locality.

Distribution: Mexico (map 5).

17. AGONOCRYPTUS AMOENUS, n. sp.

Female: Face and clypeus shiny, impunctate with face finely transstriate in the middle. Malar space granulose. Frons and vertex shiny, impunctate. Temples smooth and shiny. Pronotum shiny, impunctate. Epomia absent. Mesoscutum shiny, impunctate. Mesopleurum and metapleurum shiny with a few scattered fine and superficial punctures. Prepectal carina 0.8 the height of mesopleurum. Propodeum finely distinctly striate, striations arched and areas in between shiny. Area basad of basal carina semicircularly striate, particularly in the central basal area. Areolet pentagonal, wider than high. Second recurrent vein meeting areolet at its apical 0.4. Nervellus intercepted at its upper 0.33. Abdomen wholly smooth and shiny, without punctures. First abdominal segment 3.0 its apical width and slender, its sternite reaching up to 0.75 the length of its tergite, ovipositor about 5.3 x the length of abdomen.

Black with yellow marks and bands. Head (including face) yellow with apex of mandible, with frons, vertex, and occiput medially, black. Pronotal collar near neck and upper margin of pronotum yellow. A central mark on mesoscutum, scutellum, metascutellum, tegula, subtegular ridge, a mark below wings, mesopleurum broadly, mesosternum wholly, upper 0.75 of metapleurum, two rectangular marks at base of propodeum and a dagger-shaped mark at the apical area of propodeum, yellow speculum black. Fore leg with coxa, trochanters, and femur yellow with femur dorsally darker, its tibia and tarsus yellowish brown. Middle leg with coxa, first trochanteral segment and femur yellow, with coxa marked black ventrally and femur dorsally. Second trochanteral segment black and tibia and tarsus brown. Hind leg with coxa yellow with a dorsal black line, its trochanters black, basal 0.5 of femur and 0.25 of tibia yellow. Hind tarsus yellow. Wings hyaline.

Male: Unknown.

Length: ♀, 18 mm. Fore wing 12 mm; ovipositor 0.55 as long as abdomen.

Holotype: ♀, *Brazil:* Matto Grosso, Sinop, 12° 31' S, 55° W, Oct. 1975, M. Alvarenga (Townes).

Distribution: Brazil (map 5).

18. AGONOCRYPTUS BISPOTUS, n. sp.

Female: Face and clypeus shiny, impunctate, area just below antennal socket with a few striations. Malar space granulose. Frons and vertex shiny, impunctate. Occipital carina erased for a distance equal to the basal width of mandible. Pronotum with scattered fine punctures (especially in the sulcus). Pronotal collar impunctate. Epomia present. Mesoscutum smooth and shiny, impunctate. Scutellum, mesopleurum and metapleurum mostly shiny, with a few fine punctures. Prepectal carina 0.8 the height of mesopleurum. Propodeum trans-striate, striations not very close, interspaces with punctures. Area basad of basal carina with weak striations. Striations in the petiolar area rather strong. Areolet wider than high. Second recurrent vein meeting areolet in its apical 0.33. Nervellus intercepted at its apical 0.33. First abdominal segment slender, nearly 3.0 its apical width, its sternite extending up to 0.75 the length of its tergite. First abdominal tergite shiny, impunctate. Second tergite medially finely punctate. Third also finely punctate. The following tergites smooth. Ovipositor about 0.5 the length of abdomen.

Black with yellow marks and bands. Head yellow except frons, vertex and occiput medially and apex of mandibles black. Pronotum along its collar and upper margin broadly, propleurum except basally, a rectangular mark on mesoscutum, scutellum, metascutellum, tegula, subtegular ridge, speculum, a mark below wings, mesopleurum, mesosternum and metapleurum broadly, two squarish marks at base of propodeum and a dagger-shaped mark at apical slope of propodeum, yellow. Legs yellow, except all coxae dorsally black marked. Hind trochanters black. Hind femur and tibia apically darker. Hind tarsus whitish-yellow. Fore and middle trochanters and femur dorsally darker. Wings hyaline. First abdominal tergite with a sub-basal bifurcated mark. All abdominal tergites except the eighth, with yellow apical bands. Second and third tergites with sub-basal lateral spots. Sides of third to seventh tergites yellow. Eighth tergite wholly black.

Male: Unknown.

Length: 18 mm; fore wing 12 mm.

Holotype: ♀, *Mexico:* Oaxaca, Metate, 85.5 kilometers south of Tuxtepec, 900 m, Oct. 16, 1962, H. & M. Townes (Townes).

Distribution: Mexico (map 5).

19. AGONOCRYPTUS TRICOLOR, n. sp.

Female: Face and clypeus shiny. Face submedially, below antennal socket with a few striations. Clypeus with a few fine striations laterally. Malar space granulose, equal to the basal width of mandible. Frons and vertex smooth and shiny. Occipital carina erased for a distance, away from the hypostomal carina (as in *bispotus*), but this carina not deflected inwards; this area appears rough. Pronotum with minute scattered punctures, its hind margin smooth. Epomia absent. Mesoscutum shiny, impunctate. Scutellum, mesopleurum and metapleurum shiny and with a few scattered punctures. Prepectal carina 0.8 the height of mesopleurum. Propodeum smooth with area apicad of basal transverse carina with weak striations interposed with shallow punctures. (Sides of propodeum smooth, and whole propodeum much more shiny than in other species of this group.) Areolet much wider than high. Second recurrent vein meeting areolet at its middle. Nervellus intercepted at its upper 0.33. First abdominal segment smooth and shiny, slender, 3.0 its apical width, its sternite extending to 0.75 the length of its tergite. Second tergite with minute scattered punctures sub-basally. Third tergite finely punctate. The following tergites smooth and shiny. Ovipositor about 0.6 the length of abdomen.

Black, marked with yellow. Abdomen reddish-brown. Head yellow except frons, vertex and occiput medially, malar space and apex of mandible, black. A rectangular mark on mesoscutum, two marks on pronotal collar, upper edge of pronotum, propleurum apically, scutellum, metascutellum, tegula, subtegular ridge, an oblong (somewhat L-shaped) mark on mesopleurum, a mark on mesosternum, an oblong mark on metapleurum, two oval marks on basal area of propodeum, and a dagger-shaped mark on apical slope of propodeum, yellow. Speculum black. Fore and middle legs yellow except their coxae, trochanters and femora dorsally black. Hind coxa, trochanters, and femur reddish-brown with coxa dorsally with a yellow oval mark and a black line on its outer side. Hind tibia and tarsus yellowish, with last tarsal segment black. Wings hyaline. First tergite yellow basally and apically, medially black. The rest of the tergites reddish-brown with tip of abdomen black.

Male: Generally similar to the female. Flagellar segments with long hairs. Each segment also with long spine-like stout seta amongst the hairs at the apex of the segment, particularly from fourth onwards. Frons punctate. Pronotum shiny, Propodeum centrally with strong striations. Apical transverse carina present.

Color black with yellow marks. Antenna black, without yellow mark dorsally. Head yellow. Apex of mandible, frons, vertex and occiput medially black. Pronotal collar, pronotum, propleurum, a mark on middle lobe of mesoscutum, scutellum, metascutellum, tegula, subtegular ridge, speculum, a mark behind hind wing, mesopleurum broadly, mesosternum wholly, metapleurum except basal region, and petiolar area of propodeum, yellow. Fore and middle coxae and first trochanters yellow, their femora and tibiae yellowish-brown. Fore tarsus basally yellow, apically black. Middle tarsus black. Hind coxa black, apically yellow, their trochanters, femur, tibia and tarsus black. Hind femur lighter in color basally. Wings hyaline, apically a little fuscous. Abdomen black, with first tergite basally and apically, and the following tergites apically and apicolaterally yellowish-brown.

Length: 10-18 mm. Fore wing 10-14 mm.

Holotype: ♀, *Brazil*: Matto Grosso, Sinop, 12° 31' S. 55° 37' W., Oct. 1975, M. Alvarenga (Townes).

Allotype: ♂, same data as the type (Townes).
Paratypes: 27♂, same data as the type, collected October 1974, 75, and 1976 (Townes).
Distribution: Brazil (map 5).

VI. THE RUGIFRONS GROUP

The Rugifrons Group is characterized by having the frons rugose; face trans-striate to somewhat rugoso-striate; vertex punctate or sparsely so; whole of thorax including propodeum strongly punctate; and whole of abdomen punctate. In males the first tergite is smoother or more sparsely punctate and thorax also somewhat less strongly punctate.

This group includes *A. rugifrons* with clear wings, reddish abdomen and body with yellow marks; *A. mulleus,* with clear wings, and body wholly reddish-brown; and *A. fumosus* and *A. infuscatus* with blackish-brown wings, reddish thorax and black abdomen. They occur in Argentina and Brazil.

20. AGONOCRYPTUS RUGIFRONS, n. sp.

Male and *Female:* Face trans-striate with punctures in between. Face centrally strongly rugose. Clypeus granulose, with scattered punctures, its apical margin with a minute median tooth. Malar space as long as the basal width of mandible, granulose. Temple smooth and shiny. Frons rugose. Vertex punctate. Occipital carina incurved and meeting hypostomal carina. Pronotum punctate, its sulcus trans-striate. Epomia absent. Mesoscutum closely and deeply punctate. Scutellum sparsely, superficially punctate. Mesopleurum, metapleurum, and propodeum wholly strongly punctato-rugose. Apical transverse carina of propodeum interrupted medially. Areolet rectangular, about 1.4 as wide as high. Second recurrent vein meeting areolet distad of its middle. Nervellus intercepted about the middle. First tergite less than 2.0 its apical width, its sternite ending at the level of spiracle, about 0.5 the length of its tergite. All tergites punctate. First tergite closely and deeply punctate medially. Second and third tergites mat and with compact deep punctures. The following tergites with finer punctures and finely mat. Ovipositor equal to the length of abdomen.

Black with yellow marks. Abdomen reddish-brown to blackish-brown. Head black, except for basal half of clypeus, labrum, and a broad interrupted orbital ring. A mark on mesoscutum, scutellum, metascutellum, interrupted pronotal collar, upper part of pronotum, tegula, subtegular ridge, speculum, a small mark on apical corner of mesopleurum, a mark below wings, an elongate mark on dorsal part of metapleurum, and an inverted T-shaped mark on petiolar area of propodeum (which does not reach the basal transverse carina), whitish-yellow. Wings hyaline. Legs reddish-brown, with fore coxa and first trochanter dorsally and hind tarsal segment yellowish-white. Abdomen largely brown. First tergite apically with a broad white band. The following tergites with a narrow white band. Color of abdominal tergites variable from reddish-brown to blackish brown.
Length: 16 mm. Wings 10 mm.
Holotype: ♀, *Argentina:* Jujuy, Jan. 16, 1966, H. & M. Townes (Townes).
Allotype: ♂, same data as the type (Townes).
Paratypes: 32♀, 33♂, *Argentina:* from Horco Molle (near Tucuman), Calalao, San Pedro de Tucuman, Jujuy, El Pintado, Al Tola Vina, Salta,

Yacochuya (Tafayata), Tacuil, Angastaco, San Miguel de Tucuman, Amaicha del Valle, Aguas Blancas, Trances Tacanas, and Arnau. Collected on various dates during January, February, March, April, September, October, November, and December (Townes, Cambridge and Ottawa).

Distribution: Various localities near Tucuman, Argentina (map 6).

21. AGONOCRYPTUS MULLEUS, n. sp.

Female: Face rugose. Clypeus rough, with a few striations apicolaterally, its margin with a median apical tooth. Malar space aciculate, 0.8 as long as the basal width of mandible. Temple shiny, with a few superficial punctures close to malar space. Frons rugose. Vertex punctate medially, punctures not deep. Occipital carina not meeting hypostomal carina. Pronotum deeply and closely punctate, punctures running into striations along the sulcus and along hind margin. Epomia absent. Mesoscutum and scutellum closely and deeply punctate, appearing rugose at places. Mesopleurum, metapleurum and propodeum wholly strongly punctato-rugose. Apical transverse carina of propodeum broadly interrupted medially. Areolet pentagonal, about 1.4 as wide as high. Second recurrent vein meeting areolet at its apical 0.4. Second intercubitus unpigmented. Nervellus intercepted at its upper 0.4. First tergite not stout, 2.0 as long as its apical width. First sternite ending at the level of spiracle, about 0.5 as long as the tergite. First tergite shiny, with a few scattered punctures medially. Second and third tergites finely, closely punctate. The following tergites finely mat. Ovipositor 0.5 as long as the abdomen.

Reddish-brown. Scape and basal flagellar segments reddish-brown. Narrow (interrupted) inner and outer orbital rings, labrum, pronotal collar and propleurum centrally, and tegula, white. Wings hyaline. Basal tarsal segments of middle and hind legs white. Apical tarsal segments blackish.

Length: 15 mm. Fore wing 10 mm.

Male: Unknown.

Holotype: ♀, *Argentina:* Misiones, Leandro N. Alem., Inst. Alberdi, Nov. 17-19, 1969, C. Porter (Porter).

Distribution: Argentina (map 6).

22. AGONOCRYPTUS INFUSCATUS, n. sp.

Female: Face and clypeus transversely striate, with punctures in between. Clypeal margin with a median minute tooth. Malar space strongly granulose, as long as the basal width of mandible. Temple smooth and shiny, near malar space finely striate. Frons rugose. Vertex sparsely punctate, punctures rather shallow. Occipital carina not meeting hypostomal carina. Pronotum deeply and closely punctate. Pronotal sulcus without any striations. Epomia absent. Mesoscutum and scutellum closely punctate. Mesopleurum punctate. Area below subtegular ridge a little striate. Speculum shiny, with fine scattered punctures. Metapleurum punctate to striato-punctate. Basal area of propodeum sparsely punctate. Apical transverse carina present only on sides. Area between basal and apical transverse carinae striato-punctate. Petiolar area shiny, with a few longitudinal striations. Areolet rectangular, 1.5 as wide as high. Second recurrent vein meeting areolet at its apical 0.3. Nervellus intercepted above its middle. Petiole stout, nearly 2.0 as long as its apical width. First sternite reaching spiracle. Abdominal tergites punctate. First tergite punctate strongly on apical half. Second, third and fourth tergites closely punctate. The following tergites mat. Ovipositor 0.7 as long as the abdomen, its tip pointed, narrow and tapering, with seven teeth.

Reddish-brown with black abdomen. Head reddish-brown. Face reddish-brown, without yellow orbital stripes. Malar space and mandibles black. Wings strongly infuscate. Legs black except coxae. Fore femur ventrally and tibia dorsally, brownish. Hind femur black. Second and third hind tarsal segments yellow. Abdomen black with only the first tergite basally brown.

Length: 18 mm. Fore wing 12 mm.
Male: Unknown.
Holotype: ♀, *Brazil:* Nova Teutonia, Santa Catarina, Oct. 1970, F. Plaumann (Townes).
Paratype: ♀, same data as the holotype, but Nov. 1970 (Townes).
Distribution: Brazil (map 6).

This species is rather close to *A. rufithorax* in general appearance, but can be differentiated by the structure of frons, first tergite, nature of ovipositor teeth, absence of apical transverse carina of propodeum, and also by the color of malar space, orbital rings, hind tarsus and wings.

23. AGONOCRYPTUS FUMOSUS, n. sp.

Female: Face finely transversely striate, with a few rugosities in the middle. Clypeus finely striate, its apical margin with a median tooth. Malar space granulose, with a few longitudinal striations, about as long as the basal width of mandible. Temple similar to that of *infuscatus.* Frons rugose. Vertex closely and finely punctate. Occipital carina meeting hypostomal carina. Epomia absent. Pronotum punctate. Pronotal sulcus punctato-striate. Mesoscutum closely and deeply punctate. Scutellum sparsely punctate. Mesopleurum finely and closely punctate, area below subtegular ridge punctato-striate. Metapleurum deeply and closely punctate. Apical transverse carina of propodeum absent or merged into striations on propodeum. Propodeum punctato-striate. Petiolar area smooth and with a few longitudinal ridges. Wings and abdomen as in *A. infuscatus.* Ovipositor tip blunt, not tapering, and with only 5 teeth. Ovipositor 0.6 as long as the abdomen.

Reddish-brown with head and abdomen black. Face black, with yellow. Interrupted orbital rings. Wings infuscate. Fore leg black with femur ventrally and tibia yellowish-brown. Middle coxa reddish-brown, trochanters black, and rest of middle leg brownish. Hind coxa and femur reddish-brown; trochanters, femur apically, tibia, basal half of basitarsus, and fifth tarsal segment, black. Hind tarsus otherwise yellow. Abdomen black, with first tergite reddish-brown.

Male: Rather similar to the female, with pronotal groove smoother, fore leg lighter in color and abdomen brown rather than black.

Length: ♀, 18 mm. Fore wing 12 mm. Male 13 mm. Fore wing 8-10 mm.

Holotype: ♀, *Brazil:* Nova Teutonia, Santa Catarina, April 14, 1952, F. Plaumann (Townes).

Allotype: ♂, *Argentina:* La Plata, Dec. 18, 1965, H. & M. Townes (Townes).

Paratypes: 4♂, 3♀. *Argentina:* La Plata, 1♂, Dec. 18, 1965, H. & M. Townes (Townes). Berisso, 1♂, Dec. 8, 1965, H. & M. Townes. Mar del Plata, Prov. Buenos Aires, 1♀, Nov. 1, 1949 (Townes). Buenos Aires, Abra de la Ventana, 1♀, Feb. 6, 1947 (Porter). *Brazil:* Nova Teutonia, Santa Catarina, 1♂, Nov. 11, 1952, 1♂, Nov. 1955, F. Plaumann (Townes); 1♀,

April 26, 1960 (Porter).

Distribution: Brazil and Argentina (map 6).

Acknowledgments: I am grateful to Dr. Henry Townes for giving me the opportunity to study at the American Entomological Institute and for his guidance and assistance at every stage of the investigation. I am also thankful to the curators of the various museums from where the types and other collections were borrowed for the present study. The respective locations of the types and other material received for study are given in the text after the specimens examined.

REFERENCES

Brues, Charles T. 1912. Brazilian Ichneumonidae and Braconidae Obtained by the Stanford Expedition. Ann. Ent. Soc. America 5(3): 193-228.

Brullé, M. A. 1846. *In* Lepeletier: Histoire Naturelle des Insects. Hyménoptères 4: 1-680.

Carlson, R. W. 1979. Family Ichneumonidae. *In* Krombein *et al.*: Catalog of Hymenoptera in America North of Mexico. 1: 315-741.

Cresson, E. T. 1873. Description of Mexican Ichneumonidae. Proc. Acad. Nat. Sci. Philadelphia 1873: 104-176.

Cushman, R. A. 1929. A Revision of the North American Ichneumon-flies of the genus *Mesostenus* and Related Genera. Proc. U. S. Natl. Mus. 74(16): 1-58.

Taschenberg, E. L. 1876. Einige neue tropische, namentlich Südamerikanische. Ztschr. f. die Gesam. Naturw. Halle 48: 61-104.

Townes, Henry K. 1946. The Generic Position of Neotropic Ichneumonidae (Hymenoptera) with Types in the Philadelphia and Quebec Museums, Described by Cresson, Hooker, Norton, Provancher and Viereck. Bol. Ent. Venezolana 5: 29-63.

Townes, Henry K. & Marjorie 1962. Ichneumon-flies of America north of Mexico 3: subfamily Gelinae, Tribe Mesostenini, U. S. Natl. Mus. Bull. 216(3): 1-602.

Townes, Henry & Marjorie 1966. A Catalogue and Reclassification of Neotropic Ichneumonidae. Mem. Amer. Ent. Inst. 8: 1-367.

Viereck, H. L. 1905. Notes and Descriptions of Hymenoptera from the Western United States in the Collection of the Univ. of Kansas. Trans. Kansas Academy Sci. 19: 264-326.

39

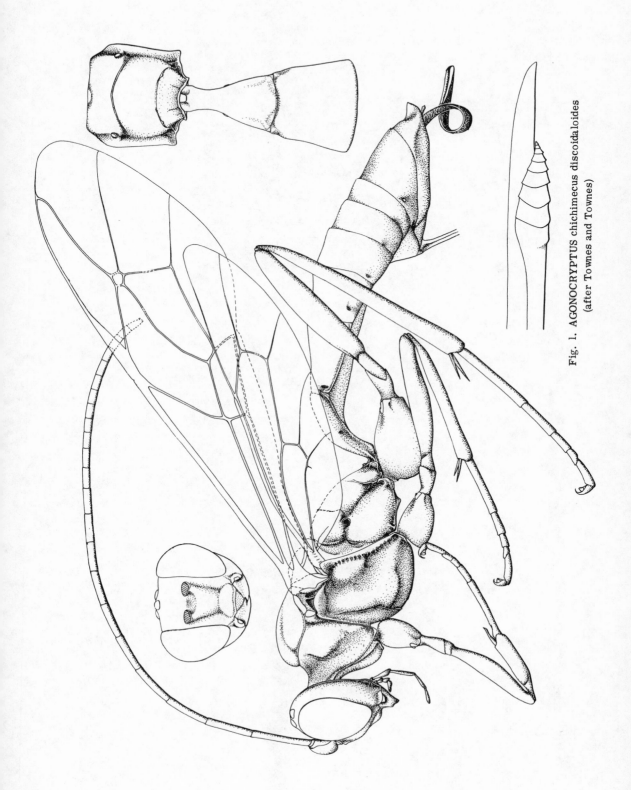

Fig. 1. AGONOCRYPTUS chichimecus discoidaloides
(after Townes and Townes)

▲ A. chichimecus
   chichimecus

● A. chichimecus
   discoidaloides

■ A. ruficrus

✶ A. bicolor

Map 1.   AGONOCRYPTUS - Chichimecus Group

*   A. rufithorax (Chichimecus Gr.)

■   A. russulus   (,,   )

○   A. heathi   (Heathi   Group)

▲   A. violascens (Group ?)

Map 2. AGONOCRYPTUS - Chichimecus Group, Heathi Group

▲ A. lioneli lioneli

△ A. lioneli coxinota

● A. argentinus argentinus

✳ A. argentinus tucumanus

■ A. physocnemis physocnemis

○ A. physocnemis nigristernum

Map. 3. AGONOCRYPTUS - Physocnemis Group

A. gossypii

A. erugatus

A. rufigaster

A. adustus adustus

A. adustus paulus

A. varus varus

A. varus nigrifemur

A. leurosus flavosternum

A. leurosus leurosus

Map. 4. AGONOCRYPTUS - Varus Group

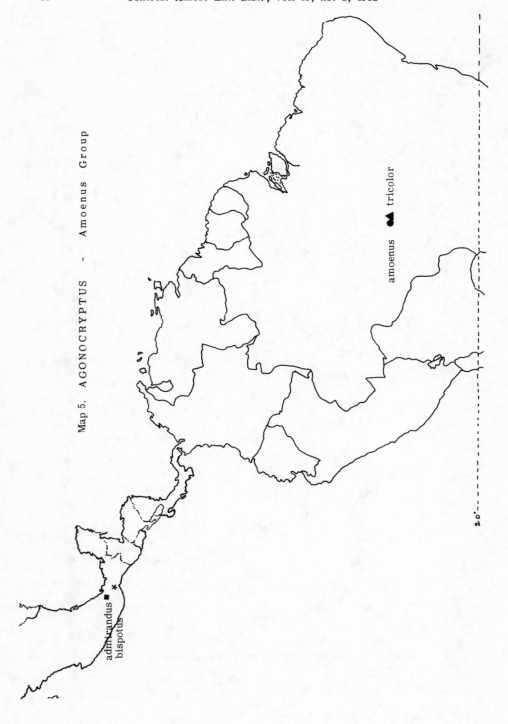

Map 5. AGONOCRYPTUS - Amoenus Group

● A. rugifrons

▲ A. fumosus

■ A. mulleus

○ A. infuscatus

Map 6. AGONOCRYPTUS - Rugifrons Group

A REVIEW OF THE GENUS *CRYPTOHELCOSTIZUS*

(HYMENOPTERA: ICHNEUMONIDAE)

by
Santosh Gupta *

American Entomological Institute
5950 Warren Road, Ann Arbor, Michigan 48105

The genus *Cryptohelcosti zus* was described by Cushman (1919) with
C. rufigaster (from California) as the type-species. In 1922, he synonymized
rufigaster with *Cryptus alamedensis* Ashmead (1890). Viereck (1921) described
C. dichrous from North Carolina. Cushman (1940) added two more species:
C. chrysobothridis and *C. ornatus* from Oklahoma and California, respectively.
Townes and Townes (1962) added six more species: *genalis*, *leiomerus*,
caudatus and *maculosus* from California, and *fumipennis* and *nigricans* from
Arizona. Thus a total of ten species are at present known under the genus,
all from the U.S.A., in the Nearctic Region.

In the Townes Collection there were a number of unstudied specimens
from Northwestern United States, which prompted me to make a restudy of
the genus. All the types were kindly loaned to me by the curators of the
various museums. No new species turned up in the collections, but a new
subspecies of *C. genalis* is described from Oregon. New distributional
records are given for the various species, and a key is provided to distinguish
them. In the treatment that follows, only the significant characters of the
species are mentioned, and the records of specimens studied. For fuller
descriptions refer to Townes and Townes (1962). Synonymical references to
the species can be seen in the catalogues of Townes (1944), Townes and Townes
(1951) and Carlson (1979).

The genus *Cryptohelcostizus* has been traditionally put under the old sub-
family Cryptinae close to *Helcostizus* Foerster and *Xylophrurus* Foerster.
Townes and Townes (1962) placed it under the subtribe Echthrina of tribe
Mesostenini, subfamily Gelinae. In 1970, Townes chanted the name of the
subtribe to Gabuniina. With the raising of the tribe Mesostenini to subfamily
Mesosteninae by Gupta (1970), the Gabuniina becomes the tribe Gabuniini.

Genus CRYPTOHELCOSTIZUS Cushman   (fig. 1)

> *Cryptohelcostizus* Cushman, 1919. Proc. U. S. Natl. Mus., 55: 534.
>   Type-species: (*Cryptohelcostizus rufigaster* Cushman) = *alamedensis*
>   (Ashmead); original designation.
> Taxonomy: Townes and Townes, 1962: 504. Carlson, 1979: 478.

The salient features of the genus are: Head transverse. Clypeus short,
convex, its apical margin broadly truncate, without a median tooth. Sternaulus
absent. Pleural carina and apical transverse carina of propodeum absent.
Propodeum usually short and rugoso-punctate. Basal transverse carina always
present. Areolet pentagonal, large, 0.5 to 0.9 as high as second recurrent

* Present address: Center for Parasitic Hymenoptera. Dept.
of Entomology & Nematology, Univ. of Florida, Gainesville, FL 32611.

vein. Nervulus basad of basal vein. Nervellus intercepted below its middle. Fore femur in female concave below. First tergite without a lateral triangular tooth at base, its dorsolateral carina weak or absent, ventrolateral carina present. Ovipositor 0.5 to 1.0 the fore wing length.

Distribution: Nearctic Region.

Host: Buprestid larvae boring in branches and twigs of trees and shrubs.

Two species groups can be recognized based on the nature of the basal transverse carina of the propodeum and the coloration of the wings:

I. *The Alamedensis Group,* characterized by having the wings clear hyaline, and the basal transverse carina of propodeum complete, uniformly arched or sometimes narrowly interrupted medially, but never turned basad medially. This group includes *C. alamedensis, nigricans, maculosus, ornatus, caudatus,* and *chrysobothridis.*

II. *The Dichrous Group,* characterized by having the wings black or strongly infuscate, and basal transverse carina of propodeum interrupted medially and each end turned basad. This group includes *C. dichrous, genalis, fumipennis,* and *leiomerus.*

KEY TO THE SPECIES OF *CRYPTOHELCOSTIZUS*

1. Wings clear hyaline. Basal transverse carina of propodeum complete, uniformly arched or sometimes narrowly interrupted medially, but never turned basad in the middle (Alamedensis Group). 2
Wings black or strongly infuscate. Basal transverse carina of propodeum interrupted medially and its ends turned basad medially (Dichrous Group). 7

2. Hind femur on its front side with very sparse hairs, their sockets separated by more than 2.0 the length of hairs. 3
Hind femur on its front side with closely set hairs, their sockets separated by 0.7 to 2.0 x the length of hairs. 6

3. Hind femur polished, black. Nervulus basad of basal vein by 0.5 to 0.66 its length (by more than half its length). Propodeum dorsally flat, not very convex. Trochanters of fore leg black ventrally.
                              2. nigricans Townes and Townes (p. 4)
Hind femur granulose, subpolished. Nervulus basad of basal vein by less than half its length. Propodeum dorsally convex, sloping. Trochanters of fore leg with white marks ventrally. 4

4. Basal transverse carina of propodeum complete and uniformly arched. Propodeum and hind coxa black, without any spot or mark.
                              1. alamedensis (Ashmead) (p. 4)
Basal transverse carina of propodeum narrowly interrupted medially with the ends not turned basad medially. Propodeum and hind coxa with yellow marks. 5

5. Mesopleurum and metapleurum each with a large yellow mark. Vertex deeply and closely punctate. Second intercubitus weak medially.
                              3. maculosus Townes and Townes (p. 5)

Mesopleurum and metapleurum without any yellow marks. Vertex with sparse, moderate sized punctures. Second intercubitus uniformly strong. (Ovipositor sheath 0.4 the length of front wing.)

6. Propodeum and hind coxa marked with white. Ovipositor sheath 0.75 the length of front wing. Legs lighter in color.

Propodeum and hind coxa black, without any white marks. Ovipositor sheath 0.4 the length of front wing. Legs darker in color.

7. Hairs on front face of hind femur moderately dense, their sockets separated by about 0.7 the length of the hairs. Punctures on mesopleurum separated by about 0.7 their diameter. Prepectal carina reaching subtegular ridge. Face granulose, with punctures on sides and rugulose medially. Pronotal collar black without any central white or brown mark. . . . .

Hairs on front face of hind femur sparser, their sockets separated by 2.0 the length of the hairs. Punctures on mesopleurum separated by about 0.4 their diameter. Face shiny with punctures or with a few oblique striations. Pronotal collar with a white or brown mark medially (except in *leiomerus*). Prepectal carina usually reaching 0.6 to 0.75 the height of mesopleurum. 8

8. Hind margin of temple, and its occipital carina, strongly bulging outward at the lower corner of the eye. Face strongly punctate. Vertex moderately closely punctate close to ocelli.

Hind margin of temple, and its occipital carina, not bulging outward, sometimes a little convex. Side of face punctate. Vertex sparsely punctate close to ocelli. 10

9. Hind femur reddish-brown. Coxa brownish-black. Propodeum rugulosopunctate. First tergite shiny and grooved subapically.

Hind coxa and femur black. Propodeum punctato-striate. First tergite punctate and not grooved. .

10. First tergite sparsely and finely punctate. Clypeus and upper part of pronotum marked with white. Temple weakly convex. Second abscissa of cubitus about 1.4 as long as the third abscissa.

First tergite moderately punctate dorsally. Clypeus and upper part of propodeum marked with white. Temple moderately convex. Second abscissa of cubitus about 1.15 as long as the third abscissa.

I. THE ALAMEDENSIS GROUP

1. CRYPTOHELCOSTIZUS ALAMEDENSIS (Ashmead)
    Cryptus alamedensis Ashmead, 1890. Proc. U. S. Natl. Mus., 12: 409.
        ♀. Type: ♀, Alameda, California (Washington).
    Cryptohelcostizus rufigaster Cushman, 1919. Proc. U. S. Natl. Mus.,
        55: 534. ♂, ♀. Type: ♀, Harold, California (Washington).

The characteristic features of this species are: Mesopleurum with fine
rugosities. Propodeum dorsally convex and sloping. Basal transverse carina
of propodeum complete and evenly arched. Nervulus basad of basal vein by
less than 0.5 its length. Hind femur granulose with very sparse hairs. Ovipositor 0.9 as long as fore wing. Ovipositor sheath about 0.65 as long as fore
wing.
    Black with reddish abdomen. A mark on clypeus, base of mandible, a mark
on pronotum, a narrow interrupted orbital mark and metascutellum, white.
Propodeum black, sometimes with two very small marks on apical area.
Wings hyaline. Legs brownish. First trochanter of fore leg white ventrally.
Hind coxa black. Basal 0.5 of first abdominal tergite brownish-black.
    Specimens: 61♀, 26♂. U. S. A.: *California:* Leevining, 4♂, June 22,
1948. Portero in San Diego Co., 11♀, 5♂, April 8-17, 1974. Lake Wohford,
San Diego Co., 11♀, 1♂, April 21-May 1, 1974. Julian, 24♀, 10♂, May 6-30,
1974. North of Leggett, 1♀, May 18, 1978 (All above collected by H. and M.
Townes and in Townes Collection.). California: Simla Station, ♂, June 28,
1922, R. D. Hartman, on Apricot, parasitic on *Chrysobothris mali.* Los
Gatos, 1♀, May 23, 1918, R. D. Hartman, on *Osmaronia,* parasitic on
Chrysobothris mali. Walter Sqr., Napa Co., 1♀, May 26, 1951, E. I.
Schlinger (Townes). Elk Grove, Sacramento Co., 1♀, Sept. 24, 1949, H. A.
Hunt (Townes). W. Walker R., 6000 ft., ♀, C. D. Michner, on *Prunus*
(Townes). Idaho: near Stanley, 6000 ft., ♀, Aug. 8, 1978 (Townes). Low-
man, 6000 ft., ♀, April 14, 1978, H. & M. Townes (Townes). *Nevada:*
Tuscarora, ♂, June 5, 1978, H. & M. Townes (Townes). *Oregon:* Selma, 2♀,
3♂, May 20-21, 1974, H. & M. Townes (Townes). Hyatt Reservoir, 3♀, May
2, 20, and 22, 1974, H. & M. Townes (Townes). Pinehurst, 3♀, June 23 and
29, 1974, H. & M. Townes (Townes). *Utah:* Strawberry Daniel Pass, ♀, June
19, 1948, H., M., G. & D. Townes (Townes).
    Variations: Two females from California: Poterero, April 8, 14, 1974 are
lighter in color with body thin and slender. Other specimens also show considerable variations in the punctation of face, interocellar area, and mesopleurum, color of legs and base of abdomen, and the length of ovipositor
sheath as compared to the fore wing length.
    Distribution: Townes and Townes (1962) reported this species from
British Columbia, California (widely distributed), Oregon, Texas and Utah.
It is here reported from Idaho and Nevada for the first time.
    Hosts: Chrysobothris mali, Myrmex arizonicus, Argilus angelicus.

2. CRYPTOHELCOSTIZUS NIGRICANS Townes and Townes
    Cryptohelcostizus nigricans Townes and Townes, 1962. U. S. Natl. Mus.
        Bull. 216(3): 508. ♂, ♀. Type: ♀, Arizona: Sierra Ancha, Parker
        Creek (Townes). Examined, 1980.

Some of the salient features of this species are: Mesopleurum punctate, without striations or rugosities, but punctures close and deep. Propodeum dorsally flat. Basal transverse carina of propodeum complete and evenly arched. Nervulus basad of basal vein by 0.5 to 0.66 its length. Hind femur polished and with very sparse hairs. Ovipositor about 0.7 as long as fore wing. Ovipositor sheath about 0.45 as long as fore wing.

Black with reddish abdomen. Clypeus, base of mandible, and pronotum each with a small white mark. Metascutellum and propodeum wholly black. Wings hyaline. Legs black. First trochanter of fore leg without any white mark. Apical two tarsal segments of hind leg brownish. First tergite black.

Specimens: 1♂, 2♀. *Arizona:* Sierra Ancha, Parker Creek, ♀ (type), May 7, 1947, H. & M. Townes (Townes). Workman Creek, ♂, ♀ (paratypes), April 28, 1947, H. & M. Townes (Townes).

Distribution: U. S. A.: Arizona.

3. CRYPTOHELCOSTIZUS MACULOSUS Townes & Townes

Cryptohelcostizus maculosus Townes and Townes, 1962. U. S. Natl. Mus. Bull., 216(3): 512. ♀. Type: ♀, California: Death Valley (Washington). Reared from *Prosopis*. Examined, 1980.

The salient features of this species are: Mesopleurum closely and deeply punctate with a few rugosities medially. Propodeum convex and sloping. Basal transverse carina of propodeum interrupted medially, but its ends not turned basad. Nervulus basad of basal vein by less than 0.5 its length. Hind femur granulose with very sparse hairs. Ovipositor sheath about 0.5 as long as fore wing.

Black, marked with yellow. Abdomen reddish-brown with whitish marks. Face, clypeus, labrum, base of mandible, complete orbital stripe, temple, pronotal collar, pronotum broadly above, a Y-shaped mark on middle lobe of mesoscutum a crescentic mark on lateral lobe of mesoscutum, scutellum, metascutellum, tegula, subtegular ridge, a broad oblong mark on mesopleurum, metapleurum dorsally, a mark behind hind wing and propodeum apically, yellow. Wings hyaline. Legs reddish-brown. Coxae black, dorsally marked with white. First trochanter of fore leg ventrally white. First tergite basally brownish and apically marked with white.

Specimen: California: Death Valley, ♀ (type), Sept. 3, 1957, R. C. Hall, "Reared from *Prosopis*" (Washington).

Distribution: U. S. A.: California.

4. CRYPTOHELCOSTIZUS ORNATUS Cushman

Cryptohelcostizus ornatus Cushman, 1940. Proc. U. S. Natl. Mus. 88: 359. ♀. Type: ♀, Death Valley, California (Washington). Examined, 1980.

The salient features of this species are: Mesopleurum closely and deeply punctate. Propodeum convex. Basal transverse carina of propodeum interrupted medially but ends not turning basad. Nervulus basad of basal vein by less than 0.5 its length. Hind femur granulose with very sparse hairs. Ovipositor 0.6 the length of fore wing, its sheath 0.4 the length of fore wing.

Black with yellow marks. Abdomen reddish-brown. Face black. Clypeus base of mandible, broad complete orbital stripes, an interrupted mark on pronotal collar, upper part of pronotum, scutellum, metascutellum, tegula, and subtegular ridge, yellow. Propodeum with two small white marks on either side and a small mark basad of them. Wings hyaline. Legs reddish-brown. Fore coxa dorsally and ventrally, and first trochanter ventrally white. Middle coxa dorsolaterally and trochanter ventrally white. Hind coxa dorsally white marked. Basal 0.7 of first tergite black.

Specimens: ♀, ♂. *Arizona:* Sahuarito, ♂, April 11, 1947, H. & M. Townes (Townes). *California:* Death Valley, ♀ (type) Feb. 23, 1939, M. F. Gilman, reared from *Chrysobothris deserta* in desert holly (Washington).

Distribution: U. S. A.: Arizona, California.

Host: Chrysobothris deserta.

5. CRYPTOHELCOSTIZUS CAUDATUS Townes and Townes

Cryptohelcostizus caudatus Townes and Townes, 1962, U. S. Natl. Mus. Bull., 216(3): 511. ♀. Type: California: Lake Tahoe (Davis). Examined, 1980.

The characteristic features of this species are: Mesopleurum moderately closely punctate, punctures arranged in transverse rows or forming fine rugosities at places. Basal transverse carina of propodeum complete and evenly arched. Hind femur with closely set hairs. Ovipositor as long as the fore wing. Ovipositor sheath 0.75 as long as fore wing.

Black with reddish abdomen. Clypeus except its apical margin, base of mandible, interrupted orbital ring, temple towards lower corner of eye, upper margin of pronotum, two lateral marks on scutellum (meeting at apex), metascutellum, tegula, subtegular ridge, and two sublateral marks on apical slope of propodeum, whitish-yellow. Wings hyaline. Legs reddish-brown. All coxae black with white dorsal marks. First trochanter of fore and middle legs ventrally white marked. First tergite basally black.

Specimens: 2♀. *California:* Lake Tahoe, ♀ (type), Aug. 15, 1950, R. M. Bohart (Davis). *Oregon:* Pinehurst, ♀, July 2, 1978, H. & M. Townes (Townes).

Variation: The color of mesosternum in the specimen from Oregon is somewhat lighter than that in the type. Otherwise it agrees with the same.

Distribution: U. S. A.: California, Oregon.

6. CRYPTOHELCOSTIZUS CHRYSOBOTHRIDIS Cushman

Cryptohelcostizus chrysobothridis Cushman, 1940. Proc. U. S. Natl. Mus., 88: 358. ♂, ♀. Type: ♀, Stillwater, Oklahoma (Washington). Paratype examined, 1980.

This species is characterized as: Mesopleurum with fine trans-rugosities. Propodeum a little flat. Basal transverse carina of propodeum complete and evenly arched. Hairs on front face of hind femur close together. Ovipositor 0.6 as long as fore wing. Ovipositor sheath 0.4 as long as fore wing.

Black. Abdomen reddish-brown. A mark on clypeus, a mark on pronotum and an interrupted narrow orbital stripe, white. Metascutellum brownish. Propodeum wholly black. Wings hyaline. Legs brownish, with all coxae blackish. First trochanter of fore leg ventrally not marked with yellow. Third and fourth tarsal segments of hind leg white. In male middle and hind femora reddish-brown.

Specimen: Oklahoma: Stillwater, ♀ (paratype), April 4, 1936, Myron

Maxwell (Townes).
>   *Host: Chrysobothris* sp. in *Malus pumila.*
>   *Distribution:* U. S. A.: Oklahoma.

II. THE DICHROUS GROUP

7. CRYPTOHELCOSTIZUS DICHROUS Viereck

Cryptohelcostizus dichrous Viereck, 1921. Psyche, 28: 73. ♂, ♀.
>   Type: ♀, Southern Pines, N. C. (Cambridge).

This species is characterized by: Face granulose with punctures. Punctures below antennal sockets obliquely striate. Hind margin of temple and its occipital carina not bulged out at the lower margin of eye. Punctures on mesopleurum deep and well separated. Prepectal carina reaching subtegular ridge. Basal transverse carina of propodeum interrupted medially and turned basad. Hind femur granulose with moderately dense hairs. Ovipositor 0.45 as long as fore wing, shorter than in *fumipennis* and *leiomerus*. Ovipositor sheath about 0.35 as long as fore wing.
>   Black, abdomen reddish-brown. Orbital ring interrupted and white. Pronotum without any yellow mark. Wings black to strongly infumated. Legs black. In male hind femur red.
>   *Specimens:* 4♀. *North Carolina:* Lumberton, ♀, Oct. 27, 1949, Rabb and Townes. Council, ♀, May 16, 1940, D. L. Wray. Page Lake, ♀, Oct. 27, 1949, Rabb and Townes. Southern Pines, ♀, Oct. 1955, H. Townes (all in Townes collection).
>   *Distribution:* Townes and Townes reported it from Georgia, Missouri, New Jersey, North Carolina, Texas and Virginia.

8. CRYPTOHELCOSTIZUS GENALIS Townes and Townes

Female: Face shiny, punctate; punctures arranged in oblique striations just below the antennal sockets. Vertex close to ocelli moderately closely punctate. Hind margin of temple and its occipital carina strongly bulged outwards at the level of lower corner of the eye. Mesopleural punctures deep and separated by 0.4 their diameter. Prepectal carina reaching 0.6 the height of mesopleurum. Propodeum basad of basal carina rugoso-punctate; centrally punctato-striate to punctato-rugose. Basal transverse carina of propodeum interrupted medially and its ends turned basad. Hairs on front face of hind femur sparse. First abdominal tergite subapically grooved or not so; groove punctate or impunctate. Ovipositor 0.5 as long as fore wing.
>   Black. Abdomen reddish-brown. Face black. A mark on clypeus, a small mark on upper part of pronotum, a narrow interrupted orbital ring, and a mark on pronotal collar, white. Wings infuscate. Coxa and trochanters brownish-black. Legs reddish-brown. Hind tibia and tarsus brownish.
>   Two subspecies are recognized: *C. genalis genalis* Townes and Townes from California and *C. genalis niger,* n. subsp. from Oregon.

8a. CRYPTOHELCOSTIZUS GENALIS GENALIS Townes and Townes

Cryptohelcostizus genalis Townes and Townes, 1962. U. S. Natl. Mus.
>   Bull., 216(3): 507. ♀. key, des. Type: ♀, California (Davis).

Female: Propodeum centrally rugoso-punctate. First abdominal tergite subapically longitudinally grooved, groove shiny, impunctate. Hind coxa brownish-black. Hind femur reddish-brown.

Specimen: California: Mendocino, Capella, ♀ (type), May 20, 1955, E. I. Schlinger (Davis).

Distribution: U. S. A.: California.

8b. CRYPTOHELCOSTIZUS GENALIS NIGER, n. subsp.

Female: Face deeply and densely punctate, below antennal sockets punctures sometimes appear semicircularly arranged. Malar space 0.75 the basal width of mandible. Temple shiny with a few fine punctures, its hind margin and occipital carina strongly bulged outwards at the lower corner of the eye. Frons ruguso-striate. Vertex closely and deeply punctate with a few transstriae in the groove. Pronotal collar finely and closely punctate. Mesoscutum deeply and closely punctate. Notaulus broad, with trans-striae. Scutellum closely and deeply punctate, but punctures a little sparse in the middle. Metascutellum shiny. Mesopleurum with moderately close and deep punctures. Prepectal carina reaching 0.7 the height of mesopleurum. Metapleurum punctato-striate. Basal transverse carina of propodeum interrupted medially and its ends turned basad. Area basad of basal carina punctate submedially and with a few trans-striations. Basolateral area of propodeum densely and deeply punctate, central area punctato-striate, petiolar area rugose. Nervulus basad of basal vein by less than 0.5 its length. Nervellus intercepted below the middle. Hind coxa punctate. Hind femur with sparse hairs. First tergite stout, closely and finely punctate. First sternite reaching up to spiracle. Dorsolateral carina absent. Ventrolateral carina present and stout at base. Second and following tergites weakly punctate to mat. Ovipositor 0.5 as long as fore wing.

Black. Abdomen reddish-brown. Pronotal collar without white mark. Small rounding mark on clypeus, and upper edge of pronotum, and a thin broken orbital ring, white. Wings infuscate. Legs black, except the second trochanter of each leg, which is reddish.

Male: Similar to the female in structure, but legs brownish-black and first abdominal tergite a little darker in color. Wings not as dark as those of females.

Length: ♀, 16 mm. Fore wing 11 mm. Ovipositor 6 mm. ♂, 13 mm. Fore wing 8 mm.

Holotype: ♀, Oregon: Selma, May 28, 1978, H. & M. Townes (Townes).

Allotype: ♂, Oregon: Selma, May 20, 1978, H. & M. Townes (Townes).

Paratypes: 1♀, 2♂. Oregon: Pinehurst, ♀, June 29, 1978, H. & M. Townes; 2♂, same locality, dates, and collector as allotype (Townes).

Distribution: U. S. A.: Oregon.

9. CRYPTOHELCOSTIZUS LEIOMERUS Townes and Townes

Cryptohelcostizus leiomerus Townes and Townes, 1962. U. S. Natl. Mus. Bull., 216(3): 506. ♀. key, des. Type: ♀, California (Berkeley).

The characteristic features of this species are: Face punctate, punctures a little sparse and not forming striations. Hind margin of temple and its occipital carina not bulged outward at the level of lower corner of eye. Vertex sparsely punctate as compared to *genalis*. Punctures on mesopleurum close and forming trans-striations. Prepectal carina reaching 0.7 the height of mesopleurum. Basal transverse carina of propodeum interrupted medially and turned basad. Second abscissa of cubitus about 1.4 as long as the third abscissa. Hairs on front face of hind femur sparse. First abdominal tergite sparsely punctate. Ovipositor 0.6 as long as fore wing.

Black. Abdomen reddish-brown. Orbits narrowly white. Wings infuscate. Legs brownish-black.

Specimen: California: Los Angeles, Tanbark Flats, ♀ (type), June 14, 1952, W. V. Garner (Berkeley).

Distribution: U. S. A.: California.

10. CRYPTOHELCOSTIZUS FUMIPENNIS Townes and Townes

Cryptohelcostizus fumipennis Townes and Townes, 1962. U. S. Natl. Mus. Bull., 216(3): 507. ♀. key, des. Type: ♀, Arizona (Washington).

This species can be recognized as follows: Face shiny, with distinct punctures. Hind margin of temple and its occipital carina faintly bulged outward. Vertex sparsely punctate as compared to *genalis*. Punctures on mesopleurum close and deep. Prepectal carina reaching 0.6 the height of mesopleurum. Propodeum rugoso-striate, its basal transverse carina interrupted medially and the ends turned basad. Second abscissa of cubitus about 1.1 as long as third abscissa. Hairs on front face of hind femur sparse. First tergite closely punctate (sparsely punctate in leiomerus). Ovipositor 0.6 as long as fore wing.

Black. Abdomen reddish-brown. Clypeus except its apical margin, a mark at the base of mandible, interrupted orbital ring, upper edge of pronotum, pronotal collar medially, and tegula, white. Wings infuscate. Legs brownish-black (including femur). Fore and middle legs a little lighter.

Specimen: Arizona: Sabino Canyon, "Resting on blossom", ♀ (type), Nov. 24, 1917, W. D. Edmonston (Washington).

Distribution: U. S. A.: Arizona.

ACKNOWLEDGMENTS

I am thankful to Dr. Henry Townes for providing me with facilities for work and for his guidance during the course of the present research work. Thanks are also due to the curators of various museums who loaned the types for the present study.

REFERENCES

Ashmead, W. M. 1890. Description of New Ichneumonidae in the Collection of the U. S. National Museum. Proc. U. S. Natl. Mus. 12: 387-451.

Carlson, R. W. 1979. Family Ichneumonidae. *In* Krombein *et al.*: Catalog of Hymenoptera in America North of Mexico. 1: 315-741.

Cushman, R. A. 1919. Description of New North American Ichneumon-flies. Proc. U. S. Natl. Mus. 55: 517-543.

Cushman, R. A. 1922. New Species of Ichneumon-flies with Taxonomic Notes. Proc. U. S. Natl. Mus. 60(21): 1-28.

Cushman, R. A. 1940. New Genera and Species of Ichneumon-flies with Taxonomic Notes. Proc. U. S. Natl. Mus. 88: 355-372.

Gupta, V. K. 1970. Ichneumon Hunting in India pp. 1-80.

Townes, Henry K. 1944. A Catalogue and Reclassification of the Nearctic Ichneumonidae (Hymenoptera). Part I. The Subfamilies Ichneumonidae, Tryphoninae, Cryptinae, Phaeogeninae and Lissonotinae. Mem. Amer. Ent. Soc., 11(1): 1-477.

Townes, Henry K. 1970. The Genera of Ichneumonidae, Part 2. (Gelinae). Mem. Amer. Ent. Inst. 12: 1-537.

Townes, H. & M. 1951. *In* Muesebeck *et al.*: Hymenoptera of America North of Mexico - Synoptic Catalog. U. S. Dept. Agr. Agr. Monog. No. 2: 1-1420.

Townes, Henry K. and Marjorie 1962. Ichneumon-flies of America North of Mexico: 3. Subfamily Gelinae. Tribe Mesostenini. U. S. Natl. Mus. Bull. 216(3): 1-602.

Vireck, H. L. 1921. Description of New Ichneumonidae in the Collection of the Museum of Comparative Zoology Cambridge, Mass. Psyche 28(3): 70-83.

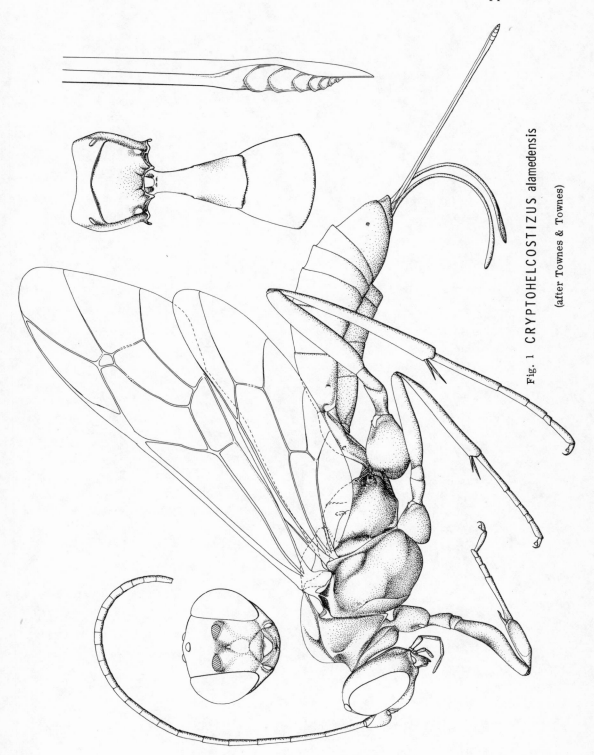

11

Fig. 1 CRYPTOHELCOSTIZUS alamedensis

(after Townes & Townes)

A REVIEW OF THE GENUS *PERITHOUS*, WITH DESCRIPTIONS of NEW TAXA

(HYMENOPTERA: ICHNEUMONIDAE)

by
Virendra Gupta

Center for Parasitic Hymenoptera
Department of Entomology and Nematology
University of Florida, Gainesville, FL 32611

Perithous Holmgren is a small genus belonging to the tribe Theroniini, subfamily Pimplinae (= Ephialtinae). It is mainly distributed in the Holarctic Region. Hosts are aculeate Hymenoptera (larvae of Sphecidae, Vespidae, and Chrysididae) in stems and twigs, particularly of *Rosa* and *Rubus*. There are reports, however, of attacks on larvae of Xiphydriidae, Cerambycidae and some Lepidoptera.

In this paper the world species are reviewed, with some new diagnostic characters. The distributional ranges of the American and European subspecies of *P. mediator* are analyzed. Three new species *(digitalis, kamathi* and *sundaicus)* and two new subspecies *(digitalis taiwanensis* and *divinator himalayensis)* are described from Kashmir and Himachal Pradesh in India (areas of Palaearctic affinities) and Java and Taiwan (typically Oriental), thus extending the known distribution of the genus to the Oriental Region.

Baltazar (1961) described *Hybomischos* as a subgenus of *Perithous* from the Philippines, distinguishing it mainly in having lateral spine-like teeth at the base of first tergite, its spiracle distant from lateral carina, ovipositor tip sinuate and thickened before apex, and by the absence of notauli in the female. Aubert (1969) synonymized it with *Perithous,* while Constantineanu & Pisică (1977) treated it as a distinct genus. *Hybomischos* is here considered a separate genus. It is treated in an adjacent paper.

Genus PERITHOUS Holmgren (figs. 1-4)

> *Perithous* Holmgren, 1959. Öfvers. Svenska Vetensk.-Akad. Forh.,
> 16: 123. Type-species: *Ephialtes albicinctus* Gravenhorst; designated
> by Viereck, 1914.
> Taxonomy: Oehlke, 1966: 279. Aubert, 1969: 100. Townes, 1969: 126.
> Constantineanu & Pisică, 1977: 86.

Body moderately long and slender. General coloration black with mesoscutum and mesopleurum often reddish and abdominal tergites narrowly margined with yellow (some species with black thorax). Legs pale brown. Face of male whitish and of female black with whitish orbital borders.

Antenna moderately long and slender to shorter and a little thickened subapically. Clypeus flattened apically and with a deep median apical notch. Mandibular teeth equal in length and similar in shape. Face a little convex, sparsely to moderately punctate. Eyes only slightly emarginate just above antennal sockets (cf. *Hybomischos*). Occipital carina complete, without a median apical dip. Epomia short. Prepectal carina dorsad to 0.7 the height of mesopleurum. Notauli weakly impressed anteriorly. Propodeum convex, with only the apical transverse carina present, which is semicircularly arched and usually complete. Fore wing with stub of ramellus distinct and continued to some distance as a brownish or unpigmented groove. Areolet triangular, receiving second recurrent vein near its outer corner. Nervellus intercepted above the middle. Tarsal claws of female without a lobe or an enlarged spatulate bristle. First abdominal tergite convex dorsally, with short median dorsal carinae enclosing a basal declivity, its dorsolateral carina present, and baso-lateral corner not projecting as a tooth. Spiracle just below the dorsolateral carina and touching or almost touching it. First sternite with a carina-like median fold, the basal end of which may or may not be projecting out. Second to fifth abdominal tergites with low smooth tubercles. Those on second often fused into a rhomboidal area. Ovipositor 0.8 to 1.5 x as long as the body, compressed, weakly upcurved, its tip with lower valve occupying almost the entire depth, teeth compressed and close together or spaced out to an area about 5.0 x the depth of ovipositor (figs. 1-3). Upper valve with slanting ridges, but not appearing saw-like in profile view (cf. *Hybomischos*).

Some recent studies on *Perithous* are: Townes & Townes (1960), on North American species; Townes, Momoi & Townes (1965), A Catalogue of Eastern Palaearctic species; Oehlke (1966), on Taxonomy of Haupt species; Oehlke (1967), Catalogue of Western Palaearctic species; and Constantineanu & Constantineanu (1968) and Constantineanu & Pisică (1977), on Rumanian species. Aubert (1969) also catalogued the West Palaearctic species, with notes on hosts. These papers should be consulted for fuller bibliographic references and synonymy of the species. Biological references mainly mention host records and are given in the above references as well as in Carlson (1979). Short (1978) figures the larval head of *Perithous divinator divinator*. The larva has a large tooth-like projection at the junction of the base and blade of the mandible and has three sensilla on maxillary and labial palpi (fig. 4). In other genera of Theroniini, there are only 2 sensilla on these palpi.

Species that have been described under *Perithous* or those that belong to that genus (in the strict sense) are listed below, with their present status. Taxa that have been described as subspecies and are no longer considered valid and other synonyms not originally described under *Perithous* are omitted. These can be found in the various catalogues mentioned above. Species falling under *Hybomischos* are also not included as they are being treated in a separate paper that follows.

The following species have so far been reported under *Perithous*:

1. *albicinctus* (Gravenhorst), 1829. Europe, USSR, Japan.
    Originally described under *Ephialtes* but transferred to *Perithous* by Holmgren. Valid species.
2. *brunnescens* Koornneef, 1951. Europe.
    Synonymized by Oehlke (1967) under *Perithous (Hybomischos) septemcinctorius* (Thunberg).
3. *divinator* (Rossi), 1790. Europe, USSR, North Africa, North China, North America.
    Originally described under *Ichneumon* and first transferred to *Perithous* by Marshall, 1872. Valid species. *P. pimplarius* Haupt is a synonym of it (cf. Oehlke, 1966).

4. *exiguus* Haupt, 1954. Europe.
    Synonymized by Oehlke (1966) under *P. (Hybomischos) septemcinctorius*.
5. *japonicus* Uchida, 1928. Japan.
    Reduced to a subspecies of *P. mediator* (Fabricius) by Townes & Townes (1960). Valid subspecies.
6. *longiseta* Haupt, 1954. Europe.
    Synonymized with *P. mediator* by Oehlke (1966). Constantineanu & Constantineanu (1968) considered it a valid species, represented by two subspecies *longiseta longiseta* Haupt and *longiseta moldavica* Const. & Const. from Rumania. Not seen.
7. *mediator* (Fabricius), 1804. Holarctic.
    Originally described under *Pimpla* and transferred to *Perithous* by Holmgren, 1860. Townes & Townes (1960) recognized four subspecies under it from Europe, North America, and Japan. Townes, Momoi & Townes (1965) included *nigrinotum* Uchida from China as another subspecies. Valid subspecies.
8. *neomexicanus* (Viereck), 1903. North America.
    Originally described under *Pimpla* and first placed as a synonym of *P. pleuralis* Cresson by Townes (1944) and subsequently (1960) treated as a subspecies of *P. mediator* (Fabricius). Valid subspecies.
9. *nigrinotum* Uchida, 1942. China.
    Townes, Momoi & Townes (1965) reduced it to a subspecies of *P. mediator*. Valid subspecies.
10. *nigrigaster* Constantineanu & Constantineanu, 1968. Europe: Rumania.
11. *pleuralis* Cresson. North America.
    Townes (1960) considered it as a subspecies of *P. mediator*. Valid subspecies.
12. *pimplarius* Haupt, 1938. Europe.
    Oehlke (1966) synonymized it with *P. divinator* (Rossius). Constantineanu & Constantineanu (1968), however, considered it distinct. Not examined.
13. *speculator* Haupt, 1954. Europe.
    Oehlke (1966), who examined the type considered it distinct. Constantineanu & Constantineanu (1968) reported it from Rumania, with two subspecies: *speculator speculator* and *speculator transsylvanicus*. Not seen.

In addition the following species and subspecies are described here as new from the Orient.

1. *divinator himalayensis*, n. subsp. Himachal Pradesh and Uttar Pradesh, India.
2. *digitalis*, n. sp., with three subspecies:
    (a) *digitalis digitalis*, n. subsp. Kashmir, India
    (b) *digitalis nepalensis*, n. subsp. Nepal.
    (c) *digitalis taiwanensis*, n. subsp. Taiwan.
3. *sundaicus*, n. sp. Java, Indonesia.
4. *kamathi*, n. sp. Himachal Pradesh, India.

PART I. THE HOLARCTIC SPECIES

Since all the European species so far described were not available for study, a key to the world species is not attempted. Notes on all species known to me are given below, incorporating new diagnostic features observed on the metasternum, posterior mesosternal carina and the nature of the submetapleural carina. These characters helped in the recognition of the new taxa described in Part II of this paper.

1. PERITHOUS ALBICINCTUS (Gravenhorst)

Ephialtes albicinctus Gravenhorst, 1829. Ichneumonologia europaea,
   3: 259. ♀. des. Type ♀, Germany: Hannover (Wroclaw).
Perithous albicinctus: Holmgren, 1860. Öfvers Svenska Vetensk.-Akad.
   Forh., 16: 123. Uchida, 1928. J. Faculty Agr. Hokkaido Imp. Univ.,
   25: 91.
Taxonomy: Townes et al., 1965: 69. Oehlke, 1967: 35.
Biology: Mirek, 1963: 419. Aubert, 1969: 100.

The posterior metasternal carina is curved and only slightly raised in the middle, appearing like two low tubercles between the hind coxae. Median metasternal furrow is deep and wide. Submetapleural carina is almost complete to (but not quite touching) metasternal carina and flattened and wider just above middle coxa (projection above middle coxa broadly rounded and its anterior margin reflexed). Antennal flagellum composed of 37 segments, slender. Ovipositor longer than body and its tip with teeth which are widely spaced and occupying an area about 4.0 x its depth. First abdominal sternite with only a moderately projecting basal tooth.

This species is distinctive in having a black thorax without reddish parts, which is also the case in *mediator japonicus* from Japan, and *sundaicus*, n. sp. described from Java.

Hosts: Ectemnius nigritarsus (H.-S.) (Hymenoptera: Sphecidae). Mirek (1963) has published on the biology of this species.

Distribution: Europe (England, Germany, Poland, Belgium, Finland, Sweden, France, Austria, Czechoslovakia, Switzerland, Rumania, U. S. S. R.), Japan. Specimens from Germany seen in Townes Collection.

2. PERITHOUS MEDIATOR (Fabricius)

The posterior metasternal carina is better developed and more prominent in this species than in the preceding species. Tubercle-like prominences in the middle are low but distinct. Median metasternal furrow narrow and sometimes partly obliterated. Submetapleural carina complete to metasternal carina and forming a short conical projection above middle coxa (in American subspecies this projection without a reflexed margin and somewhat broader and thin; in Chinese subspecies also a little broader but reflexed). Ovipositor longer than body and its tip with widely spaced teeth occupying an area about 5.0 the apical depth of ovipositor. First sternite with a distinct conical or obtuse projection at base, better developed than in *albicinctus*. Antennal flagellum longer and slender (cf. *divinator*).

This species is widespread in Eurasia and North America. Five subspecies are recognized (*mediator, japonicus, nigrinotum, pleuralis* and *neomexicanus*). The non-reflexed nature of the submetapleural projection in the North American subspecies readily separate them from the Eurasian subspecies. The coloration of the various subspecies is diagnostic, although the subspecies do show variations and perhaps intergrade in zones of overlap. Some of the new subspecies described by Constantineanu & Constantineanu have been synonymized with this subspecies by Aubert (1969) but this aspect could not be verified as specimens of those subspecies were not available.

Key to the subspecies of *Perithous mediator*

1. Thorax wholly black except for yellow lines along upper margin of pronotum, submetapleural ridge, mesepimeron, and apices of scutellum and meta-scutellum. Hind tibia and tarsus blackish-brown and femur reddish-brown, without any distinct black marks. Antenna black. Flagellar segments 33±2. Japan, Korea, Sakhalin.

    mediator japonicus Uchida (p. 7)

    Thorax with mesopleurum and often also the mesoscutum reddish-brown. Mesepimeron brown or black. Hind tibia and tarsus black marked. Hind femur with or without black marks. 2

2. Mesopleurum with an oval yellow spot at dorsal end of prepectal carina. Mesoscutum black. Scutellum largely yellow. Hind tibia and tarsus pale yellow with blackish marks. Hind femur without black mark. Face broadly yellow on sides. Antenna brown. Flagellar segments 28 (only 1♀ known). China: Manchuria.

    mediator nigrinotum Uchida (p. 6)

    Mesopleurum without a yellow spot. Mesoscutum usually wholly reddish-brown, but sometimes brownish to blackish with reddish lines. Specimens from North America with black mesoscutum also have apex of hind femur black. 3

3. Submetapleural projection above middle coxa conical and with a reflexed margin. First abdominal tergite largely smooth with scattered minute punctures. Second and following tergites also largely smooth in apical half. All legs of female reddish-brown with only inconspicuous fuscous marks on hind tibia and tarsus. Male fore and middle coxae may be yellow. Antenna brown. Flagellar segments 31±3 (less in specimens from Italy, which are comparatively smaller). Europe.

    mediator mediator (Fabricius) (p. 6)

    Submetapleural projection above middle coxa broader and without a reflexed margin. First abdominal tergite with denser punctation; other tergites also with crowded punctures, particularly in basal half. Hind tibia and tarsus of female yellowish to brownish with conspicuous dark fuscous marks. Fore and middle coxae partly whitish. Antenna black in female, brownish black in male. 4

4. Hind femur without a black apical mark (rarely faintly infuscate). First abdominal tergite with punctures well-separated. Punctures on second tergite crowded in its basolateral areas, but centrally (including swellings) smooth and with widely scattered and minute punctures. Mesoscutum usually reddish brown. Hind tibia and tarsus pale with black marks. North America in Rocky mountains and west of them.

    mediator neomexicanus (Viereck) (p. 8)

Hind femur always with an apical black ring. First and second abdominal
tergites with crowded punctures, those on second extending over the
basal half of swelling (except rarely particularly specimens from N. W.
parts of range). Mesoscutal color varying from reddish-brown to
black. Hind tibia and tarsus often dark. North America east of Rocky
mountains and Alaska and also in mountains of British Columbia and
Oregon (zones of overlap). . . . <u>mediator</u> <u>pleuralis</u> (Cresson) (p. 8)

2a. PERITHOUS MEDIATOR MEDIATOR (Fabricius)

Pimpla mediator Fabricius, 1804. Syst. Piez., 2: 117. ♀. des.
 Type ♀, Czechoslovakia: Mähren (Kiel, on deposit in Copenhagen).
Perithous mediator mediator: Townes & Townes, 1960. U. S. Natl.
 Mus. Bull., 216(2): 214. ♂, ♀. England, Germany.
Taxonomy: Oehlke, 1967: 35 (full synonymy).
Biology: Borries, 1897: 153-159. Brocher, 1926: 393-410. Aubert,
 1969: 102.

Characterized by having punctures on abdominal tergites rather fine and
sparse. Submetapleural carina with a short conical projection above middle
coxa with its apical margin reflexed. Antenna brownish. Mesopleurum,
mesoscutum, scutellum (except apically), and often the upper dorsal margin
of pronotum reddish-brown (extreme apical margin of pronotum yellow); tegula
brown. Subtegular ridge yellow. Head black with inner orbital borders
yellow. All legs orange-brown with hind tibia and tarsus with rather inconspicuous light fuscous marks. Propodeum with a crescentic yellow mark in
the middle along apical transverse carina. Abdominal tergites narrowly
yellow along their apical margins.
 Specimens from Italy are more yellow: face is largely yellow with black
mark confined in the middle (one female normal); clypeus is also largely
yellow; fore coxa is lighter in color and is yellowish; and the apical margins
of abdominal tergites are more distinctly yellow. Specimens from U. S. S. R.
almost lack the yellow mark on propodeum and the yellow apical bands on
abdominal tergites are rather narrow.
 The males have the face wholly yellow, fore and middle coxae yellow,
their tibiae and tarsi also yellow but marked with orange, particularly femora,
and hind coxa and femur orange, their tibia and tarsus yellow with light fuscous markings. In one male from Italy there is a yellow faint spot on mesopleurum as seen in female of *nigrinotum*.
 Hosts: Aubert (1969) gives a long list of hosts. They belong to Coleoptera
(Cerambycidae) and Hymenoptera (Xiphydriidae, Cynipidae, Chrysididae,
Eumenidae, Sphecidae, and Megachilidae).
 Distribution: Widely distributed in Europe, England, USSR. Specimens
from USSR, Germany, Czechoslovakia, Italy, Denmark, Sweden, and England
seen in Townes and Ottawa Collections.

2b. PERITHOUS MEDIATOR NIGRINOTUM Uchida

Perithous medinator (!) nigrinotum Uchida, 1942. Ins. Matsumurana,
 16: 118. ♀. des. Type ♀, China: Kaiyuan in Manchuria (Sapporo).
 Examined, 1980.
Perithous (Perithous) mediator nigrinotum: Townes et al., 1965. Mem.
 Amer. Ent. Inst., 5: 69.

This subspecies is known only from the female type-specimen from Kaiyuan (= Kaigen), Manchuria. It is rather similar to the typical subspecies but the propodeum is shiny, impunctate, and the second and third abdominal tergites are more definitely punctate, the punctures separated from each other by about 1.5 to 2.0 their diameter. Facial punctures somewhat larger but sparser than in *mediator mediator*. Submetapleural projection above middle coxa not very acute, its margin narrowly reflexed. Face more yellowish on sides, median black mark narrow towards antennal sockets (which are yellow). Antenna brownish, yellowish ventrally towards base. Pronotum black with a yellow line along its upper margin. Mesoscutum and metapleurum black. Mesoscutum with faint brownish-yellow lines. Mesopleurum largely orange-brown with a yellow spot near upper end of prepectal carina. Scutellum, metascutellum, tegula, and subtegular ridge yellow. Propodeum black with a broad crescentic yellow mark along apical transverse carina. Apices of all abdominal tergites yellow. Fore and middle legs largely yellow with their femorae yellowish orange. Hind leg orange-brown up to femur, its tibia and tarsus yellowish and with fuscous marks subbasally and apically on tibia and at apices of tarsal segments.

Distribution: China: Manchuria.

2c. PERITHOUS MEDIATOR JAPONICUS Uchida

Perithous japonicus Uchida, 1928. J. Fac. Agri. Hokkaido Imp. Univ., 25: 91. ♀, ♂. des., fig. Lectotype ♀, Japan: Yamagata (Sapporo). Japan: Sapporo, Nikko; Sakhalin.

Perithous mediator japonicus: Townes & Townes, 1960. U. S. Natl. Mus. Bull., 216(2): 214. key, des. Japan.

Taxonomy: Townes et al., 1965: 69.

Similar to the typical subspecies in having an acute submetapleural projection above middle coxa with its margin reflexed. Propodeum punctate as in *mediator mediator*. Facial punctures comparatively coarser and abdominal tergites, particularly the first and second more densely punctate (first tergite smoother with scattered punctures in *mediator mediator* and *m. nigrinotum*).

Facial orbits broadly yellow. Black mark on face parallel-sided (variable in width). Thorax black except for yellow marks along upper margin of pronotum, tegula (brownish-yellow), subtegular ridge, apical half of scutellum, metascutellum, and a crescentic to circular mark on propodeum along apical transverse carina. Apical margins of abdominal tergites yellow. Fore and middle coxae yellow. Hind coxa reddish-brown. Fore and middle femora yellowish-brown and their tibiae and tarsi infuscate. Hind femur reddish-brown. Hind tibia and tarsus blackish-brown.

Distribution: Japan, Korea and Sakhalin. Specimens from Hokkaido, Kamikochi and Tanigumi Prov., Japan, examined in Townes Collection.

Hosts: Unknown.

Alaskan specimens of *mediator pleuralis* somewhat approach this subspecies in body punctation and color of hind leg. *Japonicus* also resembles *albicinctus* in coloration, but submetapleural carina is complete and the basal conical projection on first sternite is distinct. The submetapleural projection above middle coxa is short, conical and with a reflexed margin.

2d. PERITHOUS MEDIATOR NEOMEXICANUS (Viereck)

Pimpla neomexicana Viereck, 1903. *In* Skinner: Trans. Amer. Ent.
     Soc., 29: 88. ♀. des. Type ♀, Beulah, N. Mexico (Philadelphia).
Perithous mediator neomexicanus: Townes & Townes, 1960. U. S. Natl.
     Mus. Bull., 216(2): 213.
Perithous (Perithous) mediator neomexicanus: Carlson, 1979. *In*
     Krombein et al.: Catalog of Hymenoptera in America North of Mexico,
     1: 348.
Biology: Parker and Bohart, 1966: 94.

This and the other American subspecies differ from the typical subspecies
in having a thin and broader submetapleural projection above middle coxa, the
apical margin of which is not reflexed. The punctation on abdominal tergites
is also denser. The fore and middle coxae are often partly yellowish-white
and hind tibia and tarsus with distinct black markings. The males are not
very distinctive, except that in males of *neomexicanus,* the antenna tend to be
dark brown to black and black markings on hind leg are more conspicuous.
     First abdominal tergite with scattered punctures. Second tergite with
punctures confined mostly in basolateral areas. Swellings on it largely smooth
(cf. *pleuralis* and *japonicus*). Hind leg brown with tibia and tarsus broadly
infuscate (usually only the apices of tarsal segments black and tibia largely
yellowish-brown along inner side). Hind femur without black apical mark
(except rarely faintly so). Mesoscutum usually entirely reddish-brown.
Yellow markings on face, propodeum, and abdominal tergites as in the typical
sub-species.
     Host: *Pemphredon confertim.*
     Distribution: North America in Rocky mountains and west of the same.
Specimens from various localities in   Alberta, Arizona, Idaho, Colorado,
California, Montana, Oregon, British Columbia, New Mexico and Washington
State seen in Townes and Ottawa Collections.

2e. PERITHOUS MEDIATOR PLEURALIS (Cresson)

Perithous pleuralis Cresson, 1868. Canad. Ent., 1: 46. ♀. des.
     Type ♀, Ontario: Grimsby (lost).
Perithous mediator pleuralis: Townes & Townes, 1960. U. S. Natl.
     Mus. Bull., 216(2): 211.
Perithous (Perithous) mediator pleuralis: Carlson, 1979. *In* Krombein
     et al.: Catalog of Hymenoptera in America North of Mexico, 1: 348.
Biology: Champlain, 1922: 97. Rheinhard, 1929: 155. Krombein, 1960:
     31.

Readily distinguished from *neomexicanus* as well as from *mediator*
mediator by having crowded punctures on first and second abdominal tergites,
those on second tergite extending a little over the swellings, at least in basal
half (except rarely), hind leg brown with femur always having an apical black
ring which is usually wide, about as wide as the apical depth of femur, hind
tibia and tarsus yellowish with black apical marks, but often tibia and tarsus
with extensive black marks, color of mesoscutum varying from reddish-
brown to black, mesopleurum also often black dorsally, abdominal tergites
sometimes only with faint yellow margins, and fore and middle legs also
sometimes tending to be infuscate.

Specimens from Alaska and British Columbia are generally darker, with mesoscutum black, with or without two small reddish lines, hind femur with wider black apical band, hind tibia and tarsus almost wholly black in female and with white markings in male.

This subspecies intergrades with *neomexicanus* in the northern parts of its distributional range and specimens of both the subspecies have been collected in Oregon (same locality) as well as in British Columbia and Idaho.

Hosts: Pemphredon concolor, P. inornatus, and *P. harbecki?*.

Distribution: North America east of Rocky mountains and Alaska. Also occurring in British Columbia mountains, Idaho, and Oregon. Specimens from most eastern states north to Michigan as well as Quebec seen in Townes Collections.

3. PERITHOUS DIVINATOR (Rossi)

Ichneumon divinator Rossi, 1790. Fauna Etrusca,2: 48. ♀. des.
    Type ♀, Italy: Etrusca (location?).
*Perithous divinator:*Marshall, 1872. Catalog of British Hymenoptera,
    p. 86. England.
Taxonomy: Townes & Townes, 1960: 215. Townes et al., 1965: 69.
    Oehlke, 1967: 35. Carlson, 1979: 348.
Biology: Horstmann,1964: 193. Thomas, 1964: 199. Horstmann,1967: 95.
    Aubert, 1969: 101.

In this species the median portion of posterior mesosternal carina forms a posteriorly directed pair of closely set tubercles, which are also hairy, metasternal furrow shallow to obliterated, submetapleural carina narrow, complete to metasternal carina or slightly short of it, not widened above middle, midventral fold of first sternite not forming a conical projection at base, antenna short and a little thickened subapically, about 0.75 the length of body, flagellum composed of 28±2 segments, ovipositor short, about 0.8 the body length, its tip with teeth occupying an area which is only about 2.5 its apical depth, teeth comparatively closely set, scape and pedicel combined a little longer than second flagellar segment (14: 12), and nervellus intercepted at its upper 0.33.

Head black with inner orbits, base of clypeus, mandible, and scape and pedicel beneath, yellow. Upper margin of pronotum, subtegular ridge, apex of scutellum, metascutellum and an arched mark on propodeum, yellow. Mesoscutum, mesopleurum, mesosternum, and legs reddish-brown. Thorax otherwise black. Hind tibia and tarsus often lightly infuscate. Abdominal tergites black with apices of second and the following tergites narrowly to faintly yellow. Sometimes these yellow lines partly obliterated.

In males, face wholly yellow and tegula and fore and middle legs largely yellowish-white with hind tibia and tarsus blackish dorsally.

This species apparently has a wide range of distribution, having been recorded from Europe, USSR, North Africa, North America and Northern China. I have examined specimens of it from England, Sweden, Germany and Eastern North America in Townes Collections and Canadian National Collections. According to Townes & Townes (1960) it is an introduced species in North America.

A new subspecies, *divinator hymalayensis*, is described here from Himachal Pradesh, India. The European and American populations should therefore be referred to as *divinator divinator*.

Regional color variations occur in the various populations of *divinator divinator* studied. The females from Sweden have two small yellow spots on the face just below antennal sockets, connected to the yellow orbital stripes (as is also seen in American specimens). In males the hind coxa is lightly infuscate on the inner side. In all European specimens the tegula is yellowish-brown, and hind tibia and tarsus faintly infuscate.

Specimens from England tend to be darker, with mesopleurum and meso-scutum darker reddish-brown, fore coxa partly yellow, hind coxa largely infus-cate, hind tibia and tarsus brownish-black, tegula dark brown in female and brownish-yellow in male, and mesopleurum often black along prepectus and along mesopleural groove (varying in extent in different specimens). The yellow marks on abdomen and propodeum are extremely narrow and often absent.

The North American specimens have bright reddish-brown mesopleurum, mesoscutum, and all coxae of female (as in European specimens), except fore coxae often partly yellow, the hind tibia and tarsus of female lightly infuscate, tegula yellow, male fore coxa yellow, and yellow lines on abdomen and pro-podeum more prominent. Even the orbital yellow stripes are a little wider and often there are two small yellow spots below antennal sockets which may be or may not be connected with the orbital stripes.

One female from Bulgaria: Sozopol, in Townes Collection, is rather different and fits the description of *Perithous pimplarius* Haupt. The first abdominal segment is short with a deep declivity at base and second and third abdominal tergites are transverse (wider than long) rather than long as is usual in the genus. The flagellum is short (27-segmented), scape and pedicel combined not longer than second flagellar segment, abdominal tergites includ-ing postpetiole more densely punctate, tubercles on abdominal tergites more prominent and oblong, nervellus intercepted almost at middle, tegula yellowish-white, scutellum largely yellow and apex of first tergite white (never seen in the specimens of *divinator* with me), and yellow lines on abdominal tergites more distinct.

Whether or not this specimen represents one of the species described by Haupt or Constantineanu could not be decided without reference to the type-specimens. Horstmann (1967) has incidentally shown that the nature of the abdomen in *divinator* can vary with reference to its hosts, and therefore he synonymized *P. pimplarius* with *divinator*.

Hosts: Chiefly parasitic on species of *Pemphredon, Passoloceus, Psenulus, Trypoxylon,* and other Sphecidae in Europe and North America. Aubert (1969) lists several hosts belonging to Coleoptera (Cerambycidae), Lepidoptera (Pyralidae), and Hymenoptera (Cynipidae, Chrysididae, Eumenidae, and Sphecidae).

Distribution: Europe (Italy, Germany, France, England, Finland, Rumania, Czechoslovakia, Spain, USSR), North Africa, China, Canada, U.S.A. In North America distributed mainly from Quebec south to Connecticut and Pennsylvania and west to Wisconsin.

II. DESCRIPTIONS OF NEW TAXA FROM THE ORIENT

Key to the Oriental Species of *Perithous*

1. Mesopleurum and mesoscutum black with yellow spots or stripes, finely polished. Metapleurum and metasternum also polished. Submetapleural carina flattened out and forming a triangular or rounded projection over base of middle coxa (fig. 1).   2
    Mesopleurum and mesoscutum reddish-brown, subpolished. Metapleurum with minute punctures or smoother *(digitalis)*. Metasternum smooth or hairy. Submetapleural carina narrow and not flattened or widened above middle coxa (fig. 2, 3) more reflexed.   3

2. Metasternal carina forming conical projections in the middle between hind coxae (fig. 1). Median metasternal groove absent and represented by a small pit only. Submetapleural carina distinct on anterior 0.8 of metapleurum and forming a triangular projection above middle coxa. First sternite conically produced basomedially. Mesoscutum with two lateral yellow stripes near middle. Propodeum smooth dorsally, with only scattered punctures laterally. (In general thorax sparsely hairy and highly polished). Java (♀ only known).
    5. <u>sundaicus</u>, n. sp. (p. 15)
    Metasternal carina not forming conical projections in the middle. Median metasternal groove distinct and widely open behind between hind coxae (metasternal carina curving around hind coxae). Submetapleural carina short, extending only in anterior 0.5 of metapleurum and forming a broadly rounded flange above middle coxa. First sternite not conically produced basomedially, only with a short tubercle. Mesoscutum wholly black. Propodeum more punctate dorsolaterally. (In general thorax hairy and subpolished.) India: Himachal Pradesh. (♂ only known.)
    6. <u>kamathi</u>, n. sp. (p.16)

3. Metasternum shiny and with a finger-like projection between middle and hind coxae (fig. 3). Submetapleural carina broadly reflexed above middle coxa and forming a short triangular projection. Antenna 0.8 as long as the body, slender. Ovipositor 0.8 to 1.0 as long as body, its tip with very closely spaced oblique teeth (fig. 3). Base of first sternite with a forwardly directed projection. (Female only known.) India, Nepal, Taiwan. . . . .   4. <u>digitalis</u>, n. sp. (p. 13)
    Metasternum apically rough and hairy, with a posteriorly directed pair of closely set and blunt tubercles between hind coxae (fig. 2). Submetapleural carina narrowly reflexed and broadly triangular above middle coxa. Antenna shorter, about 0.65 to 0.75 the body length and somewhat thickened preapically. Ovipositor about 0.8 as long as the body, its tip with moderately spaced vertical teeth (fig. 2). Base of first sternite without any projection. . . . .   3. <u>divinator</u> . . . 4

4. Tegula brown or with a brown spot (in American populations yellow). Side of scutellum and middle coxa reddish-brown. Abdominal yellow bands very narrow, often incomplete or absent on some or most of the tergites. Body slender, 7-9 mm. long. Submetapleural carina complete to hind coxal cavity, straight. Metapleurum not strongly striate near hind coxa.

Male without yellow marks on mesoscutum or mesopleurum. Widely
   distributed in Holarctic Region. 3a. <u>divinator divinator</u> Rossi (p. 9)
Tegula, sides of scutellum, and middle coxa largely yellow. Abdominal
   yellow bands always prominent. Body stouter, 10-11 mm long. Submetapleural carina a little obliterated posteriorly and bent inwards.
   Metapleurum obliquely striate near hind coxa (fig. 2). Male with
   yellow marks in the region of notauli and on mesopleurum. Hind tibia
   and tarsus tending to be darker. India: Himalaya.
                            3b. <u>divinator himalayensis</u>, n. subsp. (p. 12)

3b. PERITHOUS DIVINATOR HIMALAYENSIS, n. subsp. (fig. 2)

Female: Antenna a little thickened preapically, about 0.65 to 0.75 as long
as the body. Face punctate in the center, smooth along orbital borders, its
upper edge a little cleft between antennal sockets. Frons, vertex, temple and
occiput smooth and shiny, but hairy, especially the temples. Mesoscutum subpolished, beset with short hairs. Pronotum shallowly wrinkled in the median
groove, otherwise subpolished. Mesopleurum polished, sparsely hairy. Metapleurum polished, with minute punctures dorsoapically, area close to hind
coxa ruguloso-striate (fig. 2). Submetapleural carina narrow and not widened
above middle coxa, a little obliterated apically, where it is bent inward and
meets metasternal carina which forms two closely set posteriorly directed
teeth-like projections (fig. 2). This area also conspicuously hairy. Median
groove on metasternum distinct in apical 0.75. Propodeum finely punctate
dorsolaterally, smooth basomedially. Apical transverse carina complete or
slightly obliterated in the middle, bounding a smooth and shiny petiolar area.
Propodeum with a shallow median groove up to the transverse carina. Abdominal tergites subpolished with scattered punctures, those on first distributed
all over except in the apical portion, those on second weaker in apical half,
those on third and the following tergites confined to basal 0.25 with tubercles
and apical areas mat to subpolished. Sides of tergites densely punctate.
Tubercles rather weak and more distinct only on tergites 3-5. Ovipositor
about 0.8 the body length, its tip with parallel, slightly arched teeth confined
to an area about 2.5 its depth (fig. 2).
   Black with mesopleurum, mesosternum, mesoscutum and scutellum
reddish-brown. Scape and pedicel yellow ventrally. Flagellum brown. Basal
two to three flagellar segments also yellowish ventrally. Face laterally and
just below antennal sockets, upper orbital borders, clypeus along basal and
lateral margins, mandible in basal half, upper margin of pronotal collar,
tegula, subtegular ridge, sides and apex of scutellum, metascutellum, mesepimeron, and a semicircular mark on propodeum along apical transverse
carina, yellow. Legs reddish-brown except fore and middle coxae yellow,
their femora yellow marked and hind tibia and tarsus blackish-brown. Second
to fifth abdominal tergites yellow apically. Margins of sixth and seventh tergites broadly yellow laterally and very narrowly so dorsally.
   Male: Similar to female with face and clypeus wholly yellow. Mesoscutum
with yellow marks along notaular areas. Scutellum largely yellow. Mesopleurum with indistinct one or two yellow marks. First tergite may be yellow
apically in the middle.
   This subspecies differs from the typical subspecies in having the submetapleural carina almost obliterated apically and metapleurum more punctate.
The tubercles are mat. The male has faint to distinct yellow marks on
mesoscutum and mesopleurum. The body coloration is more like that of the

American population of the typical subspecies. The European populations tend to have less yellow on propodeum and abdomen. The abdomen is comparatively strongly punctate. Specimens from Britain examined are rather dark and largely devoid of yellow markings.

Length: 10-11 mm. Fore wing 7.5-8.5 mm. Ovipositor 7.5-8.0 mm. Antenna 7.0-7.5 mm.

Holotype ♀, *India:* Himachal Pradesh: Manali, 1828 m., 19-V-1970, M. K. Kamath (Gupta).

Allotype ♂: Same data as the holotype.

Paratypes: 26♀, 4♂: Same locality as the type, but collected from 2 to 30-V-1970 by various Collectors (Gupta). H. P.: Sangla, 2743 m., Kalpa Valley, 1♀, 16 - VI-1972, Gupta. U.P.: Gangotri, 3000 m., 1♀, 20-VI-1977, G. Singh (Gupta).

4. PERITHOUS DIGITALIS, n. sp. (fig. 3)

Readily distinguished by having a finger-like median projection on metasternum between middle and hind coxae, submetapleural carina narrow, but rather broadly reflexed and triangular above middle coxa, this carina a little curved near metasternal carina, first abdominal sternite with a median basal projection, and ovipositor tip compressed and with teeth spaced close together (fig. 3).

Female: Antenna slender, about 0.8 the body length. Face minutely to closely punctate, raised near antennal sockets and with a faint median carina extending in the upper half of face. Clypeus smooth, with setiferous punctures and concave in apical half. Frons and vertex smooth and polished. Vertex posteriorly and temple smooth but hairy. Pronotum polished. Mesoscutum subpolished and hairy. Scutellum a little more convex than in other species and subpolished. Mesopleurum polished, sparsely hairy. Metapleurum polished. Submetapleural carina narrow and forming a short triangular projection anteriorly above middle coxa, just short of reaching posterior metasternal carina. Metasternal carina irregular and forming a characteristic posteriorly directed long finger-like projection (fig. 3) in the middle, with its apex knob-like. Metasternum without a median longitudinal groove. Propodeum shiny, hairy or punctate on sides, with or without a median basal depression; its apical transverse carina complete or incomplete. First tergite smooth or punctate, raised in the middle. First sternite with a strong basomedian forwardly directed projection. Abdomen with protuberances on second and the following tergites, which may be smooth or punctate. Ovipositor rather strongly compressed, about 0.8 to 1.0 the body length, teeth on its lower valve compressed into a small area, occupying about 2.5 its maximum depth, teeth circular and close together, with basal 2-3 teeth divergent (fig. 3).

This species is known only from three females, one each from India, Nepal and Taiwan. They are essentially similar to each other, but exhibit slight differences in puncation and coloration. I prefer to consider them as allopatric subspecies until the extent of variability of each population is known, when they may turn out to be distinct species. Among the three specimens, the specimen from Kashmir has stronger punctation, while the specimens from Nepal and Taiwan are closely related. They can be separated by the following key:

Key to the subspecies of *Perithous digitalis*

1. Inner eye orbits narrowly yellow. Face closely punctate. Propodeum
without a white apical semicircular mark. Apical transverse carina
distinct and semicircularly arched. Median groove on propodeum
extending to its middle. Propodeum punctate laterally. Hind coxa
yellowish-brown with a pale yellow dorsal spot. Scape and pedicel
black. Bands on abdominal tergites narrow, those on apical tergites
rather inconspicuous. India: Kashmir.
                    4a. digitalis digitalis, n. subsp. (p.14)
   Inner orbits broadly yellow—these stripes meeting on face below antennal
sockets. Face with minute scattered punctures. Propodeum with a
complete semicircular yellow stripe in the region of apical transverse
carina, or with two lateral spots, the carina itself indistinct and
represented by two lateral tubercles. Median groove of propodeum
indistinct to absent. Propodeum smoother. Scape and pedicel white
marked. Bands on abdominal tergites wider and conspicuous on second
and the following tergites. 2

2. Face with scattered punctures, smooth medially. Propodeum smooth and
shiny, with two apicolateral yellow spots. Hind coxa and femur yellow.
Hind leg largely yellow with fuscous marks on tibia and tarsus.
Nepal. . . . .   4b. digitalis nepalensis, n. subsp. (p. 15)
   Face with minute punctures, without a smooth medial area. Propodeum
with a semicircular yellow mark in the region of apical transverse
carina, with minute setiferous punctures laterally. Hind coxa and
femur yellowish-brown. Hind leg darker yellowish-brown, with tibia
extensively black marked. Taiwan.
               4c. digitalis taiwanensis, n. subsp. (p. 15)

4a. PERITHOUS DIGITALIS DIGITALIS, n. subsp.

Female: Face with distinct well separated punctures. Metapleurum
wrinkled near hind coxa, otherwise smooth and polished. Propodeum shallowly
punctate laterally, with a median groove which extends to its middle; its apical
transverse carina distinct and semicircularly arched. Petiolar area mat.
First tergite punctate with its apex smoother. Protuberances on second and the
following tergites smooth medially, otherwise abdomen sparsely punctate and
hairy. Sides of tergites with denser punctures.
    Black. Mesoscutum, scutellum (except its apex), and mesopleurum
(except narrowly below wings), reddish-brown. Clypeus along epistomal
groove, mandible basally, inner orbits narrowly up to the ocellar level, hind
corner of pronotum, tegula, subtegular ridge, apex of scutellum, metascutellum
wholly, and apices of second to sixth abdominal tergites narrowly, yellow.
Mesepimeron brownish-black. First tergite faintly yellow in the middle and
seventh tergite yellow laterally. Scape and pedicel without yellow marks,
black. Fore and middle legs largely yellowish with yellowish-brown patches
on femora. Hind coxa yellowish-brown with a pale yellow dorsal oval spot.
Hind femur yellowish-brown with a black subapical ring and its apex yellow.
Middle and hind tibiae with black dorsal lines, and their bases and apices also
black. Middle and hind tarsal segments yellow with their apices black.
    Male: Unknown.

Length: ♀, 8.5 mm. Fore wing 7 mm. Ovipositor 8.5 mm.

Holotype: ♀, *India:* Kashmir, Pahalgam, 7200 ft., 29-VI-1966, M. K. Kamath (Gupta).

4b. PERITHOUS DIGITALIS NEPALENSIS, n. subsp.

Female: Face with scattered punctures, smooth medially. Metapleurum smooth and polished. Propodeum smooth and shiny, with a faint indication of a basomedian groove, its apical transverse carina absent, but represented laterally by weak tubercles. Petiolar area smooth and shiny. First tergite largely smooth and shiny. Protuberances on second and the following tergites impunctate and shiny. Second to fourth tergites punctate laterally.

Black. Similar to *digitalis digitalis* but inner eye margins broadly yellow —these marks meeting on face below antennal sockets. Scape and pedicel with ventral whitish-yellow marks. Upper margin of pronotum yellow. Upper margin of mesopleurum broadly black. Mesepimeron yellow. Apex of propodeum with two small lateral yellow marks. Hind leg largely yellow with lateral black marks on femur, tibia and tarsus. Hind coxa faintly marked laterally with yellowish-brown. Bands on second and the following abdominal tergites wider and conspicuous on all tergites.

Male: Unknown.

Length: ♀, 8 mm. Fore wing 7 mm. Ovipositor 6.6 mm.

Holotype ♀, *Nepal:* Kathmandu, Godavari, 1500 m., 29-IX-1970, Gupta (Gupta).

4c. PERITHOUS DIGITALIS TAIWANENSIS, n. subsp. (fig. 3)

Female: Face with minute punctures, without a smooth smooth median area. Metapleurum smooth and polished. Propodeum with minute setiferous punctures laterally, with a faint indication of a basomedian groove, without an apical transverse carina, though with lateral weak tubercles. Petiolar area mat. Abdomen smooth as in *nepalensis,* with smooth tubercles and second to fourth tergites punctate laterally.

General coloration similar to that of *digitalis nepalensis,* with scape, pedicel and abdominal tergites broadly yellow. Yellow on face more extensive below antennal sockets. Propodeum with a semicircular yellow mark in the region of apical transverse carina. Hind leg somewhat darker, yellowish-brown, with femur almost wholly yellowish-brown and tibia more extensively black marked.

Male: Unknown.

Length: ♀, 8.5 mm. Fore wing 7.5 mm. Ovipositor 7.5 mm.

Holotype: ♀, *Taiwan:* Taipei, ex Chiu Colln., 1971 (Gupta).

5. PERITHOUS SUNDAICUS, n. sp. (fig. 1)

Superficially resembling the Japanese *Perithous mediator japonicus* in coloration, but distinct in having a yellow spot on mesopleurum near upper end of prepectal carina and also by having a very polished metapleurum and metasternum, with the posterior metasternal carina forming distinct conical processes in the middle and the median metasternal groove absent and represented only by a small pit. The submetapleural carina is absent in apical 0.25 and the projection above middle coxa is conical and longer than in *japonicus.* The propodeum is shiny and not punctate as in *japonicus.*

Female: Antenna slender, about 0.8 as long as the body. Face with scattered shallow punctures. Frons, vertex, temple and occiput smooth and shiny. Temple with scattered hairs. Mesoscutum shiny, with minute setiferous punctures. Pronotum, mesopleurum, metapleurum, mesosternum and metasternum shiny and polished. Mesopleurum with scattered minute setiferous punctures. Metasternal carina forming conical projections in the middle between hind coxae. Median metasternal groove indistinct and represented by a pit only. Submetapleural carina distinct in anterior 0.8 of metapleurum, forming a triangular projection above middle coxa, its apical margin reflexed. Propodeum smooth and shiny, with scattered and a little elongate punctures laterally, with a shallow median groove. Apical transverse carina erased medially, although two triangular white marks at this place demarcate a semicircular petiolar area. Tergite shiny, with scattered punctures laterally. First sternite forming a conical projection basomedially. Punctures on second to fourth tergites moderate and well separated. The following tergites smooth with minute setiferous punctures. Teeth on ovipositor tip arched and parallel, extending over to nearly 5.0 the width of ovipositor (fig. 1).

Black. Antenna wholly blackish-brown, without white marks on scape and pedicel. Face except narrowly in the middle, clypeus except in the middle, teeth except apically, upper orbital borders, two linear marks on mesoscutum, upper margin of pronotum, pronotal collar largely, subtegular ridge, an oval spot on mesopleurum at upper end of prepectal carina, mesepimeron, two lateral triangular marks on propodeum in the region of apical transverse carina, scutellum laterally and apically, and metascutellum except basally, yellow. Tegula brown. Fore and middle coxae and fore trochanters yellow. Hind coxa reddish-brown with an oval dorsal yellow spot at base. Legs otherwise brownish-yellow, with fore legs lighter in color and middle tarsus and hind tibia and tarsus brownish-black. Apices of first (narrowly), 2nd and 3rd tergites brownish, those of fourth laterally, fifth completely and sixth and seventh tergites laterally, yellow.

Male: Unknown.

Length: ♀, 12 mm. Fore wing 10 mm. Ovipositor 15 mm.

Holotype ♀, Java: Tjibodas, 1400 m., 27 to 29 - VII - 1930, M. A. Lieftinck (Townes).

Paratype ♀, W. Java: Mt. Gedeh, 28 - XI - 1935, ex Betrem Coll. (Gupta).

6. PERITHOUS KAMATHI, n. sp.

Similar to the Javanese *sundaicus*, n. sp., and mainly distinguished by the absence of conical tubercles on metasternum and first sternite, complete metasternal groove, which is widely open behind between hind coxae, and short submetapleural carina extending in anterior 0.5 to 0.6 only and forming a broadly triangular flange above middle coxa. The propodeum is punctate dorsolaterally and the apical transverse carina is distinct throughout. The mesoscutum does not have yellow stripes.

Male: Antenna long and slender, about 0.8 as long as the body. Face punctate. Frons, vertex, temple and occiput smooth, with temple and vertex posteriorly sparsely hairy. Mesoscutum subpolished, with short hairs. Pronotum, mesopleurum, metapleurum, and metasternum shiny, polished, with mesosternum subpolished and more hairy (as in mesoscutum). Metasternal carina almost absent, faintly represented in the middle, not forming conical projections. Median metasternal groove distinct and widely open behind.

Submetapleural carina extending in anterior 0.5 to 0.6 only and forming a broadly triangular flange above middle coxa. Propodeum with distinct well separated punctures except in the central region, its median groove faint. Apical transverse carina distinct. First tergite with scattered punctures, its sternite with a blunt projection at base. Second to fifth tergites more punctate, punctures well separated. The following tergites progressively minutely punctate.

Black. Scape and pedicel yellow beneath. Flagellum black. Face and clypeus wholly, upper orbital borders, pronotal collar medially, upper margin of pronotum, subtegular ridge, an oval spot on mesopleurum at the end of prepectal carina, mesepimeron, apical half of scutellum, metascutellum, two triangular marks on propodeum near apical transverse carina, and apices of all abdominal tergites, yellow. Tegula yellowish-brown. Fore and middle coxae and trochanters yellow, their femora, tibia and tarsi yellowish-brown. Hind coxa and femur orange brown, coxa yellow basodorsally, tibia and tarsus blackish. Apical tarsal segment of middle tarsus also black.

Female: Unknown.

Length: ♂, 11-14 mm. Fore wing 8-11 mm.

Holotype: ♂, *India:* Manali, Himachal Pradesh, 1828 m., 20-V-1970, M. K. Kamath (Gupta).

Paratypes: 7♂, same locality, collected during 17 to 23-V-1970 (Gupta).

REFERENCES

1. Aubert, Jacques -F. 1969. Les Ichneumonides Ouest-Palearctiques et leurs hôtes. 1. Pimplinae, Xoridinae, Acanitinae. Ouvrage Publee avec le Concours du CNRS Pp 1-304.

2. Baltazar, C. R. 1961. The Philippine Pimplini, Poemeniini, Rhyssini, and Xoridini (Hymenoptera: Ichneumonidae, Pimplinae). Monogr. Natl. Inst. Sci. & Technol. Manila 7: 1-130.

3. Borries, H. 1897. Om *Perithous mediator* og *Omalus auratus.* Vidensk. Medde. Dansk. Naturh. Foren. 1897: 153-159.

4. Brocher, Fr., 1926. Observations sur la *Perithous mediator* Gr. Ponte, Oeuf, larve, Nymphe et Imago. Etude anatomique de la tariere, de ses muscles et de son fonctionnement. Ann. Ent. Soc. France 95: 393-410.

5. Carlson, R. W. 1979. Family Ichneumonidae. *In* Krombein, *et. al.:* Catalog of Hymenoptera in America north of Mexico 1: 315-740.

6. Champlain, A. B. 1922. Records of Hymenopterous Parasites in Pennsylvania. Psyche 29: 95-100.

7. Constantineanu, M. I. & Constantineanu, R. M. 1968. Contributions à l'etude du genre *Perithous* (Hym., Ichneum.) de la R. S. Romania. Zool. Anz. 180(3-4): 228-258.

8. Constantineanu, M. I. & Pisică, Constantin, 1977. Fauna Republicii Socialiste Romania. Insecta, Hymenoptera, Ichneumonidae 9(7): 1-310.

9. Danks, H. V. 1971. Biology of Some Stem-Nesting Aculete Hymenoptera. Trans. R. Ent. Soc. London 122: 323-399.

10. Haupt, H. 1938. Die Pimplinen der Schlupfwespen-Fauna von Bellinchen (Oder). Märk. Tierwelt 3(3): 181-221.

11. Haupt, H. 1954. Fensterfänge bemerkenswerter Ichneumonen. Deutsch. Ent. Ztschr (n.f.) 1: 99-116.

12. Horstmann, K. 1964. Zur Biologie der Holzanbohrenden Schlupfwespen, *Perithous divinator* Rossi (Hym., Ichneum.).Fauna Mitt. Norddeutschl. 2: 193-197.

13. Horstmann, K. 1967. Untersuchungen uber eine Wirtsbedingte Modification bei der Schlupfwespen *Perithous divinator* (Rossi) (Hym., Ichneum.). Zool. Anz. 178: 95-102.

14. Krombein, K. V. 1960. Biological Notes on some Hymenoptera that Nest in Sumach Pith. Ent. News 71(2-3): 29-36; 63-69.

15. Mirek, J. 1963. Zweiter Fund von *Perithous albicinctus* (Gravenhorst) in Polen Samt Angaben über andere Arten der Gattung Perithous Holmgren (Hymenoptera, Ichneumonidae). Fragm. Faunist. 10: 419-423.

16. Oehlke, J. 1966. Revision der Ephialtinae-Typen von H. Haupt (Hymenoptera, Ichneumonidae). Reichenbachia 6(32): 279-285.

17. Oehlke, J. 1967. Westpaläartische Ichneumonidae. I. Ephialtinae. *In* Junk: Hymenopterorum Catalogus *(nova editio)*, part 2: 1-48.

18. Parker, F. D. & Bohart, R. M. 1966. Host-Parasite Associations in some Twig-Nesting Hymenoptera from Western North America. Pan-Pacific Ent. 42: 91-98.

19. Rheinhard, E. G. 1929. Pemphredon and her enemies. Nature Mag. 13: 155-157.

20. Short, J. R. T. 1978. The final larval instars of the Ichneumonidae. Mem. Amer. Ent. Inst. 25: 1-508. *(Perithous* on pages 25 and 173).

21. Thomas, S. J. 1964. *Perithous divinator* Rossi and its host, *Pemphredon lethifer* Shuckard. Michigan Acad. Sci., Arts & Letters 49: 199-201.

22. Townes, H. 1944. A Catalog and Reclassification of the Nearctic Ichneumonidae (Hymenoptera), Part I. The subfamilies Ichneumoninae, Tryphoninae, Cryptinae, Phaeogeninae and Lissonotinae. Mem. Amer. Ent. Soc. 11(1): 1-477.

23. Townes, H. & M. 1960. Ichneumon-flies of America North of Mexico: 2: Subfamilies Ephialtinae, Xoridinae, Acaenitinae. U. S. Natl. Mus. Bull. 216(2): 1-676.

24. Townes, H., Momoi, S. & Townes, M. 1965. A Catalog and Reclassification of Eastern Palearctic Ichneumonidae. Mem. Amer. Ent. Inst. 5: 1-661.

25. Townes, H. 1969. The Genera of Ichneumonidae, Part I. Ephialtinae to Agriotypinae. Mem. Amer. Ent. Inst., 11: 1-300.

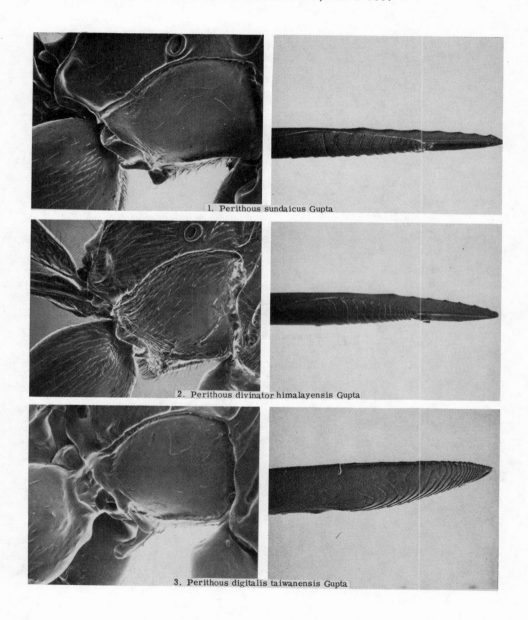

1. Perithous sundaicus Gupta

2. Perithous divinator himalayensis Gupta

3. Perithous digitalis taiwanensis Gupta

Figs. 1-3

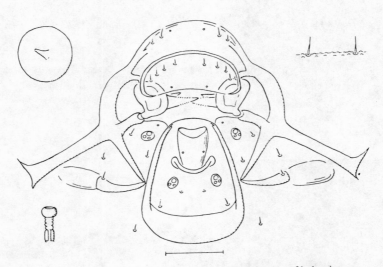

4. Larval head of Perithous divinator divinator

(After Short, 1978)

A STUDY OF THE GENUS *HYBOMISCHOS*

(HYMENOPTERA: ICHNEUMONIDAE)

by
Virendra Gupta

Center for Parasitic Hymenoptera
Department of Entomology and Nematology
University of Florida, Gainesville, FL 32611

Hybomischos was originally described by Baltazar (1961), as a subgenus of *Perithous*. However, the differences in the nature of the ovipositor tip, notauli, first tergite, etc., justify its generic separation. Baltazar described from the Philippines two species: *Perithous (Hybomischos) galbus* and *virgulatus*. Townes et al. (1965) placed in it a Eurasian species, *septemcinctorius* Thunberg. Constantineanu and Constantineanu (1968) described five additional taxa from Rumania *(septemcinctorius rufatus, romanicus romanicus, romanicus rufigaster, transversus transversus* and *transversus trapezoidalis)*. *Perithous exiguus* Haupt, *P. brunnescens* Koornneef, and *P. septemcinctorius meridionator* Aubert were synonymized with *semptemcinctorius* by Oehlke (1966, 1967). In the Townes Collection, there are specimens of *septemcinctorius* from Europe and North America, type of *galbus* from the Philippines, and specimens of an undescribed species from Japan, which is described below as *Hybomischos townesorum*, n. sp. The type of *virgulatus* Baltazar has been examined in the U. S. National Museum, Washington. The species from Rumania are unknown to me.

Genus HYBOMISCHOS Baltazar

Perithous (Hybomischos) Baltazar, 1961. Monogr. Natl. Inst. Sci. & Tech. Manila, 7: 49. Type-species: *Perithous (Hybomischos) galbus* Baltazar; original designation.
Taxonomy: Townes et al., 1965: 69. Constantineanu & Constantineanu, 1968: 250. Aubert, 1969: 101 (as a syn. of *Perithous*). Constantineanu & Pisică, 1977: 92 (as a separate genus).

Rather similar to *Perithous*, but lower mandibular tooth acute. Upper tooth with a wider cutting edge. Eyes emarginate just above antennal sockets. Notauli absent. Apical propodeal carina absent to faintly visible. Propodeum strongly convex, granulose to sparsely or distinctly punctate, usually with a weak or fairly distinct median groove. Submetapleural carina forming a flange along anterior half of metapleurum, absent posteriorly (type-species), or bent inwards toward posterior metasternal carina, which is, however, weak and not fully formed. Sometimes submetapleural carina appears forked posteriorly due to a carina-like rugosity connecting it to the hind coxal cavity. Nervellus intercepted at or above the middle. First abdominal tergite flat or convex dorsally, granulose or punctate, its median dorsal carinae absent, but at base forming lateral tooth-like projections. Spiracle of first tergite below and

and distant from the dorsolateral carina. Ovipositor about 1.4 to 2.0 the body length, its tip strongly compressed and sinuate; teeth on lower valve weaker, subvertical to semicircular; upper valve with saw-like teeth on upper edge.

Hybomischos has the same type of ovipositor as does *Atractogaster i.e.*, the tip being sinuate. The latter genus, however, is distinct in having the lower mandibular tooth longer than the upper, clypeus with a median subapical point, spiracle of first tergite situated just above the dorsolateral carina and touching the same, eyes only very slightly emarginate (as in *Perithous*), and propodeum with a semicircular carina separating a smooth petiolar area (as in *Perithous*). *Perithous* is further distinguished from *Hybomischos* by the presence of notauli, absence of lateral tooth-like projections at base of first tergite, spiracle of first tergite touching or almost close to the dorsolateral carina, and ovipositor tip different and not sinuate.

Biological references on *Hybomischos* are few and deal mainly with host records of the European *H. septemcinctorius*, which are larvae of aculeate Hymenoptera, similar to those parasitized by species of *Perithous*. Nothing is known about the larval morphology.

Species of *Hybomischos* appear to form two species groups: In *septemcinctorius* and *townesorum*, the nervellus is intercepted at the middle (or just above it), antennal flagellum is shorter and thicker, about 0.5 the body length, male flagellum is with tyloids, teeth on ovipositor tip angled at 60°, and submetapleural carina appears forked posteriorly. In *galbus* and *virgulatus* (both from Philippines), the nervellus is intercepted above the middle, flagellum is as long as the body and slender, submetapleural carina is not forked posteriorly, and teeth on ovipositor tip angled at 30° from the horizontal. The males are unknown. The Rumanian species are unknown to me.

1. HYBOMISCHOS GALBUS (Baltazar) (figs. 1-3)

> *Perithous (Hybomischos) galbus* Baltazar, 1961. Monogr. Natl. Inst.
>   Sci. & Tech. Manila, 7: 50. ♀. Key, des., fig. Type ♀, Philippines:
>   Luzon, Nueva Ecija, Sierra Madre Mts. (TOWNES). Examined, 1980.
>   Paratype ♀, Luzon, Laguna, Mt. Bahanao.

This and the other Philippine species *(virgulatus)* are readily distinguished from *H. septemcinctorius* and *townesorum*, n. sp., by the characters mentioned under species groups. Baltazar (1961) has provided a key to distinguish the two.

H. galbus is characterized by having a yellowish body color with abdomen a little brownish, without any black marks. First tergite convex dorsally (fig. 1) and granulose. Flagellum long and slender, about as long as the body. Teeth on lower valve of ovipositor semicircularly arched and the basal tooth very much slanting, making an angle of 30° with the horizontal axis. Nervellus intercepted above the middle. Propodeum granuloso-punctate, with a distinct basomedian groove. Abdominal tergites 2-5 coarsely punctate.

Distribution: Philippines: Luzon. Known so far by the type-specimens only.

2. HYBOMISCHOS VIRGULATUS (Baltazar)

> *Perithous (Hybomischos) virgulatus* Baltazar, 1961. Monogr. Natl. Inst.
>   Sci. & Tech. Manila, 7: 50. ♀. key, des., fig. Type ♀, Basilan
>   (Washington). Examined, 1980.

Readily distinguished from *galbus* by having three black stripes on meso-scutum. Frons in middle, ocellar region, tubercles on abdominal tergites, and middle of first tergite also black. First abdominal tergite flatter dorsally and shiny. Propodeum with only a few scattered punctures. Abdominal tergites punctate only laterally. Ovipositor 1.8 x the body length. Otherwise related to *H. galbus*.

Distribution: Philippines: Basilan. Known by the type specimen only.

3. HYBOMISCHOS SEPTEMCINCTORIUS (Thunberg)

Ichneumon septemcinctorius Thunberg, 1922 (1824). Mem. Acad. Imp.
    Sci. St. Petersburg, 8: 280; 9: 363. ♀. des. Lectotype ♀, Sweden(Uppsala)
Taxonomy: Oehlke, 1967: 36. Torgersen, 1972: 99. Townes et al., 1965: 70.

This species, originally described under *Ichneumon* from Sweden and first placed under *Perithous* by Roman (1912) is fairly well distributed in Europe and USSR. Constantineanu & Pisică mention Japan also in its distributional range, but I could not find a reference to that effect. Togerson (1972) reported it from North America for the first time by a single specimen found by him in Ann Arbor, Michigan, in a spider web in his garage. Subsequently, only three more females from North America have been caught: One in Ann Arbor by H. Townes in his garage, one in Ottawa, Canada, at light by him, and another in Canada by Mike Sanborne at Stillsville, Ont., in a Malaise trap.

This species is characterized by having flagellum short and a little thick in the middle. Male flagellum with tyloids on segments 9-15. Face sparsely to minutely punctate. Thorax reddish with yellow markings. Propodeum granulose, with a weak median groove. Nervellus intercepted at the middle or at upper 0.4. First abdominal tergite granulose, flat dorsally. Abdominal tergites moderately punctate, punctures well separated from each other. Teeth on lower valve of ovipositor weak and angled at 65° from the horizontal.

Distribution: Europe, Russia, North America.

Hosts: (From Aubert, 1969): *Andricus quercus-tozae* (Cynipidae); *Omalus auratus* (Chrysididae); *Pemphredon lugubris, P. morio, Psen (Mimumesa) dahlbomi,* and *Psenulus fuscipennis* (Sphecidae). These hosts have been reported in Europe.

4. HYBOMISCHOS TOWNESORUM, n. sp.

Female: Face with scattered punctures, smoother laterally along orbital borders. Frons, vertex, temple, and occiput smooth and shiny, with scattered hairs, particularly on temple. Mesoscutum subpolished and with small hairs. Pronotum, mesopleurum, mesosternum, and metapleurum polished. Mesosternum more hairy than other parts. Submetapleural carina short and flange-like in anterior half and forked posteriorly, one arm of which is continuous with striations along hind coxal cavity. Propodeum strongly convex, apical half almost vertical, punctate; punctures evenly separated and mostly in the middle; its apical transverse carina incomplete and poorly defined; petiolar area smooth. Nervellus intercepted near middle. First abdominal tergite with scattered punctures, which are a little more crowded subapically. Second tergite with sparse scattered punctures. The following tergites progressively less punctate and smoother. Sides of second to fourth tergites with crowded punctures. Ovipositor about 1.35 x as long as the body, its tip sinuate and strongly compressed. Lower valve with faint or very faint teeth, which are slanting at an angle of 60° with the horizontal. Upper valve with 6-7 saw-like teeth on upper edge.

Black. Scape and pedicel yellow ventrally, flagellum brownish-black. Face along orbits, upper orbital borders up to slightly beyond the ocellar area, clypeus wholly, mandible except apically, basolateral corners of mesoscutum, hind corner of pronotum, tegula, subtegular ridge, scutellum except basally, metascutellum, a crescentic mark on propodeum in the region of apical trans- verse carina, and apices of all abdominal tergites, yellow. Apical mark on first tergite short. Fore and middle coxae and trochanters yellow. Rest of fore and middle legs yellowish-brown, with their femora with brownish patches. Middle femur darker and apices of its tarsal segments blackish. Hind coxa brownish with black mark basodorsally and yellow apically; femur orange-red; tibia and tarsus blackish-brown with basal and apical tarsal segments yellowish- white in basal 0.5 to 0.75.

Male: Essentially similar to the female except that face is wholly yellowish- white. Flagellum pale brownish, segments 9-17 with linear tyloids. Second to fourth abdominal tergites more densely punctate. Hind leg lighter in color: coxa yellow apically, femur light orange, tibia whitish medially.

Length: ♀, 9 mm. Fore wing 6-7 mm. Ovipositor 11.5-12 mm. ♂, 7.5 mm. Fore wing 5.5 mm.

Holotype: ♀, *Japan:* Mt. Norikura, 1600 m, 31-VII-1954, Townes family (Townes).

Allotype ♂, *Japan:* Sapporo, 9-VII-1954, Townes family (Townes).

Paratype ♀, *Japan:* Kamikochi, 22-VII-1954, Townes Family (Townes).

Hybomischos townesorum is related to *H. septemcinctorius* in the nature of nervellus, antennal flagellum and submetapleural carina, but can be readily distinguished by its black thorax, punctate propodeum without a distinct median groove, sparsely punctate frist tergite, and the following tergites with scattered shallow punctures and smoother on apical half.

LITERATURE CITED

Aubert, Jacques - F. 1969. Les Ichneumonides Ouest-Palearctiques et leurs hôtes. I. Pimplinae, Xoridinae, Acaenitinae. Ouvrage Publee avec la Concours du CNRS Pp 1-304.

Baltazar, C. R. 1961. The Philippine Pimplini, Poemeniini, Rhyssini and Xoridini (Hymenoptera, Ichneumonidae). Monogr. Natl. Inst. Sci. Tech. Manila 7: 1-130.

Constantineanu, M. I. and R. M. 1968. Contributions à l'etude du genre *Perithous* (Hym., Ichneum.) de la R. S. Romania. Zool. Anz. 180 (3-4): 228-258.

Constantineanu, M. I. and Pisică, Constantin, 1977. Fauna Republicii Socialiste Romania. Insecta, Hymenoptera, Ichneumonidae, 9(7): 1-310.

Oehlke, J. 1966. Revision der Ephialtinae-Typen von H. Haupt (Hymenoptera, Ichneumonidae). Reichenbachia 6(32): 279-285.

Oehlke, J. 1967. Westpaläartische Ichneumonidae I. Ephialtinae. *In* Junk: Hymenopterorum Catalogus *(nova editio)*, part 2: 1-48.

Townes, H., Momoi, S., and Townes, M. 1965. A Catalog and Reclassification of Eastern Palaearctic Ichneumonidae. Mem. Amer. Ent. Inst. 5: 1-661.

Torgersen, T. R. 1972. A *Perithous* (Hymenoptera Ichneumonidae) introduced from Europe. Great Lakes Ent. 5(3): 99.

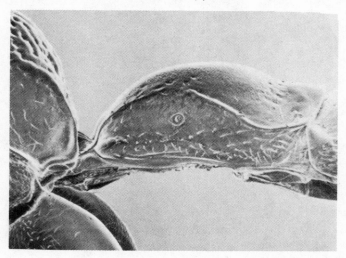

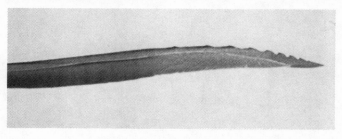

Hybomischos galbus (Baltazar)

Fig. 1

REVISION OF THE NEW WORLD GENUS HADROMEROPSIS PIERCE
(Coleoptera, Curculionidae, Tanymecini)
by
Anne T. Howden

Department of Biology, Carleton University,
Ottawa, Ontario, K1S 5B6, Canada

ABSTRACT

The genus Hadromeropsis Pierce is revised; included are a review of the nomenclature, taxonomy, a key to the species, illustrations, distribution maps, and a discussion of the morphology of the male and female genitalia.

The genus is divided into two subgenera: the nominate subgenus Hadromeropsis, type-species nobilitatus (Gyllenhal), and the new subgenus Hadrorestes, type-species pectinatus n. sp. Of the 52 species assigned to the genus 35 are described as new. The species in the nominate subgenus are as follows: opalinus (Horn) from Arizona and Mexico; fulgens (Champion) from Mexico; crinitus n. sp. from Mexico; flagellatus n. sp. from Mexico; dejeanii (Boheman) from Mexico; amoenus n. sp. from Mexico; scintillans (Champion) from Guatemala and Mexico; micans (Champion) from Guatemala; brevicomus n. sp. from Mexico; aureus (Blanchard) from Bolivia; cretatus (Champion) from Costa Rica and Panama; rufipes (Champion) from Costa Rica; superbus (Heller) from Argentina; gemmifer (Boheman) from Colombia, Costa Rica, Guatemala, Panama, and Venezuela; meridianus n. sp. from Brazil and Colombia; batesi n. sp. from Bolivia, Brazil, and Peru; togatus (Boheman) from Brazil and Paraguay; argentinensis (Hustache) from Argentina, Brazil, Paraguay, and Uruguay; plebeius n. sp. from Brazil; pulverulentus n. sp. from Brazil; pallidus n. sp. from Argentina, Brazil, Paraguay, and Uruguay; speculifer n. sp. from Brazil; beverlyae n. sp. from Brazil; nobilitatus (Gyllenhal) from Argentina and Brazil; atomarius (Boheman) from Brazil; excubitor n. sp. from Brazil; and fasciatus (Lucas) from Brazil. The species in the subgenus Hadrorestes are as follows: alacer n. sp. from Colombia and Ecuador; inconscriptus n. sp. from Peru; nebulicolus n. sp. from Colombia; silaceus n. sp. from Colombia; exilis n. sp. from Bolivia; impressicollis (Kirsch) from Colombia; institulus n. sp. from Ecuador; bombycinus n. sp. from Peru; brachypterus n. sp. (locality unknown); transandinus n. sp. from Ecuador; nanus n. sp. from Ecuador; apicalis n. sp. from Ecuador; dialeucus n. sp. from Venezuela; contractus n. sp. from Bolivia; pectinatus n. sp. from Bolivia, Colombia, and Peru; conquisitus n. sp. from Ecuador; scambus n. sp. from Ecuador; magicus (Pascoe) from Brazil (?) and Colombia; nitidus n. sp. from Ecuador; mandibularis n. sp. from Colombia; spiculatus n. sp. from Bolivia and Peru; picchuensis n. sp. from Peru; and cavifrons n. sp. from Peru. The new species earinus from Ecuador and striatus from Colombia are not assigned to a subgenus. Species synonymized are: Hadromerus scabricollis Faust, a synonym of gemmifer (Boheman); Hadromerus brachispinosus Boheman, a synonym of togatus (Boheman); Hadromerus herbaceus Lucas, a synonym of fasciatus (Lucas); Hadromerus ruficrus Kirsch and Hadromeropsis subaeneus Voss, synonyms of impressicollis (Kirsch).

Lectotypes are designated for opalinus (Horn), fulgens (Champion), scintillans (Champion), micans (Champion), superbus (Heller), brachispinosus (Boheman), argentinensis (Hustache), fasciatus (Lucas), and herbaceus (Lucas).

TABLE OF CONTENTS

Fig. 1.   *H. (Hadromeropsis) nobilitatus* (Gyllenhal),
type-species of *Hadromeropsis*. Male, dorsal habitus.

REVISION OF THE NEW WORLD GENUS HADROMEROPSIS PIERCE
(Coleoptera, Curculionidae, Tanymecini)
by
Anne T. Howden

HISTORICAL REVIEW

There has been considerable nomenclatural confusion involving the use of Hadromerus Schoenherr, Hadromeropsis Pierce, and Siderodactylus Schoenherr, partly due to various interpretations of the literature and partly because the original very obscure description of Hadromerus has apparently been overlooked by many authors, including me.*

In 1823, column 1141, Schoenherr first published "96. Hadromerus nob. Typ.: Curc. sagittarius Oliv." This bare reference, without description, is nonetheless a valid indication according to Art. 16a(v) of the International Code (1964). Thus sagittarius Olivier, 1807, from Senegal, is fixed as the type-species of Hadromerus Schoenherr, 1823, by original designation and monotypy. The characters of the genus are first described and discussed by Schoenherr in 1826 (pp. 136, 137) where he lists nobilitatus from Brazil for the first time as "Typus" of Stirps 1a and sagittarius Olivier from Senegal as "Typus" of Stirps 2a. However, there is no description of nobilitatus here, other than the description of Stirps 1a; this is interpreted as Stirps 1a being a subgenus but invalid here since nobilitatus is a nomen nudum. Thus, in this second publication, nothing was changed nomenclaturally.

In 1833 (p. 12), Schoenherr in his "Tabula synoptica," lists Hadromerus much as he did in 1823: "Genus 102. Hadromerus. Nob. Typus: Hadr. sagittarius. Curc. id. Oliv."

In 1834, Schoenherr (p. 125) first published the name Siderodactylus citing, "Hadromerus Schh. olim (1).-Curculio Oliv.," "(1) Schh. Curc. Disp. Meth. p.137. Hadromerus, stirps 2.," and listing sagittarius as the type. Thus, Siderodactylus is a synonym of Hadromerus Sch., 1823. Added to the genus Siderodactylus at this time were adstringatus Gyll. n. sp. and rhodinus Gyll. n. sp., both from Senegal. On p. 127, Schoenherr (1834) describes Hadromerus citing, "Schh. Curc. Disp. Meth., p. 136, no. 69 [sic]" as the original description. On p. 128 (loc.cit.) nobilitatus is first described by Gyllenhal, and no other species added.

Pierce, 1913, p. 400, tried to resolve the confusion by stating, "For the genus with nobilitatus as type we may take a new name, Hadromeropsis." It is fortunate that Pierce discussed the problem and stated the type-species clearly, because he accidentally cited Schoenherr, 1834, p. 125, the page for the sagittarius-based genus instead of the nobilitatus-based genus (p.127), to receive the new name.

---

* I thank Richard T. Thompson, London, for bringing this to my attention.

Marshall (1952:261-262) disagreed with Pierce, but Marshall's reasoning would be disallowed by the International Code published subsequently in 1964.

Since Pierce's publication of the name Hadromeropsis in 1913, the name has been applied to the New World genus in most of the literature including Günther and Zumpt (1933), Blackwelder (1957), Hustache (1938), and Van Emden (1944). Exceptions to this usage are Heller (1921), Hustache (1928), Marshall (1952), and Kuschel (1955). Based on taxonomic characters, all Hadromerus are African and all Hadromeropsis are New World.

In both Günther and Zumpt (loc. cit.: 102-103) and Blackwelder (loc. cit.: 800), six nomina nuda are listed as synonyms of Hadromeropsis. I have not repeated these names with the valid species but record them here. They are as follows: "Hadromerus irroratus Klug" and "Hadromerus pygalpis Germar" listed by Boheman as nomina nuda of his atomarius (Boheman, 1840:292); "Hadromerus micronychus Germar" listed by Boheman as nomen nudum of his brachispinosus [= togatus Boheman] (Boheman, 1840:291); "Hadromerus lepidopterus Klug" listed by Boheman as a nomen nudum of his dejeanii (Boheman, 1840:293); "Hadromerus splendidus Salle" listed by Champion as a nomen nudum of his fulgens (Champion, 1911:184); and "Hadromerus schonherri Chevrolat" listed by Champion as a nomen nudum of his scintillans (Champion, 1911:182).

Taxa originally associated with Hadromeropsis but which have been subsequently removed include the following:
Hadromerus porosus Boheman, 1840:294, transferred to Pandeleteius by Kuschel (1955:280).
Hadromerus tuberculifer Boheman, 1840:295, transferred to Pandeleteius by Kuschel (1955:280) then to Airosimus by Howden (1966:202).
Pandeleteinus Champion as a subgenus of Hadromeropsis, sensu Voss (1954:232), reinstated as genus by Howden (1969:76).

Taxa originally associated with Hadromeropsis HERE REMOVED from the genus are the following:
Hadromeropsis distinctus Voss, 1954:232 here removed to Pandeleteius.
    Pandeleteius distinctus (Voss), NEW COMBINATION.
Hadromeropsis griseus Voss, 1954:233 here removed to Pandeleteius.
    Pandeleteius griseus (Voss), NEW COMBINATION.

MATERIAL STUDIED

This study was based on over 2000 specimens in 35 institutional and private collections. Institutional collections are referred to in the text by the city in which they are located, and private collections by the owner's name.

Institutional collections studied and the persons assisting are as follows:

Auckland   New Zealand Arthropod Collection, DSIR.
                        G. Kuschel.

Basel   Naturhistorisches Museum. W. Wittmer.

Berkeley	California Insect Survey. E. Sleeper.
Berlin	Museum für Naturkunde an der Humboldt-Universität. F. Hieke.
Buenos Aires	Museo Argentino de Ciencias Naturales. M. J. Viana.
Cambridge	Museum of Comparative Zoology, Harvard University. J. Lawrence and M. Thayer.
Chicago	Field Museum of Natural History. H. Dybas and R. Wenzel.
Curitiba	Universidade Federal do Paraná. G. H. Rosado Neto.
Dresden	Staatliches Museum für Tierkunde. R. Hertel and R. Krause.
Eberswalde	Deutsche Akademie der Landwirtschaft-swissenschaften zu Berlin. G. Morge.
Hamburg	Universität Hamburg. H. Strümpel.
Ithaca	Cornell University. L. L. Pechuman.
Leiden	Rijksmuseum van Natuurlijke Historie J. Krikken.
London	British Museum (Natural History). R. Thompson.
Los Angeles	Natural History Museum of Los Angeles County. R. Snelling.
Maracay	Universidad Central de Venezuela. F. Fernández-Yepez and L. Joly.
Mexico	Instituto de Ecologia. G. Halffter.
New York	American Museum of Natural History. P. Vaurie.
Ottawa	Canadian National Collection. D. Bright.
Oxford	Hope Department of Entomology. University Museum. E. Taylor and M. W. R. Graham.
Paris	Muséum National d'Histoire Naturelle. H. Perrin.

Pittsburgh	Carnegie Museum. G. Wallace.
Rosario de Lerma	Instituto de Investigaciones Entomologicas Salta, INESALT, Argentina (formerly INESAN, San Miguel). A. Martínez, M. J. Viana, G. J. Williner.
Sacramento	California Department of Food and Agriculture. A. Hardy and T. Seeno.
San Francisco	California Academy of Sciences. H. Leech and D. Kavanaugh.
San José	University of Costa Rica. A. Wille.
São Paulo	Universidade de São Paulo. S. A. Vanin.
Stockholm	Naturhistoriska Riksmuseet. L. Janzon and P. I. Persson.
Tucson	University of Arizona. F. Werner.
Tucumán	Instituto Miguel Lillo. A. Willink.
Washington	United States National Museum of Natural History. D. R. Whitehead.

The following individuals generously loaned material from their private collections:

Henry Hespenheide, Los Angeles, California.
Charles O'Brien, Tallahassee, Florida.
Warren Steiner, College Park, Maryland.
Guillermo Wibmer, Tallahassee, Florida.

In addition, there are specimens in the Henry and Anne Howden collection. Types in our collection are deposited on loan in the Canadian National Collection, Biosystematics Research Institute, Ottawa.

All types were examined, and lectotypes were selected where appropriate.

Old museum specimens are often not duplicated in modern material. In this genus where some species appear to be rare - at least in collections - locality data are often very helpful and internal parts are sometimes essential for identification. When the old specimens have the setae destroyed by psocids and the internal parts consumed by dermestids and psocids, when the locality is incorrect, or incomplete or absent entirely, and when there are no modern specimens that match them, the taxonomist must decide if it is better to ignore the old specimens or include them in the study. In the interest of completeness, I have attempted to include such faulty material in this revision.

METHODS AND TERMINOLOGY

Species related by apparent synapomorphies are assembled into groups. No nomenclatural status is intended for these groups. The characteristics of a group are not repeated in species descriptions except to qualify them.

The Diagnosis gives the optimum number of characteristics which distinguish a species within the species group.

Where described species are represented by fewer than 10 specimens, all the information on labels, sex, and collection are given for each specimen. This information is summarized for described species represented by longer series.

Previous authors have used various terms to describe the characteristics of the scales of Hadromeropsis. I have used "scintillating," "iridescent," and "metallic lustre" which I distinguish as follows. Scintillating scales are very smooth and flat, like discs of thin metal, uniform in color but each with brilliant green, gold and red reflections. Such scales leave only a faint, pin-prick scar or a marginal scar as well as the central hole and are more easily abraded than iridescent scales. They are found on scintillans, gemmifer, dejeanii, etc.

Iridescent scales are flecked with multiple colors like the iris of the eye or like an opal and may be slightly convex or at least have the edges deflected. Iridescent scales are basically green, blue or white and may be strongly metallic or not. The scar of an iridescent scale consists of the outer perimeter as well as the central hole. Iridescent scales are charactertistic of fulgens, opalinus, etc. The scales of species in the nominate subgenus can be distinctly categorized as iridescent or scintillating; these distinctions break down in Hadrorestes and in the species earinus.

A metallic lustre that is not visible with a microscope but is visible macroscopically in a glancing light may be present on species with iridescent scales as in brevicomus, or non-iridescent scales as in nobilitatus and fasciatus.

Setae are of several diverse forms. On the sides of the elytra and the tibiae some species groups may have stiff, dark wiry setae. In the opalinus and scintillans groups there are erect setae on the dorsal surface. The characteristics of these erect setae are evident only on specimens in very good condition because the setae are prone to matting and the delicate filamentous tips break off readily leaving no obvious evidence of having been there. Shorter and arcuate setae are less vulnerable.

Another type of seta which I refer to as "long, wispy" is particularly noticeable on the prosternum and inner surface of the fore legs and is more highly developed in males than females. A similar but probably not analagous, long, wispy seta occurs often on the base of the elytra and prothorax. Such a "wispy" seta is pale, crinkled, and filamentous apically; when the tip is accidentally broken off, the remaining base resembles an ordinary erect seta. These setae seem to be associated with species with a waxy coating on the dorsum as in many Hadrorestes species. This wax conceals details of the sculpture and vestiture. It can be removed by gently stroking the surface with a fine camel hair brush dipped in ammonia solution to soften the wax and then further stroking with the brush frequently rinsed in water to remove the globs of wax. This can be tedious but often completely restores the color of the scales and position of setae without matting as can happen in an ultrasonic cleaner. See incompletely cleaned prothorax in Fig. 239, uncleaned surface in Fig. 240.

The "interantennal line" is a transverse line across the dorsum of the rostrum approximately opposite the insertion of the antennae. In Hadromeropsis it is usually unmarked.

The relative length of the segments in the funicle and antennal club are difficult to assess uniformly and have not been used in the descriptions unless they are extreme.

The elytra in most species are approximately 2x longer than their width at the humeri. This measurement is not given in the descriptions unless it differs by more than 0.2x.

Where the tibia rests against the femur the surface of the femur may be slightly smoother and concave (Fig. 257). This is referred to as the "tibial groove," although a groove is not present in many species. The contiguous edges of the tibia and femur are referred to as the "inner edge" and the opposite surface is referred to as the "outer" (Fig. 209). The remaining surfaces are referred to as anterior (Fig. 258) and posterior (Fig. 257) as they are when the leg is held at right angles to the body.

The "distal tooth" refers to the elongation of the extreme inner apical angle of the fore tibia. According to the species, the distal tooth provides many useful characteristics, ranging from short to long and curved, and from simple to divided into two.

To dissect the genitalia, the abdomen was usually removed from the beetle and the contents were treated with hot KOH. In the few cases where specimens recently killed in ethyl acetate were available, no treatment was necessary. For specimens dead several weeks, hot water with a bit of detergent was sufficient treatment before extracting the internal sac. On the other hand, many specimens of the argentinensis and nobilitatus groups were never softened sufficiently for satisfactory dissections of the male or female genitalia. In these cases I suspect the specimens had been collected or stored in formalin or alcohol which so hardened the tissues that they did not respond to rigorous KOH treatment. After extraction of the genitalia, the abdomen was glued to a paper point on the same pin as the beetle and the contents put in glycerin in a microvial on the same pin.

The aedeagus is relatively conservative in characters. Although it is a heavily sclerotized structure, its dorsum may appear slightly flattened or sunken in dried specimens compared to specimens in glycerin. This is not apparent in the lateral view I used for illustrations.

The internal sac was dissected from every species where it was possible to do so, except species where a flagellum was obviously extending from both ends of the aedeagus. The best technique was to insert a fine, hooked pin a short distance into the apical orifice of the aedeagus in very soft condition. Care was necessary to keep the parts soft with a drop of water during the dissection. Repeated short pulls were more satisfactory than a single deep effort, especially when working with a curved aedeagus. After dissection the internal sac often distended naturally after a day or so in glycerin. Sometimes an air bubble captured in the sac could be manipulated into the various lobes to inflate them.

Because of their extremely long genitalia, males suspected of belonging to the Mexican species flagellatus and dejeanii require a different dissection technique. The most successful dissections of old, dry specimens were obtained by the following procedure: keep specimen in humidor for 12 to 24 hours; lift left elytron and wing out of the way; with razor blade or scalpel slit the metathorax longitudinally to the left of the median line and across to the side; slit or cut integument of abdominal terga at extreme side nearly to apex; with fine brush apply drop of ammonia or warm water to tissues within integument until softened; observe position of genitalia and lift out, or remove entire abdomen after freeing proximal end of genitalia.

Measuring the spermathecal duct is a delicate operation. Measurements of the duct given here should be considered close estimates instead of exact measurements.

Areas of the spermatheca are referred to as in Howden (1976:7-8). The junction of the spermathecal duct with the body of the spermatheca is the nodulus; the junction of the gland with the spermatheca is the ramus; and the apex is the cornu. See Fig. 130.

MORPHOLOGY OF THE MALE AND FEMALE GENITALIA

Three rather distinct types of male and female genitalia are found in the genus. Each type of male genitalia corresponds to an equally distinct type of female genitalia. It is thus possible to examine one sex of a species and predict some characteristics of the genitalia of the other sex.

The genitalia of both sexes are known for only 36 of the 52 species (70%), and there are some anomalies that cannot be explained until both sexes of all species are known. This must be remembered in the following discussion which is based on these incomplete data.

Description of the Types of Genitalia

The first type of genitalia is found in the nobilitatus and argentinensis groups of the nominate subgenus. In the male genitalia of these groups (Figs. 159-180), the internal sac of the aedeagus when everted is seen to consist of a membranous tube with various patches of spicules and inflatable lobes on the proximal portion; a heavily sclerotized internal structure toward the distal end; a dorsal, inflatable, cupped, membranous lobe externally; the whole sac terminating distally in a slightly deflected membranous tube spiculate on the sides but with only a very fine, smooth membrane at the apex. In dorsal or ventral view the internal sclerite* is shaped

---

* The proximal end of the internal sclerite is always concealed by membranes, and I was never able to satisfy myself whether one or two sclerites (fused or movable) are involved.

proximally like a condyle, i.e., like the end of a vertebrate femoral bone; in lateral view, as I have drawn them, this is not particularly evident. Good specific characters are found in the position of the lobes as well as in the additional sclerotization of the distal deflected portion.

Females of the nobilitatus and argentinensis groups (Figs. 181-183, 187-190) have a large complex sclerite within the bursa copulatrix, sometimes almost filling the bursa. This sclerite is basically funnel shaped with fins and appears very different at various angles (Figs. 188-190). The vagina (or genital chamber) (Figs. 182, 188) is faintly testaceous and has vaguely sclerotized longitudinal streaks or flat rods its entire length. The spermathecal duct is moderate in length. Below the distal end of the common oviduct is an inflatable membranous cupped lobe.

The second type of genitalia is exhibited in the remaining H. Hadromeropsis: the gemmifer, scintillans, and opalinus groups. In the males of these groups the internal sclerite is elongated into a flagellum of various lengths (Figs. 73-108), in its extreme development longer than the beetle itself. The proximal end of the flagellum is always stiff and in extreme forms, such as flagellatus, forms a large loop or figure 8 in the mesothorax (Fig. 80). The length of the flagellum is species specific. In addition, the manubrium of the spiculum ʻgastrale, the aedeagal apodemes, and the tegminal strut may be individually or collectively elongated. The relative length of these rods is also species specific. The tegmen is much wider in species with a longer flagellum (Figs. 73, 75, 78). The ejaculatory duct ends distally (the gonopore) in a membrane encasing the proximal end of the ventral portion of the internal sclerite. This junction is readily visible at the anterior end of the aedeagus in species with the flagellum longer than the aedeagus.

The male genitalia of gemmifer (Figs. 101-103) are extreme with only a minute, free, needle-like sclerite representing the flagellum. There is also a sclerite on the ventral surface of a sort of dorsal cupped flap. In dorsal and ventral views these two sclerites are very similar in shape to the proximal end of a flagellum. The dorsal cupped lobe of gemmifer seems to be homologous with the dorsal inflatable cupped lobe of the argentinensis and nobilitatus groups. With this in mind, and in view of the location of the gonopore in gemmifer (Fig. 102), it seems possible that the gonopore in the argentinensis and nobilitatus groups is immediately beneath the dorsal inflatable cupped lobe that is located over the proximal end of the internal sclerites, but I was not able to establish this. If this is the case, the large ventral inflatable lobe of gemmifer could be comparable to the apex of the internal sac of the argentinensis and nobilitatus groups.

In females of the flagellate species there is no sclerite in the bursa copulatrix, nor any sclerotization of the vagina. The length of the spermathecal duct is proportionate to the length of the flagellum and is species specific*. The basic or unmodified length of the spermathecal duct is 1-1.5 mm in the nominate subgenus.

---

* This is not unique to Hadromeropsis. Thompson (1977:196) showed

(next page)

In the new subgenus <u>Hadrorestes</u>, an abruptly different modification of the apex of the internal sac and a gradual change in the internal sclerites constitute the third type of male genitalia. In the <u>alacer</u> group (Figs. 310-316) the internal sclerite is condyle-shaped proximally as in the <u>argentinensis</u> and <u>nobilitatus</u> groups of the nominate subgenus, but the distal end of the internal sac instead of being a thin membrane is sclerotized in the form of a short, slightly spiculate tube which is always curved and directed ventrad. The membranous parts of the internal sac are usually only faintly spiculate. The genitalia of <u>exilis</u> (Fig. 366) resembles that of the <u>alacer</u> group, but the sclerotized distal end is straight and directed distad, the membranous tube is simple and has two distinct patches of spicules, and the proximal end of the internal sclerite is evanescent and not condyle-shaped. In <u>impressicollis</u> (the only male representative of its group of four species) the sclerotized distal end of the internal sac is almost straight and directed distad; the condyle-shaped proximal end of the internal sclerite has a small dorsal elongation (Figs. 317, 319). In the remainder of <u>Hadrorestes</u> (Figs. 320-334) the internal sclerite at the proximal end is enlarged and acutely angled, and the distal end of the internal sac is finely spiculate, at least moderately sclerotized, curved, and directed dorsad.

Females of <u>Hadrorestes</u> (Figs. 336-348, 350-357) have no sclerotization in the vagina and inside the bursa have no complex funnel-shaped structure, but there may be a small, faint cup, ring, arc, or plate at the junction of the spermathecal duct. The ventral inflatable cupped pouch is larger and tougher than in the other groups; the ends of this inflatable pouch are attached to a pair of lateral sclerites, and this constitutes the third type of female genitalia. The caudal end of the common oviduct was occasionally seen to also be connected to the lateral sclerites. To study the female genitalia more thoroughly, it would be necessary to stain them. These lateral sclerites are somewhat variable in size and shape intraspecifically; they are absent in <u>cavifrons</u> (male unknown) and are presumed to be secondarily lost in that species.

Intraspecific Variation in Genitalia

Variation in the genitalia is incorporated in the species descriptions and remarks, but the following female parts warrant special note.

The spermatheca varies greatly in some species. Four examples of the spermatheca of <u>nobilitatus</u> are shown in Figs. 191-194. Three examples of variation in the spermatheca of <u>excubitor</u> are shown in Figs. 196-198.

---

that in <u>Apirocalus</u> (Otiorhynchinae, Celeuthetini) the length of the flagellum was "reflected in the length of the spermathecal duct of the females". Sharp (1918: Pl.IX, fig. 7; 1920:76) made similar observations on Celeuthetini.

In H. (Hadrorestes) spiculatus the spermathecal duct of the allotype (Fig. 357) is highly modified, but the only female paratype of this species has a normally slender duct. See Remarks on the species.

The lateral sclerites of the vagina of Hadrorestes vary in detail intraspecifically, as noted in previous section. Two examples of the lateral sclerite of one species are shown in Fig. 354.

In the subgenus Hadrorestes where a distinctive bursal sclerite is apparently not indicated by the morphology of the male genitalia, the bursal sclerite is variable in development intraspecifically. An extreme example is cavifrons where a large disc is present in the bursa copulatrix of one specimen but absent in the other.

Interpretations of Male and Female Genitalia

The implications of the morphology of the corresponding male and female genitalia are: a) in the flagellate species the flagellum enters the spermathecal duct which is an appropriate corresponding length; this would bring the gonopore into close proximity with the bursa; b) in Hadrorestes the tough, strongly supported pouch below the common oviduct accomodates either the dorsal spiculate lobe or the acutely angled internal sclerite, anchoring it while the distal end enters the bursa copulatrix; this would bring the gonopore into close proximity with the spermathecal duct in the bursa; c) in the nobilitatus and argentinensis groups, the faintly sclerotized vagina withstands the heavily spiculate proximal portion of the internal sac, the cupped lobe beneath the common oviduct does not need the extra reinforcement provided by lateral sclerites to receive the dorsal cupped lobe of the internal sac, and the funnel-like receptacle within the bursa copulatrix is of an appropriate design for the broad, expandable spiculate sides of the distal extension of the internal sac.

Thus, if the gonopore must be brought into proximity with the spermathecal duct in the bursa copulatrix to effect fertilization, then the morphology of the genitalia gives visual evidence of a possible reproductive isolating mechanism. At the very least the morphology of the male and female genitalia of Hadromeropsis has proved to be very useful in associating the sexes of dimorphic species as well as in relating species sometimes represented by only one sex when the external properties of those species are variable and sexually dimorphic.

BIOLOGY

The 15 species of Hadromeropsis which colleagues and I have collected and observed were all diurnal. Three species are brachypterous, but of the fully winged species at least superbus, alacer, and pectinatus were observed to be strong, fast fliers.

Many species in the nominate subgenus were taken on trees or shrubs of species of Leguminoseae, but not exclusively. Species in the subgenus Hadrorestes, which occurs in the Andes outside the range of most of these Leguminoseae, were taken on alder, grass, bamboo, and

miscellaneous unidentified plants. There was evidence of adult food
preference but not of specificity.

It is presumed the larvae are general root feeders. Bruch
(Hayward, 1960:18) reports the larvae of argentinensis attacking "los
espinelles" (Acacia caven (Mol.) and Mimosa scabrella Benth.).

There was physical evidence of predation on eight specimens of
Hadrorestes but in only six specimens of the much more common nominate
subgenus. Possibly the species with thinner integument or slower
movements were consumed outright by the predator whereas the very hard
or faster species escaped with only a cracked or nipped elytron. See
Remarks under Hadrorestes.

Genus Hadromeropsis Pierce

Hadromerus Schoenherr, 1834:127. Type-species Hadromerus nobilitatus
     Gyllenhal, in Schoenherr, 1834:128, by original designation.
Hadromeropsis Pierce, 1913:400. Replacement name for Hadromerus
     Schoenherr, 1834, nec Hadromerus Schoenherr, 1823.

Diagnosis.-Small to large, 5 to 20 mm in length. Scales never
sculptured. Posterior margin of epistoma never carinate or keeled,
although margin elevated distally in a few species. Mandible with
vestiture of lateral and ventral surface similar to that of rostrum.
Pronotum never produced anteriorly over vertex. Female with caudal
surface of ventrites 2, 3, and 4 conspicuously elevated, often
perpendicular or slanted anteriorly or posteriorly, edge of caudal
surface usually sharply delimited; character less developed in male.

Description.-Glabrous, squamose, or with both squamose and
glabrous areas arranged in a pattern. Scales with or without metallic
lustre, in many species strongly iridescent or scintillating. In
lateral view ventral surface of rostrum almost as long as dorsal
surface. Rostrum with interantennal line faintly marked at most.
Scar of mandibular cusp not situated on a process of the mandible.
Anterior constriction of prothorax absent dorsally in most species.
Fore leg larger than other legs; fore femur of male often greatly
enlarged (1.3-2.7x wider than hind femur) and often with a flange on
inner edge distally, flange situated anterior to tibial groove,
surface posterior to tibial groove at most weakly modified; fore femur
of female less modified. Fore coxae narrowly separated, the
separation sometimes concealed by the convex inner surface of coxae.
Hind tibia with corbel open. Inner surface of elytra with
well-developed apical carina between suture at end of interlocking
flange and side at apex of epipleural fold. Last tergite of male
simple.

Remarks.-The larger size, smooth scales, and presence of scales on
the mandible distinguish Hadromeropsis from Pandeleteius which is
probably the most closely related New World genus. Airosimus Howden
differs from Hadromeropsis in the presence of a "platform" on the apex
of the rostrum, the ventral surface of rostrum much shorter than the
dorsal surface, and the hind corbels semi-enclosed. The South
American Macropterus Schoenherr is the only other New World Tanymecini
with which Hadromeropsis might be confused. Macropterus has a thin
integument, extremely elongate elytra and ventrites 1 and 2, and lacks

the apical carina on the inner surface of the elytra.

The African Hadromerus differs from Hadromeropsis in the following characters: mandibular scar situated on a process of the mandible; prothorax with basal constriction adjacent to the perpendicular margin; elytra very narrow across base which is perpendicular and has the edge carinate. Hadromerus shares the following characters with Pandeleteius but not with Hadromeropsis: scales sculptured; inner edge of fore femur in some species distally modified with spur or teeth on posterior side of tibial groove; last tergite of male variously modified with carina or concavity.

The antennal scape of Hadromeropsis reaches the eye or exceeds it (especially in the subgenus Hadrorestes) but does not reach the prothorax (except striatus). The length of the scape is useful in field recognition of similarly sized species of Naupactini with enlarged fore legs, such as Naupactus and Macrostylus.

Three brachypterous species of Hadromeropsis are described from uniques, two from males and one from a female. The proportions of the elytra and prothorax look wrong for the genus, but this is a consequence of brachyptery. In all three the wing is only 0.6-0.7x as long as the elytra, the elytra are very narrow across the humeri, the apical umbone is absent (elytra flat at the apical termination of striae 4-6), and the mesepimeron is very narrow.

The spiculum gastrale and the spiculum ventrale are conservative in form and only one example of each is figured in Figs. 111 and 349 respectively.

In some males I observed a pair of very small sclerites (Fig. 109) set on a flexible stalk arising from the first connecting membrane between the spiculum gastrale and the 8th sternites. These sclerites are apparently the same as "spicule plates", sensu Gilbert (1952:636; Clark, 1977:105). The sclerites were very similar in all species but not all were surveyed for this character.

All species examined have a rectal ring (Fig. 172).

The gender of Hadromeropsis is masculine since the ending "opsis" is adjectival and modifies the masculine noun "Hadromerus."

Key to the Species of Hadromeropsis

1.   Occurring in South America. 2
        Occurring in Panama and northward. 48

2.   Eye flattened, scarcely exceeding outline of head.
          Prothorax in dorsal outline strongly rounded on
          sides, much wider than long (Fig. 2). Elytra
          clothed with patches of flat, imbricate, green
          or lavender scales alternating with irregularly
          placed rectangular glabrous areas. Length 7.5-
          9.5 mm. Argentina. . . . . 13. superbus (Heller)
        Eye convex. Prothorax with sides less strongly
          rounded. Elytra not so patterned. 3

3(2)   Eye small (Fig. 289). Pronotum (Fig. 290) flattened,
          surface irregularly sculptured. Base of elytra

between striae 5 abruptly perpendicular and
slightly anteriorly produced. Male unknown.
Length 13 mm. Ecuador . . . . 51. earinus n. sp.
Eye larger. Pronotum not so sculptured. Base
of elytra rounded. 4

4(3)   Brachypterous, wing only 0.6-0.7x length of
elytron. Mesepimeron very slender (Fig. 226).
Elytra without apical umbone, i.e., flat at
apical termination of striae 4 to 6. 5
Wing much longer than elytron. Mesepimeron
normal (Fig. 225). Elytra usually with
distinct apical umbone. 7

5(4)   Elytra with intervals conspicuously convex;
striae distinct, straight, strial punctures
large. Female unknown. Colombia . . . 52. striatus n. sp.
Elytra with intervals not convex; striae
indistinct or confused, strial punctures small. 6

6(5)   Elytra without tubercles. Disc of prothorax
smooth, sides of prothorax with obsolete
tubercles. Clothed with large scintillating
green scales. Female unknown. Bolivia . 32. exilis n. sp.
Elytra and prothorax with crowded, small, uniform
tubercles. Clothed with small white setae and
seta-like scales. Male unknown. Locality
unknown. . . . . .36. brachypterus n. sp.

7(4)   Dorsal surface densely clothed with scales of
one color, the only pattern (if any) formed by
shiny, bare areas confined to elytral intervals
between distinct striae, forming random dark
spots; or if entirely glabrous (some
gemmifer) with distinct tubercle on fore
coxa. Vestiture of male like that of female.
Elytral margin always smooth around ventrite
5. Many species with a tubercle on inner
distal edge of fore coxa opposite trochanter
(Figs. 57, 58). 8
Dorsal surface glabrous or squamose. If
clothed with scales of one color, then usually
striae not evident or bare areas forming a
fasciate or other pattern. Males often
glabrous, females squamose or colored differently. Elytral margins smooth, tuberculate,
or denticulate around ventrite 5. Never with
a coxal tubercle. 12

8(7)   Rostrum relatively long and narrow, 1.4-1.6x
longer than wide (Fig. 63). Male fore femur
at distal end with ventral margin expanded
into a flange (Fig. 65). Scintillating
green. Bolivia (Pando), Brazil (Amazonas, Pará),

Peru. . . . . 16. _batesi_ n. sp.
Rostrum moderate in width and length, usually
less than 1.2x longer than wide (Figs. 55,
56). Male fore femur unmodified distally.
Scintillating green or not. 9

9(8)   Prothorax in dorsal outline strongly rounded
on sides, much wider than long (Fig. 49).
Unicolorous white (mature specimens) or
slightly yellowed (teneral specimens).
Length 11-14 mm. Bolivia. . . . . . 10. _aureus_ (Blanchard)
Prothorax in dorsal outline not strongly
rounded on sides. Seldom white. 10

10(9)   Anterior edge of rostrum at margin of epistoma
produced upwards and outwards (Fig. 156).
Elytral interval 9 with the glabrous spots
tuberculate. Fore tibia in end view (Fig. 154)
with a ventral lobe as long as the distal
tooth. Brazil . . . . 23. _beverlyae_ n. sp.
Anterior edge of rostrum not produced. Elytral
interval 9 with glabrous spots not elevated.
Fore tibia in end view with ventral lobe
absent or much smaller than the distal tooth. 11

11(10)   Female with apex of ventrite 5 emarginate
(Fig. 59). Aedeagus longer, 3.1-4.2 mm, with
a short endophallic structure (Figs. 101, 102).
Elytra of male with sides tapering from humeri.
With a glabrous black form (mostly males).
Guatemala to northern Colombia and northern
Venezuela . . . . 14. _gemmifer_ (Boheman), in part
Female with apex of ventrite 5 rounded (Fig. 60).
Aedeagus shorter, 2.4-3.0 mm, with flagellum
as long as aedeagus (Figs. 104, 105). Elytra
of male with sides not tapered from humeri.
Colombia (Cundinamarca, Tolima) . . . 15. _meridianus_ n. sp.

12(7)   Striae 9 and 10 contiguous, confused or otherwise
indistinct beneath apical umbone; if condition
uncertain, then with long, wispy setae across
base of elytra and usually with tubercles or
denticles around apex of elytra. Occurring in
the Andes, including the Sierra Nevada de Santa
Marta. . . . . subgenus _Hadrorestes_ . . 23
Striae 9 and 10 separated, distinct beneath
apical umbone; never with long, wispy setae
on base of elytra, only _fasciatus_ with
distinct tubercles around apex of elytra.
Occurring in South America other than in the
Andes. 13

13(12)   Epistoma wider than long, occupying approximately

0.5 of anterior edge of rostrum (Fig. 149).
Size small, length 5.0-8.0 mm. Strial
punctures often foveate, especially in male.
Brazil (Minas Gerais). . . . . . 20. pulverulentus n. sp.
Epistoma as wide as or (more frequently) longer
than wide, occupying 0.2-0.4 of anterior edge
of rostrum. Length 4.5-16.5 mm. Strial
punctures not foveate. 14

14(13)   Edge of elytra around ventrite 5 with distinct
acute setiferous tubercles when viewed from
above (Fig. 157). Fore femur without distal
flange. Rostrum of female with distinct
pterygia (Fig. 158), apex of rostrum also
produced slightly in a flange on either side
of epistoma, epistoma thus long and narrow and
anterior edge of rostrum bisinuate. Elytra
with a complete postmedian dark fascia (Fig. 27),
very conspicuous in female and often appearing to
be composed of three contiguous diamonds; male
with blue or green scales, fascia weak, outlined
in white. Length 9.2-16.5 mm. Southeastern
Brazil. . . . . 27. fasciatus (Lucas)
Edge of elytra around ventrite 5 smooth when
viewed from above, or if appearing somewhat
tuberculate or crenulate, then without all
above remaining characters. 15

15(14)   Fore femur with distinct (but very weak in some
females) distal flange when viewed from behind.
Elytra with tubercles on lateral intervals
usually distinct in dorsal view. With
metallic lustre of gold or red. Fore tibia
abruptly bowed inwardly at distal fourth (Fig. 1). . . . 16
Fore femur without distal flange or swelling
above tibial groove when viewed from behind.
Elytra with or without tubercles on lateral
intervals. With or without a metallic lustre.
Fore tibia straight or slightly curved inwardly
distally. 17

16(15)   Male with distal tooth of fore tibia as long as
width of tibia at apex. Female with distal
tooth of fore tibia single, 0.5x as long as
width of tibia. Apex of elytra of male
produced into a brief divergent tooth. Length
of male 7-8.6 mm, female 7.5-10.7 mm. Southeastern Brazil and adjacent Argentina. . . . . .
. . . . 24. nobilitatus (Gyllenhal), in part
Male with distal tooth of fore tibia 0.5x as long
as width of tibia (Fig. 141). Female with a
pair of distal teeth on fore tibia, the most
distal tooth equal to or shorter than the

second (Fig. 142). Apex of elytra of male not
at all produced. Length of male 8.2-10.5 mm,
female 9.4-12.8 mm. Brazil (Rio de Janeiro,
Santa Catarina). . . . . 26. <u>excubitor</u> n. sp.

17(15) Large, length of male 9.5-11.4 mm; female
11-14.2 mm. Elytra without tubercles. Elytra
of female with anchor-shaped dark or glabrous
area (Fig. 13). Elytra of male green, blue, or
tan with pattern of female faintly marked with
white. Southeastern Brazil and Paraguay. . .
. . . . 17. <u>togatus</u> (Boheman)
Smaller, length of male less than 9 mm, of female
less than 10.7 mm. Elytra with or without
tubercles, not patterned as above. 18

18(17) Dorso-lateral edges of rostrum poorly defined,
broadly rounded (Fig. 151). Argentina
(Misionés), Brazil (Minas Gerais), Paraguay,
Uruguay . . . . 21. <u>pallidus</u> n. sp.
Dorso-lateral edges of rostrum distinct. 19

19(18) Elytra short and thick, convex in profile,
declivity almost perpendicular, apical umbone
absent (Figs. 7-10). Elytra with a large dark
postmedian spot on intervals 5, 6, and 7; without
metallic lustre. Northern Argentina, Brazil
(Santa Catarina), Paraguay, Uruguay. . . . . .
. . . . 18. <u>argentinensis</u> (Hustache)
Elytra not shaped as above. If elytra with a dark
postmedian spot, then also with either glabrous
tubercles on apical umbone or with a distinct
golden lustre. 20

20(19) Elytra with some glabrous setiferous tubercles on
apical umbone and usually on declivity and
lateral intervals as well. Without metallic
lustre. Color pattern similar to that of
argentinensis. Brazil (Paraná, Santa
Catarina, São Paulo). . . . . 19. <u>plebeius</u> n. sp.
Elytra without distinct tubercles, but glabrous
spots may be slightly convex. With or without
metallic lustre. 21

21(20) Female ventrite 4 with caudal margin straight in
ventral view, arcuate in caudal view (Fig. 143).
Elytra of male long and flat in profile. Fore
tibia almost straight, only weakly curved inward
distally (Fig. 140). Brazil (Minas Gerais),
Santa Catarina, São Paulo). . . . . . 22. <u>speculifer</u> n. sp.
Female ventrite 4 with caudal margin pointed
medially (Fig. 155). Elytra of male shorter,
more convex. Fore tibia distinctly bowed

inward distally (Figs. 1, 144). 22

22(21)   With distinct golden lustre. Ventrite 5 of
         female medially smooth and polished, without
         scales. Southeastern Brazil and adjacent
         Argentina. . . . . . 24. nobilitatus (Gyllenhal), in part
         With metallic lustre absent or very faint.
         Ventrite 5 of female with scales evenly
         distributed or sparser medially (Fig. 155).
         Brazil (Minas Gerais) . . . . . . 25. atomarius (Boheman)

23(12)   Scales when present very elongate, seta-like;
         scutellum conspicuously clothed with dense white
         elongate scales. Pronotum usually with a
         shallow depression on each side and a median
         depression as well. 24
         Scales when present rounded in shape; scutellum
         glabrous or variously clothed. Pronotum without
         both a pair of lateral depressions and a median
         depression (except some pectinatus). 26

24(23)   Head and rostrum with a shagreened microsculpture
         and scattered very fine punctures, the only
         other sculpture being a frontal fovea (Fig. 222).
         Ventrite 5 of female flat. Male unknown. Peru
         (Huánuco) . . . . 35. bombycinus n. sp.
         Head and rostrum not sculptured as above.
         Ventrite 5 of female longitudinally convex
         apically. 25

25(24)   Female with elytra (Figs. 215, 218) abruptly
         constricted beneath apical umbone, then flared
         outward in a prominent flange. Sculpture of
         rostrum and frons consisting of short,
         irregular rugae (Fig. 216). Male unknown.
         Ecuador . . . . 34. institulus n. sp.
         Female with elytra (Fig. 214) gradually tapered
         apically. Sculpture of rostrum and frons
         consisting of long, longitudinal rugae curving
         around frontal fovea (Fig. 210). Colombia. . .
         . . . . 33. impressicollis (Kirsch)

26(23)   Rostrum with extreme hemispherical concavity
         exaggerated by keeled dorso-lateral edges
         (Figs. 282, 283). Size large, length 15.8-17.4
         mm. Peru. . . . . 50. cavifrons n. sp.
         Rostrum not so modified. Size small to large. 27

27(26)   Large, length 11-20 mm. Punctures of elytra not
         aligned in striae and with many extra punctures
         of the same size. Fore tarsus with segment 2
         much wider distally, abruptly tapered proximally
         (Figs. 302, 304, 306, 309). Eye separated from

anterior margin of prothorax by its own diameter
or more, thus forming a slight neck. 28
Smaller, length 7.6-16 mm. Punctures of elytra
arranged in straight striae, usually without
extra punctures. Fore tarsus with segment 2
narrower distally, gradually tapered (Figs. 305,
307, 308). Eye closer to prothorax. 31

28(27)   Pronotum slightly, evenly convex, without lateral
flattened or depressed area. Elytral punctures
very small in glabrous areas (Fig. 206). Glabrous specimens with vitta on lateral edge of
metasternum of dense, imbricate white scales.
Female with apex of elytra broadly rounded-
truncate; ventrite 5 with deep lateral fovea.
Sutural interval at summit of declivity without
long, erect setae in either sex. Colombia. . . .
. . . . 31. silaceus n. sp.
Pronotum with lateral flattened or depressed area.
Elytral punctures larger, sometimes foveate.
Glabrous specimens never with concentration of
scales along side of metasternum. Female with
apex of elytra slightly produced, sutural interval attenuate into a brief tooth; ventrite 5
flat or convex laterally. Sutural interval at
summit of declivity with long erect setae in all
females and in males of alacer and inconscriptus 29

29(28)   Elytra sparsely (male) or more densely (female)
(Fig. 205) clothed with rounded white and pale
colored scales, both sexes with a vague postmedian "V" outlined in larger, paler scales;
some specimens with metallic green scales
ventrally. Female with, male without long erect
setae at summit of declivity. Colombia (Sierra
Nevada de Santa Marta) . . . . 30. nebulicolus n. sp.
Elytra of male and sometimes female with only
minute scales, no pattern; appearing glabrous
to naked eye. Female and male with long erect
setae at summity of declivity. 30

30(29)   Setae of disc of elytra conspicuous, 2x longer
than the diameter of a fovea (Fig. 204). Both
sexes with minute scales only. Spermatheca very
long and contorted, as in Fig. 339. Peru . . .
. . . . 29. inconscriptus n. sp.
Setae of disc of elytra as long as one fovea
(Fig. 203). Both sexes with minute scales only
or female with larger ochraceous scales forming
a pattern with 1, 2, or 3 dark fasciae. Spermatheca as in Fig. 336. Colombia, northern
Ecuador . . . . 28. alacer n. sp.

31(27) Apical umbone of elytra prominent in dorsal
        outline (Figs. 259, 260), set with acute
        tubercles each bearing a short, curled seta.
        Dorsal surface nowhere with long, wispy setae
        or erect setae except at summit of declivity.
        Black with sparse minute green scales or
        covered with full sized iridescent green
        scales with a pattern of glabrous black
        fasciae. Ecuador. 32
        Without all the above characteristics. Ecuador
        or not. 33

32(31) Fore tibia (Fig. 258) almost cyclindrical,
        smooth, nearly impunctate, with a few fine
        recumbent pale setae. Fore tibia straight or
        nearly so. Fore femur of male (Fig. 257)
        enormously swollen, with tubercles
        above and below tibial groove. . . . 43. conquisitus n. sp.
        Fore tibia (Fig. 262) with confluent punctures
        and rugulae, with numerous dark wiry hairs or
        setae. Fore tibia distinctly bowed distally.
        Fore femur of male with distal flange weak,
        without tubercles except minute ones on
        flange . . . . 44. scambus n. sp.

33(31) Prothorax (Fig. 266) with very uniform, rounded,
        evenly distributed tubercles. Male: posterior
        face of fore femur with 6 or more well-
        developed shelf-like tubercles; fore tibia with
        3 or more extremely long teeth (Fig. 267).
        Female: caudal surface of ventrites 2, 3, and
        4 rugulose (Fig. 269, 270). . . . . . 42. pectinatus n. sp.
        Prothoracic sculpture various, not as above. Male
        with or without shelf-like tubercles on
        posterior face of fore femur; teeth of fore
        tibia smaller, more nearly uniform in size.
        Female with caudal surface of ventrites 2, 3,
        and 4 smooth and shiny (finely sculptured in
        contractus) 34

34(33) Pronotum broadly flattened (Figs. 271, 288). Base
        of elytra with long, wispy setae. 35
        Pronotum not broadly flattened, may be concave or
        narrowly depressed along median line. Base of
        elytra with or without long, wispy setae. 36

35(34) Elytra long and slender, in male 2.2x longer than
        width across humeri (Fig. 38), in female 2.4x
        longer than width across humeri. Fore femur
        of male 2.4x, of female 1.2-1.4x wider than
        hind femur; distal flange distinct, its edge
        denticulate . . . . 48. spiculatus n. sp.
        Elytra shorter and broader in female (Fig. 31)

(male unknown), 2.2-2.3x longer than width
across humeri. Fore femur 1.2x wider than
hind femur; distal flange obsolete, not
denticulate. . . . . 49. picchuensis n. sp.

36(34)   Males. 37
         Females. 42

37(36)   Posterior face of fore femur with shelf-like
            tubercles. Basal third of elytral interval
            10 with setiferous tubercles approximately as
            large as the punctures of adjacent striae 9
            and 10. Posterior half of hind coxa covered
            with rounded scales. Venezuela. . . . .
            . . . . 40. dialeucus n. sp., in part
         Posterior face of fore femur smooth, without
            tubercles. Basal third of elytral interval
            10 smooth or almost, setae arising from smooth
            surface, punctures on basal third of striae 9
            and 10 often large or even foveate. Hind
            coxa with some rounded scales or not. 38

38(37)   Disc of elytra smooth and polished, with large
            white scales densely clustered to form three
            spots (Fig. 36). Mesepisternum densely
            covered with large white scales (prone to
            abrasion). Elytral interval 10 opposite
            metasternum wider than adjacent interval 9
            (Fig. 251). Colombia . . . . 45. magicus (Pascoe), in part
         Disc of elytra with minute scales assembled in
            brief, transverse depressions (depressions
            present whether scales abraded or not);
            without large scales. Elytral interval 10
            opposite metasternum not conspicuously wider
            than adjacent interval 9 except briefly in
            mandibularis. 39

39(38)   Small, length 7.6 mm. Fore femur very slender,
            as wide as rostrum. Pronotum (Fig. 238) with
            distinctly impressed median line. Ecuador. . . . .
            . . . . 38. nanus n. sp.
         Larger, length 8.5-11.7 mm. Fore femur enlarged,
            0.3 or more wider than dorsum of rostrum. Pro-
            notum with or without distinctly impressed
            median line. 40

40(39)   Pronotum (Fig. 234) with median line deeply,
            broadly impressed, ending in distinct basal
            and apical constrictions. Size smaller, 8.5-
            10.5 mm. Ecuador. . . . . 37. transandinus n. sp., in part
         Pronotum with median line unmarked, apical
            constriction unmarked dorsally. Size larger,
            over 11 mm. 41

41(40) Elytra in dorsal view with sides gradually tapered
        from about middle to apex. Elytra in cross
        section convex dorsally. Apices of elytra almost
        vertical, only very briefly individually rounded
        (Fig. 241). Ecuador. . . . . . 46. nitidus n. sp., in part
        Elytra in dorsal view with sides parallel or nearly
        so to apical umbone. Elytra in cross section
        flattened dorsally. Apices of elytra oblique,
        broadly individually rounded (Fig. 247).
        Colombia . . . . 47. mandibularis n. sp., in part

42(36) Mesepisternum with large, rounded white scales.
        Elytra smooth and polished between conspicuous
        white markings, interval 2 densely squamose
        forming a vitta from base to summit of
        declivity. Colombia. . . . . 45. magicus (Pascoe), in part
        Mesepisternum never with large white scales.
        Elytra variously patterned, never with a vitta
        on interval 2. 43

43(42) Ventrite 5 (Fig. 278) with very deep lateral
        depression. Ventrites 3, 4, and 5 much narrower
        than ventrites 1 and 2; edges of elytra likewise
        rather abruptly convergent (Figs. 37, 277). With
        white fasciae on elytra. Bolivia . . 41. contractus n. sp.
        Ventrite 5 with surface not depressed below
        lateral edge. Ventrites 3, 4, and 5 gradually
        narrowed to apex; edges of elytra gradually con-
        vergent. Elytra with or without white markings. 44

44(43) Dorsal outline of elytra serrulate from acute
        tubercles of intervals 8 and 9, apical umbone,
        and apex (Fig. 279). Posterior half of hind
        coxa covered with rounded scales. Elytra with
        minute white or blue scales in transverse
        depressions, also usually marked with larger
        white scales in a maximum pattern of basal
        ring, median and apical fasciae (Fig. 28).
        Venezuela . . . . 40. dialeucus n. sp., in part
        Dorsal outline of elytra serrulate on apical half
        at most. Hind coxa without rounded scales.
        Elytra not so patterned. 45

45(44) Elytra marked with two long, white, elliptical
        loops. Elytra shorter, 3.6x longer than pro-
        thorax. Colombia. . . . . . 46. nitidus(?) n. sp., in part
        Elytra without white markings. Elytra longer,
        3.8-4.0x longer than prothorax. Colombia,
        Ecuador. 46

46(45) Pronotum medially not depressed, surface rather
        smooth with punctures (Fig. 245). Mandible with

deep, sharply defined transverse dorsal groove
(Fig. 244). Length 16.2 mm. Colombia . . . . .
. . . . 47. mandibularis n. sp., in part
Pronotum medially depressed or flattened, median
line with transverse sculpture. Mandible with
dorsal setae not set in groove. Length 11.0-11.5 mm . . 47

47(45)   Elytra with an aeneous lustre. In profile apex of
elytra dorsad of apical umbone. Occurring at
3200 m. Ecuador. . . . . 39. apicalis n. sp.
Elytra shiny black without metallic lustre. In
profile apex of elytra ventrad of apical umbone
(Figs. 231, 232). Occurring at 1500-2000 m.
Ecuador. . . . . 37. transandinus n. sp., in part

48(1)   Fore tibia straight (Figs. 66-68). Scales iridescent or not, never scintillating (as defined
in Introduction). 49
Fore tibia curved inward distally (Figs. 69-72).
Scales scintillating or not. 52

49(48)   Eye large and only feebly convex (Figs. 40, 42).
All dorsal setae both sexes erect, long, dense.
Mexico (Michoacan) . . . . 3. crinitus n. sp.
Eye more convex. Setae of elytra erect; setae of
prothorax and head erect or not. 50

50(49)   Very uniformly and densely clothed; scales of
elytra not overlapping; discrete or contiguous.
Glabrous spots not larger than one scale. Mexico
(Morelos to Chiapas). . . . . 2. fulgens (Champion)
Scales more irregularly distributed; scales of
elytra overlapping in some areas or with larger
glabrous areas or both. 51

51(50)   Fore femur without distal flange. Postocular
vibrissae lacking or very poorly developed.
Setae of head and prothorax not erect in either
sex. North of Tropic of Cancer in Mexico
(Durango, Sonora, and Zacatecas) and U.S.A.
(Arizona). . . . . 1. opalinus (Horn)
Fore femur with brief but distinct distal flange,
often with small shiny granules on edge of
flange. Postocular vibrissae moderate. Setae
of head and prothorax erect in both sexes.
Mexico (central highlands). . . . . . 4. flagellatus n. sp.

52(48)   Fore coxa with a tubercle on inner distal edge
opposite trochanter (Figs. 57, 58). Prothorax
with low tubercles on sides. Ventrite 5 of
female distinctly notched at apex (Fig. 59).
Guatemala, to northern Colombia and northern
Venezuela . . . . 14. gemmifer (Boheman), in part

Fore coxa without a tubercle on inner distal
    edge. Prothorax usually without tubercles.
    Ventrite 5 of female rounded. 53

53(52)   Densely clothed with imbricate pure white scales;
    elytra with sparse (8-10 per interval) glabrous
    spots not at all elevated and nowhere with
    tubercles (Fig. 52). Base of elytra arcuate
    between striae 5. Length 10.0-12.5 mm. Panama
    (Chiriquí), Costa Rica. . . . . . . 11. cretatus (Champion)
Usually clothed with scintillating or iridescent,
    colored scales. If clothed with white scales
    (occasional amoenus, brevicomus, dejeanii,
    and scintillans), scales with some opalescence
    or iridescence under magnification. Base of
    elytra usually straight. 54

54(53)   Female over 13 mm in length (male unknown). Costa
    Rica, Guatemala. 55
Female usually less than 10 mm in length, male
    under 9.5 mm. 56

55(54)   Ventrite 5 flat, glabrous medially. Elytra with
    a vitta on interval 4 from base to summit of
    declivity and a vitta on interval 8 continuous
    with one on prothorax; vittae formed of larger,
    overlapping scales. Costa Rica. . . 12. rufipes (Champion)
Ventrite 5 convex, evenly squamose. Scales of
    elytra all of one size, not forming a pattern,
    except of random glabrous spots gradually dimin-
    ishing in size from base to apex. Guatemala. . . . . . .
    . . . . 8. micans (Champion)

56(54)   (Examination of genitalia may be necessary for
    positive identification of next four species).
    Setae of disc of prothorax both sexes short (as
    long as 1-1.5 scales), appressed. Elytra
    shorter; in male 2.5-2.7x longer than prothorax,
    in female 2.8-3.2x longer than prothorax. 57
Setae of prothorax both sexes longer (as long as
    1.5-6.0 scales), not appressed, usually erect.
    Elytra longer; in male 2.7-3.1x longer than
    prothorax, in female 3.1-3.7x longer than
    prothorax. 58

57(56)   Scales scintillating green, sometimes with irides-
    cent scales interspersed. Elytral glabrous areas
    evenly distributed. Mexico, usually north of
    Isthmus of Tehuantepec. . . . . 5. dejeanii (Boheman)
Scales iridescent green, often with golden or
    coppery sheen in reflected light, or white (one
    female). Elytral glabrous areas largest on inter-
    vals 1, 2, and 3, often coalescing to form short,

sinuous fasciae. Mexico (Chiapas) . . 9. <u>brevicomus</u> n. sp.

58(56)   Flagellum shorter than aedeagus, 1.3-2.0 mm long
         (Figs. 84-86). Spermathecal duct 1.5-2.1 mm
         long. Occurring in Mexico north of Isthmus of
         Tehuantepec. . . . . 6. <u>amoenus</u> n. sp.
         Flagellum measured within aedeagus 4.1-6.0 mm
         (Figs. 87-89). Spermathecal duct 4.3-7.0 mm.
         Occurring in Guatemala and Mexico south of
         Isthmus of Tehuantepec. . . . . . 7. <u>scintillans</u> (Champion)

Subgenus <u>Hadromeropsis</u>, new status

<u>Hadromeropsis</u> Pierce, 1913:400. Type-species, <u>Hadromerus</u> <u>nobilitatus</u>
      Gyllenhal, by original designation.
      Diagnosis.-Clothed with scales (except some <u>gemmifer</u> glabrous).
Color pattern of male similar to that of female or not. Prothorax
more or less evenly convex, never with depressions and elaborate
sculpture as in many <u>Hadrorestes</u>. Elytra with apical edge around
ventrite 5 smooth or weakly tuberculate (<u>nobilitatus</u> group). Striae 9
and 10 discrete, usually well-separated below apical umbone. Internal
sac of aedeagus in the form of a membranous tube with lobes and
patches of spicules on the surface and internally with sclerites; in
some groups membranous tube variously reduced and internal sclerites
in the form of a flagellum. Female never with paired sclerites in
vagina caudad of bursa copulatrix; spermatheca in one plane or nodulus
slightly angled (<u>fasciatus</u>).
      Remarks.-The species assigned to this nominate subgenus on
morphological characters occur from Arizona in the United States to
Argentina, and usually not in the Andes. Exceptions are <u>aureus</u>, which
occurs in semi-tropical mountain valleys of Bolivia at elevations up
to 1500 m, and two species of the <u>gemmifer</u> group, <u>gemmifer</u> and
<u>meridianus</u>, which occur in the Andes of Colombia and Venezuela at
elevations up to 1800 m.

The <u>opalinus</u> Group

1. <u>opalinus</u> (Horn)   3. <u>crinitus</u> n. sp.
2. <u>fulgens</u> (Champion)   4. <u>flagellatus</u> n. sp.
      Characteristics of Group.-Scales iridescent or not, never
scintillating. Without a sharply defined color pattern. Without
tubercles or pustules on prothorax or elytra. Setae of elytra uniform
or of randomly varying lengths, not conspicuously longer on alternate
intervals. Fore femur with distal flange of inner edge weak to
moderate in <u>flagellatus</u>, weak or absent in others. Fore tibia
straight (Figs. 66-68), distal tooth weak. Endophallic structure a
flagellum (but not known for <u>crinitus</u>); ring of tegmen wide (except
<u>crinitus</u>). Female genitalia without bursal sclerite; spermathecal
duct proportionate in length to that of flagellum.
      Remarks.-These are the northernmost representatives of the genus,

ranging from the Isthmus of Tehuantepec to Arizona.

In this group the flagellum reaches its maximum development (flagellatus) and the postocular vibrissae their maximum reduction (opalinus).

1. Hadromeropsis (Hadromeropsis) opalinus (Horn)

Figs. 39, 67, 74-76, 112, 153; Map 1

Hadromerus opalinus Horn, 1876:85. Champion, 1911:183. LECTOTYPE, HERE DESIGNATED, male, labelled "Ariz. 49," "Type 326" (Cambridge). See Type Material.
Hadromeropsis opalinus (Horn); Pierce, 1913:400.
Pandeleteius viridissimus Van Dyke, 1943:108. Type, female, labelled "Montezuma Pass, Huachuca Mountains, Arizona, VIII-19-1940," "Van Dyke Collection", "Holotype No. 5342 Pandeletius virisissimus [sic] Van Dyke" (San Francisco). Synonymized by Howden, 1959:419.

Diagnosis.-Clothed with blue, blue-green, green, white or coppery scales; scales not scintillating, most not strongly iridescent. Elytra often with an elongate pattern on intervals 2, 3, and 4 or 3, 4, and 5 formed of imbricate, slightly paler scales. Postocular vibrissae (Fig. 39) absent or very poorly developed (maximum length 0.3 mm). Setae of prothorax not erect in either sex, parallel to surface but usually not touching it. Fore femur with distal flange absent or obsolete. Apical edge of fore tibia (Fig. 67) with narrow, acute lobe next to inner edge; lobe approximately as long as distal tooth. Flagellum 4-4.5 mm long. Spermathecal duct 4.5-6 mm long.

Description.-Male, length 6.4-8.2 mm, width 2.3-3.0 mm. Female, length 7.1-9.5 mm, width 2.8-3.8 mm. Color as in Diagnosis, the green and blue occurring in 86% of the specimens, white, pinkish white and coppery in 14%. Green or blue specimens with greenish or golden sheen or metallic lustre, others with or without rosy lustre. Prothorax sometimes faintly vittate. Color of venter same color as dorsum or infrequently pink in specimens which are green dorsally. Naturally glabrous areas of prothorax and elytra irregular in size and placement, ranging from the size of a scale to width of interval, a variety of sizes present on any specimen. Rostrum flattened or slightly concave; dorso-lateral edges parallel or slightly divergent basally; median line usually finely impressed and narrowly glabrous from between eyes to interantennal line where it is often not impressed, more broadly glabrous. Epistoma as long as wide, occupying 0.3 of anterior edge of rostrum; with 1 or more scales on epistoma in 6% of males, 3% of females, remainder without scales. Prothorax of male averaging 1.1x, of female 1.2x wider than long; surface with some fine punctures; tubercles or granules at most very weak, usually absent. Postocular vibrissae as in Diagnosis. Setae of disc as long as 2 (usually) or 3 scales, anteriorly directed, not erect; setae may be perpendicular to surface on sides at basal and apical constrictions. Elytra across humeri in male usually 1.2x, in female 1.3x wider than prothorax. Elytra of male 2.8-3.1x, of female 3.1-3.2x longer than prothorax. Humeral angles prominent. Sides of elytra parallel or almost; apex of male rounded, without a tooth. Apical edge of elytra often thickened both sexes. Female with apex slightly attenuate or not beyond weak apical umbone, sutural interval ending in a tooth no longer than 3 scales, usually much shorter. Widest intervals of disc with 4-6 scales abreast. Setae of elytra erect in both sexes, stiff, as long as 2-3 scales, uniform in length. Female at summit of declivity on sutural interval with line of 4-6 elongate setae of increasing then decreasing size, the longest seta 2-2.5x longer than those of disc. Fore femur of male 1.5-2x, of female 1.4-1.6x wider than hind femur, usually without a trace of flange on inner edge. Fore tibia as in Diagnosis. Venter evenly squamose. Ventrite 5 of male 1.9x wider than long, flat to distinctly convex, apex emarginate. Ventrite 5 of female flat or almost flat, triangular with apex very narrowly rounded, averaging 1.6x wider than long; scales sparser medially and apically in some. Male genitalia as in Figs. 74-76; aedeagus 2.4-2.7 mm long, aedeagal apodeme 1.8 mm long, tegminal strut 1.5-1.8 mm long, flagellum 4.0-4.5 mm long. Spermatheca as in Fig. 112; spermathecal duct 4.5-6.0 mm long.

Type Material.-PARALECTOTYPES of Hadromerus opalinus, HERE DESIGNATED, 1 male, 1 female: 1 male, labelled "49," "5/1 73," "Lectotype 2830" on red paper, "Horn Coll, H 8313", "Hadromerus opalinus N" in pencil on white paper [Note: this specimen had been isolated in the Academy of Natural Sciences, Philadelphia, collection of types as the lectotype by a technician before the collection went to the Museum of Comparative Zoology (J. Lawrence, in litt.); the specimen is in very poor condition.] (Cambridge); 1 female, "Ariz," "5/1 73," "49," "169," "Type 326," "Hadromerus opalinus Horn" on white paper ruled with pale blue lines (Cambridge).

Distribution.-Map 1. MEXICO. Durango: Canelas, Durango, 23 mi S Durango, 24 mi NE Durango, El Salto, 6 mi NE El Salto, 28 mi E El Salto, Navahos (20 mi E El Salto), La Ciudad, Francisco I. Madero, Otinapa, Palos Colorados. Sonora: 6 mi NW Cananea. Zacatecas: 8 mi S Fresnillo, 27 mi NW Fresnillo. UNITED STATES. Arizona: Canelo, Chiricahua, Copper Canyon, Greaterville (Sta. Rita Mts.), Montezuma Pass, Parker Canyon (Huachuca Mts.), Tex Canyon (Cochise County).

Specimens were collected in July and August.

Specimens examined: 108 males, 96 females. Specimens in Berlin, Cambridge, London, New York, Ottawa, Sacramento, San Francisco, Tucson, Howden, O'Brien.

Remarks.-Of the 31 specimens in the British Museum labelled opalinus by Champion, only two are actually opalinus. Hence, the description and discussion of opalinus offered by Champion (1911:183-184) should be disregarded. He notes that he did not see an example from Arizona, the type locality. The male illustrated by Champion in Tab. 7, Fig. 29, is from Las Vigas, Veracruz; it is a teneral specimen of flagellatus with the scales incompletely colored. The specimen depicted in Fig. 30 is a teneral female labelled "Ciudad, Mex., 8100 ft., Forrer." There is a male, also teneral, with identical data to this female and these are the only two specimens of opalinus in Champion's series. Champion refers to this locality as "Ciudad in Durango" and not in the state of Mexico; Selander and Vaurie (1962) place La Ciudad at 148 km WNW of Durango. This female is robust and exceptionally large, 9.5 mm long, 0.5 mm longer than the next largest female. The setae of the elytra are as long as about 4 scales, but the scales appear to be incompletely developed.

Other specimens labelled opalinus by Champion are 6 scintillans from Chiapas, 3 amoenus, 21 flagellatus (1 in Dresden).

In opalinus the postocular vibrissae and the femoral flange are the least developed of any Hadromeropsis. This combined with the more northern range are usually sufficient to recognize the species. However, the single specimen from 27 mi NW Fresnillo has rather well-developed postocular vibrissae; this specimen also differs from typical opalinus in having the setae of alternate intervals of several lengths, and the rostrum more concave. Since the spermathecal duct is 4.5 mm long and the spermatheca is within the range of variation of opalinus, I presume that in this specimen the vibrissae are exceptional.

Arizona specimens are less often patterned than those from Mexico. Adults of opalinus seem to prefer Mimosoidea. A long series from 24 mi NE Durango (Aug. 20, O'Brien and Marshall) were taken on Acacia schaffneri, and the Arizona specimens were often on Calliandra eriophylla. Other plants recorded are Acacia sp., pine (one specimen) and Ceanothus depressus.

Various Durango and Zacatecas localities are described by Spieth (1950) in his account of the David Rockefeller Mexican Expedition of the American Museum of Natural History.

2. Hadromeropsis (Hadromeropsis) fulgens (Champion)

Figs. 41, 46, 66, 73, 116; Map 1

Hadromerus fulgens Champion, 1911:184, Tab. 7, Figs. 32, 33. LECTOTYPE, HERE DESIGNATED, male, labelled "Amula, Guerrero, 6000 ft., Sept. H.H. Smith," "♂," "Sp. figured," "B.C.A. Col., IV, pt. 3, Hadromerus fulgens, Ch." (London). See Type Material.

Diagnosis.-Densely, evenly clothed with scales of one size and one color, green, blue, or white (4%); scales iridescent (as defined in Methods) or not, never scintillating. Scales not imbricate on dorsal surface of elytra. Usually (96%) with some scales on epistoma. Natural bare areas, when present, the size of one scale, convex. Flagellum 4.0-4.4 mm long. Spermathecal duct 4.3-4.5 mm long.

Description.-Male, length 7.0-8.7 mm, width 2.6-3.4 mm. Female, length 7.5-9.0 mm, width 3.1-3.8 mm. Color as in Diagnosis, usually with a strong sheen; scales of elytral intervals 7 and 8 often less iridescent. Epistoma (Fig. 46) as in Diagnosis; averaging 1.2x longer than wide; occupying 0.3 of anterior edge of rostrum. Rostrum (Fig. 41) flat (especially female) or feebly concave; median line finely impressed from between eyes to interantennal line which may be weakly elevated in maximum development. Prothorax averaging 1.2x wider than long (range 1.1-1.29x); in dorsal view much wider than basal and apical constrictions, sides rounded or parallel in male, in female often widest posteriorly. Setae of prothorax erect in male, not erect in female (one exception), as long as 1.5-3 scales. Postocular vibrissae moderate. Elytra across humeri in male usually 1.1x, in female 1.2x wider than prothorax. Elytra of male 2.5-2.8x, of female 2.9-3.3x longer than prothorax. Sides of elytra of male parallel or slightly wider behind middle, apex truncate, without a tooth. Sides of elytra of female widest at about middle, apex slightly attenuated beyond weak apical umbone or not, ending in a minute tooth or not. Widest intervals of disc with up to 8 scales abreast. Setae of elytra erect, as long as 2-5 scales in males, 2-4 scales in females. Female usually with 4 (extreme 10) long, erect setae on sutural interval on summit of declivity. Strial punctures not larger than a scale, may be completely concealed by scales. Fore femur of male averaging 1.8x, of female 1.6x wider than hind femur; with a weak, short distal flange. Apical edge of fore tibia (Fig. 66) with narrow, rounded to acute lobe next to inner edge; distal tooth of male approximately as long as teeth on inner edge, shorter in female. Ventrite 5 of male moderately to slightly convex, apex narrowly truncate, usually 2x wider than long. Female sometimes with a few scales on perpendicular caudal surface of ventrites 2, 3, and 4; ventrite 5 across base averages 1.7x wider than long, evenly squamose or scales somewhat sparser medially. Male genitalia similar to Fig. 73 (see Remarks); aedeagus 2.2-2.7 mm long, aedeagal apodeme 1.1-1.5 mm long, tegminal strut 1.3-1.5 mm long, flagellum 4.0-4.4 mm long (1.7x longer than its aedeagus). Spermatheca as in Fig. 116; spermathecal duct 4.3-4.5 mm long.

   Type Material.-PARALECTOTYPES, HERE DESIGNATED, 10 males, 14 females: 1 male, 1 female, same data as lectotype but "Aug." (London); 1 male, 3 females, "Cuernavaca, Mor., Mex., Wickham" (London, Washington); 2 males, 3 females, "Puebla, Mexico, Salle Coll." (London); 2 males, 1 female, "Parada, Mexico, Salle Coll." (London, Howden); 1 male, 1 female, "Istepec, Mexico, Salle Coll." (London, Washington); 2 males, 2 females, "Boucard, Mex., Fry Coll." (London); 1 female, "1278, Mexico, Bowring 63.47" (London); 1 male, 2 females, "Oajaca, 58.135 Mex." (London).

   Distribution.-Map 1. MEXICO. Chiapas: Finca Guatimoc. Guerrero: Amula, Chilapa, Taxco. Mexico: Ixtapan de la Sal. Morelos: Cuernavaca, Tepoztlán. Oaxaca: Istepec, Monte Alban, Parada. Puebla.
   Specimens were collected in July, August, September, and October.
   Specimens examined: 64 males, 47 females. Specimens in Auckland, Basel, Berlin, Cambridge, Chicago, Dresden, Eberswalde, Leiden, London, Los Angeles, Mexico, Paris, Stockholm, Washington, Howden, O'Brien.

   Remarks.-In the white specimen of fulgens figured in Tab. 7, Fig. 33, by Champion (1911), the prothorax is actually 1.14x wider than long, the figure being somewhat misleading. Champion mentions a "claw" on the intermediate tibia of the male; this is the small mucro which is present on the hind and middle tibia of all Hadromeropsis males.
   There are three syntypes from Amula including the male lectotype (not dissected) and a similar female. The third specimen is a male which differs considerably from the above description of typical fulgens. The discrepancies are as follows: postocular vibrissae more numerous and longer; elytra 3.3x longer than prothorax, 1.3x wider across humeri than across prothorax; elytral setae sparser and a few setae on alternate intervals extremely long (as long as 8 scales); distal tooth of fore tibia long; apex of ventrite 5 emarginate; flagellum in curved position within aedeagus 6 mm long, 1.9x longer than aedeagus (Fig. 73).
   Most specimens of fulgens can be readily recognized

macroscopically by the uniform color with a sheen. In addition to the diagnostic characters, white specimens or worn specimens can be separated on external characters from flagellatus by the lack of punctures on the rostrum and head and by the weaker postocular vibrissae; from dejeanii by the shape of the epistoma and smaller scales; from opalinus by the very evenly distributed scales of intervals 2 and 3 especially; from crinitus by the distal flange on the fore femur and convex eye.

3. Hadromeropsis (Hadromeropsis) crinitus n. sp.

Figs. 40, 42, 43, 47, 77, 117, 131; Map 1

Diagnosis.-Densely, uniformly squamose as fulgens, but all dorsal setae both sexes erect, longer, and denser than in fulgens. Eye large and only feebly convex (Figs. 40, 42). Postocular vibrissae poorly developed, no longer than adjacent setae. Fore femur without distal flange. Aedeagus (Fig. 77) arcuate proximally only; remainder straight, flattened dorsally and with dorso-lateral edges carinate. Spermatheca as in Fig. 117, duct 1.3 mm long.

Description.-Holotype, male, length 9.1 mm, width 2.9 mm. Densely clothed with scales of pinkish tan, obsoletely iridescent, without a sheen; scales discrete, imbricate only on pronotum either side of median line. Scales of ventral surface faintly opalescent whitish. Naturally glabrous areas of elytra the size of a scale or less, except sutural interval where scales are sparser. Setae of head and rostrum as long as 3-4 scales; setae of prothorax as long as 5-7 scales, setae of elytra as long as 5-8 scales. Rostrum approximately as wide as long, dorso-lateral edges parallel, sides broadly visible in dorsal view; flattened dorsally, apex rather strongly deflected, epistoma thus almost vertical. Median line finely impressed between eyes only; dorsal surface of rostrum with scattered fine punctures. Epistoma (Fig. 43) occupying 0.4 of anterior edge of rostrum, 1.3x wider than long, without scales. Head thick in profile (Fig. 40); eye large, elongate, scarcely exceeding head in outline (Fig. 42). Prothorax 1.08x wider than long, sides slightly rounded between constrictions; in profile, disc feebly arcuate, constrictions obsolete. Prothorax with a very few small punctures, no pustules. Postocular vibrissae as in Diagnosis, Fig. 40. Scutellum with a few scales and a few appressed setae. Elytra across humeri 1.3x wider than prothorax. Elytra 3.0x longer than prothorax. Humeri prominent, sides of elytra very slightly convergent from humeri to apical 0.3, thence more strongly convergent to apex, apical umbone weak in dorsal outline; apices briefly individually rounded, without a tooth. Base of elytra including scutellum abruptly perpendicular. Strial punctures small and striae poorly defined, especially apically. Widest intervals 5 scales abreast; interval 5 narrower, as narrow as 3 scales abreast. Setae very numerous on elytra (Fig. 47); uniform in length except occasional setae of alternate intervals slightly longer. Fore femur 1.9x wider than hind femur, strongly but gradually swollen. Apical edge of fore tibia with inner lobe approximately equal to distal tooth, distal tooth 0.2 of width of apical edge. Last tergite in apical view scarcely convex. Abdomen long, convex; ventrite 5 (Fig. 131) convex, apex truncate, 1.7x wider than long. Genitalia as in Diagnosis, Fig. 77; aedeagus 3.3 mm long, aedeagal apodeme 1.3 mm long, tegminal strut 1.8 mm long. Endophallic structure not seen, internal sac damaged.

Allotype, female, length 9.0 mm, width 3.4 mm long. Differs from type as follows. Rostrum with median line foveate between eyes. Prothorax 1.2x wider than long; sides of prothorax with a few glabrous pustules. Elytra across humeri 1.3x wider than prothorax. Elytra 3.6x longer than prothorax. Sides of elytra very slightly divergent to apical 0.3, thence gradually, slightly rounded to apex, apical umbone weaker than in male; sutural interval ending in a blunt tooth about as long as a scale. Intervals and striae slightly more evident, partly because intervals are feebly convex, especially basally and laterally. Widest intervals up to 8 scales abreast. Setae of summit of declivity on sutural interval little different from other setae of same interval, but 9 setae slightly longer. Fore femur 1.5x wider than hind femur. Ventrite 3 with caudal margin arcuate, posteriorly directed. Ventrite 5 across base 1.7x wider than long, elongate-triangular, apex very narrowly rounded, slightly convex longitudinally; scales on sides basally only. Spermatheca as in Fig. 117; spermathecal duct 1.3 mm long.

Type Series.-Holotype, MEXICO, [Michoacan], Patzcuaro, Koebele, Koebele Collection (San Francisco). Allotype, same data as type (San Francisco). No paratypes.

Remarks.-The wide epistoma is equalled or exceeded in flagellatus and some opalinus. The elytral setae are more dense than on any other species seen. The eyes may be as feebly convex in female flagellatus and, less frequently, in female opalinus.

The internal sac was damaged in dissection and its form is consequently unknown, but no obvious flagellum or sclerites were observed. The tegmen is narrower in crinitus than in the other three species in the group. The aedeagal apodemes are short as in aureus, brevicomus, superbus. The length of the spermathecal duct is "normal", i.e., not modified for a flagellum. These characters all indicate an internal sac with a short flagellate structure similar in size to that of aureus, brevicomus, superbus.

H. crinitus occurs in an area from which I saw no other specimens of Hadromeropsis. It is closest in characters and range to opalinus to the north and fulgens from Guerrero. The diagnostic characters listed will separate it from both.

The Latin word crinitus means hairy or with long hair.

4. Hadromeropsis (Hadromeropsis) flagellatus n. sp.

Figs. 45, 48, 68, 78-81, 113-115, 132, 133; Map 2

Diagnosis.-Often appearing dusty or slightly tessellate because of arrangement of scales and long, erect setae. Clothed with white, white and blue, iridescent green (especially in Veracruz), or coppery scales. Scales often of several sizes, elytra often vaguely vittate because of denser scales on intervals 2, 4, 6, and 8; vittae often interrupted with irregular glabrous areas; with a vague V-shaped mark before declivity in 14% of specimens; never evenly squamose as in fulgens. Setae of head, prothorax and elytra long, fine, erect in both sexes. Fore femur with weak to moderate distal flange (Fig. 48), flange with at least a few shiny granules or tubercles in males and in some females. Last tergite of male convex (especially northern part of range), or flattened medially (especially center of range) or with a longitudinal concavity (southern end of range). Genitalia extreme (Figs. 78-81). Aedeagus 2.5-3.2 mm long, aedeagal apodeme 1.1-1.4 mm long, tegminal strut 2.8-3.3 mm long, thus aedeagal apodeme less than half the length of aedeagus, tegminal strut averaging as long as aedeagus; ring of tegmen very wide; flagellum 9-11 mm long (3-4x longer than aedeagus), longer than entire beetle. Spermathecal duct 7.5-12.0 mm long.

Description.-Holotype, male, length 7.8 mm, width 2.8 mm. Shiny black, sparsely clothed with white scales interspersed with gray-blue scales, scales varying greatly in size and shape. Scales especially sparse on pronotum, scales of elytra forming a few vague clusters on intervals 2 and 4. Dorsal setae fine, with long filamentous tips mostly abraded in type; setae of male topotypes as long as 3 large scales on head and rostrum, as long as 3-6 large scales on prothorax, as long as 4-7 large scales on elytra. Head and rostrum (Fig. 45) punctate, each seta arising from a puncture approximately as large as a scale. Punctures of rostrum confluent apically, each puncture bearing a scale or seta. Rostrum 1.2x longer than wide, dorso-lateral edges distinct, parallel; sides of rostrum narrowly visible in dorsal view. Median line marked with elongate pit between anterior half of eyes; basal half of rostrum weakly concave. Epistoma approximately as wide as long, occupying 0.4 of anterior edge of rostrum; epistoma without scales, sides weakly carinate anteriorly. Segments 1 and 2 of funicle equal; club approximately 3x longer than wide. Prothorax 1.13x wider than long, slightly convex, sides slightly rounded between constrictions; constrictions obsolete on disc. Disc of prothorax with a few fine punctures, sides of prothorax with base of some setae weakly granulate. Postocular vibrissae consisting of about 8 moderately long, fine, white setae. Scutellum with a few small scales, mostly glabrous. Elytra across humeri 1.3x wider than prothorax. Elytra 3.1x longer than prothorax. Humeri prominent, right-angled. Sides of elytra parallel to approximately apical 0.3, apical umbone weak, apex beyond umbone broadly rounded, without a tooth. Strial punctures uniform in size, moderate. Widest intervals 4 scales abreast, but scales not contiguous. Setae of ventral surface and legs much more numerous than on dorsal surface, otherwise similar. Fore femur (Fig. 48) 1.7x wider than hind femur; distal flange brief but distinct, with shiny, sharp tubercles. Fore tibia (Fig. 68) cylindrical; apical edge with inner lobe obsolete; distal tooth very small, only 0.2 of width of apical edge. Ventrite 5 (Fig. 132) slightly convex, 2.2x wider than long, apex broadly emarginate. Last tergite convex with a central flattened area; surface punctate. Genitalia as in Fig. 78-81. Aedeagus flattened dorsally, dorso-lateral edge sharp, almost carinate basally. Aedeagus 2.7 mm long, aedeagal apodeme 1.1 mm long, tegminal strut 2.8 mm long, flagellum more than 9 mm long.

Allotype, female, length 8.7 mm, width 3.3 mm. Differs from type as follows. Scales similar in color, but more dense; prothorax with wide almost glabrous median and dorso-lateral vittae. Setae of head and rostrum more numerous, most directed toward median line, as long as 2-3 adjacent scales. Head more robust, eyes much less prominent. Segment 2 of funicle 0.8x as long as segment 1; club 2.3x longer than wide. Prothorax 1.16x wider than long, flat in profile; disc indistinctly flattened laterally. Elytra across humeri 1.4x wider than prothorax. Elytra 3.1x longer than prothorax. Elytra slightly wider across middle; widest intervals 5 scales abreast; setae of various lengths, as long as 3-8 scales, all setae of declivity longer; sutural interval at summit of declivity with 8 very long, erect setae. Fore femur 1.4x wider than hind femur. Caudal margin of ventrites 2 and 3 anteriorly arcuate and caudal surface slanted anteriorly; caudal margin of ventrite 4 straight, surface perpendicular. Ventrite 5 (Fig. 133) across base 1.8x wider than long, elongate-triangular, apex narrowly rounded, slightly convex longitudinally; scales on sides basally only, medially with only sparse, short setae. Allotype not dissected; female topotypes with spermatheca as in Fig. 113, spermathecal duct 8-8.5 mm long; duct scarcely wider or more sclerotized at bursa.

Type Series.-Holotype, MEXICO, Tlaxcala, 21 mi W Apizaco, Aug. 20, 1958, E. Mockford, on juniper (Howden). Allotype, same data as type (Howden). Paratypes, 43 males, 49 females. MEXICO. 1 female, Hoege, Coll. Kuschel (Auckland); 2 males, 1888 [1], Coll. Kuschel (Auckland); 1 male, 1 female (Basel); 1 male, 2 females, F.C. Bowditch Coll. (Cambridge); 6 males, 6 females, Koltze Coll. or Boucard [1], J. Faust, Ankauf 1900 (Dresden*); 1 female, Hoege, Samml. K.F. Hartmann (Dresden); 1 male, 1 female, Coll. Kraatz, Hadromerus opalinus Horn, Hust. det. 1938 [1], Kuschel det. [1] (Eberswalde); 2 males, 1 female, Sharp Coll. 1905-313 (London*); 1 female, Boucard, Fry Coll. 1905-100 (London); 1 female, Tylden Coll. (Oxford); 1 female, coll. de Bonneuil, Ex. Coll. Clerc (Paris); 2 females, Hustache Coll. (Paris); 1 male, 1 female, ex Coll. Oberthur (Paris); 3 males, 1 female, Salle 1859 (Paris). Guerrero: 1 male, 1 female, Texquitzin bei Chilapa, X.29, L. Schultze S.G. (Berlin). Hidalgo: 3 males, 1 female, 15 mi NE Huichapan, 6900 ft, 8-18-1971, C & L O'Brien & Marshall (O'Brien); 1 male, 6 mi E Tulancingo, VI.24.1962, J.M. Campbell (Howden). Jalisco: 1 female (Mexico); 1 male, 5 mi NW Lagos de la Moreno, VII-16-1974, R.L.Mangan & D.S.Chandler (O'Brien). Mexico: 1 female, D.F., J. R. Inda Collector (Washington); 1 male, Ixtapan la Sal, 5500 ft, 9.VIII.1954, J. G. Chilcott (Ottawa); 2 males, 1 female, 5 mi SE Texcoco, 30.VII.67, on Mimosopsis aculeaticarpa Br. & Rs.(Howden); 1 female, 10 km de Villa Morelos, VIII.26.1956, G. Halffter (Howden). Puebla: 1 male, 3 females, Salle Coll. (London*, Howden*); 1 female (Stockholm). San Luis Potosí: 1 male, Hacienda de Bleados, Dr. Palmer (London*). Tlaxcala: 3 males, 5 females, same data as type, or H.F. Howden [2] (Ottawa, Howden). Veracruz: 1 female, Deyr. (Cambridge); 1 male, 3 females, Jalapa, Hoege (London*, Washington*); 1 female, 2.3 mi W Acultzingo, 7000 ft, Aug.24, 1951, J.E.Mosimann, T.M. Uzzell, G.B. Rabb (Howden); 4 males, 3 females, El Camaron, Salle Coll. (Dresden*, London*); 1 female, Cordova, Hoege (London*); 1 female, Playa Vicente, Hoege (London*); 1 male, Las Vigas, Hoege, Sp. figured [the male figured in Tab. 7, fig. 29, B.C.A., Col., IV, pt.3] (London*). No data: 5 males, 5 females (Cambridge, Dresden, London*, San Francisco, Stockholm, Washington*).

* Bearing Champion label "B.C.A., Col., IV, pt. 3 Hadromerus opalinus Horn."

Remarks.-Paratypes vary as follows. Males vary in length from 7.4-9.0 mm and in width from 2.5-3.1 mm. Females vary in length from 8.1-10.4 mm and in width from 3.2-4.0 mm. Champion's illustration of "opalinus" (1911, Tab. 7, Fig. 29) accurately depicts the form and dusty appearance of many specimens of flagellatus. The vittate appearance of some specimens may be emphasized by the fact that intervals 3 and 5 are often very slightly more elevated and scales there are more frequently abraded from them leaving scales more numerous in shallower intervals 2 and 4. Macroscopically most specimens appear white or whitish, 11% are green and only 3% are distinctly coppery. No specimens were seen with scintillating scales as in dejeanii, amoenus, etc.

The filamentous tips of dorsal setae are readily damaged; when apparently entire, they may be of several lengths on the elytral intervals and, infrequently, as much as twice as long on alternate intervals. Setae of elytra are in single but irregular rows on each interval on disc, more numerous on sutural and lateral intervals. Females may have as many as 14 long setae on the sutural interval on the declivity. The epistoma has 1 or more scales in 7% of the males, 18% of the females. The antennal club is variable in length. Widest elytral intervals of females are up to 7 scales abreast. Apex of elytra in females often with sutural interval produced in oblique overlapping knobs or teeth, these teeth directed posteriorly in only 10% of the females. Fore femur of females often lack tubercles on the inner distal edge. Inner lobe of the apical edge of the fore tibia may be moderately developed; inner edge of fore tibia with as many as 16 teeth. Variation in genitalia is listed in the Diagnosis; the spermatheca varies in shape as in Figs. 114, 115.

The flagellum is always longer than the entire length of the beetle and its proximal end is always very stiff. Within the beetle the proximal end is found in one or two large loops or a figure "8" over the proventriculus in the metathorax (Fig. 80). In one instance the proximal end seemed to extend into the prothorax. A special technique for dissecting these extreme genitalia is outlined in Methods and Terminology.

No special technique was needed to dissect females; the extremely long spermathecal duct was often found neatly and compactly folded back and forth beside the spermatheca.

There is some indication of characters grading from the southern, lower elevation part of the range to the central and northern part of the range and higher elevations. "Southern" specimens are from El Camarón (7), Córdoba (1), Playa Vicente (1), and "Mexico" (10 males, 4 females). They are characterized as follows: 55% clothed with iridescent green scales (compared to only 7% from elsewhere); last tergite of male with distinct longitudinal concavity (flattened or convex elsewhere); flagellum 10, 10.5 and 11 mm in the three dissected (9-10 mm in the four flagella measured of 10 others dissected); spermathecal duct 8.7 and 10 mm in the two dissected (8-12 mm, average

9.3 mm in the 10 others dissected). Males from the "center" of the range have the last tergite flattened medially; this includes the type locality of Apizaco in Tlaxcala, Chilapa in Guerrero, Puebla, Texcoco (1 of 2 males) and Ixtapan la Sal in Mexico, and Las Vigas in Veracruz. Males from the states of Hidalgo, San Luis Potosí, and some from Mexico have the last tergite evenly convex.

H. flagellatus is most closely related to opalinus which differs from flagellatus in the greatly reduced femoral flange, and consequently the absence of tubercles or granules on the area of the femoral flange, in the further reduction of the postocular vibrissae, in the thickened apical edge of the elytra, in the prothoracic setae not erect in either sex, and in the different genitalia. Apparently opalinus is restricted to the Sierra Madre Occidental, extending as far south as the Tropic of Cancer, whereas flagellatus occurs in the Sierra Madre Oriental south of 22° as in Map 2.

Many specimens of fulgens have labels identical to those of flagellatus. H. fulgens is most readily distinguished externally by its more strongly rounded sides of prothorax, lack of punctures on head and rostrum, denser scales, fine strial punctures, and longer, slightly curved fore tibia.

The name flagellatus refers to the extremely long flagellum of the males of this species.

The scintillans Group

5. dejeanii (Boheman)
6. amoenus n. sp.
7. scintillans (Champion)
8. micans (Champion)
9. brevicomus n. sp.

10. aureus (Blanchard)
11. cretatus (Champion)
12. rufipes (Champion)
13. superbus (Heller)

Characteristics of Group.-Scales scintillating, iridescent or neither, without a sharply defined color pattern, often with a pattern formed by size and arrangement of small glabrous areas. Without tubercles or pustules on prothorax or elytra but aureus, and to a lesser extent brevicomus, with a tendency for glabrous areas to be slightly elevated. Setae of elytra in some species much longer on alternate intervals. Fore tibia at least slightly curved and at least slightly bowed distally (except some scintillans). Endophallic structure a flagellum, but not known for micans and rufipes. Female without bursal sclerites; spermathecal duct proportionate in length to that of flagellum.

Remarks.-Of the other two flagellate species groups, the opalinus group differs from the scintillans group in the straight fore tibia and wide tegmen and the gemmifer group differs in the presence of a tubercle on the fore coxa on the inner distal edge.

5. Hadromeropsis (Hadromeropsis) dejeanii (Boheman)

Figs. 44, 71, 82, 83, 118; Map 1

Hadromerus dejeanii Boheman, 1840:293. Champion, 1911:183; Tab. 7,

Fig. 27. Type, male, labelled "34" mechanically printed on white square, "Vera Crux in Mexico, Chevr. 675" hand written on pale blue coated paper, "9" printed on white paper, "Typus" on red paper rectangle outlined in black, "dejeani [sic] Chevr." penciled on yellow paper, "163, 53" on salmon paper (Stockholm).

Diagnosis.-A short, stout species uniformly clothed with large, flat, scintillating green scales, or with smaller iridescent scales interspersed, or (1 female) with iridescent scales only. Epistoma long and narrow (Fig. 44), often connected to median line by glabrous line; occupying only 0.1-0.3 of anterior edge of rostrum. Setae of pronotum in both sexes appressed, as long as 1-1.5 scales. Fore femur stout, with strong distal flange on inner edge. Fore tibia (Fig. 71) often evenly arcuate, or bowed apically only. Apical edge of fore tibia of male with weak to obsolete lobe near inner side; in female lobe obsolete to absent. Aedeagal apodeme very long, 0.76 as long as aedeagus; flagellum 4.9-5.5 mm long.

Description.-Male, length 6.8-8.5 mm, width 2.6-3.1 mm. Female, length 8.0-9.6 mm, width 3.1-4.2 mm. Scales imbricate or discrete. Glabrous areas of pronotum rather evenly distributed, approximately the size of a scale, occasionally confluent, weakly convex; often with median line partly glabrous. Glabrous areas of elytra ranging in size from less than the diameter of a scale to the size of several scales; areas flat to weakly convex. Rostrum (Fig. 44) flat to concave, median line often narrowly glabrous and connected to apex of epistoma, thus emphasizing the length of the epistoma. Epistoma usually approximately 2x (range 1.3-2.0x) longer than wide; epistoma occupying 0.19-0.26 in male, 0.23-0.38 in female of anterior edge of rostrum; epistoma with 1 or more scales in 17% of males, 15% of females. Antennal club short, length 0.6-0.8 mm in male, 0.6-0.7 mm in female. Prothorax convex, sides strongly rounded, averaging 1.1x wider than long. Setae of pronotum as in Diagnosis. Postocular vibrissae moderate. Elytra across humeri 1.2x wider than prothorax. Elytra of male 2.5-2.7x, of female 2.8-3.2x longer than prothorax. Sides of elytra of male parallel, elytra of female wider apicad of middle. In both sexes apical umbone weak, distinct; apex truncate, without a tooth. Widest intervals of disc with 5 or 6 scales abreast. Setae of elytra in both sexes stiff, erect, as long as 2-3 scales, those of alternate intervals may be slightly longer than those of other intervals. Female with an average of 6 (extreme 12) long, erect setae on sutural interval at summit of declivity. Fore femur as in Diagnosis; in male 1.8-2.2x, in female 1.4-1.7x wider than hind femur. Fore tibia as in Diagnosis and Fig. 71, with some small, shiny pustules on posterior surface near teeth. Ventrite 5 of male 2.0-2.2x wider than long, moderately convex, apex broadly truncate-emarginate. Last tergite of male convex. Ventrite 5 of female 1.8-2.0x wider than long, sparsely squamose, almost flat. Male genitalia as in Figs. 82, 83; aedeagus 2.8-3.2 mm long, aedeagal apodeme 2.3-2.6 mm long, tegminal strut 2.0-2.2 mm long, flagellum within aedeagus 4.9-5.5 mm long. Spermatheca as in Fig. 118; spermathecal duct 5.9-6.0 mm long.

Distribution.-Map 1. MEXICO. Chiapas [see Remarks]. Nuevo Leon: Santiago. San Luis Potosí: 6 mi E Ciudad del Maiz, 4400 ft. Veracruz: Córdoba, El Camarón, Jalapa, Orizaba, Playa Vicente.

Specimens were collected in July and August. A series of specimens from Santiago, Nuevo Leon, was taken on pear trees.

Specimens examined: 31 males, 18 females. Specimens in Auckland, Berlin, Dresden, Eberswalde, Leiden, London, Paris, San Francisco, Washington, Howden, O'Brien.

Remarks.-Champion's description and illustration of dejeanii are accurate. H. dejeanii superficially resembles brevicomus in the characters of the prothoracic setae, body shape, etc., but differs in the more evenly distributed glabrous areas of the elytra (concentrated medially in brevicomus), scintillating scales (iridescent in brevicomus), and very different male and female genitalia. The aedeagal apodomes are longer in dejeanii than in any other species in the nominate subgenus.

Only a pair of specimens of dejeanii labelled "Chiapas, ex. Col. Salle" (Dresden) indicate that dejeanii possibly overlaps the range of scintillans and brevicomus. H. scintillans, although often similarly clothed with green scintillating scales, is not as closely related and can be readily distinguished by its elongate body, long dorsal setae, etc.

6. Hadromeropsis (Hadromeropsis) amoenus n. sp.

Figs. 84-86, 121, 134, 135; Map 3

Diagnosis.-Strongly resembling a Guatemalan scintillans but flagellum and spermathecal duct measuring approximately only one-third as long and occurring north of the Isthmus of Tehuantepec. Sympatric with dejeanii and similar to it in color but fore tibia usually less bowed, elytra longer, female with tooth on apex of sutural interval, genitalia different. Flagellum shorter than aedeagus, 1.3-2.0 mm long; proximal end with a heavy collar (Figs. 85, 86). Spermathecal duct 1.5-2.2 mm long.

Description.-Holotype, male, length 8.5 mm, width 2.7 mm. Elytra and sides and dorsum of prothorax clothed with large, green scintillating scales; remainder of body and legs clothed with pale, iridescent pinkish scales with occasional scintillating green scales. Scales contiguous or not, nowhere forming a vitta; glabrous areas smooth, polished black, irregular in shape on pronotum, glabrous areas of elytra of various sizes, sometimes as wide as intervals and continuous with adjacent glabrous areas, thus forming vague, short, wavy fasciae; glabrous areas smaller towards apex, on declivity the size of one or more scales. Rostrum approximately as wide as long, dorsal surface very slightly inclined towards median line; median line glabrous, very finely impressed between eyes only. Epistoma occupying 0.3 of anterior edge of rostrum, without scales, 1.4x longer than wide. Segment 2 of funicle 1.3x longer than segment 1; club 3.6x longer than wide, 0.8 mm long. Prothorax 1.08x wider than long, strongly convex, sides strongly rounded between constrictions, basal constriction complete, apical constriction obsolete on disc; postocular vibrissae well developed. Setae of disc of pronotum as long as 1.5-2 scales, setae of sides of prothorax as long as 2-3 scales, setae matted except a few basal setae erect. Scutellum covered with scales and a few fine, appressed setae. Elytra across humeri 1.2x wider than prothorax. Elytra 2.75x longer than prothorax. Elytra shaped as in scintillans. Strial punctures subfoveate basally, becoming smaller towards apex. Widest interval 4 scales abreast. Setae of elytra obviously worn on type; erect; in male paratypes some setae of alternate intervals longer, shorter setae as long as 2.5 scales, longest setae as long as up to 6 scales. Fore femur 1.9x wider than hind femur, distal flange well developed, its edge with a few acute pustules lower than the diameter of one scale. Fore tibia as in scintillans, distal tooth as long as 0.5 of width of apical edge. Ventrite 5 (Fig. 134) slightly convex, 2x wider than long, apex broadly truncate, shallowly emarginate. Last tergite convex. Aedeagus as in Fig. 84; apex cupped, not depressed ventrally; dorsally testaceous and flexible from orifice to apical 0.28 of total length of aedeagus. Aedeagus 2.4 mm long, aedeagal apodeme 1.2 mm long, tegminal strut 1.4 mm long, flagellum 1.4 mm long.

Allotype, female, length 10.1 mm, width 4.0 mm. Differs from type as follows. All scales scintillating green. Glabrous areas smaller, seldom the width of one elytral interval. Epistoma occupying 0.26 of anterior edge of rostrum, 2.1x longer than wide. Segment 2 of funicle 1.1x longer than segment 1; club 3.3x longer than wide, 0.7 mm long. Prothorax 1.1x wider than long, much less convex. Elytra across humeri 1.3x wider than prothorax. Elytra 3.3x longer than prothorax. Alternate intervals of elytra with a few very long setae basally and apically; most setae as long as 2 scales, the very long setae as long as 4 scales. Widest intervals 6 scales abreast. Sutural interval at summit of declivity with 6 or more long, erect setae. Apex of sutural interval with short, distinct tooth. Fore femur 1.3x wider than hind femur. Ventrite 5 (Fig. 135) as in scintillans female. Spermatheca as in Fig. 121; spermathecal duct broken in allotype, 1.5-2.2 mm in paratypes.

Type Series.-Holotype, MEXICO, Veracruz, Jalapa, 9/28-X/3/61, R & K Dreisbach (Howden). Allotype, same data as type (Howden). Paratypes, 13 males, 16 females. MEXICO. [Oaxaca?] : 1 male, Juquila, Flohr S. (Berlin). Puebla: 1 female, Crawford [syntype of Hadromerus micans Champion] (London); 1 male, 1 female, 6 mi. W Teziutlan, Aug. 18, 1958, R.B. Selander (Howden). Veracruz: 6 males, 6 females, same data as type, or VIII/1-6/61 [1 female] (Paris, Washington, Howden, O'Brien); 2 females, no additional data (San Francisco); 1 female, 10 km N Fortin, July 21-29, 1976, E. Giesbert Coll (Los Angeles); 1 male, Las Vigas, Hoege [opalinus Horn, det. Ch.] (London); 1 male, 1 female, 13 mi E Las Vigas, VI-29-1962, J.M. Campbell (Ottawa, Howden); 1 male, Orizaba, HHS & FDG, Dec. 1887 [syntype of Hadromerus scintillans Champion] (London); 1 female, Orizaba, Flohr S. (Berlin); 1 male, Orizaba, 25.VII.1978, G & M Wood (Howden); 1 female, Playa Vicente, Hoege [opalinus Horn, det. Champion] (London); 1 male, Playa Vicente, Hoege [syntype of Hadromerus scintillans Champion] (London). No data: 1 female, S. Mex. [opalinus Horn, det. Champion] (London); 1 female, Mex., F.C. Bowditch Coll. (Cambridge).

Remarks.-Males vary in length from 7.7-9.1 mm and in width from 2.8-3.1 mm. Females vary in length from 8.1-11.4 mm and in width from 3.1-4.1 mm. All but four specimens were completely clothed with scintillating green scales. One male from Las Vigas was entirely clothed with slightly convex iridescent pinkish scales but with

occasional flat scintillating scales, the iridescent scales of the
middle and hind femora more green than pink. In the other Las Vigas
male there are iridescent scales intermingled everywhere with the
scintillating green. In one female from Playa Vicente the scales are
about one-half scintillating green and most of the remainder are
slightly convex non-iridescent whitish. In the female from "S. Mex."
iridescent green scales outnumber the occasional opalescent scales.
In amoenus as in scintillans, a distal portion of the epistomal edges
are shiny and weakly carinate. Scales encroached on the epistoma in
two males and four females. The antennal club of the male ranges from
3-4x longer than wide. Glabrous spots on the prothorax were slightly
elevated in one male from Las Vigas and one male from Orizaba. Setae
seem to be particularly delicate in this species, but even in the
freshest specimens only occasional setae were erect on the pronotum.
In the largest females the widest intervals occasionally were 7 scales
abreast. The female elytral tooth ranges from scarcely perceptible to
as long as 4 scales. The fore tibia is definitely bowed in the male
from 13 mi E Las Vigas. Male genitalia varies as follows: aedeagus
2.1-2.5 mm long, aedeagal apodeme 1.0-1.3 mm long, tegminal strut
1.2-1.5 mm long, flagellum 1.3-2.0 mm long. Spermathecal duct is
1.5-2.2 mm long.

Vague, wavy, slightly elevated transverse glabrous lines were
present on the elytra in half the male paratypes and may be helpful in
tentative identification of males. I did not see such elytral
markings in any specimens of the sympatric dejeanii, but they are
distinct in brevicomus described here from Chiapas.

In addition to dejeanii, H. amoenus is also sympatric with
flagellatus and is readily separable from the latter by the convex
prothorax, narrower epistoma, bowed apex of fore tibia and characters
of the genitalia.

The locality is helpful, but examination of the male and female
genitalia is usually necessary for positive identification of amoenus
among other species of the scintillans group.

Particular care should be exercised regarding a locality of
"Jalapa" not associated with a state. Jalapa is the name of a
medicinal plant and towns of that name abound outside the known range
of amoenus.

This species is named amoenus, meaning "pleasant, delightful",
partly in reference to the physical attractiveness of the specimens
and partly in memory of the late R.R. and Kay Dreisbach, a delightful
couple who thoroughly enjoyed collecting insects and sharing their
catch with others.

7. Hadromeropsis (Hadromeropsis) scintillans (Champion)

Figs. 69, 87-89, 120, 136, 137; Map 3

Hadromerus scintillans Champion, 1911:182; Tab. 7, Figs. 25, 26.
    LECTOTYPE, HERE DESIGNATED, male, on card with female, labelled,
    "Quiche Mts., 7-9000 ft., Champion," "♂♀," "B.C.A., Col. IV,
    pt.3, Hadromerus scintillans, Ch." (London). See Type Material.
    Diagnosis.-Very variable, the range (Guatemala to Chiapas and

Tabasco in Mexico) and the characters of the genitalia often necessary for recognition. Setae (see caution in Methods) of entire body erect in male; in female, setae may or may not be erect on head and prothorax. Elytra with humeri prominent; setae of alternate intervals distinctly longer. Fore femur with flange moderately to well developed. Apical edge of fore tibia (Fig. 69) divided into two lobes subequal in length and often in width. Flagellum within aedeagus measures 4.1-6.0 mm long. Spermathecal duct measures 4.3-7.0 mm long.

Description.-Male, length 7.5-9.1 mm (Guatemala specimens), 6.8-9.3 mm (other localities); width 2.7-3.2 mm (Guatemala specimens), 2.5-3.1 mm (other localities). Female, length 8.9-10.0 (Guatemala specimens), 7.6-10.0 mm (other localities); width 3.2-3.8 mm (Guatemala specimens), 2.0-3.8 mm (other localities). Typical (Guatemala) specimens densely clothed with green scintillating scales, scales randomly discrete or overlapping. Specimens from other localities with scales as in Guatemala specimens or one of the four following patterns. Pattern 1, scales of elytral intervals 3, 8, and 9 more dense, forming a tone-on-tone vitta. Pattern 2, hind legs and often venter tan; scales slightly convex, whitish, tan or slightly pinkish, not at all scintillating or iridescent (Montebello, Teopisca), or with varying amounts of metallic reflections (Bochil, Las Margaritas, San Cristóbal, Teopisca). Pattern 3, bicolored with green and pinkish scales, e.g., dorsum green, venter pinkish or the reverse, or head and prothorax pinkish, elytra green; or elytral intervals 4 to 7 green and remainder pinkish (San Cristóbal). Pattern 4, venter and legs black, scales sparser exposing shiny black integument; all scales macroscopically silvery, rosy or tan, under magnification appearing brilliant, scintillating colors, especially red, green, and gold, each scale one color but all colors on each specimen; these specimens usually with vittae on elytral intervals 3, 8 and 9 and scales often very sparse between vittae (Montebello, San Cristóbal, San Felipe, Simojovel, Villahermosa). Densely squamose specimens with random glabrous spots on prothorax and elytra; glabrous spots largest on basal portion of elytra, becoming much smaller on declivity; females more densely squamose than males of all patterns. Rostrum slightly concave medially and median line usually narrowly glabrous (Guatemala specimens); rostrum flattened (patterns 1, 2, 3) or slightly concave (pattern 4) and median line occasionally glabrous; median line not impressed except between eyes. Epistoma occupying 0.25-0.3 of anterior edge of rostrum in male, 0.2-0.4 in female. Epistoma with 1 or more scales in 83% of specimens from Guatemala, 13% of other specimens. In Guatemala specimens, prothorax of male convex, sides strongly rounded, averaging 1.08x wider than long; prothorax of female slightly less convex dorsally, sides less rounded, averaging 1.14x wider than long. Prothorax of pattern 1, 2, and 3 specimens averaging 1.1x wider than long in male, 1.14x in female. Prothorax of pattern 4 specimens averaging 1.08x wider than long in male, 1.13x in female. In profile prothorax of Guatemala specimens with basal constriction distinct, apical constriction slightly less so; often with vague dorso-lateral depression; other specimens similar or not; glabrous area of disc flat to slightly pustulate. Median line often partially faintly impressed or glabrous or both. Setae of disc of prothorax erect in male; female with no setae on disc erect (Guatemala specimens) or erect on basal third (most specimens of patterns 1, 2, and 3), or all setae erect (specimens of pattern 4 and some others). Postocular vibrissae moderate to well developed. Scutellum covered with scales and a few appressed setae or (pattern 4) setae only. Elytra across humeri average 1.3x (range 1.2-1.4x) wider than prothorax in male, 1.4x (range 1.3-1.5x) in female. Elytra average 2.9x (range 2.7-3.1x) longer than prothorax in male, 3.4x (range 3.1-3.7x) in female. Humeri prominent; in male elytra widest across base, sides parallel or gradually tapered, in female elytra widest apicad of middle; apical umbone distinct in both sexes. Apex of elytra slightly longer than in dejeanii, that of female usually with a short tooth. Widest intervals of disc with 5 or 6 scales abreast (extreme, 8). Setae of alternate elytral intervals much longer on apical portion and occasionally longer on base. Female with an average of 8 long, erect setae on sutural interval at summit of declivity. In densely squamose specimens, strial punctures the size of a scale or smaller; in more sparsely squamose specimens, strial punctures may be foveate, especially near base of elytra. Fore femur of male averages 1.8x wider than hind femur (Guatemala specimens) or 2x (pattern 4) or 2.2x (others). Fore femur of female averages 1.4x wider than hind femur in Guatemala specimens, 1.5x in all others. Fore femur with distal flange moderately to well developed, weaker in female, its edge often with a few small acute pustules. Fore tibia (Fig. 69) weakly arcuate distally or not, apical edge as in Diagnosis, distal tooth acute, usually longer than teeth of inner edge. Ventrite 5 of male (Fig. 136 of Quiché specimen) moderately convex, averaging 2x wider than long, apex broadly, shallowly emarginate. Last tergite of male convex. Ventrite 5 of female averaging 1.7x wider than long (Fig. 137 of Guatemala specimen) (Guatemala and patterns 1, 2, and 3), 1.5x in pattern 4 specimens; ventrite very slightly convex, squamose as remainder of abdomen. Male genitalia as in Figs. 87-89. In Guatemala and pattern 1, 2, and 3 specimens, aedeagus 2.4-2.5 mm long, aedeagal apodeme 1.6-1.9 mm long, tegminal strut 1.3-1.7 mm long, flagellum 4.1-4.8 mm long; in pattern 4 specimens, aedeagus 2.3-3.2 mm long, aedeagal apodeme 1.7-2.0 mm long, tegminal strut 1.7-2.0 mm long, flagellum 4.7-6.0 mm long. Spermatheca as in Fig. 120. In Guatemala and patterns 1, 2, and 3 specimens, spermathecal duct 4.0-5.5 mm long, in pattern 4 specimens, 5.5-7.0 mm long.

Type Material.-None of the specimens in the type series bears the usual "Sp. figured" label. There is a "Type" label on the pin of one card bearing a male and female and labels like the lectotype, but this "Type" label is thought to have been affixed by Arrow or someone other than Champion (R. Thompson, in litt.).

PARALECTOTYPES, HERE DESIGNATED, 7 males, 14 females. GUATEMALA. 2 males, 6 females, no additional data (London); 3 males, 5 females, same data as type (London, Washington); 1 male, 2 females, Tepan (London, Howden). MEXICO. 1 female (London). Locality in doubt, 1 male, Costa Rica, "locality doubtful" in Champion's handwriting (London).

Two of Champion's syntypes are actually amoenus. These are 1 male labelled "Orizaba, H.H.S. & F.D.G. Dec. 1887" (London), and 1 male, labelled "Playa Vicente, Mexico, Höge" (London). One male syntype with no data is dejeanii.

Distribution.-Map 3. GUATEMALA. Chimaltenango: Tecpán (="Tepan"), Zaragoza. El Quiché: Quiché Mts. Huehuetenango: 25 mi S Huehuetenango. MEXICO. Chiapas: 8 mi N Bochil, Comitán, 12 mi NW Comitán, 14 mi W Comitán, Dolores, Las Margaritas, 8 mi W P N Montebello, 10 mi W P N Montebello, Rancho Nuevo (8.6 mi E San Cristóbal), San Cristóbal de las Casas, 5 mi SE San Cristóbal, 35 mi E San Cristóbal, San Felipe, 2 mi S Simojovel, Teopisca, 2 mi SE Teopisca, 8 mi SE Teopisca, 9 mi SE Teopisca, 28 mi E Tuxtla Gutiérrez. Tabasco: Villahermosa.

In addition to the "Costa Rica" paralectotype, a locality which Champion questioned, there are two specimens in Berlin labelled "Costa Rica, Wagner" and one specimen in Stockholm labelled "Costa Rica, D. Geminger". I suspect these refer to something other than the country now known as Costa Rica.

Specimens were collected in May (1), June, July, August (1), September (4), and December (1).

Specimens examined: 81 males, 63 females. Specimens in: Berkeley, Berlin, Cambridge, Dresden, Eberswalde, Leiden, London, New York, Ottawa, Paris, Stockholm, Washington, Howden, O'Brien.

Remarks.-The figure in the color plate in Champion (1911: Tab. 7, Fig. 25) shows the elytra too short.

The black legged specimens (pattern 4) are primarily from the lower elevations and are often distinguishable on morphological characters. However, no characters appear to be exclusive to one area and the mingling of characters is such that only field work could justify a different treatment.

At 9 mi SE Teopisca H. Howden collected scintillans on Acacia, an unidentified legume, and on oak. At 8 mi W P. N. Montebello, O'Brien and associates collected five scintillans on Juniperus mexicana; at 12 mi S Comitán they took a series of scintillans on Quercus.

8. Hadromeropsis (Hadromeropsis) micans Champion

Figs. 70, 119, 139; Map 3

Hadromerus micans Champion, 1911:183; Tab. 7, Figs. 28, 28a.
LECTOTYPE, HERE DESIGNATED, female, labelled, "Purula, Vera Paz, Champion," "♀," "Type" on orange-circled disc, "Sp. figured," "B.C.A., Col., IV. pt. 3, Hadromerus micans, Ch.," "Lectotype" on purple-circled disc (London). See Type Material.

Diagnosis.-Known only from the female lectotype. Very large and robust. Resembling Guatemalan scintillans but apex of ventrite 5 pointed, spermatheca (Fig. 119) with ramus not at all produced, spermathecal duct 7.0 mm long.

Description.-Unique female, length 13.4 mm, width 5.4 mm. Habitus as in Champion (1911, Tab.7, Fig. 28). All scales scintillating green, dense but overlapping only around eye and on specific areas of ventral surface such as coxae. Glabrous areas as in Champion (1911:183 and Tab. 7, Fig. 28). Rostrum with dorso-lateral edges slightly divergent apically where rostrum is approximately as wide as long. Median line glabrous, impressed between eyes, thence obsoletely elevated to apical 0.6 at a vague transverse elevation; epistoma reaching almost to this transverse mark. Epistoma occupying 0.3 of anterior edge of rostrum; epistoma 1.4x longer than wide, its distal emargination extending 0.3 of length of epistoma; 2 scales on epistoma. Segment 2 of funicle 1.2-1.4x longer than segment 1, club 3.2x longer than wide. Prothorax 1.15x wider than long, sides only moderately rounded, slightly flattened dorso-laterally; in profile scarcely convex with basal and apical constrictions faint. Setae of prothorax very fine, erect basally only, as long as 2-3 scales. Postocular vibrissae well developed. Scutellum with a few scales and a few setae. Elytra across humeri 1.5x wider than prothorax. Elytra 3.3x longer than prothorax. Humeri very prominent, interval 5 at base depressed thus further emphasizing humeri. Elytra with post-scutellar area flattened. Tooth at apex of sutural interval approximately as long as 3 scales. Widest interval with extreme of 9 scales abreast, usually 5-7 scales abreast. Setae of alternate intervals longer basally and apically, the longest as long as 9 scales, shorter setae as long as 2 or more scales. Sutural interval on declivity with 9-12 very long erect setae. Strial punctures smaller than a scale. Fore femur 1.36x wider than hind femur, distal flange weak. Fore tibia as in Fig. 70, distal tooth acute, 0.5 as long as width of apical edge. Ventrite 5 (Fig. 139) convex as in scintillans, evenly covered with scales as remainder of abdomen; 1.7x longer than wide, sides convergent from base to apex in a blunt point. Spermatheca as in Fig. 119; spermathecal duct 7 mm long.

Type Material.-The syntype from Puebla is described as a paratype of amoenus in this paper; its spermathecal duct is only 1.8 mm long.

Distribution.-Map 3. GUATEMALA. Baja Verapaz: 1 female, Purulá (London).

Remarks.-This is possibly only an extremely large and robust specimen of scintillans, but the different spermatheca seems to be sufficient justification to let it stand until field observations or more material can elucidate the situation.

The large size and somewhat similar spermatheca bring to mind cretatus, but in that species the base of the elytra is distinctly arcuately emarginate instead of straight.

9. Hadromeropsis (Hadromeropsis) brevicomus n. sp.

Figs. 90, 91, 122, 123; Map 2

Diagnosis.-Densely clothed with white (1 female) or iridescent green scales, dorsum usually with a slight coppery or golden lustre in reflected light; glabrous areas of elytra often as wide as one interval, these areas larger on intervals 1, 2, and 3 where they often coalesce to form short, vague, sinuous fasciae. Setae of prothorax in both sexes appressed, very short, as long as one scale on center of disc, on sides maximum length of 2 scales. Endophallic structure (Figs. 90, 91) less than half the length of the aedeagus, bent and very stout proximally, with short flagellum. Spermathecal duct very short, 0.9-1.4 mm long.

Description.-Holotype, male, length 9.0 mm, width 3.2 mm. Scales and glabrous areas as in Diagnosis. Scales especially dense and overlapping on side of rostrum and head, side of prothorax, side of fore coxa, side of meso- and metasternum. Dorsum of head, rostrum and prothorax with glabrous areas irregular in size and placement, as large as 1 to many scales. Glabrous areas of elytra forming 5 irregular sinuous lines, the longest reaching stria 4, the glabrous areas polished and slightly elevated. Rostrum 1.12x longer than wide; dorso-lateral edges distinct, parallel, glabrous. Median line glabrous entire length, impressed between eyes. Dorsal surface of rostrum with faint transverse interantennal elevation at about middle, very slightly concave caudad of elevation; basal glabrous surface punctate, slightly rugose. Epistoma occupying 0.27 of anterior edge of rostrum, 1.3x longer than wide; without scales. Funicle with segment 2 scarcely longer than segment 1; antennal club 3.8-4x longer than wide. Prothorax 1.12x wider than long, convex; sides rounded, much wider than constrictions. In profile pronotum strongly convex basally, less so anteriorly, anterior constriction distinct though weak. Setae of prothorax as in Diagnosis; on disc glabrous base of setae with a faint crescent-shaped elevation, elevation less pronounced on sides. Postocular vibrissae well developed. Elytra across humeri 1.2x wider than prothorax. Elytra 2.6x longer than prothorax. Base of elytra slightly arcuately emarginate between striae 5. Sides of elytra parallel to approximately apical third, thence slightly rounded, apical umbone distinct in outline, apex elongate-rounded, without a tooth. Elytra in profile thick as in dejeanii, but declivity more gradual. Widest interval with 7 scales abreast, most intervals 5-6 scales abreast. Strial punctures moderate; those of sutural interval and interval 2 basally foveate, gradually becoming smaller apicad. Elytral setae of two types; short curved setae on disc as long as 1 scale, becoming longer, straighter on declivity and on sides; alternate intervals, including interval 7, with much longer, straight, erect setae in addition; sutural interval with intermediate length, slightly arcuate setae. Fore femur 1.9x wider than hind femur, strongly swollen, with moderate arcuate flange on distal edge. Fore tibia weakly bowed, apical edge with distinct small lobe near inner edge; distal tooth as long as 0.5 of apical width of tibia; ventral surface with some indistinct sculpturing and near inner edge with very small, shiny tubercles. Metasternum protuberant compared to many species, this especially evident in the broadly concave posterior median pit and the abrupt ledge adjacent to the hind coxa. Ventrite 5 convex, 1.9x wider than long, apex emarginate. Last tergite convex but with high point slightly flattened. Genitalia as in Diagnosis and Figs. 90, 91. Aedeagus 2.4 mm long, aedeagal apodeme 1.1 mm long, tegminal strut 1.1 mm long, endophallic structure 1.0 mm long.

Allotype, female, length 9.3 mm, width 3.7 mm. Differs from type as follows. Scales sparser on mesepisternum and metepimeron (but this not a sexual expression in paratypes). Rostrum more robust, dorso-lateral edges weakly bowed outward. Epistoma 1.4x longer than wide. Prothorax less convex, apical constriction not marked on disc; 1.09x wider than long. All prothoracic setae as long as 1 scale except at extreme anterior and basal edges on sides. Elytra across humeri 1.36x wider than prothorax. Elytra 3x longer than prothorax. Sides of elytra slightly divergent to apical third thence gradually rounded, forming apical umbone, apex triangular with arcuate sides; sutural interval with very weak distal tooth. Sutural interval at summit of declivity with 5 longer setae of graduated lengths. Ventrite 5 slightly convex apically, 1.8x wider than long, apex narrowly rounded; medially with short setae, few scales. Spermatheca (Fig. 122) 0.8 mm long, spermathecal duct 0.9 mm long.

Type Series.-Holotype, MEXICO, Chiapas, 7 mi SW Ocozocoautla, 2500', Aug 1 1974, O'Brien & Marshall, on Acacia pennatula (O'Brien). Allotype, same data as type (O'Brien). Paratypes, 5 males, 2 females. MEXICO. Chiapas: 3 males, 1 female, same data as type (Howden, O'Brien); 2 males, El Sumidero, 14 Sept. 1974, G. Bohart, W. Hanson (O'Brien); 1 female, 14 km N. Tuxtla Gutiérrez [Sumidero, approx. 4000'], VII.14.1962, J. M. Campbell (Howden).

Remarks.-Males vary in length from 7.9-9.5 mm and in width from 2.7-3.4 mm. The female paratypes are 9.0-9.5 mm in length and 3.8 mm in width. The female from the type locality is clothed with white scales only, otherwise paratypes exhibit little variation. The green female paratype and one male from El Sumidero are more densely and more evenly squamose, the elytral glabrous areas rarely as wide as an interval. No specimens have scales on the epistoma. The antennal club of males is 3.0-3.8x longer than wide, of females 3.1-4.0x longer than wide. In dejeanii the antennal club of males is 2.7-4.0x longer than wide, in females 2.2-3.0x longer than wide.

Five males and three females were dissected. Variation in the male genitalia is as follows: aedeagus 2.3-2.4 mm long, aedeagal apodeme 1.0-1.3 mm long, tegminal strut 1.1-1.3 mm long, endophallic structure (3 specimens) 0.9-1.0 mm long. The spermatheca of the paratype from Tuxtla Gutiérrez is shown in Fig. 123; the duct in this specimen is 1.3 mm long and in the white female is 1.4 mm long.

H. brevicomus appears to be most closely related to aureus. With aureus, brevicomus shares the elevated, often contiguous glabrous areas of the elytra (but in aureus they are evenly distributed dorsally and laterally), elevated bases of prothoracic setae (but in aureus they are much more tuberculate), only 5 long erect setae on sutural interval on declivity of female (but these longer and straighter in aureus), and similar male genitalia (but in aureus the endophallic structure is straight proximally). H. aureus is known only from Bolivia. D'Orbigny recorded aureus from "les mimosas" (see Type Material of aureus); brevicomus was collected on Acacia. Both sexes of aureus have a distinct tooth at the apex of the sutural interval; in brevicomus there is no trace of a tooth in the male and only a weak tooth in the female.

The name "brevicomus" means "short setae", one of the most distinctive external features of this species.

10. Hadromeropsis (Hadromeropsis) aureus (Blanchard)

Figs. 49, 51, 53, 92-94, 124; Map 4

Hadromerus aureus Blanchard, 1846:201. Type, female, labelled "7247, 34" on underside of green disc; "2248"; "Pl.16 Fig.7" on vertical label on pin; "Museum Paris, De Chiquitos, A. Mojos, D'orbigny

1834"; and on green label in box "H. aureus Blanch. Bolivia, M.J. 'Orbigny" (Paris). See Type Material.

Diagnosis.-Clothed with scales which are yellowish with a metallic lustre in teneral specimens (as the type), becoming white without any metallic lustre in mature specimens; prothorax (Fig. 49) with random glabrous spots ranging from elevated to pustulate; elytra (Fig. 51) with larger glabrous areas, each area usually the entire width of the interval, slightly convex, some areas confluent forming slender sinuous subfasciae. Apex of elytra of male slightly produced into a pair of blunt, divergent teeth (Fig. 53).

Description.-Male, length 8.9-9.2 mm, width 3.2-3.5 mm. Female, length 8.7-10.8 mm, width 3.5-4.3 mm. Color as in Diagnosis. Scales especially dense around eye, in depressed areas of pronotum and on elytra, often imbricate on the latter. Metepisternum sparsely scaled and with a row of long, fine, white prostrate setae, the tip of one overlapping the base of the succeeding one by 0.3-0.5. Setae of head, rostrum, and prothorax appressed, as long as 1 scale. Rostrum 1.2x longer than wide, dorso-lateral edges slightly elevated, parallel or weakly convergent caudally. Epistoma occupying less than 0.3 of anterior edge of rostrum, its apex narrow, rounded or acute. Interantennal line faintly convex at most, rostrum feebly concave caudad of interantennal line. Funicle with segment 2 averaging 1.3x longer than segment 1. Prothorax (Fig. 49) 1.1x wider than long, almost as wide at apical constriction as at basal constriction; disc vaguely flattened or depressed either side of midline, these areas densely squamose, remainder of disc with random bare spots as in Diagnosis, occasional spots confluent. In profile, disc feebly convex or flattened and apical constriction weakly to not at all impressed. Elytra across humeri 1.2x wider than prothorax in male, 1.3-1.4x wider in female; elytra 3.0x longer than prothorax in male, 3.2x longer in female. Sides of elytra parallel basally, thence subparallel, very weakly rounded to the weak apical umbone; apices of elytra of male (Figs. 51, 53) divergent, sutural interval produced in a brief, blunt tooth; apices of female produced in a pair of longer, parallel teeth. Base of elytra straight. Declivity of male oblique, summit gradual; declivity of female more abrupt, suture at summit with row of 4 or 5 much longer setae. Elytra with elevated glabrous areas as in Diagnosis; each glabrous area with a long, fine seta near its caudal edge, the setae of the apical half of elytral intervals 1, 3, and 5, and the declivity longer and more erect than the setae of the basal half. Striae distinct, punctures smaller than a single scale and separated by width of 1-3 scales. Fore coxa without a tubercle but in 2 of 4 males examined with a feeble thickening at inner distal margin. Fore femur of male 2-2.3x wider than hind femur; distally with a moderate, thick flange on inner edge and a few weak tubercles below tibial groove. Fore femur of female 1.5-1.7x wider than hind femur, flange much less developed, tubercles absent. Fore tibia bowed distally; distal tooth as long as 0.6-0.8 of apical width of tibia, with a row of 8-10 small, blunt denticles on inner edge, these obscured by long setae, especially in males. Ventrite 5 of male 2.3x wider than long, rather evenly, sparsely squamose; apex broadly emarginate. Ventrite 5 of female 1.8-2.0x wider at base than long; triangular with apex briefly rounded; almost flat. Male genitalia as in Figs. 92-94; aedeagus 2.4-2.6 mm long; aedeagal apodeme 1.1 mm long (short in relation to length of aedeagus); tegminal strut 1.3-1.4 mm long; flagellum 1.2-1.3 mm long, approximately 0.5 the length of the aedeagus, stiff, heavily sclerotized, gradually tapering from base to apex. Female with spermatheca as in Fig. 124, 0.8-0.9 mm long; spermathecal duct approximately 1.0 mm long, moderately sclerotized at junction with bursa copulatrix.

Type Material.-Apparently described from the unique type in Paris, a teneral female in excellent condition. The numbers on the labels of the type refer to the manuscript catalogue of the d'Orbigny collection: "Chiquitos, sur les mimosas à Santa Ana (Compicon?) 1834" (H. Perrin, in litt.).

Distribution.-Map 4. BOLIVIA. La Paz: Chulumani, Coroico, Nigrillani. Santa Cruz: Santa Ana in Chiquitos Prov.

Specimens were collected in January and August.

Specimens examined: 5 males, 9 females. Specimens in Auckland, Dresden, London, Paris, Rosario de Lerma, Washington, Howden.

Remarks.-One female has some scales very pale green, especially on intervals 6, 7, and 8, as if a vague remnant of an elytral vitta, and indicating the possibility of all-green or green maculate specimens.

In addition to the diagnostic characters, aureus differs from meridianus and gemmifer in the longer segment 2 of the funicle, more strongly developed femoral flange, and mostly imbricate scales of elytra. From cretatus, another all white species, aureus differs in the prothoracic sculpture, strial punctures larger, setae of alternate intervals longer, base of elytra straight, and apex of elytra produced. See Remarks on brevicomus for additional comparisons.

The name aureus is appropriate for the type specimen which is teneral and consequently golden, but "argenteus" which appears on the Dresden specimens would better describe mature specimens.

11. Hadromeropsis (Hadromeropsis) cretatus (Champion)

Figs. 50, 52, 54, 95-97, 125; Map 3

Hadromerus cretatus Champion, 1911:184, Tab. 7, Fig. 31. Type, female, labelled "Caldera, 1200 ft, Champion" (London). See Type Material.

Diagnosis.-Densely clothed with imbricate, pure white scales, with small glabrous spots randomly scattered over entire dorsum (Fig. 52); the glabrous spots not elevated above scales and nowhere with tubercles. Apices of elytra of male individually briefly rounded (Fig. 52); apices of female with a sharp tooth concealed by long setae (Fig. 54). Fore femur with very weak flange on inner edge distally. Flagellum approximately twice as long as aedeagus, evenly sclerotized, tapered to apex. Spermathecal duct 6.0-6.5 mm long.

Description.-Male, length 10.1-11.2 mm, width 3.9-4.4 mm. Female, length 10.8-12.5 mm, width 4.6-5.0 mm. Color as in Diagnosis. Rostrum broad, 1.1x or less longer than wide; obsoletely concave (male) or flat (especially female); median line only obsoletely impressed in one male, otherwise without sculpture. Dorso-lateral edges of rostrum abrupt, prominent, convergent caudally; in dorsal view sides of rostrum including scrobe clearly visible. Epistoma long, narrow, its apex vague, rounded; occupying less than 0.3 of anterior edge of rostrum. Funicle with segment 2 averaging 1.4x longer than segment 1. Prothorax in dorsal view 1.0-1.3x wider than long; sides strongly, evenly rounded between constrictions; squamose as in Diagnosis. In profile (Fig. 50) disc weakly convex, apical constriction absent dorsally. Elytra across humeri 1.3x wider than prothorax in male, 1.4x wider in female. Elytra 2.9x longer than prothorax in male, 3.2x longer in female; sides parallel basally, thence subparallel, very weakly rounded to the weak apical umbone; apices of elytra as in Diagnosis. Base of elytra arcuately emarginate between striae 5. Declivity very gradual; in female sutural interval at summit of declivity with a row of 4-14 setae approximately 3x as long as adjacent setae. Glabrous spots of elytra each with a single seta not conspicuously longer or straighter on alternate intervals. Strial punctures (Fig. 52) very small and inconspicuous; in abraded areas a puncture seen to be smaller than the scar of a scale and very shallow; punctures separated by 2-5 scales. Fore femur 2.1x wider than hind femur in male, 1.6-1.8x wider in female; with very feeble flange on inner edge distally. Fore tibia weakly bowed, similar to that of rufipes (Fig. 72), distal tooth as long as width of apex in male, slightly shorter in female; with a row of small teeth on inner edge, with only a few minute tubercles on posterior surface distally. Ventrite 5 of male 2-2.2x wider at base than long, rather evenly, sparsely squamose; apex broadly truncate. Last tergite of male slightly transversely flattened. Ventrite 5 of female 1.8-2.0x wider at base than long, apex narrowly rounded; feebly convex. Male genitalia as in Figs. 95-97 and Diagnosis; aedeagus 3.5-3.6 mm long; aedeagal apodeme 1.9 mm long; tegminal strut 1.9-2.3 mm long; flagellum 5.4-6.5 mm long. Female with spermatheca as in Fig. 125; spermathecal duct 6.0-6.5 mm long.

Type Material.-Described from a unique with the head slightly deformed or injured in life over left eye.

Distribution.-Map 3. COSTA RICA. Alajuela: 1 male, Grecia, 14 May 1977, R. Campos (Howden). San José: 1 male, San José, Feb. 28, 1924, Nevermann Coll. (Washington); 1 male, Monte Rey, Río Candelaria, 1946 (San José). No locality given: 1 male, 3 females, Septiembre, Paul Serre (Paris, Howden); 1 male, 1 female (Cambridge). PANAMA. Chiriquí: 1 female, Caldera, 1200 ft., Champion (the type) (London).

Remarks.-Champion's description, "the rostrum canaliculate and moderately excavate" might appear to contradict my description. His "canaliculate" means the median line is finely impressed and his "excavate" refers to the caudal margin of the epistoma. Except in two males, the median line is not modified in any way; it is usually completely covered with scales as well, though the scales may not always be contiguous at the median line. Another point of confusion is found in the "flattened, tuberculiform" glabrous spots of the elytra; these spots are indeed flattened; in none of the specimens examined did I see anything I would call tuberculiform.

On the abdomen and fore coxae of one female there are individual scattered iridescent green scales which are smaller than the white scales.

The very distinctive habitus should make this species immediately recognizable. Within Central America, the only other white species is gemmifer which has the fore coxa with a tubercle, fore tibia tuberculate on posterior surface, female ventrite 5 notched apically and genitalia very different.

12. Hadromeropsis (Hadromeropsis) rufipes (Champion)

Figs. 62, 72, 138, 367; Map 2

Hadromerus rufipes Champion, 1911:185; Tab. 7, Figs. 34, 34a. Type, female, labelled "Costa Rica, P. Biolley," "Arcangeles, 1500-1700 m.," "♀," "Arcangeles 1200 m [in pencil] 1192 [in ink]" on folded piece of paper, "Type" on orange-ringed disc, "Sp. figured.," "B.C.A., Col., IV. pt. 3, Hadromerus rufipes, Ch." (London).

Diagnosis.-Based on female only. Very large. Apparently normally covered with two sizes of scintillating green and some opalescent scales, the larger scales forming a vitta on elytral interval 4 as depicted by Champion (1911, Tab. 7, Fig. 34) as well as a vitta on interval 8; both elytral vittae may be continuous with narrow vittae on prothorax. Ventrite 5 (Fig. 138) smooth and polished medially, rounded apically. Spermathecal duct 2.2 mm long.

Description.-Female, length 13.0-13.4 mm, width 5.0-5.2 mm. Black, legs and antennae rufous. Normally covered with scales as in Diagnosis; ventrally scales dense on sides of abdomen, mostly absent on central half of ventrites 1 and 2, sparser and smaller on ventrites 3 and 4 centrally, sparse basally on ventrite 5, apical half of ventrite 5 with only 1 scale; legs very sparsely clothed with small scales. Rostrum with dorso-lateral edges basally indistinct, rounded; edges apicad of interantennal line well defined, approximately parallel. Surface of rostrum almost flat, median line finely impressed between eyes; distal 0.3 of rostrum deflected, here very slightly wider than long. Surface of head and rostrum with scattered fine punctures which may be concealed by scales in fresh specimens. Epistoma occupying 0.3 of anterior edge of rostrum, approximately as wide as long, its distal emargination extending 0.3 of length of epistoma. Segment 2 of funicle 1.1x longer than segment 1; club 2.6-2.8x longer than wide. Prothorax 1.2x wider than long, sides only moderately rounded, slightly flattened dorso-laterally, large scales thus protected from abrasion by the depression consequently forming a vitta; in profile prothorax scarcely convex, apical constriction absent, basal constriction very faint. Disc of prothorax medially apparently naturally glabrous, smooth and polished. Setae of prothorax very fine, as long as 1-2 scales, appressed(?). Scutellum mostly glabrous with some small scales and setae. Elytra across humeri 1.4-1.6x wider than prothorax. Elytra 3.8x longer than prothorax. Base of elytra weakly produced or not between striae 5, thence slightly obliquely deflected. Sides of elytra parallel for basal 0.15, thence 1.1x wider than across humeri and parallel to middle, thence very gradually convergent to apex, the apical umbone scarcely to distinctly evident in outline. Elytral sutures briefly divergent on declivity below erect setae, ending in weak to prominent pair of diagonally directed teeth whose apices overlap (Fig. 62).

Edge of elytra opposite ventrite 5 thickened and prominent, interval 10 depressed in this area and striae 9 and 10 here indistinct but ending apically in several foveae which further emphasize the distal tooth. Strial punctures very fine, smaller than a small scale. Setae of elytra difficult to assess because of abrasion; apparently setae of basal portion of intervals 3, 5, and 7 longer, as long as 2-3 of the larger scales; all setae very fine and erect. Fore femur 1.3x wider than hind femur, distal flange arcuate but weak. Fore tibia as in Fig. 72 or apical edge with inner lobe as in micans (Fig. 70). Ventrite 5 (Fig. 138) 1.8x wider than long, flat or feebly convex, medially smooth and polished; rufous apex punctate. Spermatheca as in Fig. 367; spermathecal duct 2.2 mm long (type not dissected).

Distribution.-Map 2. COSTA RICA. 1 female, "Arcangeles", 1200-1700 m (London). 1 female, [San José] Escazú, Coll. Kuschel (Auckland).

Extensive questioning in Costa Rica and elsewhere was not successful in locating "Arcangeles", but it was frequently suggested that it is the name of a farm or estate.

Remarks.-Both specimens are abraded and the glabrous areas of the elytra are covered with scars of small scales. The elytral vittae should still be evident in fresher specimens because of the different nature of the scales in the vittae - much larger and overlapping compared to the scales of adjacent intervals.

H. cretatus also occurs in Costa Rica and resembles rufipes in the very fine elytral punctures and occasionally bears scintillating green scales on the ventral surface. Compared to cretatus, rufipes is elongate, has a shorter second funicular segment, more attenuate elytral apex, and much shorter spermathecal duct.

From gemmifer, another large species possibly occurring in Costa Rica, rufipes differs in the absolutely smooth, non-tuberculate surface of the prothorax and elytra and ventrite 5 rounded apically instead of emarginate. The coxal tubercle is absent in rufipes but the other characters are often more easily viewed.

13. Hadromeropsis (Hadromeropsis) superbus (Heller)

Figs. 2, 3, 98-100, 126; Map 4

Hadromerus superbus Heller, 1921:29. LECTOTYPE, HERE DESIGNATED, male, labelled "Rep. ARGENTINA, Prov. Tucuman, II 1906, C. Bruch," "K.M. Heller," "Typus" on green rectangle with black border, "Hadromerus superbus Heller i.l.," "Hadromerus superbus m. i. l." handwritten (Bruch Coll., Buenos Aires). See Type Material.

Diagnosis.-Clothed with iridescent green scales only or with both green and lavender iridescent scales; prothorax with random glabrous areas not elevated; elytra with random glabrous areas rectangular, usually the entire width of an interval, scarcely elevated. Eye flattened (Fig. 2). Prothorax (Fig. 2) much wider than long, sides very strongly rounded.

Description.-Male, length 7.2-8.1 mm, width 2.8-3.0 mm. Female, length 7.4-9.6 mm, width 3.1-3.8 mm. Vestiture as in Diagnosis. Integument and scales very shiny, sometimes with golden lustre. Lavender scales when present randomly interspersed with green scales on head and prothorax; on elytra lavender scales concentrated on disc and intervals 9 and 10 leaving intervals 7 and 8 entirely green; often with a vague green pattern basally and medially. Rostrum with dorso-lateral edges abrupt from apex to middle of eye, parallel or slightly bowed outward; sides perpendicular or concave between edge of dorsum and scrobe and thus not visible from above; sides convex below ventral margin of scrobe. Rostrum 1.2x longer than wide; dorsal surface shallowly concave; frons with a small interocular fovea; median line glabrous. Epistoma occupying 0.3 of anterior edge of rostrum. Funicle with segments 1 and 2 usually equal in length. Eye as in Diagnosis and Fig. 2. Prothorax as in Diagnosis and Figs. 2, 3, averaging 1.2x wider than long, convex, the apical constriction weak dorsally. Postocular vibrissae arising from a small knob. Elytra across humeri averaging 1.2x wider than prothorax in male, 1.3x wider in female. Elytra of male and female averaging 3.0x longer than prothorax. Shape of elytra as in Figs. 2, 3. Apices of elytra of male conjointly rounded, with a minute tooth in some; in female with small distinct tooth. Elytra with a single, fine seta arising from each glabrous spot, the setae of alternate intervals erect on apical half of elytra and much longer; setae much more numerous on sutural interval on declivity, but not clustered there, all setae slightly longer in female than in male. Strial punctures approximately as large as a scale, separated by approximately one diameter. Fore femur of male 2.0x wider than hind femur, fore femur of female 1.4-1.6x wider than hind femur; inner edge both sexes with a brief arcuate flange sometimes separated from femur by a crease, edge of flange with a few minute tubercles, a few tubercles of same size below tibial groove. Fore tibia bowed distally; distal tooth approximately 0.5 as long as width of apex in male, shorter in female; inner edge with an average of 10 small blunt denticles. Ventrite 5 of male 2.1x wider than long, apex truncate, edges deflected. Ventrite 5 of female 1.7-1.8x wider at base than long, flattened, side edges deflected, apex narrowly rounded. Male genitalia as in Figs. 98-100; aedeagus 1.5-2.1 mm long, aedeagal apodeme 0.9 mm long, tegminal strut 0.9-1.2 mm long, flagellum 1.6 mm long. Spermatheca as in Fig. 126; spermathecal duct 1.0-1.3 mm long.

Type Material.-PARALECTOTYPES, HERE DESIGNATED, 6 females: 1 female, labelled "Rep ARGENTINA, Prov. Tucuman, I 1906, C. Bruch" (Buenos Aires); 1 female, labelled "Rep. ARGENTINA, Prov. Catamarca, 7.III.1907, C. Bruch," "Typus" on green rectangle, "Hadromerus superbus Heller" (Buenos Aires); 1 female, labelled "Rep. ARGENTINA, Prov. Tucuman, 190 , C. Bruch," "1906, II," "superbus TYPUS" on red rectangle (Dresden); 2 females, labelled "Argentina, Prov. Salta, mir 1905, Steinbach" (Dresden); 1 female, labelled "Rep. ARGENTINA, Prov. Cordoba, 189, C. Bruch", "Cotypus" (Paris). Heller does not list the province of Catamarca but the specimen from there is considered a paralectotype because it bears a type label.

Distribution.-Map 4. ARGENTINA. Catamarca. Córdoba: Alta Gracia, La Granja, Quiscate. Jujuy: Ledesma. Salta: Cerrillos, Cerro San Bernardo, La Merced, Salta, Viñaco (15 km S. El Carril). Santiago del Estero. Tucumán.

Specimens were collected from December through March.

Specimens examined: 11 males, 72 females. Specimens in Auckland, Berlin, Buenos Aires, Curitiba, Dresden, Paris, Washington, Howden.

Remarks.-This is a very distinctive species, readily recognized by the flattened eye alone. The green and lavender vestiture is likewise unique but six specimens appeared oily and no color was present at all, the specimens being a uniform dark brown.

Specimens which I collected were very lively and appeared quite colonial. The five specimens taken at Viñaco were on one branch of Mimosa farinosa Grisebach and landed on the beating sheet simultaneously. The 17 specimens from Cerrillos were taken on four different days, all from one very large Prosopis chilensis (Md.) Stuntz, mostly on one branch. At Cerro San Bernardo six specimens were taken on Piptadenia macrocarpa Benth. and one very teneral specimen on herbaceous weeds.

The gemmifer Group

14. gemmifer (Boheman)   16. batesi n. sp.
15. meridianus n. sp.

Characteristics of Group.-Glabrous or densely clothed with scales, elytra without color pattern except that formed by random bare spots or more dense scales. Fore coxa with a tubercle on inner distal edge opposite trochanter (Fig. 57); tubercle less prominent in female (Fig. 58); tubercle most readily observed from a caudal or oblique angle as in the illustrations. Fore tibia of male on inner edge with a row of cylindrical tubercles, in female these are more acute and flattened like a tooth. Male with endophallic structure variable, flagellum well developed or indistinct. Female without bursal sclerites.

Remarks.-The distinct coxal tubercle which characterizes this group is sometimes present in a greatly reduced state in aureus (scintillans group) and in beverlyae (argentinensis group). The tubercle is most strongly developed in gemmifer. This and the absence of both a bursal sclerite and paired lateral sclerites in the female definitely place gemmifer with the flagellate species in spite of the indistinct flagellum.

In the three species in this group the fore tibia is bowed distally, the distal tooth is long and acute, and the inner apical lobe is moderate.

14. Hadromeropsis (Hadromeropsis) gemmifer (Boheman)

Figs. 6, 55-59, 61, 101-103, 127; Map 4

Hadromerus gemmifer Boheman, 1845:418. Faust, 1892:1; Champion, 1911:184. Type, female, labelled "Typus" on a red rectangle outlined in black; "129" printed on white; "762" written on white; "Hadromerus gemmifer, Reiche, Venezuela, Mannerheim" written on white (Stockholm).

Hadromerus scabricollis Faust, 1892:1. LECTOTYPE, HERE DESIGNATED, male, labelled "Colonia Tovar, E. Simon, I.II.88," "Museum Paris, Venezuela, E. Simon 1897," "J. Faust det. 1897," "type" handwritten on white, "Hadromerus scabricollis n. sp.," "TYPE" printed on red (Paris). See Type Material. NEW SYNONYMY.

Diagnosis.-Glabrous, shiny black (84% of males, 20% of females) or densely squamose with scales of iridescent or scintillating green (12% of males, 38% of females), iridescent cupreous or mixed colors (1% of males, 31% of females), or non-iridescent white or greenish (3% of males, 11% of females). Elytra of male (Fig. 6) slightly wider across humeri, tapering gradually to declivity, or parallel-sided. Female with ventrite 5 distinctly notched at apex (Fig. 59). Aedeagus with granules of dorsum extending onto sides distally. Internal sac (Figs. 101-102) with 2 pairs of inflatable lobes (only the dorso-lateral pair spiculate), and one long ventral lobe, internally with ejaculatory duct leading to a heavily sclerotized structure which forms the ventral surface of a dorsal cupped lobe and beneath this with a short, needle-like, free sclerite like a rudimentary flagellum. Spermatheca with nodulus distinctly shorter than cornu (Fig. 127).

Description.-Male, length 8.5-12.3 mm, width 2.8-4.0 mm. Female, length 9.0-15.0 mm, width 3.6-5.3 mm. Color as in Diagnosis, naturally glabrous (i.e, not glabrous because of abrasion) specimens with integument shiny and smooth, lacking scars of scales, but often with a few minute iridescent scales especially in the depressions of apex of elytra, and with occasional full-sized scales ventrally. All with a dense patch of scales on hind coxa and in a metasternal spot anteriorad to hind coxa. All sculpture stronger and more conspicuous in glabrous specimens. Rostrum moderately broad (Figs. 55, 56) ranging from flat (some females) to deeply triangularly concave; median line varying from unimpressed to distinctly impressed between eyes to their caudal margin. Epistoma long, narrow, occupying approximately 0.3 of anterior edge of rostrum. Eye moderately large, convex. Antennal funicle with segment 2 equal to or slightly longer than segment 1. Prothorax of male almost spherical, as wide as long, wider in female; evenly sculptured on disc and sides with faint to strong, dense tubercles each with an inconspicuous recumbent seta at its anterior edge, setae all directed toward middle of anterior margin of prothorax. Scutellum with or without scales; densely clothed with long, often broad, setae which may conceal the scales under them. Elytra of male as in Diagnosis and Fig. 6. Elytra of male 1.1-1.2x wider across humeri than across prothorax; elytra 2.5-3.0x longer than prothorax. Elytra of female 1.26-1.4x wider across humeri than across prothorax; elytra 3.0-3.3x longer than prothorax. Elytra of female slightly wider at about middle. Strial punctures small, striae regular but somewhat indistinct on declivity, this especially evident in glabrous specimens. Intervals each with a sparse row of fine setae, those of alternate intervals becoming longer and straighter towards apex of elytra. Apical umbone weak. Declivity of male in profile oblique, slightly concave or not; of female perpendicular or almost, usually concave. Sutural interval at summit of declivity in female only with very long, stout, horizontal setae. Apices of elytra briefly individually rounded in male; in female, produced in a small tooth approximately as long as a tarsal claw. Tubercle of fore coxa well developed (Figs. 57, 58). Fore tibia of male on inner edge with dense row of small, blunt teeth or elongate tubercles; posterior surface (Fig. 61) with similar multiple blunt tubercles, especially distally, the tubercles rapidly becoming smaller away from inner edge; fore tibia of female with denticles of inner edge sparser, sharp, without multiple tubercles posteriorly. Ventrally with very long, fine setae on prosternum restricted to small median area. Ventrite 5 of male 1.8-2.0x wider across base than long, almost flat, apex broadly truncate-emarginate, finely punctate apically. Ventrite 5 of female as in Diagnosis and Fig. 59; 1.8-1.9x wider across base than long, almost flat. Aedeagus ranging in length from 3.1-4.2 mm, averaging 3.6 mm. Internal sac as in Figs. 101-103 and Diagnosis; Spermatheca as in Diagnosis, Fig. 127; spermathecal duct 1-1.5 mm long.

Type Material.-The lectotype designated for scabricollis Faust is the only specimen labelled "Colonia Tovar, E. Simon" as in the description. Three PARALECTOTYPES, HERE DESIGNATED, 1 male, labelled "Columbia, Baden" and 2 females, labelled "Venezuela," all three labelled "Type" on red paper and with a gold square in Dresden apparently complete Faust's type series. The Simon collection was acquired by the Paris Museum in 1897 which seems to explain the label "J. Faust det. 1897," 5 years after the publication of the description. The type locality was described by Simon in a discussion of his three-month trip to Venezuela (1889:169-171).

Distribution.-Map 4. Central America and northern Colombia and Venezuela between 1000 and 1800 m. GUATEMALA. Alta Verapaz: Baleu, mcpio. San Cristobal Verapaz. COSTA RICA. "locality doubtful". PANAMA. Darien. COLOMBIA. Cesar or Guajira ["Magdalena"] : Socorpa Mission, Sierra de Perija. Cundinamarca: "Bogota." Norte de Santander: Chinácota. Magdalena: Río Frío, Cerro Patron. VENEZUELA. Aragua: Choroní, Colonia Tovar, Rancho Grande. Distrito Federal: Caracas, Caracas Valley, Parque Nacional Avila. Mérida. Miranda: El Hatillo, Nucleo El Laurel. Táchira: Cordero. Zulia: Maracaibo.

Specimens were collected in every month except February and March, and have been taken on Bracatinga brasiliense, Cassia, Mimosa, and peach foliage.

Specimens examined: 157 males, 198 females. Specimens are in most collections studied.

Remarks.-This is a highly variable species superficially resembling Hadrorestes in the characters of the vestiture and the punctuation of striae 9 and 10 which are somewhat confused in glabrous specimens but distinct in squamose specimens. The variation in the concavity of the dorsal surface of the rostrum is so extreme, as is the color, that specimens of gemmifer can look extremely different. Two specimens are a deep iridescent ultramarine and one is a uniform non-metallic pastel green. One extreme male from Venezuela (Miranda) has the prothorax wider than long and with faint lateral depressions on the disc.

All the Central American and Sierra de Perija specimens (total of 25) are glabrous black and robust, but the single specimen from Magdalena, the only record between these two areas is an iridescent green male. H. gemmifer is the only predominantly black species in North or Central America (but see rufipes).

15. *Hadromeropsis* (*Hadromeropsis*) *meridianus* n. sp.

Figs. 4, 5, 60, 104-107, 128, 129; Map 4

Diagnosis.-Similar to gemmifer, but: Male with sides of elytra not tapered from humeri. Apical elytral umbone weaker in both sexes. Female with apex of ventrite 5 rounded instead of emarginate. Aedeagus with granules confined to dorsal surface; internally with flagellum (Fig. 105) approximately as long as aedeagus. Spermatheca (Figs. 128, 129) with nodulus long and straight, usually much longer than cornu.

Description.-Holotype, male, length 9.8 mm, width 3.9 mm. Black, densely clothed with scintillating green, often overlapping scales. Scales not noticeably more dense on hind coxa; prothoracic tubercles and random spots on elytra naturally glabrous. Rostrum 1.1x longer than wide, with a triangular concavity basad of interantennal line, the latter weakly but distinctly elevated; median line impressed between anterior half of eyes. Antenna with segment 2 of funicle equal to segment 1. Prothorax 1.2x wider than long, sides strongly rounded between constrictions, sculpture consisting of very weak setiferous tubercles as in gemmifer. Scutellum densely squamose with overlapping scales. Elytra (Figs. 4, 5) as in gemmifer male, except sides very slightly wider to and caudad of middle, summit of declivity in profile very broadly arcuate, not abrupt as in Fig. 5, apical umbone almost imperceptible, apices of elytra very briefly produced into a knob. Fore coxal tubercle well developed. Fore femur 2.1x as wide as hind femur. Fore tibia with tubercles of posterior surface much less numerous than average gemmifer male, but more numerous than in an extremely sparsely tuberculate gemmifer male. Caudal surface of ventrites 2, 3, and 4 perpendicular, approximately equal in height. Antero-lateral corner of ventrites 4 and 5 with a deep pit. Ventrite 5 slightly convex, 2.1x wider at base than long, shorter and apex more broadly emarginate than in gemmifer male. Aedeagus as in Diagnosis, Fig. 104, 1.9 mm long; aedeagal apodeme 1.4 mm long; tegminal strut 1.9 mm long; internally with flagellum (Fig. 105) approximately 2.5 mm long, strongly sclerotized for proximal 0.3, distal remainder filamentous.

Allotype, female, length 10.3 mm, width 4.0 mm. Like holotype, except elytra in profile with disc evenly convex, highest at middle, summit of declivity more distinct, declivity slightly oblique, not concave; with elongate setae on sutural interval graduated in length, not numerous. In dorsal view sides of elytra gradually convergent from middle to apex, apices ending in a tooth shorter than the tarsal claw. Fore leg as in gemmifer female; fore femur 1.6x wider than hind femur. Ventrite 5 (Fig. 60) rounded at apex, 1.7x wider at base than long. Spermatheca as in Diagnosis, similar to Fig. 128; spermathecal duct 1.5 mm long.

Type Series.-Holotype, COLOMBIA, Cundinamarca, Monte Redondo nr. Bogota, 1500-1600 m, VI.27.1947, L. Richter Coll., Frank Johnson Donor (New York). Allotype, same data as type (New York). Paratypes, 8 males, 7 females. COLOMBIA. Cundinamarca: 6 males, 2 females, same data as type or [1] 1400 m, VI.26.1947 (New York, Howden); 1 female, Bogota, v. Lansberge (Leiden); 1 female, Cordillera de Subia between

Bogota and Girardot, 2300 m, VII.24.1947, L. Richter Coll., Frank Johnson Donor (New York); 1 male, Quetame, IX-44 (San Francisco). Tolima: 1 male, 3 females, Ibague, Fr. Claver (Paris). BRAZIL. 1 female, Amaz., Bowring 63.47 (London).

Remarks.-The specimens from Ibague and "Amaz." are uniformly larger (male, length 10.0 mm, width 3.6 mm; females, length 11.7-12.5 mm, width 4.9-5.2 mm) than the specimens with more explicit data from Cundinamarca (males, length 8.5-10.4 mm, width 3.0-3.9 mm; females, length 9.1-11.1 mm, width 3.6-4.1 mm). All specimens are densely squamose. The three males and two females that appear to be glabrous as in the glabrous gemmifer are actually squamose when viewed under magnification, the scales being transparent due in part to a teneral condition. The elytral shape of both sexes exhibits little variation and is rather conspicuously different from that of gemmifer. The aedeagus ranges in length from 2.4-3.0 mm (6 males dissected), thus smaller and with no overlap in size with gemmifer. Other parts of the male genitalia vary as follows: aedeagal apodeme 1.4-1.5 mm long, tegminal strut 1.5-1.9 mm long, flagellum 2.0-2.7 mm long. Length of spermathecal duct is 1.3-1.5 mm long.

The spermathecal duct is scarcely modified in length, approximately the same length as in gemmifer, and seems short for the length of the flagellum. I interpret this not as a different system from other flagellate species, but as evidence of recent change, a view supported by the closeness of meridianus and gemmifer.

The name "meridianus" means southern and alludes to the range being more southern than that of its sister species gemmifer.

16. Hadromeropsis (Hadromeropsis) batesi n. sp.

Figs. 63-65, 108-111, 130; Map 4

Diagnosis.-Rostrum (Fig. 63) long and dorsally narrow, the dorso-lateral edges prominent (almost keeled) and convergent from apex to head, the area between concave. Fore femur of male distally with the posterior edge enlarged into an arcuate flange (Fig. 65). Flagellum as long as or longer than aedeagus.

Description.-Holotype, male, length 11.4 mm, width 4.5 mm. Black, densely clothed with scintillating green scales; disc and sides of prothorax with small glabrous black areas; elytral intervals with irregular, random glabrous areas occasionally confluent; all glabrous areas of dorsum slightly elevated (squamose areas consequently slightly depressed), smooth and shiny, and with a single seta. Rostrum as in Diagnosis, Fig. 63, 1.5x longer than width at interantennal line, the dorso-lateral "keel" extending beyond anterior edge of eye; in dorsal view, scrobes completely visible, dorsally rostrum only about 0.8 ventral width. Mandibular scar very prominent (Fig. 64). Epistoma occupying 0.3 of anterior edge of rostrum. Antenna with segment 2 of funicle 1.7x longer than segment 1. Prothorax 1.2x wider than long; sides strongly rounded, 1.4x wider than basal constriction; 1.2x wider at basal constriction than at apical constriction. Disc in profile with basal constriction strong, apical constriction absent, thickest at basal 0.7, feebly arcuate to apex. Glabrous elevated spots of prothorax becoming weakly tuberculate laterad, each with a very fine, short, inconspicuous, recumbent seta; no other setae on disc or sides. Postocular vibrissae very well developed, arising from a prominent knob. Scutellum densely squamose, without setae. Elytra across humeri 1.3x wider than prothorax; elytra of paratypes averaging 2.7x longer than prothorax, sides almost parallel, apical umbone weak. Declivity gradual, oblique. Apices of elytra unmodified. Glabrous spots on base of elytra with minute, recumbent setae; apicad setae gradually becoming erect, stiff, dark, but still fine, until all on declivity are erect; summit of declivity without additional or longer setae. Fore coxa with tubercle well developed, but left tubercle abraded in type. Fore femur enormously thickened, 1.2x wider than head between outer edge of eyes, 2.7x wider than hind femur; with a brief arcuate flange on inner edge distally and a few weak tubercles in tibial groove beside flange. Distal end of fore femur (Diagnosis, Fig. 65) with posterior flange, inner surface of flange with a few scales and setae. Fore tibia proximally peculiarly flattened and twisted downward (Fig. 65); inner edge with 9 small, blunt teeth (or elongate tubercles); distally with a posterior lobe similar to but smaller than that of beverlyae. Ventrite 5 broadly truncate-emarginate at apex, 2.0-2.2x wider than long in paratypes. Genitalia as in (Figs. 108-111); aedeagus 3.2 mm long; flagellum (Fig. 108, 110) as long as aedeagus. Spiculum gastrale as in Fig. 111; apex of spiculum gastrale in ventral view with stalked sclerites arising from first connecting membrane as in Fig. 109 (see Remarks under genus).

Allotype, female, length 12.3 mm, width 5.1 mm. Differs from type as follows. Occasional scales, especially on fore legs, vivid blue. Segment 2 of funicle 1.4x longer than segment 1. Prothorax 1.3x wider than basal constriction. Elytra 1.4x wider across humeri than prothorax; elytra 3.1x longer than prothorax; sides of elytra only slightly wider medially than in type, apical umbone scarcely stronger than in type; sutural interval below summit of declivity with 6 dark setae of increasing then decreasing length; apex of sutural interval attenuated into brief tooth. Fore femur 0.8x as wide as head between outer edge of eyes, 1.5x as wide as hind femur; flange of inner edge greatly reduced and tubercles absent; without flange on distal end of fore femur. Fore tibia unmodified proximally; inner edge with 5 small, sharp teeth and several denticles; posterior lobe absent. Caudal surface of ventrites 2, 3, and 4 equal in height. Ventrite 5 triangular, 2x as wide at base as long, very weakly convex, apex rounded. Spermatheca of paratypes as in Fig. 130, nodulus slightly deflected; spermathecal duct 1.7-2.0 mm long, sinuous and much wider at bursa copulatrix.

Type Series.-Holotype, BRAZIL, Pará, Santarém (Cambridge). Allotype, BRAZIL, Santarém, 151 (Washington). Paratypes, 36 males, 16 females. BOLIVIA. Pando: 1 male, 1 female, Tambopatha Hath, Lat. 13° 5', Long. 69° GR, Cap[itain]es Mailles & Vincent, 1914 Août (Paris). BRAZIL. No other data: 1 female, Bates (London); 2 males, 1 female, E.A. Klages Coll. (Ithaca); 1 male, Lower Amazon (Ithaca). Minas Gerais: 1 male, Ponte Nova, May, 1899 (São Paulo). Pará: No other data, 2 males (London); 1 female, Santarem, Coll. Kuschel (Auckland);

2 females, Santarem (Washington); 1 male, Santarem, Amazonas, Staudinger and Bang-Haas (Eberswalde); 20 males, 6 females, Santarem, Klages Coll'n Exotic Coleop. Acc. No. 2275 or 2966 (Pittsburgh, Howden); 1 male, Tapajos (London). PERU. 1 female, Chaquimayo, Coll. Kuschel (Auckland). Locality unknown or questionable: 2 males (Cambridge); 1 male, "Costa Rica"* (London); 4 males, 3 females, "Panama"** (4) F. Psota Coll. (Chicago, Howden).

* The Costa Rica label is identical to some which are labelled "Locality doubtful."

** This could be Panema, a Bates collecting area near Santarém, improperly transcribed to Panama (Bates, 1910:189). The Psota Collection locality labels are often incorrect and Panama being so far out of the documented range of batesi, the Costa Rica and Panama localities are questionable.

Remarks.-Males vary in length from 8.5-11.2 mm and in width from 3.3-4.4 mm. Females vary in length from 9.6-13.5 mm and in width from 4.1-5.5 mm. The rostrum is consistently long and narrow facilitating quick identification of batesi among the many green species. The female specimen from Peru is 13.5 mm long, 1.5 mm longer than any other female. It also differs in the color of the scales of the fore tibia which are a vivid lavender. The only other obvious variation in the paratypes is in the degree of development of the fore femur of the male. In the least developed, it is 2.3x wider than the hind femur and the distal flange is also much smaller.

This spectacular species is named in honor of Henry Walter Bates, who collected at least one of the above paratypes, and to commemorate his 11 years spent collecting specimens and recording "aspects of nature under the equator" (Bates, 1910, Title page).

The argentinensis and nobilitatus Groups

These two groups represent subdivisions of a unit and the groupings are primarily for the taxonomist's convenience. In both groups the internal sac of the aedeagus has patches of spicules, spiculate inflatable lobes, a very heavy internal sclerite which is condyle-shaped at its anterior (proximal) end, a dorsal inflatable cupped lobe near the proximal end of the internal sclerite, and the apex of the internal sac membranous. Females of the species have a complex, somewhat funnel-shaped sclerotization inside the bursa copulatrix at the end of the spermathecal duct; spermathecal duct moderate in length, spermatheca in one plane except nodulus slightly angled in fasciatus. The vagina in all species has longitudinal faint to moderate sclerotized lines (Figs. 182, 188).

The groups do not readily lend themselves to interpretation from museum specimens only. Apparently the disruption of the habitat of the coastal Brazilian forests is reflected in confusing phenotypes. In several instances in the nobilitatus and argentinensis groups, a more northern taxon has a Santa Catarina form which is morphologically different and more subdued in color.

The _argentinensis_ Group

17. _togatus_ (Boheman)
18. _argentinensis_ (Hustache)
19. _plebeius_ n. sp.
20. _pulverulentus_ n. sp.

21. _pallidus_ n. sp.
22. _speculifer_ n. sp.
23. _beverlyae_ n. sp.

Characteristics of Group.-Metallic lustre infrequent. Elytra without tubercles except _plebeius_ and _beverlyae_. Fore femur without flange on inner edge distally except _beverlyae_. Fore tibia almost straight. Female ventrite 5 smooth and highly polished in many species. Internal sac of male with ventral plate absent, or present but not free, forming a bottom to the sac, or present but not sclerotized.

Remarks.-_H_. _argentinensis_ is associated with thorn scrub, _beverlyae_ with cerrado; the remainder appear to have ranges roughly corresponding with coastal forest formations of Brazil and are often sympatric with _nobilitatus_ group species.

17. _Hadromeropsis_ (_Hadromeropsis_) _togatus_ (Boheman)

Figs. 13, 159, 160, 184; Map 5

Hadromerus _togatus_ Boheman, 1840:290. Type, female, labelled "Brasilia, Westin" printed on white paper; "Typus" on red rectangle bordered in black (Schoenherr coll., Stockholm).

Hadromerus _brachispinosus_ Boheman, 1840:291. LECTOTYPE, HERE DESIGNATED, male, labelled with small green triangle with no inscription; "Hadr. brachispinosus Chevr. Brasil. Germar" written in black on white (Schoenherr coll., Stockholm). See Type Material. NEW SYNONYMY.

Diagnosis.-Moderate to large size. Elytra of female (Fig. 13) covered with a conspicuous anchor-shaped dark mark sharply defined in white; remainder of elytra ochraceous, white, pale olive or moss green, or obscure; with or without green or rosy lustre; declivity with a dark spot on apical umbone. Elytra of male with same anchor pattern evident or confused by random clusters of scales alternating with glabrous spots creating a tessellate appearance. Apex of rostrum and epistoma "normal", i.e., anterior corners of epistoma not produced in a lobe. Glabrous spots of elytra at most weakly convex, never tuberculate. Edge of elytra smooth with no trace of tubercles.

Description.-Male, length 9.5-11.4 mm, width 3.3-4.0 mm. Female, length 11.0-14.2 mm, width 4.0-5.3 mm. Elytra patterned as in Diagnosis and Fig. 13. White or ochraceous scales dense, imbricate around eye, continuing along side of head and side of prothorax to base of elytra from interval 4 over and under humerus and continuing thence into the elytral pattern, these markings vague in male, very sharply defined in female. Elytra of male with glabrous areas often the full width of interval and sometimes confluent; apical umbone not always with glabrous spot. Dorsal surface of male and female head, rostrum, dark central area of pronotum, and elytral anchor mark sparsely clothed with scales which under the microscope appear weakly to strongly iridescent green, faintly cupreous, or non-iridescent.

Rostrum approximately as long as wide, flat or (rarely) feebly concave in male, median line briefly impressed basally; sides parallel or apically divergent; surface sparsely punctate. Setae of head and rostrum slender, as long as 1.5-2 scales, recumbent. Epistoma occupying 0.28-0.37 of anterior edge of rostrum, epistoma very slightly ogival. Prothorax 1.0-1.1x wider than long, sides weakly to moderately rounded between constrictions; apical constriction obsolete dorsally; female especially with vague flattened areas on disc on either side of median line. Pronotum smooth medially or with setiferous arcuate pustules and other low convexities; glabrous setiferous spots sparser on sides because of denser scales, spots pustulate, setae as long as 1-2 scales. Elytra across humeri 1.3x wider than prothorax in male, 1.3-1.4x wider in female. Elytra averaging 3.0x longer than prothorax in male, 3.3-3.6x longer in female. Elytra in dorsal view with sides parallel for basal 0.16, thence gradually divergent to beyond middle where they average 1.1x wider than across humeri in both sexes, gradually rounded to apical umbone which enters dorsal outline; apex usually broadly rounded in male, interval 10 at apex slightly thickened and often briefly horizontally flared posteriorly; apex of female and some males slightly elongated triangular. Female apex of each sutural interval thickened in a vertical lobe; apical modification densely pilose in both sexes. Setae of elytra fine, straight; recumbent, semi-erect or perpendicular; recumbent setae as long as 1 scale, semi-erect and erect setae up to 2 scale lengths (posture and length of setae possibly artifacts of abrasion); setae longer on base, sides and declivity; female on sutural interval at summit of declivity with a cluster of up to 12 very long setae. Legs, metasternum, and abdomen with many fine, long setae in male; in female setae much sparser, shorter, curved. Fore femur 1.7-2.0x as wide as hind femur in male, 1.6x in female; faintly pustulate and sometimes rugulose distally. Fore tibia bowed distally; distal tooth averaging 0.5x as long as apex of fore tibia; apex posteriorly with edge expanded into a lobe the size of the outer dorsal lobe; posterior surface with numerous granules or small acute tubercles near teeth of inner edge, granules most numerous in well-developed males, sometimes absent in female. Male ventrite 5 across base 1.6-1.8x wider than long; almost flat; almost evenly squamose; with moderate, sparse punctures; apex narrowly emarginate. Female caudal surface of ventrites 2, 3, and 4 equal, perpendicular, margin evenly arcuate; perpendicular surface of ventrite 2 vertically striate. Female ventrite 5 across base 1.7-1.9x wider than long, almost flat, scales sparser or absent medially. Aedeagus as in Fig. 160. Internal sac as in Fig. 159, differing from similar species in the pair of small fused sclerites at extreme apex ventrally. Spermatheca as in Fig. 184.

Type Material.-PARALECTOTYPES of Hadromerus brachispinosus Boheman, HERE DESIGNATED, 2 males: 1 male, labelled, "Brazil"; orange square with a gritty black texture on top, white underside; "Coll. Chevrol."; "Cotypus" on red rectangle with black border (Schoenherr coll., Stockholm). 1 male, with no labels but listed as "Hadromerus brachispinosis Schönherr (Brasilia)" (Chevrolat coll., Stockholm).

Distribution.-Map 5. BRAZIL. Minas Gerais: Faz. dos Campos. Rio de Janeiro: Nova Friburgo, Rio de Janeiro. Rio Grande do Sul (?).

Santa Catarina: Cauna, Corupá, Mafra, Rio Natal, Rio Vermelho.   São Paulo: Pindamonhangaba.   PARAGUAY. Alto Paraná: Hohenau.

Specimens were collected in January, February, and September through December.

Specimens examined: 13 males, 35 females.   Specimens in Auckland, Cambridge, Curitiba, Dresden, London, New York, Oxford, Paris, São Paulo, Stockholm, Washington, Howden.

Remarks.-From the few specimens available there is an indication that scales with metallic green iridescence in both sexes occur in the northern part of the range only, not in Santa Catarina.

The male labelled "Rio Grande" is very similar to the Rio Vermelho male suggesting that this locality is in the state of Rio Grande do Sul.   Both specimens are very fresh with the vestiture in good condition; all elytral setae are erect and nearly perpendicular.

The ventral lobe and granules on the posterior surface of the male fore tibia are developed about half as much as in the extreme beverlyae.

One male had an obsolete flange on the fore femur which is an important diagnostic character of nobilitatus.   The lack of tubercles on the elytra, especially the apical edge, should distinguish togatus from nobilitatus at a glance.

The anchor-shaped elytral mark is diagnostic and very uniform, the only deviation being one in which the fascia is almost bisected on interval 6 and another in which it is terminated at interval 7.

18. Hadromeropsis (Hadromeropsis) argentinensis (Hustache)

Figs. 7-10, 161, 162, 181, 182; Map 5

Hadromorus [sic] argentinensis Hustache, 1928:157; Pl.5, Fig. 2.
     LECTOTYPE, HERE DESIGNATED, female, labelled "Alta Gracia La Granja, Sierras de Córdoba, C. Bruch leg."; "Foto" printed on green; "Typus" printed on green paper; "Hadromerus argentinensis Hust." printed by hand on white rectangle with fine red border (Bruch coll., Buenos Aires). See Type Material.

Diagnosis.-Elytra with a dark, mostly glabrous postmedian spot on intervals 5, 6, and 7. Rostrum with very few rounded scales on side below scrobe, no rounded scales on ventral surface of rostrum. Elytra short and thick (Figs. 7, 8), convex in profile (Figs. 9, 10), apex not or scarcely extended in dorsal view. Elytra without an apical umbone, i.e., intervals 5 and 6 at their apical termination not at all elevated. Ventrite 4 of female with caudal surface perpendicular, straight; ventrite 5 of female glabrous and polished medially. Sexes very similar.

Description.-Male, length 5.3-7.4 mm, width 2.0-3.0 mm. Female, length 6.2-8.6 mm, width 2.6-3.6 mm. Males from Uruguay as short as 4.6 mm, females as short as 6 mm. Integument brown to black. Clothed with white, tan and obscure scales, the latter rarely with green, lavender or cupreous reflections; scales sometimes gray-blue or gray-green in teneral specimens. Side of prothorax with a vitta of dense white or white and tan scales extending from eye to humerus. Ventrally with scales tan, mostly concentrated on sides; ventrites 3,

4, and 5 with very few scales and these only on the extreme lateral edges. Rostrum with dorsal surface more or less flat, dorso-lateral edges not acute, sides broadly visible from above. Surface of head and rostrum with punctures approximately the size of a scale, these often changing to rugae on rostrum. Median line broadly, evenly impressed from between eyes almost to apex of epistoma, frontal fovea obsolete to deep. Epistoma occupying .36-.45 of anterior edge of rostrum; anterior edge of rostrum at margin of epistoma slightly produced. Outer contour of mandible rounded; scar small and elliptical, not protuberant. Ventral surface of head and rostrum usually with no rounded scales. Prothorax averaging 1.1x (1.09-1.2x) wider than long; dorsal and apical constrictions strong, equal; apical constriction obsolete dorsally; in profile, convex to slightly convex. Disc of prothorax sparsely scaled, sparsely punctate, the glabrous areas smooth and polished. Elytra across humeri averaging 1.3x wider than prothorax in male, 1.4x in female. Elytra averaging 2.7x longer than prothorax in male, 3.0x in female. Form of elytra as in Diagnosis and Figs. 7-10. Elytral intervals with random stiff blunt setae, as long as 1-3 scales; those of apical half of elytra in females especially much more nearly perpendicular and longer. Infrequently summit of declivity in female with setae up to 5 scale-lengths. Apices of elytra conjointly rounded in male, in female produced in a weak tooth. Fore femur in male averaging 1.7x, in female 1.6x wider than hind femur; with longitudinal rugae distally. Fore tibia on outer edge covered with scales smaller than those of body, anterior and posterior surfaces indistinctly rugose, with some granules near teeth on inner edge. Distal tooth of fore tibia very small, never exceeding the tuft of setae; female with a pair of distal teeth. Ventrite 5 of male 2.0-2.2x wider than long, weakly convex or not, apex truncate or truncate-emarginate. Ventrite 5 of female as in Diagnosis, 1.8x wider than long, the flat polished surface sometimes depressed below sides; apex broadly rounded. Male genitalia as in Figs. 161, 162. Female genitalia as in Figs. 181, 182 (setae omitted in drawings).

Type Material.-PARALECTOTYPES, HERE DESIGNATED, 1 male, 2 females, labelled "Alta Gracia La Granja, Sierras de Córdoba, C. Bruch leg.," "30." in faded ink on pink square (Hustache Coll., Paris), these 3 specimens mounted on individual cards on a single pin, the pin also bearing a label "TYPE" printed in red on a rectangle; 1 male, 2 females, labelled "Rep. Argentina, Pr. Santiago del Estero, 190_, C. Bruch" (1 male, 1 female, Buenos Aires; 1 female, Paris, also labelled "30." as the 3 paralectotypes from Alta Gracia).

Distribution.-Map 5. South of the Tropic of Capricorn in Argentina, Paraguay, Uruguay and Brazil. ARGENTINA. Buenos Aires: Isla Martin Garcia; San Fernando. Córdoba: Alta Gracia, Anizacate, Capilla del Monte. Entre Ríos: Gualeguaychu, 54 mi E. Paraná, Santa Elena. Jujuy: Juancito. Misiones: Loreto. Salta: 4 km S. Campo Quijano, Cerrillos, Cerro San Bernardo, Las Cañas, Metan, Rosario de Lerma, Viñaco (15 km S. El Carril). Santa Fe: La Gallareta, Tartagal. Santiago del Estero: Averías, 25 km NW Icaño. Tucumán: Chuscha, Ciudacita, Tapia. BRAZIL. Most specimens no other data. Santa Catarina. PARAGUAY. Itapuá Hepua. URUGUAY. Canelones. Colonia: Carmelo, Colonia. Montevideo. Paysando: Puerto Pepe Ají. Soriano:

Arroyo Cololó.

Specimens were collected in January through July, but were most common in February and March.

Specimens examined. 215 males, 149 females. Specimens in Auckland, Berlin, Buenos Aires, Cambridge, Curitiba, Dresden, Eberswalde, Ithaca, London, Maracay, New York, Paris, São Paulo, Tucumán, Washington, Howden, Wibmer.

Remarks.-Specimens from Sierras de Córdoba, the type locality, exhibit most of the range of variation noted within the species. The principal geographic variation seems to be the smaller size of specimens from Uruguay and contiguous areas of Argentina (Provinces of Buenos Aires and Entre Ríos and Misiones Ter.). Specimens from Brazil (21) average much larger.

Additional variation is as follows. One specimen had no dark elytral spot but the area was sharply delimited by white scales so that the pattern was still similar. In a male from Salta the dark elytral spot was reduced to interval 5 only. Interval 9 of the elytra may occasionally have some setae set on weak tubercles in males and females. Ventrite 5 of the male had up to 26 scales per side; ventrite 5 in the female usually had no scales but as many as six were seen on one side.

I did not see the specimen mentioned by Hustache (1928:157) from "Alto Paraná, Misiones (Wagner)" with the golden yellow vestiture. It is possibly only a teneral specimen.

Although apparently very distinctive in appearance, small specimens of this species can still be readily mistaken for atomarius, and larger specimens can be confused with plebeius. From atomarius, argentinensis may be distinguished by the straight fore tibia; the shorter ventrite 5 of the male; elongate, cylindrical endophallic apex; straight edge of ventrite 4 of female, and glabrous ventrite 5 of female. From plebeius, argentinensis differs in the absence of round scales on the ventral surface of the rostrum, stout fore femur, rounded exterior outline of mandible, shorter elytra without tubercles and lack of an apical umbone.

According to Bruch (Hayward, 1960:18) the larvae of H. argentinensis attack "los espinelles" (Acacia caven (Mol.) and Mimosa scabrella Benth.). Data on specimens includes Acacia caven (in Entre Ríos) and Prosopis (in Córdoba). Hustache (1938:6) lists specimens from Loreto, Misiones, on Mimosa bracatinga. In the Province of Salta, I took argentinensis by beating the foliage of Acacia macracantha H.B. ex Willd., Mimosa farinosa Grisebach, Prosopis chilensis (Mol.) Stuntz, and Mimosa strigillosa Torr. and Gr.; occasional specimens were found on tall weeds and one weevil was found buried in the flower of Mimosa strigillosa. In Córdoba at Capilla del Monte argentinensis were common on Acacia caven (Mol.), small trees in a pasture yielding 10 to 15 specimens each.

19. Hadromeropsis (Hadromeropsis) plebeius n. sp.

Figs. 15-17, 147, 164, 186; Map 5

Diagnosis.-Color pattern (Figs. 15-17) similar to that of

argentinensis in the conspicuous dark spot on intervals 5, 6, and 7. Without a metallic lustre. Elytra with setae of intervals fine, "erect", i.e., at an angle greater than 45° to surface; each seta set on glabrous, flat to tuberculate spot. Apical umbone of elytra with 1 or more glabrous tubercles. Fore femur slender; fore tibia almost straight to weakly bowed, distal tooth very small and concealed by dense setae. Apex of internal sac elongate (Fig. 164), with 2 pairs of basal lobes not present in argentinensis.

Description.-Holotype, male, length 9.0 mm, width 3.3 mm. Integument piceous, clothed with white, tan and obscure scales, the white scales usually imbricate, the tan appearing ochraceous in some lights and usually not imbricate. Head, rostrum, thorax, legs and ventral surface clothed with random tan scales. Elytra patterned as follows: with dark spots caused by combination of glabrous areas and obscure scales, (1) on base of intervals 3 and 4, (2) in conspicuous spot on intervals 5, 6, and 7 as in argentinensis, connected to suture by zigzag line, and (3) on apical umbone; confused white vitta on basal third of intervals 5 and 6 and white fascia across apical third; center of disc with mixed white and tan; elytra overall with random glabrous spots flat or elevated as in description of elytra below. Rostrum 1.1x longer than wide, almost flat, dorso-lateral edges not acute, the sides not quite vertical. Surface of head with punctures approximately the size of a scale, changing to rugae on rostrum. Median line impressed from frons almost to apex of epistoma. Epistoma occupying approximately 0.3 of anterior edge of rostrum, anterior edge of rostrum at epistoma slightly produced; epistoma 1.3x longer than wide. Scales of side of rostrum rounded, changing to elongate ventrally, on ventral surface of rostrum only a central triangular area without scales. Prothorax 1.1x wider than long, sides moderately rounded between constrictions, apical constriction obsolete dorsally; in profile disc almost flat. Setae of sides of prothorax set on weak pustules. Disc of pronotum slightly irregular, sparsely punctate, surface polished between scales. Elytra as in Diagnosis and Fig. 15. Elytra across humeri 1.4x wider than prothorax. Elytra 2.9x longer than prothorax. Setae of disc of elytra (between striae 7 from basal 0.16 to declivity) arising from flat to slightly convex surface, remainder of setae set on tubercles, those of declivity very small, some no larger than a scale. Setae of elytra straight, most held at an angle greater than 45°; longer than 2 scales. Setae on disc of elytra colored testaceous or darker; setae on extreme base of elytra and intervals 8, 9, and 10 paler or white. Apices of elytra briefly divergent, thickened, densely clothed with fine setae. Edge of elytra around ventrite 5 weakly, minutely tuberculate, appearing crenulate. Fore femur 2.0x wider than hind femur, surface indistinctly sculptured. Fore tibia slightly bowed from distal 0.25, surfaces with moderate, confused sculpture; posteriorly with granules near inner edge, inner edge with about 10 small teeth; distal tooth as small as those along edge, completely concealed in dense fine setae. Central 0.3 of ventrites 3, 4, and 5 without scales. Ventrite 5 across base 2.0x wider than long; weakly convex; apex truncate. Genitalia as in Figs. 164, 165; internal sac with spiculate part of apex especially long.

Allotype, female, length 9.4 mm, width 3.8 mm. Differs from type as follows. White scales of elytra more condensed, forming a basal arc, an oblique fascia from basal third to suture at middle, and a wider apical fascia. Rostrum more robust. Epistoma 1.1x longer than wide. Prothorax 1.2x wider than long, sides more strongly rounded, surface smoother. Elytra 3.3x longer than prothorax. Elytra as in Figs. 16, 17, 147. Apical umbone less prominent; apex more elongate, ending in short, caudally directed tooth. Setae longer, mostly perpendicular; 6 or 7 setae on sutural interval at summit of declivity much longer. Edge of elytra not crenulate. Fore femur 1.5x wider than hind femur. Caudal surface of ventrites 2, 3, and 4 straight, perpendicular; in caudal view slightly arcuate. Surface of ventrites 3 and 4 flat, the caudal elevation beginning abruptly across middle of segment. Scales of ventrites 3, 4, and 5 elongate, limited to extreme sides. Ventrite 5 flat, smooth and polished except at sides; 1.8x wider than long; apex narrowly rounded. Spermatheca as in Fig. 186.

Type Series.-Holotype, BRAZIL, Paraná, O. d'Agua, Rolando, 4-44, 1296, Coleção F. Justus Jor (Curitiba). Allotype, same data as type (Curitiba). Paratypes, 6 males, 14 females. BRAZIL. Paraná: 1 female, 5041, Gregorio Bondar Collection, David Rockefeller Donor (New York); 1 female, Guarauna, 12.40, 1295, Coleção F. Justus Jor (Howden); 1 female, Ortogueira, 1-44, 1295, Coleção F. Justus Jor Curitiba). Santa Catarina: 3 males, 9 females, Nova Teutonia, 300-500 m, 27°11'B, 52°23'L, III.1945, 18.IV.1948, IV.1951, 6.3.1951, II.1966, II.1974, II.1976 [2], III.1977 [2], V.1977, Fritz Plaumann (Auckland, Curitiba, São Paulo, Hespenheide, Howden). São Paulo: 1 male, Staudinger (Dresden); 1 male, 1 female, Cantareira, 13.3.38, Dr. Nick, Coll. Kuschel (Auckland).

Remarks.-Males vary in length from 8.9-9.1 mm and in width from 3.3-3.5 mm. Females vary in length from 7.7-10.7 mm and in width from 3.1-4.5 mm. Minimum color pattern includes the dark somewhat cordiform scutellar area, the postmedian spot on intervals 5, 6, and 7, a dark spot on apical third of interval 2, and at least one glabrous tubercle on apical umbone. The pattern can appear quite different if the postmedian mark is continuous and forms a broad common fascia as in fasciatus. Some pastel blue-green scales are found on two male paratypes on the abdomen where they are randomly interspersed with tan scales, and they are also found on two female paratypes where they replace a few or most of the white scales of the elytra. Ventrally in both sexes scales are limited to the sides of the abdomen except ventrite 2 which may be squamose across entire caudal edge; in the type, however, there is a single scale in the center of ventrite 5. The apex of the rostrum of the female may be produced at the outer edges and at the corners of the epistoma almost as much as in fasciatus, but pterygia have not developed. The prothorax of male paratypes is narrower, 1.05-1.09x wider than long. In four males the apices of the elytra are prolonged into a short tooth. Ventrite 5 of females ranges from 1.7-2.0x wider than long, but in most females is 1.8x.

Both fasciatus and nobilitatus have setiferous tubercles on the elytra as in plebeius; plebeius can be quickly distinguished from both by the lack of a metallic lustre and short distal tibial tooth as well as the other diagnostic characters.

Because it lacks the golden lustre of nobilitatus and the red lustre of female _fasciatus_, this species is called "plebeius," meaning plebeian.

20. Hadromeropsis (Hadromeropsis) pulverulentus n. sp.

Figs. 12, 148-150, 163, 183; Map 6

Diagnosis.-Small; male, length 5.0-6.3 mm, female 6.4-8.0 mm. rostrum (Fig. 149) as wide as long. Epistoma wider than long. Elytra without tubercles; strial punctures usually foveate (Fig. 150); apex of elytra rounded in male, attenuate in female (Fig. 148); setae of edge of apical half of elytra, declivity, and lateral intervals very long and conspicuous. Aedeagus (Fig. 163) almost evenly cylindrical.

Description.-Holotype, male, length 6.3 mm, width 2.2 mm. Integument red-brown, metasternum and ventrites 1 and 2 almost black, ventrites 3 and 4 paler than 1 and 2, ventrite 5 paler than elytra. Without metallic lustre. Sparsely clothed with white and pale tan scales arranged in a faint pattern of (1) a white vitta along base of interval 4 joined across suture in an arc at basal fourth, and (2) a small dark diamond on suture at summit of declivity forming a "V" with a pair of dark spots on intervals 4 and 5 at middle. Scales denser around eye, on side of head, and side of prothorax to humerus. Rostrum as in Diagnosis and Fig. 149, dorso-lateral edges parallel, dorsal surface slightly concave; median line sharply impressed from between caudal edge of eyes to interantennal line, thence shallowly concave to apex of epistoma. Epistoma 1.6x wider than long, occupying 0.47 of anterior edge of rostrum, sides of epistoma slightly arcuate, anterior corners not produced. Sculpture of head and rostrum similar to that of _nobilitatus_. Prothorax 1.04x wider than long, sides moderately rounded between equal basal and apical constrictions. Each seta of disc set in a puncture, those of sides of prothorax set on a weak granule. Scutellum with several fine appressed setae, no scales. Elytra as in Diagnosis; elytra across humeri 1.3x wider than prothorax. Elytra 2.7x longer than prothorax. Elytra with sides subparallel, very slightly wider (1.09x) at middle, apical umbone very weak, apex rounded beyond umbone. Strial punctures foveate, larger than in Fig. 150 of allotype, gradually becoming smaller on declivity; on disc foveae separated by approximately their own diameter both longitudinally and transversely. Setae of disc very fine and inconspicuous; setae suddenly longer on sides and declivity from apical 0.25, longest setae (on interval 10 apically) as long as 3 scales; only those setae caudad of apical umbone and on edge of elytra around ventrite 5 straight, erect and conspicuous; sutural interval on declivity with evenly spaced, curved setae as long as 2 scales. Fore femur 1.4x wider than hind femur, without distal flange, without sculpture except for a few rugulae distally. Fore tibia almost straight, teeth of inner edge minute, distal tooth as long as 0.32x the width of tibia at apex. Ventrites 3, 4, and 5 without scales medially; ventrite 5 across base 2.1x wider than long, moderately convex, its apex broadly truncate, emarginate. Aedeagus as in Diagnosis, Fig. 163; internal sac not extracted.

Allotype, female, length 6.1 mm, width 2.3 mm. Differs from type as follows. Elytral pattern much more distinct (Fig. 12), the basal vittae enclosing a dark postscutellar spot and the apical "V" very broad and conspicuous. Rostrum less concave. Epistoma 1.3x wider than long, occupying 0.45 of anterior edge of rostrum. Prothorax 1.05x wider than long. Elytra across humeri 1.3x wider than prothorax. Elytra 3.3x longer than prothorax. Sides of elytra gently rounded at middle where they are only 1.2x wider than across humeri, apical umbone weak, apex elongate; sutural interval attenuated into a long tooth directed slightly inward and ventrad (Fig. 148). Lateral edge of elytra somewhat constricted from ventrite 3 to apex; ventrites 3, 4, and 5 consequently narrow. Strial foveae smaller (Fig. 150). Elytral setae longer and more erect, the longest as long as 4 scales; setae of apical half of lateral intervals almost perpendicular to surface and creating a bristly effect; only 2 setae on sutural interval on declivity greatly elongate. Fore femur 1.5x wider than hind femur. Teeth of inner edge of fore tibia and distal tooth short, as long as 0.24 of width of fore tibia. Caudal surface of ventrite 4 perpendicular, its edge evenly arcuate in caudal view. Ventrite 5 across base 1.6x wider than long, triangular, apex narrowly rounded; surface almost flat, medially polished and with only a few minute setae. Genitalia as in Fig. 183 but sclerotized streaks of vagina and setae not shown.

Type Series.-Holotype, BRAZIL, Minas Gerais, Ouro Preto, Topázios, 22.II.1962, J. Bechyné col. (São Paulo). Allotype, same data as type (Howden). Paratypes, 2 males, 6 females. BRAZIL. 1 female, no other data, ex coll. Oberthur (Paris). Minas Gerais: 1 male, 1 female, Bowring 63-47, longulus (Jek.) (London); 1 female, Coll. Kuschel (Auckland). LOCALITY UNKNOWN. 1 female, Gorham Collection, acc. 68498, lentus, Jekel collection (Washington); 1 female, Bowring 63-47 (London); 1 female, Sharp Coll 1905-313, Hadromerus pulverulentus Jek n. sp. , Columbia* (London); 1 male, 1279, Bowring 63-47, pulverulentus (Jek), Bogota* (London).

*In view of the accurate data on the Minas Gerais specimens, the "Columbia" and "Bogota" on these old specimens need to be substantiated.

Remarks.-The two male paratypes are 5.0 and 6.3 mm long, 2.0 and 2.3 mm wide. Females are 6.4-8.0 mm long, 2.5-3.2 mm wide. In most paratypes the color pattern on the elytra is much less developed than in the type and allotype. In a teneral female the white scales on the sides of the prothorax are slightly opalescent. In one of the two males, the rostrum is almost flat. The elytral setae in several females are much more conspicuous, longer and more erect; on the sutural interval on the declivity, 0 to 3 setae may be greatly elongated. The strial punctures are not foveate in the largest female paratype; it is possible that the size of these punctures varies in inverse proportion to the size of the specimen. In females the perpendicular portion of the caudal surface of ventrite 4 varies from the full width of the ventrite as in the allotype to only the central 0.6, and the angle varies from perpendicular as in the allotype to slightly posteriorly projecting. The internal sac was not extracted from any of the males.

H. pulverulentus may always be distinguished by the wide epistoma and almost evenly cylindrical aedeagus. The elongate habitus, small size and bristly lateral setae will also separate the species but these characteristics are more subjective and vulnerable to wear and poor condition of specimens.

Note that the glabrous elytral spot when present is on intervals 4 and 5 or 4, 5, and 6 compared to a more lateral position in plebeius and argentinensis.

For this species, I chose the Jekel manuscript name "pulverulentus" meaning "dusty" since it describes nicely the appearance caused by the pale scales scattered over the surface.

21. Hadromeropsis (Hadromeropsis) pallidus n. sp.

Figs. 20, 21, 151, 166, 167, 187; Map 6

Diagnosis.-Densely clothed with tan or grayish and white scales, elytra with a white pattern as in Figs. 20, 21; or with a vittate pattern formed of medial and apical fasciae abbreviated and connected to basal mark. Head and rostrum robust (Fig. 151); dorso-lateral edges of rostrum ill-defined, broadly rounded, especially basally. Setae of dorsum of elytra white, parallel-sided, truncate at apex, erect, scarcely curved. Female ventrite 5 weakly convex medially with setae and sometimes scales medially. Male ventrite 5 evenly punctate. Aedeagus short, in profile thick distally (Fig. 166); internal sac as in Fig. 167 or with internal sclerite reduced.

Description.-Holotype, male, length 6.8 mm, width 2.5 mm. Integument piceous. Scales tan and off-white, the latter often with weak metallic green lustre. Most tan scales very evenly spaced, not quite touching; tan scales sparser along median line of prothorax, the glabrous area thus forming a dark vitta. Elytra with dark vitta on interval 3 basally and a small, dark "V" on intervals 1-3 before declivity; these dark areas formed by tan scales sparser and some scales replaced with obscure scales. Whitish scales often imbricate; whitish pattern of ring around eye continuous with lateral vitta on prothorax and elytra to apical umbone; disc of elytra with vittate pattern as in Diagnosis. Elytral intervals with 7-10 random glabrous areas the size of 1-2 scales, each as convex as a scale, each bearing an erect seta. Rostrum as in Diagnosis and Fig. 151, approximately as long as wide; median line impressed from between eyes to interantennal line; each longitudinal half of dorsal surface of rostrum longitudinally convex to median line. Epistoma approximately as wide as long, 0.3x as wide as anterior edge of rostrum. Scar of mandibular cusp slightly produced. Surface of head and rostrum with prostrate setae as long as 1.5-2 scales, arising from minute punctures in small glabrous spaces between scales, surface otherwise not sculptured. Prothorax almost as long as wide, sides weakly rounded between weak apical and basal constrictions; surface without evident sculpture. In profile pronotum almost flat, basal constriction narrow, anterior edge produced more than is usual in the genus. All setae of prothorax prostrate, on disc as long as 1-1.5 scales, on sides as long as 2 scales. Scutellum with 1 seta-like scale, and several minute, prostrate setae. Elytra across humeri 1.3x wider than prothorax. Elytra 2.8x longer than prothorax. Elytra in dorsal view with sides parallel for basal 0.2, thence very slightly, gradually arcuate, widest at middle, apical umbone weak but distinct in outline, apex beyond broadly rounded; apices unmodified. In profile summit of declivity unmarked. Strial punctures fine, inconspicuous. Intervals as wide as 4-5 scales, sutural interval narrower. Setae of elytra as in Diagnosis, as long as 3-4 scales. Fore femur 1.5x wider than hind femur, apparently without microsculpture. Fore tibia straight to distal 0.2 thence very weakly bowed; inner edge with 8 small teeth; distal tooth as long as 0.25 width of tibia at apex. Abdomen with scales almost evenly distributed on ventrites 1 and 2, absent medially on ventrites 3 and 4, slightly sparser medially on ventrite 5. Ventrite 5 across base 2.0x wider than long, moderately convex, its apex briefly emarginate, surface evenly punctate. Aedeagus as in Diagnosis, Fig. 166. Internal sac similar to Fig. 167 but internal sclerite greatly reduced to approximately one-third; spiculate apex of sac shorter.

Allotype, female, length 6.5 mm, width 2.8 mm. Differs from type as follows. Rostrum flatter, epistoma 1.2x longer than wide. Prothorax 1.2x wider than long. Elytra 3.0x longer than prothorax. Elytra similar to Figs. 20, 21, but humeri less prominent and sutural tooth much shorter. Fore femur 1.2x wider than hind femur. Fore tibia with distal tooth single, as holotype. Ventrites 1 and 2 almost evenly squamose, ventrites 3 and 4 medially with no round scales but with appressed setae or seta-like scales. Caudal surface of ventrite 4 perpendicular; slightly, evenly arcuate from extreme sides. Ventrite 5 across base 1.5x wider than long, scales absent on weakly convex median area. Spermatheca resembling that of pulverulentus in Fig. 183; spermathecal duct 1.2 mm long.

Type Series.-Holotype, ARGENTINA, Misiones, Oberá, 10.5.47, Wittmer leg., Kuschel Coll. (Auckland). Allotype, same data as type (Auckland). Paratypes, 4 males, 1 female. ARGENTINA. Misiones: 1 male (Buenos Aires). BRAZIL. Minas Gerais: 1 male, Vila Monte Verde, 8.III.1972, J. Halik, 12383 (Sao Paulo). PARAGUAY. Alto Paraná: 2 males, Hohenau, 12.1939, Hans Jacob leg. (Auckland, Howden). URUGUAY. Montevideo: 1 female, So Amer Paras Lab, No. 298, 6-1-43, Berry (Washington).

Remarks.-Males vary in length from 5.8-7.5 mm and in width from 2.2-2.7 mm. The female paratype is 10.0 mm long, 4.1 mm wide. The specimens from Brazil and Uruguay have the elytra patterned as in Figs. 20, 21; those from Argentina and Paraguay have the vittate pattern. The male from Minas Gerais differs from the other males as follows: integument black but appendages paler, elytra flatter, intervals as wide as 5-7 scales, setae of elytra shorter, apices of elytra developed into a pair of divergent teeth or conical tubercles, internal sac as in Fig. 167.

The internal sac in the four males dissected exhibits more than the usual amount of variation. In one male from Paraguay the internal sclerites are approximately midway between the small extreme of the type from Misiones and the large extreme of the paratype from Minas Gerais (Fig. 167). In the same male from Paraguay there is a slight sclerotization of the ventral surface of the apex reminiscent of speculifer.

It could be debated whether the blunt apex of the elytral setae is natural or a consequence of a filamentous tip breaking off. I feel it is natural because all setae on the disc of the elytra are blunt but the setae on the apical edges of the elytra are gradually tapered. This holds true for teneral as well as older specimens. Likewise the setae of the thorax, head and rostrum are almost as blunt, but their apices are protected by their prostrate position.

In addition to the diagnostic characters given, the profile of the elytra is distinctive in the unmarked summit of the declivity.

The name pallidus refers to the pale color of the specimens.

22. Hadromeropsis (Hadromeropsis) speculifer n. sp.

Figs. 11, 140, 143, 168, 169, 185; Map 7

Diagnosis.-Usually with a greenish or rosy metallic lustre. Elytra immaculate or with a maximum pattern of elongate basal ring between striae 1 and 5, an acute V-shaped fascia beginning at middle and a similar postmedian fascia parallel to it, the latter two fasciae sometimes joined laterally (Fig. 11). Elytra without tubercles or pustules. Female with lateral edges of elytra strongly constricted opposite ventrites 3, 4, and 5; in ventral view (Fig. 143) sides of elytra broadly visible. Female ventrite 4 with caudal surface perpendicular, edge straight or appearing feebly emarginate. Female ventrite 5 flat, medially mirror-like, highly polished, smooth, without scales, with several short setae at most; laterally ventrite 5 with numerous setae, with or without scales. Aedeagus slightly thicker distally; internal sac as in Fig. 168.

Description.-Holotype, male, length 6.1 mm, width 2.1 mm. Integument black. Rather densely covered with scales which are mostly contiguous on dorsal surface, the random bare areas causing a slightly tessellate appearance. Scales of side of rostrum little different from those of dorsum, continuing onto ventral surface of rostrum, absent on only central 0.4. Prothorax subvittate: scales sparser and smaller along median line, condensed on either side of median line and on sides. Scales on elytra leaving random naturally glabrous areas approximately the size of a scale on base of elytra, becoming width of interval by middle of elytra. Scales pale, opalescent, with strong lavender and rosy metallic lustre. Rostrum flat, 1.1x longer than wide; dorso-lateral edges well defined but not acute, slightly inwardly arcuate; median line sharply impressed from between eyes to interantennal line. Epistoma as wide as long, occupying .37 of anterior edge of rostrum. Head and rostrum very sparsely, finely setate-punctate. Prothorax 1.07x wider than long, sides weakly rounded between equal basal and apical constrictions; in profile prothorax almost flat. Bare spots of prothorax approximately as convex as adjacent scales, setae arising from these very inconspicuous, prostrate, shorter than an adjacent scale on disc; setae of sides only slightly longer. Scutellum with 9 seta-like scales. Elytra across humeri 1.2x wider than prothorax. Elytra 2.6x longer than prothorax. Elytra 2.3x longer than width across humeri. Elytra in dorsal view with sides subparallel, very slightly wider across middle, apical umbone weak, just entering dorsal outline, apex beyond broadly rounded; apices very briefly divergent, ending in a blunt tooth. Strial punctures fine, elongate, except larger on intervals 5 and 6 basally. Intervals basally as wide as 2-3 scales, rapidly becoming as wide as 3-4 scales. Setae of basal half of elytra very inconspicuous, as long as one scale; laterally and apically changing to arched, then straight, becoming as long as 4 scales on edge of elytra around ventrite 5. Fore femur 1.8x wider than hind femur; sculpture obsolete. Fore tibia (Fig. 140) straight except distal 0.23 slightly bowed inwards; inner edge with 7 small denticles; distal tooth small, as long as 0.3x width of apex. Metasternum 1.15x longer than ventrite 1. Ventrites 3, 4, and 5 with dense patch of scales on extreme side, scales absent medially; entire abdomen sparsely, evenly clothed with fine setae as long as those of apical edge of elytra. Ventrite 5 across base 1.8x wider than long; medially impunctate, without scales, with very few setae; apex truncate. Aedeagus and internal sac of paratypes (type not dissected) as in Figs. 168, 169. Aedeagal apodeme very short, 0.2-0.36x as long as aedeagus.

Allotype, female, length 8.1 mm, width 3.2 mm. Differs from type as follows. Scales distinctly tan and white with rosy and green lustre; elytra with maximum pattern as described in Diagnosis. Rostrum approximately as wide as long. Prothorax 1.02x wider than long. Scutellum with 3 normal scales in addition to seta-like scales. Elytra as in Diagnosis, Figs. 11, 143. Elytra across humeri 1.4x wider than prothorax. Elytra 3.1x longer than prothorax. Elytra 2.2x longer than width across humeri. Elytra in dorsal view with sides gradually divergent to middle where they are 1.3x wider than across humeri, thence rounded to interval 3, apex elongate, sutural interval elongated into a short tooth; at apex interval 10 and edge of elytra expanded before sutural tooth. Intervals much wider, from middle all intervals except sutural interval as wide as 6 scales. Setae around apical edge of elytra as long as up to 6 scales; sutural interval on declivity with 3 long, stiff setae. Fore femur 1.3x wider than hind femur. Apex of fore tibia with 2 very small teeth concealed in the setae. Ventrite 1, 1.2x longer than metasternum. Abdomen as in Diagnosis and Fig. 143. Ventrite 5 across base 1.6x wider than long. Spermatheca as in Fig. 185.

Type Series.-Holotype, BRAZIL, São Paulo, Campos Jordão, 11.IV.1962, E. Halik 2510 (São Paulo). Allotype, same data as type but 12.IV.1962, 19921, Brazil Halik 1966 Collection (Washington). Paratypes, 17 males, 14 females. BRAZIL. Minas Gerais: 1 male, 2 females, Serra do Caraça, III.1963, F. Werner, U. Martins, L. Silva (São Paulo); 1 female, Serra do Caraça, 24.II-3.III.1972, Exp. MZUSP (São Paulo); 5 males, 6 females, Vila Monte Verde, 16.IV.1960, 17 .196 [2], 18. .196 [2], I.1961, 28.II.1964 [2], 9.II.1965 [2], 15.III.1966, 17.III.1966, J. Halik (São Paulo, Washington, Howden). Santa Catarina: 1 male, Rio Vermelho, III.1947, A. Maller Coll., Frank Johnson Donor (New York). São Paulo: 9 males, 3 females, Campos Jordão, 11.IV.1962 [6], 12.IV.1962 [6], J. Halik (São Paulo, Washington, Howden); 1 male, 2 females, Campos Jordão, 12.2.52, Wittmer leg., Coll. Kuschel (Auckland).

Remarks.-Males vary in length from 5.1-7.3 mm and in width from 2.0-2.7 mm. Females vary in length from 6.3-8.5 mm and in width from

2.4-3.3 mm. The apex of the elytra is rounded and without a tooth in half the males. In males the metasternum is 1.09-1.2x longer than ventrite 1; in females ventrite 1 is 1.0-1.2x longer than the metasternum. Ventrite 5 of females sometimes has a few scales at the extreme lateral edges. Ventrite 5 of males has a few scales medially in half the specimens, none in the others; at most it is sparsely, minutely punctulate.

The internal parts of most of the type series were hard and did not respond to softening techniques (see Methods). The male genitalia shown (Figs. 168, 169) are from a specimen from Rio Natal, Santa Catarina. The genitalia of the single dissectable male from the type locality are apparently identical to that in the figures.

Hadromeropsis from the state of Santa Catarina which agree with the diagnostic characters of speculifer as listed here are consistently different from those from the states of São Paulo and Minas Gerais, an exception being one of two males from Rio Vermelho. Compared to typical specimens, these Santa Catarina specimens differ as follows. Size larger: males, length 7.3-7.7 mm, width 2.5-2.8 mm; females, length 7.8-9.0 mm, width 2.8-3.5 mm. Very densely squamose, glabrous spots smaller, prothorax only obsoletely subvittate, median line of rostrum completely or almost concealed by scales. Sometimes (both males, 1 female) immaculate; metallic lustre weaker. Rostrum 1.3x longer than wide in male, 1.2-1.4x in female, dorso-lateral edges converging apically. Prothorax as wide as long in male, 1.2x wider than long in female. Elytra across humeri 1.3-1.4x wider than prothorax in male, 1.4-1.5x wider in female. Elytra 2.9x longer than prothorax in male, 3.1-3.3x wider in female. Intervals as wide as 8 scales. Elytral setae more conspicuous in profile. Fore femur of female 1.6x wider than hind femur. Male metasternum 1.4x longer than ventrite 1; female ventrite 1, 1.03-1.14 longer than metasternum. Male ventrite 5 evenly, sparsely squamose and punctate.

The Santa Catarina specimens are labelled as follows: 3 females, no additional data, A. Maller, Gregorio Bondar Collection, David Rockefeller Donor (New York); 2 females, Cauna, III.1945, A. Maller Coll., Frank Johnson Donor (New York); 1 male, Joinville, 49.0W, 26.0S sea-level, Aug. 1926, Antonio Maller, B.M. 1931-106 (London); 1 female, Mafra, Mall. (Paris); 1 female, Mafra, 800m, 12.65 (Curitiba); 1 male, Rio Natal, II.1945, A. Maller Coll., Frank Johnson Donor (New York); 1 male, 1 female, Rio Vermelho, XII.1946, III.1947, A. Maller, Frank Johnson Donor (New York).

Field study is needed to determine if these Santa Catarina specimens represent extreme geographic variation or a sibling species.

Females of speculifer are especially recognizable by the strongly contracted sides of the elytra around ventrites 3, 4, and 5 and by the flat, mirror-like ventrite 5. Both sexes can be separated from other species within its range as follows: from atomarius by the straight fore tibia and absence of elytral tubercles; from pulverulentus by the narrower epistoma (no wider than long) and finely punctured striae; from plebeius by the lack of elytral tubercles, smaller size, and very different color pattern; from pallidus by the well-defined dorso-lateral edges of the rostrum.

The mirror-like surface of ventrite 5 in the female inspired the name speculifer meaning "with a mirror".

23. Hadromeropsis (Hadromeropsis) beverlyae n. sp.

Figs. 22-25, 152, 154, 156, 170, 171, 188-190; Map 6

Diagnosis.-Densely clothed dorsally with green scales, ventrally and laterally with cupreous or golden scales. Anterior edge of rostrum at margin of epistoma flared upwards and outwards (Fig. 156). Both sexes with the glabrous spots of the basal half of elytral interval 9 tuberculate, in dorsal view their profile more prominent than the tubercles of the sides of the prothorax. Fore femur of male with very small, shiny tubercles or granules along entire inner surface including the arcuate distal flange of inner edge (Fig. 152). Apical end of fore tibia in both sexes produced in a rounded lobe on posterior edge, the lobe at least as long as the distal tooth (Fig. 154); apical end of fore tibia of opalinus (Fig. 153) shows usual condition in other species.

Description.-Holotype, male, length 7.9 mm, width 3.3 mm. Dorsal surface of rostrum, head, prothorax, elytra, and fore femur densely clothed with green scales which exhibit a golden lustre in certain lights; remainder of body including elytral intervals 8, 9, and 10 with cupreous scales with golden lustre. Scales of basal half of elytral intervals 4 and 7 slightly more dense giving a subvittate appearance. Prothorax and elytra with random, small, glabrous spots approximately the size of a scale; the single minute seta of each glabrous spot on dorsal surface recumbent, very inconspicuous; glabrous spots of side of prothorax and basal half of elytral interval 9 tuberculate, in dorsal view those of interval 9 more prominent in profile than those of sides of prothorax. Rostrum 1.2x longer than wide, dorsal surface approximately parallel-sided, feebly depressed medially, median line impressed between anterior half of eyes. Epistoma occupying approximately 0.5x of anterior edge of rostrum, strongly concave, its apex rounded; epistoma as wide as long; sides of epistoma from about middle becoming carinate then keeled and flared outward and upward (Fig. 156). Mandibular scar scarcely exceeding outline of mandible. Prothorax 1.2x wider than long; sides strongly rounded; basal and apical constrictions equal; apical constriction weak dorsally. Basal constriction with a few white, curved setae from ventral surface evident in dorsal view. Elytra (Fig. 22) across humeri 1.2x wider than prothorax; elytra 2.8x longer than prothorax; sides very slightly divergent to middle, thence very gradually convergent to apex, apical umbone very weak, sutural interval briefly attenuate to form tooth. In profile (Fig. 23) summit of declivity unmarked, elytra gradually deflected from about middle to apex, all glabrous spots of declivity including sutural interval set with short (about as long as 2 scales), straight, obliquely deflected setae. Ventral surface of entire body (except ventrites 3, 4, and 5) and all legs with the long, wispy setae particularly dense. Fore coxa on inner distal edge with trace of tubercle as in gemmifer. Fore femur as in Diagnosis and Fig. 152, 2.6x wider than hind femur, slightly narrower than head between outer edges of eyes, scales absent on inner granular area. Fore tibia as in Diagnosis and Figs. 152, 154; inner edge with cylindrical tubercles, similar smaller tubercles covering much of inner surface. Abdomen with scales sparse on ventrites 2 and

5 medially, absent medially on ventrites 1, 3, and 4. Ventrite 5 across base 2x wider than long, apex broadly truncate, weakly convex. Aedeagus as in Fig. 171; internal sac as in Fig. 170, without basal ventral lobes.

Allotype, female, length 9.3 mm, width 3.7 mm. Differs from type as follows. Rostrum 1.1x longer than wide. Elytra (Figs. 24, 25) across humeri 1.4x wider than across prothorax, sides wider at middle, apex only slightly more attenuate and apical umbone only slightly more prominent. Long, wispy setae of ventral surface replaced by shorter, curved setae. Fore leg with no trace of tubercle on fore coxa; granules of fore femur and flange greatly reduced; tibia with granules reduced; lobe of posterior apical edge of tibia shorter. Ventrites 3 and 4 with no scales on central third, the glabrous surface roughly sculptured; surface flat, rising abruptly to caudal margin. Ventrite 5 flat, 1.9x wider than long; medially glabrous and polished, laterally bordered with short white setae and with a small sublateral patch of elongate scales; apex truncate. Genitalia of paratypes as in Figs. 188-190; vagina with 8 faint, long fins.

Type Series.-Holotype, BRAZIL, Distrito Federal, 20 km N Brasilia, III.10.1970, 1250 m, JM & BA Campbell (São Paulo). Allotype, same data as type (São Paulo). Paratypes, 9 males, 10 females, same data as type (Ottawa, Howden).

Remarks.-Males vary in length from 7.4-9.0 mm, and in width from 2.8-3.4 mm. Females vary in length from 7.4-9.3 mm, and in width from 3.1-3.9 mm. The subvittate appearance is very consistent, as is the character of the glabrous spots: mostly uniserial, larger on basal half of the elytra, becoming smaller apically until smaller than a single scale; females usually more densely squamose. The type series is very uniform except for one aberrant male which has the median line impressed on the disc of the prothorax, the head protuberant, and the basal half of interval 9 of the right elytron only raised, its scales elongate and two of the tubercles bearing moderately long, curved setae. Most specimens do not have even a trace of the coxal tubercle.

The green dorsal surface and cupreous sides and venter must lend this species particularly good color camouflage against green foliage and the reddish soil of its cerrado habitat.

It gives me great pleasure to name this species in honor of Mrs. Beverly Ann Campbell of Ottawa who collected the entire type series.

The nobilitatus Group

24. nobilitatus (Gyllenhal)   26. excubitor n. sp.
25. atomarius (Boheman)   27. fasciatus (Lucas)

Characteristics of Group.-Often with a complex fasciate color pattern on the elytra which is similar in the two sexes or reduced (weaker) in the males of some species; with a strong to weak (according to the species) red or golden lustre in reflected light. Segments 1 and 2 of antennal funicle very long, equal in length. Elytra often with setiferous tubercles. Fore tibia bowed distally; distal tooth of female double in many species. Internal sac of aedeagus ventrally with a heavily sclerotized plate which is free at its distal end (e.g., Fig. 174), this plate absent in fasciatus (Fig.

180).

24. Hadromeropsis (Hadromeropsis) nobilitatus (Gyllenhal)

Figs. 1, 14, 173, 174, 191-194; Map 8

Hadromerus nobilitatus Gyllenhal, in Schoenherr, 1834:128. Type, female, labelled "Brasilia, F....[illegible]" written on white paper; "TYPUS" on red rectangle with black border (Schoenherr Collection, Stockholm).

Hadromeropsis similis Hustache, 1938:6. Type, female, labelled "Republic Argne., Rio Parana, Territoire des Missiones," "Hadromeropsis similis m.," "TYPE" in red ink on white rectangle (Hustache Collection, Paris). Synonymized by Kuschel, 1955:278.

Diagnosis.-Body with a golden or reddish metallic lustre. Elytra (Fig. 1) marked with (a). circumscutellar half circle originating at base of interval 4, (b). oblique fascia from basal third to middle at stria 1 or suture, (c). common "V" with its apex at summit of declivity; these markings creamy white or tan in female, light to dark ochraceous in male; area between markings with scales smaller, lavender or green under microscope, this area obscure macroscopically. Prothorax with sides moderately to strongly rounded. Setiferous tubercles of intervals 8 and 9 conspicuous dorsally. Male: fore femur with distinct flange; distal tooth of fore tibia approximately as long as width of tibia at apex, acute. Female: fore femur with flange obsolete; distal tooth of fore tibia shorter, approximately 0.3-0.5x width of tibia, usually single. Fore tibia both sexes abruptly bowed at distal 0.25; of equal width to apex. Female caudal margin of ventrite 4 usually arcuate medially, projecting posteriorly. Internal sac with sclerite on ventral surface as in Fig. 174. Spermatheca with nodulus and ramus approximately equal in length, much shorter than cornu.

Description.-Male, length 6.8-8.6 mm, width 2.5-3.1 mm. Female, length 7.5-10.7 mm, width 3.0-4.5 mm. Integument medium to dark red brown. Color and elytral pattern as in Diagnosis and Fig. 1. Pale scales (white, tan or ochraceous) very dense, imbricate around eye, somewhat concentrated on sides of prothorax but not forming a vitta there, otherwise body and legs with scattered pale scales. Rostrum 1.2-1.3x longer than wide, flat or weakly concave (especially in male), median line sharply impressed from between middle of eyes to middle of rostrum. Dorso-lateral edges distinct but not acute, slightly convergent caudally. Surface of head and rostrum smooth and shiny between scales; with scattered appressed setae each set in a puncture smaller than a scale. Epistoma as long as wide, occupying approximately 0.3 of anterior edge of rostrum; sides of epistoma feebly ogival, carinate anteriorly, moderately elevated, especially in female; epistoma with an average of 3 setae on each side, usually in a single row. Sides of rostrum with several rows of rounded scales ventrad of scrobe; below this, scales usually replaced by appressed white setae. Prothorax of male 1.0-1.1x wider than long, of female 1.1-1.2x wider than long; surface in both sexes with scattered fine setae, those of disc set on smooth surface, or in a fine puncture,

those of sides on smooth surface or weak tubercle. Scutellum with or
without 2 or 3 small scales and several fine appressed setae. Elytra
across humeri averaging 1.3x wider than prothorax in male, 1.5x in
female. Elytra averaging 3.1x longer than prothorax in male, 3.4x in
female. Elytra of male (Fig. 1) parallel-sided; apical umbone weak,
touching but usually not interrupting dorsal outline; apices briefly
divergent, toothed. Elytra of female with sides parallel basally,
thence gradually divergent to just behind middle where they average
1.2x wider than across humeri (1.3x in type), then abruptly
constricted under apical umbone, apex triangular between fourth
striae, ending in a pair of slightly convergent teeth. Edge of elytra
around ventrite 5 with minute setiferous tubercles in both sexes.
Disc of elytra with random glabrous spots, these as wide as an
interval and sometimes contiguous in dark areas, smaller in areas of
pale scales, larger spots convex; each spot near its caudal edge with
a seta as long as 1-2 scales, setae appressed, weakly curved or
straight. Tubercles of intervals 8 and 9 conspicuous in dorsal view
creating a serrate outline, similar tubercles also on intervals 5, 6,
and 7 across summit of declivity; setae arising from tubercles
stouter, white, approximately as long as 2 scales, curved or straight
but parallel to surface. Female with sutural interval on declivity
with as many as 5 long, stiff setae equal in length to as many as 7
scales. Summit of declivity gradual (Fig. 14). Fore femur as in
Diagnosis; in male averaging 2.2x wider than hind femur, in female
1.7x wider. Fore tibia as in Diagnosis. Ventrite 5 of male across
base 2.0x wider than long, apex broadly rounded, scales and setae
almost evenly distributed. Ventrite 5 of female across base 1.7-2.1x
wider than long, apex narrowly rounded; with a few scales concentrated
on sides, scales absent or not medially; setae appressed, evenly
distributed. Male genitalia as in Diagnosis and Figs. 173, 174.
Female with spermatheca as in Diagnosis, Fig. 191; with a bursal
sclerite similar to that of argentinensis (Fig. 181) and beverlyae
(Figs. 189, 190).

    Distribution.-Map 8. ARGENTINA. Misiónes: Rio Paraná (the type of
similis). BRAZIL. Espírito Santo: Conceição de Barra, Santa Teresa.
Minas Gerais: Matozinhos, Viçosa, Vila Monte Verde. Paraná: Caviuna,
Curitiba. Rio de Janeiro: Corcovado, Itatiaya, Petrópolis, Rio de
Janeiro. São Paulo: Bosque de Saudo, Cantareira, Interlagos,
Jabaquara, Morumbi, Parque de Estado, São Paulo, Sitio Bananal.

    Specimens were collected in all months except the winter months of
June, July, and August.

    Specimens examined: 25 males, 44 females. Specimens in: Auckland,
Berlin, Cambridge, Curitiba, Dresden, Eberswalde, Leiden, London,
Maracay, New York, Oxford, Paris, São Paulo, Stockholm, Washington,
Howden.

    Remarks.-The above Description and Diagnosis refer to the type and
specimens which are obviously conspecific with it. I have seen a
series of 12 females which I will refer to here as "smooth
nobilitatus." These specimens all differ from typical nobilitatus in
having: the elytra smooth, i.e., the glabrous areas not or scarcely
convex and the tubercles of intervals 8 and 9 weaker; caudal margin of
ventrite 4 medially acutely pointed and posteriorly directed as in
atomarius; spermatheca variable, Figs. 192, 193, 194. Instead of the

dark preapical elytral fascia, the white scales intrude on the fascia in intervals 1, 2, and 3, and sometimes 4, reducing it to a conspicuous dark lateral spot on intervals 4 or 5 and 6 and 7. The scales are more pastel in color and the reflected lustre is paler. The apical umbone is weak in all. The rostrum is as wide as long to no more than 1.1x longer than wide. These specimens, not included in the description, range from 7.5-9.6 mm in length and 3.0-4.0 mm in width. They occur throughout the range of nobilitatus: Minas Gerais: Viçosa (2); Santa Catarina (1); São Paulo: Jundiaí (1), no other data (5); [Argentina] Est. Esp. Loreto (1); no data (1). I have seen males and females of typical nobilitatus with data identical to that of smooth nobilitatus. The pointed margin of ventrite 4 always occurs in conjunction with the smoother elytra, paler scales and different (although variable) spermatheca. These apparently consistent differences may indicate sibling species or two forms of females of nobilitatus.

The relationship of atomarius to nobilitatus has some of the attributes of a geographical cline. However, although atomarius seems to be concentrated in the Serra do Caraça, there is some apparent overlap in the range and without studying the situation in the field, it seems best to leave them as separate species. The differences between nobilitatus and atomarius which seem to be clinal are as follows: atomarius is smaller and has the elytral tubercles reduced, femoral flange less developed or even absent in male, distal tibial tooth of the male smaller, margin of ventrite 4 of female more acute, spermatheca more slender and elongate.

25. Hadromeropsis (Hadromeropsis) atomarius (Boheman)

Figs. 18, 19, 144, 145, 155, 175, 176, 195; Map 8

Hadromerus atomarius Boheman, 1840:292. Type, female, labelled "Typus" on red bordered with black, "H. pygalpis Germ. Brasil Germ" written on white (Schoenherr Collection, Stockholm). See Type Material.

Diagnosis.-Similar to nobilitatus but smaller. Metallic lustre absent or very faint. Elytral pattern (Figs. 18, 19) similar to that of nobilitatus but fainter, all scales white or shades of tan or ochraceous, none obscure, rarely with faint lavender or green reflections. Epistoma as long or longer than wide. Prothorax with sides moderately to weakly rounded. Setiferous tubercles of intervals 8 and 9 weak in dorsal view. Fore tibia of male (Fig. 144) with distal tooth 0.5 or less the width of tibia at apex; fore tibia of female (Fig. 145) with two equal teeth or distal tooth slightly longer than the second. Female ventrite 4 (Fig. 155) with caudal margin pointed medially and slightly posteriorly directed. Male apex of ventrite 5 slightly emarginate to truncate. Aedeagus in profile much thicker apically than basally (Fig. 175).

Description.-Male, length 5.5-8.0 mm, width 2.0-3.0 mm. Female, length 7.1-9.4 mm, width 2.9-3.8 mm. Integument red-brown; ventrally piceous except ventrites 4 and 5 lighter. Color of scales and pattern as in Diagnosis and Figs. 18, 19; scales almost evenly distributed except for glabrous spots which are usually larger on basal portion of elytra, usually 9-12 per interval. Rostrum and head similar to that of nobilitatus in sculpture, median line and vestiture; rostrum 1.1-1.3x longer than wide in male, 1.0-1.1x longer than wide in female. Epistoma as in Diagnosis, occupying 0.4 of anterior edge of rostrum, triangular, anterior corners scarcely elevated. Prothorax as in Diagnosis, male 1.0-1.1x wider than long, female 1.1-1.2x wider than long, almost uniformly squamose, sculpture as in nobilitatus. Scutellum with 0-8 scales. Elytra across humeri averaging 1.3x wider than prothorax in male, 1.4x wider in female. Elytra averaging 3.0x longer than prothorax in male, 3.5x longer in female. Elytra of male with sides subparallel, sides slightly constricted beneath weak apical umbone, apical edge of elytra very briefly indented immediately before suture thus forming a short tooth. Elytra of female (Fig. 18) widest at middle where they are 1.2x wider than across humeri, constricted under weak apical umbone, apex ending in a pair of slightly convergent teeth. Edge of elytra around ventrite 5 at most obsoletely tuberculate. Each glabrous spot on disc with single appressed seta as long as 1 scale in male, longer in female; towards declivity and sides setae longer, stouter, straight or slightly curved, less appressed, sometimes standing well away from surface creating a shaggy effect. Tubercles of intervals 8 and 9 weak to obsolete. Female with sutural interval on declivity with an average of 5 long, stiff setae. Fore femur without distal flange; fore femur of male 1.6-1.9x wider than hind femur; of female 1.4-1.5x wider than hind femur. Fore tibia (Fig. 144, male) bowed as in nobilitatus or less so; distal teeth as in Diagnosis, Figs. 144, 145. Female ventrite 4 with caudal margin as in Diagnosis, Fig. 155; male also often with a weak median point. Ventrite 5 of male 1.4-2.0x wider than long; scales almost evenly distributed over surface, apex as in Diagnosis. Ventrite 5 of female 1.8-2.0x wider than long; scales evenly distributed or sparser medially, apex rounded. Aedeagus as in Fig. 175, internal sac as in Fig. 176; spermatheca as in Fig. 195.

Type Material.-The remainder of the type series consists of: (1) a male and a female in the Chevrolat Collection (Stockholm) labelled "Hadromerus atomarius irroratus Klug, Brasilia," the female also labelled "Minas Gerais" and the male "Paratypus," and (2) Var. beta Boheman, 1840:292, a female, labelled "Typus" on red with black border, "Brasil mer, Schupp" written on white in the Schoenherr Collection (Stockholm). This latter specimen is teneral and has the caudal margin of ventrite 4 somewhat less pointed than in the type but is conspecific.

Distribution.-Map 8. BRAZIL: Minas Gerais: Matozinhos, Serra do Caraça, Topázios (Ouro Preto).

Specimens were collected in January, February, and March.

Specimens examined: 10 males, 22 females. Specimens in Auckland, Basel, Berlin, Cambridge, Dresden, Leiden, London, Paris, São Paulo, Stockholm, Howden.

Remarks.-Boheman's statement that the body is black below would seem to refer to speculifer, but I have no doubt that I saw the genuine type series of atomarius.

The distally thickened aedeagus separates atomarius from both nobilitatus and pulverulentus. In addition to the characters in the

Diagnosis, atomarius differs from pulverulentus in the shorter setae of the sides of the elytra and from speculifer in the thicker fore femur.

Labels on museum specimens suggest the range of atomarius apparently is concentrated in the Serra do Caraça where it is sympatric with nobilitatus and pulverulentus, but field study is needed to establish this. One female atomarius bears an identical label to the type of pulverulentus.

See discussion of nobilitatus for further comments on the relationship of nobilitatus and atomarius.

26. Hadromeropsis (Hadromeropsis) excubitor n. sp.

Figs. 141, 142, 177, 178, 196-198; Map 7

Diagnosis.-Similar to nobilitatus but more robust. Edge of elytra around ventrite 5 smooth or with minute crenulations no more prominent than the height of a convex scale. Apex of sutural interval of male not attenuated into a tooth. Fore tibia of male (Fig. 141) with distal tooth 0.5x or less the width of tibia; fore tibia of female (Fig. 142) with a pair of distal teeth, the most distal tooth shorter or equal to the second. Internal sac as in Fig. 178. Spermatheca (Figs. 196-198) with nodulus much longer than ramus.

Description.-Holotype, male, length 10.2 mm, width 3.7 mm. Integument red-brown. Scales arranged and elytra patterned as in nobilitatus (Fig. 1), but markings ochraceous to copper, scales between markings smaller, with green iridescence or opalescence under microscope; declivity also solid ochraceous; with reddish lustre in glancing light. Rostrum 1.3x longer than wide. Dorso-lateral edges of rostrum subparallel, sides of rostrum broadly visible from above; median line vaguely impressed basally; dorsum concave; sculpture and vestiture as in nobilitatus. Epistoma occupying 0.35 of anterior edge of rostrum, as in nobilitatus but with 5-7 setae on each side. Prothorax 1.1x wider than long, sides moderately rounded; sculpture and vestiture as in nobilitatus. Scutellum with 6 small scales and appressed setae. Elytra across humeri 1.4x wider than prothorax. Elytra 3.0x longer than prothorax. Sides of elytra parallel to apical third thence broadly rounded, apical umbone weak but evident in dorsal outline; apices briefly divergent, feebly knobbed. Glabrous spots and tubercles as in nobilitatus. Fore femur similar to that of nobilitatus, 2.0x wider than hind femur. Fore tibia (Fig. 141) bowed as in nobilitatus, but becoming slightly wider apicad of bend; distal tooth 0.5x as long as width of tibia at apex. Ventrite 5 across base 2.4x wider than long; moderately convex apically; apex truncate-emarginate; scales and setae slightly more numerous toward sides. Aedeagus and internal sac of paratypes as in Diagnosis and Figs. 177, 178 (type not dissected).

Allotype, female, length 11.3 mm, width 4.8 mm. Differs from type as follows. Rostrum 1.2x longer than wide. Epistoma occupying 0.4 of anterior edge of rostrum. Elytra across humeri 1.5x wider than prothorax. Elytra 3.8x longer than prothorax. Elytra wider in central third than in type; sides rounded-triangular beyond apical umbone, sutural interval terminating in distinct, medially directed tooth. Sutural interval at summit of declivity with sparse line of long, erect setae, the longest as long as 5 scales. Fore femur 2.0x wider than hind femur. Fore tibia as in Fig. 142. Caudal surface of ventrites 2 and 3 equal, perpendicular, evenly arcuate; margin of ventrite 4 slightly posteriorly directed, evenly arcuate from extreme sides. Ventrite 5 across base 2.0x wider than long, with a slight depression along middle of side, apex briefly truncate. Ventrite 2 evenly squamose, remaining ventrites slightly less densely squamose medially. Spermatheca of paratypes as in Figs. 196 and 198 (allotype not dissected).

Type Series.-Holotype, BRAZIL, Rio de Janeiro, Corcovado, GB, X.1961, M. Alvarenga, Ex-coleção M. Alvarenga (São Paulo). Allotype, BRAZIL, Corcovado, Guanabara, VIII.1965, Alvarenga & Seabra, Coleção M. Alvarenga (Curitiba). Paratypes, 12 males, 13 females. BRAZIL. 3 males, 1 female, no other data (Dresden, Oxford). Rio de Janeiro: 1 female, no other data, Fry (London); 2 males, 1 female, same data as type but IX.1961, XI.1958, VIII.1960 (São Paulo); 3 males, 5 females, same data as allotype but IX.1967, 3.XI.1958, X.1960, IX.1969 [3], 18.IX.61, J. S. Moure [2] (Curitiba, São Paulo, Howden); 1 male, Corcovado, GB, 6/X/967, Moure & Seabra (Howden); 1 male, GB, Corcovado, S.A.F. Col. No. 68 (São Paulo); 2 males, 2 females, Rio de Janeiro, Guanabara, X.1963, IX.1970 [2], XI.1970, M. Alvarenga (Washington); 1 female, Nov. Friburgo, Bescke (Oxford). No data, 2 females (Oxford, Washington).

Remarks.-Males vary in length from 8.2-10.5 mm and in width from 3.0-4.2 mm. Females vary in length from 9.4-12.8 mm and in width from 3.6-5.0 mm. The dorso-lateral edges of the rostrum range from parallel to convergent basally. Females have as many as seven long setae on the sutural interval on the declivity. The two teeth of the distal end of the fore tibia of the female are equal in size in four specimens, unequal in the remaining nine; the distalmost tooth is always the smaller when the teeth are unequal. In one extreme female the second tooth is fully twice as large as the distalmost tooth. Ventrite 5 of the female varies from 1.8-2.2x wider than long, of the male 2.0-2.4x wider than long.

Apparently this species is concentrated on and around Corcovado, although there is one historic Bescke specimen from Nova Friburgo (see Papavero 1971:88). In a mixed collection the species is conspicuously different from nobilitatus in its more robust form and reddish lustre. The form of the distal tibial teeth in both sexes and the apex of the elytra of the male will confirm the identification, and the genitalia of both sexes are quite different from those of nobilitatus.

The female of atomarius also has a pair of tibial teeth, but in that species the more distal tooth is the larger one if the two teeth are not equal. H. atomarius differs in the weak or absent metallic lustre, presence of a short tooth on elytral apex of male, pointed caudal margin of ventrite 4 of female, and absent or obsolete femoral flange.

Worn males of excubitor can be separated from the sympatric togatus by the rostrum narrower and more concave, epistoma granular, apical umbone less prominent, prothorax with scales of sides similar to those of disc and not concentrated in a vitta of dense overlapping scales, and the presence of a weak distal flange on the fore femur.

There is a sample of two males and five females from Santa Catarina which differ from typical excubitor as follows. Males, length 9.1-9.2 mm, width 3.4-3.5 mm; females, length 10.6-11.6 mm, width 4.3-4.7 mm. Color subdued, lustre almost completely absent, scales gray-green; markings faint tan or flesh colored. Scales smaller, more numerous, the surface more completely covered, glabrous spots on elytra smaller, often flat, much narrower than the width of the interval. Setae of elytral intervals more numerous, not uniserial, randomly placed; setae straight, erect. Sutural interval on declivity in female with as many as 12, averaging 10, long setae. Fore femur of male 1.7x wider than hind femur. The internal sac of the male is indistinguishable from that of Corcovado specimens; the spermatheca (Fig. 197) is very close to that of Corcovado specimens. I am uncertain whether these Santa Catarina specimens are geographic variants of excubitor or a sibling species.

Data on the Santa Catarina specimens are as follows: 1 male, no other data (Berlin); 1 male, Sao Bento, 26.0S 50.0W, 800 m, Dec. 1924, Antonio Maller (London); 1 female, Lanca, Oct. 1944, A. Maller (New York); 4 females, Nova Teutonia, XI. 1966, XI. 1967, X.1967, X.4.1962 (Curitiba, São Paulo, Howden).

The position of the Corcovado range "guarding" Rio de Janeiro suggested the name excubitor meaning "sentinel," which at the same time implies a relationship with nobilitatus.

27. Hadromeropsis (Hadromeropsis) fasciatus (Lucas)

Figs. 27, 146, 157, 158, 179, 180, 199, 200; Map 6

Hadromerus fasciatus Lucas, 1857:156. LECTOTYPE, HERE DESIGNATED, female, labelled with round green disc with "11 44" in ink on underside; vertical label "Hadromerus fasciatus, sp.n."; on green card "Museum Paris, Rio de Castelnau" (Paris). See Type Material.

Hadromerus herbaceus Lucas, 1857:157. LECTOTYPE, HERE DESIGNATED, male, labelled with round green disc with "11 44" in ink on underside; vertical label "Hadromerus herbaceous, sp. n."; on green card "Museum Paris, Rio de Castelnau" (Paris). See Type Material. NEW SYNONYMY.

Diagnosis.-Moderate to large in size. Female with conspicuous dark postmedian fascia which often appears to be composed of 3 diamonds (Figs. 27, 157), also with dark cordiform scutellar area, this area and both edges of fascia bordered conspicuously in white. Male tessellate with pattern similar to that of female but fainter. Female usually with white, tan and obscure scales; male usually with white and green or neutral scales, the latter vivid iridescent green under microscope, less frequently with pastel blue or opalescent scales. Sides of epistoma (Fig. 158) anteriorly slightly elevated and produced forward in a lobe; lateral angle of anterior edge of rostrum swollen in a distinct pterygium below scrobe and in front of antennal insertion; these characteristics of rostrum all much more pronounced in female than in male. Elytral declivity (Fig. 157) with conspicuous glabrous setiferous tubercles, otherwise without a naturally glabrous spot. Edge of elytra around ventrite 5 with acute setiferous tubercles. Internal sac without a sclerotized plate ventrally (Figs. 179, 180).

Description.-Male, length 9.2-13.6 mm, width 3.4-4.8 mm. Female, length 10.7-16.5 mm, width 4.5-6.4 mm. Color and pattern as in Diagnosis and Fig. 27. Densely squamose dorsally and ventrally except ventral surface of head and rostrum and spots of prothorax and elytra glabrous; scales often abraded from postmedian fascia of female. Female with golden red metallic lustre (as in Lucas' description of fasciatus) primarily on the tan and obscure scales; lustre never strong, most frequently present on declivity; male with golden red lustre infrequent, weaker. Scales of prothorax smaller, sparser, often darker along median third thus creating a subvittate appearance. Rostrum as in Diagnosis and Fig. 158. Rostrum flat or feebly concave basally, dorso-lateral edges parallel or feebly divergent anteriorly. Setae of head and rostrum slender, recumbent, as long as 1.5-2 scales. Epistoma 0.25-0.3x as wide as anterior edge of rostrum. Median line of rostrum impressed from between middle of eyes for basal half or less, often glabrous but not impressed for remaining distance to apex of epistoma. Prothorax 1.1x wider than long in male, almost 1.2x in female; sides moderately rounded between constrictions; apical constriction obsolete dorsally; female especially with vague flattened area on disc on either side of median line. Setae of prothorax arising from small glabrous spots, those of sides often pustulate; setae up to 3 scale lengths. Elytra across humeri averaging 1.3x wider than prothorax in male, 1.4x wider in female. Elytra averaging 3.0x longer than prothorax in male, 3.5x longer in female. In dorsal view elytra of male parallel-sided from base to beyond middle thence very gradually convergent to apical umbone which enters outline, apex broad beyond umbone, individual apices briefly divergent with a small elongation or not. Elytra of female (Fig. 27) wider at middle where they are approximately 1.2x wider than across humeri, apical umbone more prominent, apex approximately triangular; sutural interval at apex produced in acute, vertically contiguous tooth. Setiferous tubercles of intervals 8 and 9, apical umbone, and edge of elytra around ventrite 5 acute and conspicuous in dorsal view. Elytra with random glabrous spots ranging in size from scarcely more than the size of a scale (in densely squamose white areas) to full width of an interval, occasionally confluent in male especially; all glabrous spots convex, those of base of elytra and declivity more so, pustulate or tuberculate. Setae arising from glabrous spots straight, often erect on base of elytra, semi-erect on disc, erect on declivity; as long as 1.5-5 scales, shorter in female than in male, longest on declivity. Sutural interval on declivity with a row of 4-6 long setae in male, 8 in female, setae longer in female than in male. Legs, metasternum, and abdomen with many fine, long setae in male; in female setae much sparser, shorter, curved. Fore femur 1.6-1.8x wider than hind femur in male, 1.4x wider in female; without distal flange; sculpture consisting of obsolete pustules at most. Fore tibia (Fig. 146) slightly bowed distally; distal tooth 0.4-0.7x width of tibia in male, shorter in female. Male ventrite 5 across base 1.9x wider than long, almost flat, scales slightly sparser apically only, apex narrowly truncate. Female caudal surface of ventrites 2, 3, and 4 equal, perpendicular, polished, margin evenly arcuate; ventrite 5 across base 1.6-1.8x wider than long, almost flat, with or without scales. Aedeagus and internal sac as in Figs. 179, 180. Spermatheca as in Figs. 199, 200; nodulus on different plane.

Type Material.-Hadromerus fasciatus Lucas, PARALECTOTYPE, HERE DESIGNATED, 1 female, labelled like the lectotype but without the vertical label (Paris).

Hadromerus herbaceus Lucas, PARALECTOTYPES, HERE DESIGNATED, 2 males: 1 male, labelled like the lectotype but without the vertical label (Paris); 1 male, labelled with round green disc with "35-42" in ink on underside; rectangular white label "Museum Paris, Bresil, Parzudacki 1842" (Paris).

According to Mlle. Perrin (Muséum National d'Histoire Naturelle, Paris), the numbers "11 44" on the underside of the round green label of the lectotypes of fasciatus and herbaceus represent order No. 11 of the Castelnau collection of insects from "Rio", received at the Museum in 1844. The "35 42" on the underside of the label of one of the herbaceus paralectotypes refers to a collection of Coleoptera from Brazil acquired by Mr. Parzudacki, or reaching the museum, in 1842.

Papavero (1971:149-152) describes Castelnau's trip from his arrival in Brazil in 1843 until the insect collections "gathered in and about Rio" reached Paris. But there seems to be no way to define the type locality, "Bresil interieur," more specifically.

Distribution.-Map 6. BRAZIL. Bahia [1]. Minas Gerais (?): Sertão de Diamantina. Rio de Janeiro: Itatiaya, Montagnes des Orgúes (Massif de la Tijuca), Nova Friburgo, Rio de Janeiro. Santa Catarina: Corupá. São Paulo.

Specimens were collected in February primarily, also March and April.

Specimens examined: 38 males, 71 females. Specimens in: Auckland, Basel, Cambridge, Curitiba, Dresden, Eberswalde, Ithaca, London, Oxford, Paris, São Paulo, Stockholm, Washington, Howden.

Remarks.-Variation or extremes not included in the description include the following. In the Diamantina male and several others, the sides of the elytra are slightly constricted behind the humeri making the humeri extremely prominent. This same Diamantina male has the tubercles of the declivity especially long and acute. The vivid metallic green iridescence common in males is almost unique to males; only two females had any green iridescence. One female had no browns or tan but was all gray and white without the typical female golden red metallic lustre. In no specimens were the light and dark areas of the prothorax sharply defined as in togatus.

The large size combined with the distinctive postmedian fascia readily distinguish most specimens; the sharp tubercles along the edge of the apex of the elytra are helpful in distinguishing poorly marked specimens from togatus (no trace of tubercles), large nobilitatus (weak tubercles) and excubitor (no tubercles). The rostral flanges are diagnostic but not always strong; because they overhang the mandibles they are subject to abrasion and in one specimen one of the lobes was actually broken off.

This species frequently appeared in collections under the names togatus, brachispinosus, and porosus (a Pandeleteius).

This species grades into the subgenus <u>Hadrorestes</u> in the tuberculate margin of the elytra and twisted spermatheca. However, the form of the male genitalia and separated striae 9 and 10 are clearly of the <u>nobilitatus</u> lineage.

<u>Hadrorestes</u>, new subgenus

Type-species, <u>Hadromeropsis</u> (<u>Hadrorestes</u>) <u>pectinatus</u> n. sp., by present designation.

<u>Diagnosis</u>.-Dorsum of male glabrous or almost (except <u>exilis</u>); dorsum of female glabrous to densely squamose, often within the same species. Dorsal color pattern often formed of aggregations of scales into specific spots, fasciae and vittae against glabrous integument. Integument particularly thick and hard. Elytra with apical edge around ventrite 5 conspicuously dentate or tuberculate (except in brachypterous species). Striae 9 and 10 confluent or at least confused under apical umbone. Male* with internal sclerites of internal sac never flagellate; membranous parts of internal sac not distinctly spiculate in most species, at most with very pale, minute spicules. Apex of internal sac lightly to strongly sclerotized; apex directed ventrally, distally, or dorsally. Female* with paired lateral sclerites in vagina caudad of bursa copulatrix (except in <u>cavifrons</u>). Spermatheca in most species with nodulus and ramus directed in different planes.

* Male genitalia are not known for 8 of the 23 species; female genitalia are not known for 4 species.

<u>Description</u>.-Scape when positioned in the scrobe exceeding the eye in most males; in females scape exceeding eye or reaching its caudal edge. Dorsal surface of mandible smooth, punctate, or grooved. Dorsum and legs with or without elaborate sculpture, tubercles, and microsculpture according to the species; sculpture always reduced in squamose examples of same species and sculpture of legs especially reduced in females of same species. Elytra without rows of long, erect setae (except <u>institulus</u>); many species with long, wispy setae across base of elytra, these setae probably associated with production of a paraffin-like wax. Middle tibia both sexes and hind tibia in many females with a row of stout, slanting bristles.

<u>Remarks</u>.-With few exceptions, the species assigned to this new subgenus on morphological characters occur at elevations of 2000 m and above, and all are known only from the Andean cordilleras including the Sierra Nevada de Santa Marta, but excluding the coastal cordillera of Venezuela. A possible exception is <u>dialeucus</u> described here from seven specimens with no data except "Venezuela".

The species seem to be uncommon; in one 10-day collecting trip in northern Ecuador, I took five specimens representing four species. In a trip to the Sierra Nevada de Santa Marta, H. Howden and J. Campbell took only eight specimens of <u>nebulicolus</u> in a concerted effort.

The integument is so hard that the metasternum often cracks in the pinning process, but it also probably confers some protection against predators. In this subgenus, eight specimens (representing seven species) had a damaged left elytron consisting of chips out of the edge and cracks in the area of the apical umbone.

The name Hadrorestes is formed of "hadro," meaning thick, from the first part of the generic name Hadromeropsis, and "orestes," meaning mountaineer, and referring to the occurrence of the subgenus in the mountains. The name is masculine.

The alacer Group

28. alacer n. sp.   30. nebulicolus n. sp.
29. inconscriptus n. sp.   31. silaceus n. sp.

Characteristics of Group.-Large, 12-20 mm in length. Mandible with dorsal setae set in a transverse depression which is weak or strong, depending on the species. Eye separated by approximately its own diameter from anterior edge of prothorax thus forming a slight neck (Fig. 26). Postocular vibrissae weak, consisting of a few fine, short setae arising from a lobe or tooth. Elytra with punctures not aligned in striae and with many extra punctures of the same size. Elytra not flattened in profile dorsally. Multiple acute tubercles of lateral intervals visible in dorsal outline of basal half of elytra, weaker tubercles of declivity and edge of apex also visible in dorsal outline. Base of elytra with postscutellar area weakly flattened at most; stria 5 moderately depressed. In glabrous specimens fore femur with weak honeycomb or netted sculpture proximally; distally becoming coarser, irregularly rugose or foveate and with weak tubercles, those of posterior face often shelf-like in males. Fore tibia of male slightly deflected distally; inner edge with a row of many small denticles each rounded at its apex; outer surface of fore tibia strigose-punctate. Fore tarsus (Figs. 302, 304, 306, 309) with segment 2 subquadrate, abruptly tapered proximally. Female with caudal surface of ventrites 2, 3, and 4 equal in height, smooth and polished, the edge not acute on ventrites 2 and 3 and slightly anteriorly directed. Internal sac internally with a heavily sclerotized structure resembling a bone with condyle at proximal end; apex of internal sac forming a short cone, the cone directed ventrally when internal sac is evaginated.

Remarks.-Characters shared with the impressicollis group include: elytra not striate and eye remote from anterior edge of prothorax by its own diameter (male) or more (female). It is interesting that these two groups and exilis, all with the eye far removed from the prothorax, also have the postocular vibrissae greatly reduced in size. If the postocular vibrissae are used to groom the eye, excessively long vibrissae would be needed because of the length of the neck. It could be argued that instead of extreme elongation to maintain the grooming function, the vibrissae have degenerated.

In this group the apical edge of the elytra is the smoothest in the subgenus, except for the brachypterous species.

A distinct sclerotized plate was observed inside the bursa copulatrix of one nebulicolus and one alacer, but information on this condition is incomplete for the group. Larger series and staining are required for further study.

28. Hadromeropsis (Hadrorestes) alacer n. sp.

Figs. 26, 201-203, 207-209, 294, 295, 302, 310, 311, 336; Map 9

Diagnosis.-Pronotum with lateral depression or flattened area (Figs. 207, 208). Sides of metasternum and abdomen not clothed differently from rest of metasternum and abdomen, squamose or not; metasternum with a patch of imbricate white scales above hind coxa. Disc of elytra in glabrous specimens (i.e., without large scales) with foveae randomly but densely and rather evenly distributed (Fig. 203), many contiguous and forming short transverse depressions. Apex of elytra of female slightly triangular. Ventrite 5 of male distinctly emarginate (Fig. 294). Ventrite 5 of female laterally not depressed below level of margin. Apex of aedeagus simply triangular with a slight lip.

Description.-Holotype, male, length 14.2 mm, width 5.0 mm. Similar to Fig. 201. Black and various shades of reddish brown; head, abdomen, meso- and metathorax black. Appearing glabrous macroscopically but all surfaces except tarsi rather evenly clothed with minute scales, those of head and rostrum ochraceous and round, most scales elsewhere white, elongate. Ventrally with patch of dense, white imbricate scales beside middle coxa and on posterior-lateral corner of metasternum above hind coxa. With long, wispy setae as follows: a few on ventral surface of apex of rostrum; on sides and ventral surface of prothorax, on base of elytra, on remainder of ventral surface and legs except tarsi. Head and rostrum with coarse punctures rapidly converging to form rugae on dorsum of rostrum. Rostrum broadly, deeply concave. Mandibular groove weak, with 2 setae. Eye especially prominent anteriorly, perpendicular to side of rostrum. Prothorax 1.1x wider than long; sculptured as in Fig. 207 and Diagnosis; apical constriction weak dorsally. Elytra across humeri 1.3x wider than prothorax, elytra 3.0x longer than prothorax. Sculpture of elytra as in Diagnosis and Fig. 203. Setae of disc of elytra inconspicuous, approximately as long as the diameter of one fovea; setae of lateral intervals similar to those of silaceus but shorter. Apex of elytra as in Fig. 202, briefly truncate. Summit of declivity with 9 long setae on sutural interval, these long setae approximately as long as the setae on apical margin of elytra, much shorter than the long setae on summit of declivity in female. Fore femur similar to but more strongly sculptured than Fig. 109; 2.2x wider than hind femur; inner edge with strong, brief, arcuate flange; posterior face with 10 very strong and several weaker shelf-like tubercles. Fore tibia with a few granules on inner surface; outer surface of all tibiae very strongly sculptured. Fore tarsus as in Fig. 302. Ventrite 5 as in Diagnosis and Fig. 294; 2.0x wider than long. Last tergite with fine microsculpture between small punctures. Aedeagus and internal sac as in Figs. 310, 311.

Allotype, female, length 16.5 mm, width 6.7 mm. Differing from type as follows. Similar to Fig. 26. Integument reddish except black head, meso- and metathorax, ventrites 1 and 2 and median vitta on pronotum. Squamose except as follows: scales abraded from more elevated positions as dorso-lateral edges of rostrum, median line of pronotum, elytral suture, and transverse arc on metasternum, the black

or reddish color of the integument thus forming dark markings. With
round ochraceous scales dorsally and ventrally and on legs; scales
more elongate on sides and ventral surface of head and rostrum, very
densely imbricate near middle and hind coxa where male has white
spots; scales greenish white on pronotum and on disc of elytra in 3
obscure fasciae; scales darker and abraded in broad fascia at apical
third. Sculpture of head, rostrum and pronotum absent, replaced by
scales. Rostrum with shallow concavity, median line deeply foveate
between eyes. Prothorax with apical constriction unmarked on disc,
with tubercles of sides replaced by shiny granules scarcely larger
than a scale. Elytra across humeri 1.5x wider than prothorax, elytra
3.4x longer than prothorax. Sculpture of elytra reduced, the larger
concentrations of scales in depressed areas. Apex of elytra slightly
triangular, ending in a brief tooth, with slight swelling at junction
of intervals 3 and 9. Sutural interval with approximately 10 long
setae at summit of declivity (some obviously broken off). Fore femur
1.2x wider than hind femur. Ventrite 5 (Fig. 295) weakly convex
longitudinally, laterally flattened, 1.75x wider than long. Genitalia
of paratypes as in Fig. 336; spermatheca very slightly deflected
between nodulus and ramus; spermathecal duct 3-4 mm long, attached to
a faint sclerite within bursa copulatrix.

Type Series.-Holotype, COLOMBIA, Staudinger (Dresden). Allotype,
COLOMBIA, Bogotá, Fry Coll. 1905-100 (London). Paratypes, 3 males, 4
females. COLOMBIA. Valle: 1 male, 1 female, Caucathal, Coll. J.
Faust, Ankauf 1900 [1], Gehr W. Muller Vermacht 1909 [1] (Dresden).
ECUADOR. Carchi-Napo border: 1 female, Sebundoi, 11-15.IX.1977, 2600
m, L. Peña (Howden). Napo: 2 males, 7 km S Baeza, 2000 m, 21,
25.II.1979, H & A Howden, on bamboo [1] (Howden). Pichincha: 1 female,
Quito, 24362, Fry Coll. 1905-100 (London). Tungurahua: 1 female,
Runtun, 22.XI.1938, Coll. F.M.Brown (New York).

Remarks.-The males vary in length from 12.4-14.0 mm and in width
from 4.1-4.8 mm. Females vary in length from 14.0-17.6 mm and in
width from 6.1-7.0 mm. The male paratype from Colombia has coloring
as the type, and the two from Ecuador are all black except the femora
reddish. The type has many more scales on the head than the other
males. Two males lack the middle coxal spot, all have the hind coxal
spot. The apex of the elytra in all males is less truncate than in
the type. The internal sac was dissected in three of the four males;
two females were dissected, one each from Colombia and Ecuador.

The two males from Baeza are so different in habitus that special
mention is made of them here. Figures 208 and 209 illustrate their
much smoother integument. In addition they are smaller, have fewer
long wispy setae, and lack the long setae at the summit of the
declivity.

The female from Colombia is not squamose as the allotype, but
covered with minute scales as the holotype; the sculpture on this
specimen is slightly reduced from that of the holotype. All squamose
females (Fig. 26) have the median line of the pronotum dark, a dark
area bordering the postscutellar depression, and a short to wide
fascia at the apical third, the latter widest in the allotype in which
it extends the full width of the elytra.

Many specimens have the dorso-lateral edge of the elytra (interval
8 of striate species) from the humerus to the apical umbone slightly

elevated.

It is the sculpture of the elytra and form of the prothorax that help to associate the sexes of silaceus and alacer when dealing with squamose females and glabrous males. Also, in silaceus both sexes seem to lack the erect long setae on the sutural interval whereas they are present, often in both sexes in alacer.

Of the specimens from Baeza, one was collected when it flew onto a bamboo overhanging the path and the other was taken on an unidentified herbaceous plant nearby. Both specimens were very lively, hence the name alacer.

29. Hadromeropsis (Hadrorestes) inconscriptus n. sp.

Figs. 204, 298, 299, 304, 312, 339; Map 9

Diagnosis.-Similar to alacer but setae of disc of elytra (Fig. 204) as long as twice the diameter of a fovea; fore femur of male with entire distal half almost uniformly set with small shelf-like tubercles; cone-like apex of internal sac longer (Fig. 312); spermatheca very long and twisted (Fig. 339).

Description.-Holotype, male, length 14.2 mm, width 5.0 mm. Black, shading to piceous at extremities. With minute oval or elongate elliptical white scales on head, rostrum, sides of prothorax, elytra; scales very sparse ventrally. With patch of imbricate white scales on metathorax in front of hind coxa and a few scales clustered on side of ventrites 3 and 4. With long, wispy setae as follows: on sides and apex of ventral surface of rostrum, on prosternum (none on sides), a few across base of elytra, on remainder of ventral surface and legs except tarsi. Head and rostrum strongly punctate, very few punctures confluent along median line, otherwise similar to alacer. Prothorax 1.1x wider than long, lateral depressions more conspicuous than in alacer; median line convex and punctate, sculpture rapidly changing to discrete tubercles each bearing a long, curved seta; apical constriction weak dorsally. Elytra across humeri 1.2x wider than prothorax, elytra 2.5x longer than prothorax. Base of elytra without postscutellar depression. Elytral sculpture as in Fig. 204, foveae very deep and crowded, rarely confluent. Setae of elytra as in Diagnosis; setae of lateral intervals only slightly longer than those of disc; sutural interval at summit of declivity with setae much longer. Apices of elytra briefly, individually rounded. Fore femur 2.2x wider than hind femur; inner edge with strong, brief, arcuate flange. All setae of femora and tibiae long and wispy. Fore tarsus as in Fig. 304. Ventrite 5 in Fig. 298. Last tergite sculptured as in alacer but setae much more numerous and longer. Aedeagus similar to that of alacer; internal sac as in Fig. 312 and Diagnosis.

Allotype, female, length 15.3 mm, width 6.2 mm. Differs from holotype as follows. Scales more numerous; with a very small cluster of scales beside middle coxa. Long, wispy setae reduced in number except on abdomen. Sculpture of head and rostrum weaker, concavity elongate triangular from about middle to very deep fovea on median line between eyes. Prothorax with sculpture weaker. Elytra across humeri 1.4x wider than prothorax, elytra 3.3x longer than prothorax.

Elytra widest at middle, sides gradually rounded from there to apex, umbone very weak, just touching dorsal outline; apex with short tooth. Elytra with foveae more frequently confluent. Sutural intervals at summit of declivity with approximately 8 long setae. Fore femur 1.4x wider than hind femur. Ventrite 5 (Fig. 299) across base 1.7x wider than long, evenly convex from base to narrow apex, sides not at all flattened, surface coarsely but sparsely punctate, with very long setae. Spermatheca as in Fig. 339 and Diagnosis; spermathecal duct 2.5 mm long.

Type Series.-Holotype and allotype, PERU, Cuzco, Alfamayo, 40 km SE Quillabamba, Elev. 2600 m, 8.Jan.1979, W.E. Steiner, mating pair (Washington, USNM Type No. 100194).

The pair was taken on new growth on a woody composite.

Remarks.-Until it was cleaned with ammonia, the male had every elytral fovea filled with wax and all the long, wispy setae matted down with the semi-opaque wax.

The characteristics of the male and female genitalia were the prime reason for considering this a different species from alacer. As additional specimens become available, the other differences noted can be assessed for variation and possible diagnostic value: individually rounded apices of elytra in male, weaker apical umbone, slightly less prominent eye, and presence of scales on ventrites 3 and 4. In addition, the locality may be significant.

A female from Yungas de La Paz (Kuschel Coll., Auckland) may be this species. The specimen is 12.7 mm long, 5.0 mm wide and differs from the allotype as follows: prothorax rather strongly sculptured; elytral tooth obsolete; sutural interval at summit of declivity with only 3 long setae; abdomen densely squamose except for flat circular glabrous spots from which setae arise; scales more elongate centrally on abdomen; surface of ventrite 5 with sides flat, punctuation weaker; spermathecal duct 4.0 mm long. The specimen is missing both fore legs. The spermatheca and lateral sclerites are as in Fig. 339.

Hadromeropsis impressicollis males bear a superficial resemblance to alacer and inconscriptus especially in the prothorax and coxal spots, but impressicollis differs in the base of the elytra not impressed at stria 5, elytral foveae much more distant, rostrum more nearly perpendicularly sided, scutellum white-scaled, and other group characteristics. Note in Fig. 312 the proximal end of the internal sclerite is split and elongated dorsally, in this respect being intermediate between impressicollis and the other alacer group species.

The name "inconscriptus" means "unarranged" and refers to the appearance of the elytral punctures.

30. Hadromeropsis (Hadrorestes) nebulicolus n. sp.

Figs. 205, 296, 306, 313, 314, 338; Map 9

Diagnosis.-Elytra sparsely (in male) or densely (in female) clothed with rounded white and pale colored scales, both sexes with a vague postmedian V or U-shaped fascia outlined in larger, usually paler scales; some specimens with metallic green scales ventrally and on sides of elytra. Female with, male without long, erect setae at summit of declivity. Elytral striae often faintly discernible. Ventrally with an area of imbricate scales on metasternum above hind coxa, often with other patches as well. Last tergite of male very finely, sparsely, punctate; setae short. Cone-shaped apex of internal sac evenly sclerotized (Fig. 313).

Description.-Holotype, male, length 14.6 mm, width 5.0 mm. Black, extremities shading to piceous. Patterned dorsally with scales of various sizes from small to medium, mostly rounded, white and pale shades of gray, ventrally some vivid blue-green as well. Scales of head and rostrum very sparse, not arranged in a pattern except a partial border above eye; scales of pronotum slightly more numerous and larger in a vague vitta between eye and stria 4; scales of elytra with larger white scales vaguely defining a basal semicircle and a postmedian "V". Ventrally with blue-green scales in a small patch on fore coxa and beside it, on mesepisternum anteriorly, on side of middle coxa, on metepisternum (very small cluster), on metasternum laterally adjacent to hind coxa and continuing onto coxa itself, and in small clusters on side of every ventrite. With long, wispy setae as in alacer. Head and rostrum densely, finely punctured; some punctures confluent in concavity. Mandibular groove weak but distinct. Eye moderately prominent anteriorly. Prothorax almost as long as wide, sides widest caudad of middle, lateral flattened area obsolete. Pronotum with median line very finely, partially impressed, otherwise sculpture of disc similar to that of alacer. Elytra across humeri 1.3x wider than prothorax, elytra 2.8x longer than prothorax. Elytral sculpture with fewer extra-strial punctures or foveae than in alacer, striae partly discernible amongst the other sculptural details; with various irregular, rectangular, depressed areas containing aggregations of scales. Setae of elytra compared to those of alacer slightly longer, more conspicuous, white; setae of sutural interval at summit of declivity scarcely longer, less curved; setae of lateral intervals as in alacer. Apices of elytra very slightly individually rounded. Fore femur 1.7x wider than hind femur; posterior face with about 10 weak shelf-like tubercles; distal flange weak, with several small tubercles on edge. Sculpture of tibia weaker than in alacer. Fore tarsus as in Fig. 306. Ventrite 5 (Fig. 296) broadly truncate-emarginate. Last tergite as in Diagnosis. Aedeagus as in Fig. 314; internal sac not evaginated in type, in paratypes as in Fig. 313 and Diagnosis.

Allotype, female, length 17.8 mm, width 7.1 mm. Differs from type as follows. Densely covered dorsally and ventrally with ochraceous scales; integument naturally glabrous only briefly along median line of pronotum and broadly medially on metasternum and abdomen where scales are distant. Scales particularly dense ventrally where type has patches of scales. Pattern of elytra similar to that of type but encircled areas filled with darker scales; interval 9 from base to middle and interval 8 from middle to apex paler, forming an indistinct vitta. Sculpture of head, rostrum and prothorax replaced by scales; rostrum only weakly concave, median fovea between eyes extremely deep. Prothorax 1.2x wider than long. Elytra across humeri 1.5x wider than prothorax, elytra 3.6x longer than prothorax. Disc of elytra (Fig. 205) almost smooth, weakly undulating between very faintly impressed striae; glabrous areas of postscutellar region subtuberulate. Sutural interval at summit of declivity with 5 very long setae and 5 or more slightly shorter ones. Apex of elytra scarcely constricted under apical umbone, gradually and slightly rounded to short tooth. Fore femur 1.1x wider than hind femur. Ventrite 5 feebly convex apically, apex narrowly truncate, 2x wider than long. Genitalia of paratypes (two dissected) as in Figs. 338; spermatheca in one plane; spermathecal duct 2.5-3.0 mm long, attached to a large vertical sclerotized plate at caudal end of the bursa copulatrix. Lateral sclerites connected caudally by a large U-shaped sclerotization in one.

Type Series.-Holotype, COLOMBIA, Magdalena, San Lorenzo, 41 km S Sta. Marta, 7000 ft, V.5.1973, Howden and Campbell (Howden). Allotype, same data as type but V.9.1973 (Howden). Paratypes, 3 males, 3 females. COLOMBIA. Magdalena: 1 male, 2 females, same data as type but V.3.1973 [1 female], V.7.1973 [1 female], V.9.1973 [1 male]; 2 males, 1 female, San Lorenzo Area, 44 km S Sta. Marta, 8000 ft, V.11.1973, Howden and Campbell. Paratypes in Ottawa, Washington, Howden.

Remarks.-Males vary in length from 11.8-15.0 mm and in width from 3.9-5.3 mm. Females vary in length from 17.8-19.0 mm and in width from 6.7-6.8 mm. The most striking variation is in the coloring of the females; one is similar to the allotype, but the other two are iridescent-scintillating green on all ventral surfaces and legs, sides of head and rostrum, most of prothorax and base of elytra; in one specimen the postmedian fascia is green also. In one male there are a few large scales scattered along the metepisternum. In two of the males the shelf-like tubercles of the fore femur are weaker than in the type; in the third male they are stronger than in the type.

The left elytron of the type has a nip out of the margin and both hind tibiae nicked as if by predators. One of the females has a similar nip and a parallel nip on the opposite elytron with the edge cracked, as well as a longitudinal crack on stria 5 near the base.

In the other three species in this group the apical fascia (when present) is transverse; in nebulicolus it is always distinctly V- or U-shaped.

The species is endemic to the Sierra Nevada de Santa Marta where it was taken in the cloud forest, hence the name nebulicolus, meaning "living in the clouds". The collecting site is described and illustrated by the collectors in Howden and Campbell (1974).

31. Hadromeropsis (Hadrorestes) silaceus n. sp.

Figs. 206, 297, 309, 315, 316, 337; Map 9

Diagnosis.-Glabrous specimens with a vitta on lateral edge of metasternum formed by dense, imbricate, oval white scales; vitta continuing down sides of abdomen almost to apex and anteriorly present on mesepimeron. Pronotum without lateral depression. Disc of elytra of male (Fig. 206) with strial punctures small, surrounding surface smooth or minutely sculptured. Apex of elytra of male and female broadly rounded. Ventrite 5 of male broadly truncate-emarginate (Fig.

297), 1.9-2.2x wider than long. Ventrite 5 of female laterally depressed below level of side margin. Apex of aedeagus very elongate (Fig. 315).

Description.-Holotype, male, length 18.7 mm, width 6.3 mm. Black to piceous. Dorsally without scales except a few minute ones on apex of elytra; small, rounded scales on legs; imbricate oval scales ventrally as in Diagnosis; all scales white. With conspicuous, numerous, long, wispy setae on sides and ventral surface of rostrum; on prothorax dorsally on basal constriction, on sides and on ventral surface; on elytra across base; on entire ventral surface, but sparse on abdomen and on all legs. Head and rostrum irregularly punctate; rostrum moderately longitudinally concave; apex of rostrum and mandibles with many extra, long setae. Mandible foveate-punctate, dorsal 5 or 6 setae contained in foveae in a broad, sharply marked transverse depression. Prothorax 1.1x wider than long; disc slightly evenly convex, basal and apical constrictions equal on sides, apical constriction unmarked on disc, median line unmarked in any way. Disc of pronotum smooth, only very finely, sparsely punctate; sides of prothorax with moderate, discrete tubercles, the long setae arising from punctures as well as from tubercles. Elytra across humeri 1.4x wider than prothorax, elytra 3.2x longer than prothorax. Elytra sculptured as in Fig. 206 and Diagnosis. Setae of lateral intervals, declivity, and edge of elytra short, fine, curved, numerous, 2 or more rows per interval. Apex broadly rounded with brief emargination before suture. Fore femur 1.8x wider than hind femur, distally with a moderate flange with crenulate edge; posterior face with 12 or more weak shelf-like tubercles. Inner surface of fore tibia with several granules. Fore tarsus as in Fig. 309. Ventrite 5 as in Fig. 297 and Diagnosis. Last tergite shiny between large punctures. Aedeagus and internal sac as in Figs. 315, 316.

Allotype, female, length 18.4 mm, width 7.5 mm. Differs from type as follows. All surfaces dorsally and ventrally including legs (except tarsi) evenly covered with small, round, convex, ochraceous scales; scales mostly contiguous except at follicles and granules. Elytra also with pale pinkish scales forming an obscure pattern of a pair of basal rings and various zigzag fasciae. Erect setae apparently as numerous as in type but shorter, those of sides of prothorax especially conspicuous, perpendicular to surface. All sculpture less evident, mostly replaced by scales. White scales of ventral surface of male replaced by densely imbricate ochraceous scales. Prothorax 1.2x wider than long; tubercles of sides replaced by shiny granules averaging diameter of 1 scale. Elytra across humeri 1.5x wider than prothorax, elytra 3.6x longer than prothorax. Postscutellar depression concave. Sides of elytra slightly (1.1x) wider at middle, apex similar to that of male. In profile elytra evenly convex from about basal third, summit of declivity scarcely perceptible; setae of sutural interval on declivity of uniform length, not longer at summit of declivity. Fore femur 1.1x wider than hind femur. Ventrite 5 as in Diagnosis, 2x wider than long. Genitalia of paratype as in Fig. 337; spermathecal duct 2.5 mm long; spermatheca in one plane.

Type Series.-Holotype, COLOMBIA, Baden, coll. J. Faust, Ankauf 1900, "Hadromerus hirtipes J Faust" (Dresden). Allotype, COLOMBIA, same data as type except "Hadromerus silaceus J. Faust" (Dresden). Paratypes, 4 males, 2 females. COLOMBIA. [Antioquia?] : 1 female, Frontina (Oxford). Valle: 1 female, Cauca, ex coll Clerc (Paris). No data, 2 males (Oxford, Howden). Questionable data, 2 males, "Brasilien," Coll. C. Felsche, Kauf, 20, 1918 (Dresden).

Remarks.-Males vary in length from 14.6-17.0 mm and in width from 5.0-6.1 mm; one female is 20.5 mm in length, 7.0 mm in width. The female from Cauca is mostly glabrous, with the elytra sharply patterned with dense, overlapping, small, round scales forming a basal ring, median zigzag fascia, and reverse zigzag fascia across apical third, the latter fascia on side continuing along interval 8 beneath apical umbone in an oblique line to suture; punctures in the naturally glabrous areas slightly larger than in Fig. 206. The paratypes otherwise vary little from the type. The semi-opaque paraffin-like wax is common on the males. None of the females has the long setae commonly found on the summit of the declivity in females.

The species is readily distinguished by the diagnostic characters listed.

Faust's manuscript name, "hirtipes", would be very apt for the species, but the stem has been used often in Tanymecini, so I have chosen the name he attached to the female, "silaceus," which aptly describes the ochraceous color of the squamose females.

The _exilis_ Group

32. _exilis_ n. sp.
Characteristics of Group.-Head somewhat elongate, eye separated by 0.7x its own diameter from anterior edge of prothorax. Postocular vibrissae weak. Fore femur slender, without sculpture. Internal sac with spiculate lobes; sclerotized apex long, straight, distally directed.

Remarks.-H. exilis is intermediate between the _alacer_ and _impressicollis_ groups in the characters above. The species, and hence the group, is represented by a single rather abraded male and it is therefore difficult to assess the characters of punctuation, setae, etc.

32. _Hadromeropsis_ (_Hadrorestes_) _exilis_ n. sp.

Figs. 363-366

Diagnosis.-Based on unique male. Slender, elytra elongate oval. Clothed with round, iridescent green scales. Brachypterous: wing 0.7x length of elytron, mesepimeron narrow, elytra narrow across humeri. Edge of elytra smooth. Fore femur very slender.

Description.-Holotype, male, length 10.1 mm, width 3.8 mm. Integument black, legs and antennae reddish. Dorsally apparently [specimen abraded] evenly, sparsely clothed with round, iridescent green scales the same size as those of _gemmifer_. Ventrally scales concentrated on sides, only occasional scales in central third. Apparently without long, wispy setae. Head and rostrum (Figs. 363,

364) robust, rostrum approximately as long as wide. Dorso-lateral edges well marked on distal half, gradually weakening and convergent basad. Dorsum of rostrum gently sloped towards median line, median line deeply impressed between anterior half of eyes only. Lateral angle of anterior edge of rostrum slightly produced, extending beyond apex of scrobe more than usual in the genus. Epistoma approximately 1.5x longer than wide, apical edge deeply excised (worn?) (Fig. 363); epistoma occupying 0.26 of anterior edge of rostrum. Surface of head and rostrum smooth except for shallow punctures, some punctures contiguous on base of rostrum. Mandibles positioned so that their dorsal surface not visible. Eye large, round, strongly convex. Prothorax 1.05x wider than long, sides slightly rounded between equal basal and apical constrictions, basal constriction continuous across disc of pronotum, apical constriction absent on disc. Pronotum (Fig. 364) without depressions, weakly convex; sculpture consisting of minute punctures medially, obsolete tubercles laterally. Elytra across humeri 1.3x wider than prothorax, elytra 3.0x longer than prothorax. Elytra as in Diagnosis, summit of declivity and apical umbone unmarked. Elytral striae with distinct, slightly irregularly spaced punctures, some glabrous areas indistinctly elevated on intervals, otherwise with no sculpture. Edge of elytra smooth, without any trace of denticles or tubercles. Apices of elytra briefly divergent, produced in a short, blunt tooth. Setae abraded, but intervals 8 and 9 with prostrate white setae as long as 2 scales. All legs long, slender, with scattered minute punctures, otherwise without sculpture; fore femur 1.3x wider than hind femur, without distal flange; fore tibia almost straight, with 8 small teeth on inner edge, distal tooth only slightly larger, concealed in setae. Middle coxae more protuberant than usual for genus, mesepisternum consequently convex. Prosternum and metasternum finely strigulate. Abdomen long and narrow; ventrite 5 across base 1.4x wider than long, slightly, evenly convex, surface smooth and polished with sparse fine punctures; apex truncate. Genitalia as in Figs. 365, 366; apex of aedeagus cupped and minutely spiculate.

Type Series.-Holotype, male, BOLIVIA, Chapare, 400 m, 3.1951, Zischka leg., Coll. Kuschel (Auckland).

Remarks.-H. exilis is rich in characters as discussed in the group Remarks and species Diagnosis.

It is surprising to find this flightless species at such a low altitude as 400 m. Of the other flightless Hadromeropsis, one occurs at a high altitude of 3760 m in Colombia and the other has no data.

The word exilis means thin, weak, and refers to the slender habitus and inbility of this species to fly.

The impressicollis Group

33. impressicollis (Kirsch)   35. bombycinus n. sp.
34. institulus n. sp.   36. brachypterus n. sp.

Characteristics of Group.-All scales greatly elongate (Figs. 213, 215, 219). Both sexes with scutellum very densely clothed with very long, imbricate white scales; with small patches of similar scales on anterior face of fore femur proximally, on mesosternum beside coxa, and on metasternum anterior to hind coxa. With long, wispy setae on base of elytra except in brachypterus. Pronotum with a pair of lateral depressed or flattened areas as well as median line depressed (Figs. 212, 213, 219) except in brachypterus (Fig. 227). Setae on tubercles of intervals 8, 9, and 10 not strongly curved or erect, prostrate or nearly so. Posterior face of fore femur of male (known only for impressicollis) with shelf-like tubercles. Anterior face and/or outer edge of all tibiae of both sexes covered with conspicuous crowded grooves or strigae.

Remarks.-Three of the four species in this group are represented by females only, and the female genitalia could not be studied for the only species represented by both sexes. H. impressicollis males share the peculiar shelf-like tubercles of the posterior face of the fore femur with the alacer and pectinatus groups.

In the characteristics of the male genitalia and long neck, this group is intermediate between the alacer and exilis groups on the one hand and the remaining Hadrorestes on the other. In impressicollis the apex of the internal sac (Fig. 317) is almost straight, only very slightly curved ventrad, thus beginning the transition from a ventrally directed to a dorsally directed apex; in this respect the distally directed apex of exilis is more advanced. The proximal end of the internal sclerite has a small dorsal spur, transitional between the condyle-shaped end and the strongly reflexed form of the remainder of Hadrorestes.

33. Hadromeropsis (Hadrorestes) impressicollis (Kirsch)

Figs. 210-214, 303, 317-319; Map 11

Hadromerus impressicollis Kirsch, 1867:233. Type, unique male, labelled "Bogota, Kirsch" in India ink on green paper (Dresden).
Hadromerus ruficrus Kirsch, 1867:232. Type, unique female, labelled "Bogota, Kirsch" in India ink on green paper (Dresden). NEW SYNONYMY.
Hadromeropsis subaeneus Voss, 1953:61. Type, destroyed (Strümpel, in litt.). Paratype, female, labelled "Socorro Col. 3500" handwritten, "Sig. C. Rudel, Eing. Nr. 1/49" mechanically printed, "Hadromeropsis subaeneus m." handprinted on white, "Paratype" mechanically printed on orange, "Coll. E. Voss, Eing. 3-75" mechanically printed on white (Hamburg). NEW SYNONYMY.

Diagnosis.-Rostrum with conspicuous longitudinal parallel rugae from interantennal line to head behind eyes, each groove infrequently interrupted and curving around the elongate frontal pit (Fig. 210). Prothorax (Figs. 212, 213) tuberculate, more strongly so on sides and elevated areas, the depressed areas of pronotum and space between tubercles may be smooth and shiny, or transversely rugulose, or minutely granulate; any scales present extremely long and narrow (Fig. 213).

Description.-Male, length 11.0-13.0 mm, width 4.0-4.8 mm. Female, length 14.0-15.0 mm, width 5.6-6.1 mm. Black; antennae, legs and apex of rostrum reddish becoming piceous in older specimens, sometimes with subaeneous lustre in depressed areas of prothorax where glabrous.

Scales white or off-white. Maximum vestiture of females: short, sparsely set scales on head and rostrum as in Fig. 210; side of head with denser, longer scales; prothorax on side (Fig. 213) with very densely placed, very long scales giving a shaggy, furry appearance, these scales sparser in lateral depressions and replaced medially by sparse, shorter scales directed anteriorly; elytra with long scales moderately closely placed except for glabrous trifasciate pattern of indistinct arc at base, oblique fascia before middle, and parallel oblique fascia halfway to apex. Minimum vestiture of female: prothorax with scales absent on disc, reduced to vitta on side; elytra with only a few scales medially, remainder of surface with minute scales like the setae of strial punctures. Vestiture of male like minimum vestiture of female. Long, wispy setae present on ventral surface including ventral surface of femora and tibiae, on base of elytra, and, at least in female, on sides of prothorax. Rostrum as in Diagnosis and Fig. 210. Eye separated from anterior edge of prothorax by its own diameter. Prothorax 1.1-1.3x wider than long in male, 1.25-1.3x wider in female. Prothorax as in Diagnosis and Figs. 212, 213; when glabrous, sculpture much more evident; median depression weaker in female, absent in subaeneus paratype. Elytra across humeri 1.2x wider than prothorax in male, 1.3x wider in female; elytra 2.8-3.0x longer than prothorax in male, 3.3-3.6x longer in female. Apical umbone scarcely distinguishable. Apices of elytra briefly individually rounded in male; in female with a very brief tooth (Fig. 214). Sutural interval at summit of declivity without additional long setae in male; in female with at least 1 very long straight and several long curved setae. Elytral intervals except on disc with a row of tubercles, each tubercle approximately the diameter of a strial puncture, female in addition with a few weak transverse rugosities, intervals 7 and 8 "with less regular rows of tubercles, so that the side contour of the elytra appears crenulate" as in the description of subaeneus by Voss (1953:62). Each tubercle with a single fine seta, those of extreme base long, erect and wispy, those elsewhere curved; setae present the length of each interval whether arising from tubercle or not. Strial punctures ranging from nowhere aligned to perfectly aligned in rows (1 male); on declivity striae confused to absent in all specimens. Fore femur of male 1.8-2.0x wider than hind femur; inner edge with abrupt conspicuous, short, thick flange, its edge crenulate or smooth; anterior face of fore femur rugulose on distal half; posterior face with shelf-like setiferous tubercles moderately to strongly developed. Fore femur of female 1.3-1.4x wider than hind femur; posterior face without shelf-like tubercles. Fore tibia on inner edge with 8-18 nearly uniform equidistant teeth, proximally with a slight flange opposite the femoral flange. Fore tarsus of male as in Fig. 303. Metasternum granulate or tuberculate, the sculpture much weaker in female. Ventrite 5 of male 2.0x wider at base than long, apex broadly truncate; surface flat, punctate, the punctures much denser apically. Male genitalia as in Figs. 317-319. Ventrite 5 of female (Fig. 211) 1.8x wider at base than long. Female genitalia unknown - see Remarks.

Distribution.-Map 11. COLOMBIA. Cundinamarca: 1 male, 1 female, Bogota, Kirsch (the types of ruficrus and impressicollis) (Dresden). Valle: 1 female, Socorro, 3500 m (the paratype of subaeneus)

(Hamburg). No additional locality: 1 male, Felipe Ovale, Q. Ac. 33501 (New York); 1 male, Pascoe Coll. 93-60 (London); 1 female, Muro (Paris); 1 male (Pittsburgh).

Remarks.-The abundant variation observed in the short series is incorporated in the description. Some scales have an iridescent green reflection in certain lights in the densely squamose female.

The female genitalia are unknown. Of the three females seen, two are types and were not dissected. The third specimen had apparently been crushed while alive and the entire contents of the abdomen exuded and removed; when the specimen was dissected the abdomen was empty.

Both the names ruficrus and impressicollis are equally descriptive and appropriate for the species, but neither is uniquely distinctive. Although ruficrus has page priority, the name impressicollis is chosen for this species since it is more distinctive among the names already in the genus.

The high, cold Monte Socorro site is described in Fassl (1915:57-58).

34. Hadromeropsis (Hadrorestes) institulus n. sp.

Figs. 29, 215-218, 340; Map 11

Diagnosis.-Based on female only. Similar to impressicollis female but pattern of elytra tessellate, not at all fasciate; elytra strongly constricted beneath apical umbone; apex of elytra conspicuously flared outward (Figs. 29, 215, 218); and ventrite 5 narrow basally.

Description.-Holotype, female, length 12.2 mm, width 5.0 mm. Black; antennae, legs and rostrum reddish; elytra with faint subaeneous lustre. With patches of long, imbricate, white scales ventrally and on scutellum as in group, but all other scales ochraceous. Scales of head and rostrum as in Fig. 216, evenly distributed on sides. Sides of prothorax with very long scales in a dense vitta; scales of disc of prothorax as in Fig. 217. Elytra with dense patches of very long scales irregularly alternating with glabrous areas of approximately the same size; scales in vicinity of scutellum, humeral angles and apical umbone half the size of the other scales (Fig. 215). Rostrum as in Fig. 216, but weakly concave. Eye separated from prothorax by approximately its own diameter. Prothorax (Fig. 217) scarcely 1.2x wider than long; disc flattened on sides, depressed along median line; in dorsal view long, wispy setae arising from lateral and ventral surfaces particularly conspicuous. Elytra (Fig. 29) across humeri 1.4x wider than prothorax; elytra 3.4x longer than prothorax. Sides of elytra gradually divergent to about apical third where they are 1.3x wider than across humeri, thence broadly rounded, strongly constricted under very prominent apical umbone, expanded apex thus forming a very conspicuous flounce (Figs. 215, 218). With long, wispy setae on at least apical third of intervals 1-4, a few on interval 5; sutural interval with 8-10 long, wispy setae distributed down center of interval, not concentrated on edge of interval at summit of declivity. Elytral tubercles as described for impressicollis female. Strial punctures in slightly irregular rows, disappearing before declivity. Fore femur only 1.15x wider than hind femur, with slight flange on inner edge distally. Fore tibia straight; with 12 or 13 small, equidistant teeth. Caudal margins of ventrites 2, 3, and 4 as in Fig. 218. Ventrite 5 narrow, 1.5x wider at base than long. Genitalia of paratype as in Fig. 340; spermatheca almost in one plane, spermathecal duct 2.0 mm long.

Type Series.-Holotype, ECUADOR, Sabanilla, Dr. Ohaus, 1.1907 (Dresden). Paratype, 1 female, ECUADOR, Loja-Zamora, 26-28.X.1977, 2800 m, L. Peña (Howden).

Remarks.-The paratype differs from the type in the following respects. Length 10.7 mm, width 4.6 mm. Scales of sides of prothorax similar to those of disc of pronotum (Fig. 217); the two sizes of scales of elytra almost the same, the larger size white instead of ochraceous and located only on intervals 1, 2, and 3 medially where they form a conspicuous streak. Rostrum flat dorsally, lateral edges more rounded. Long, wispy setae present on elytral intervals 1 and 3 only. Elytra 3.8x longer than prothorax. Fore femur 1.2x wider than hind femur. Fore tibia with 11 teeth on inner edge.

The species is very close to impressicollis of Colombia. Series of both sexes are needed to define the relationships.

The name is based on the Latin word for flounce and refers to the expanded apex of the elytra.

35. Hadromeropsis (Hadrorestes) bombycinus n. sp.

Figs. 219-222, 225, 341, 342; Map 11

Diagnosis.-Based on unique female (Fig. 220). Entire beetle with a microsculpture that is very uniform, very minutely granular except on elytra where it is irregular, coarser, more alutaceous. Head and rostrum smooth and flat with no other sculpture than the frontal pit (Fig. 222).

Description.-Holotype, female, length 11.4 mm, width 4.9 mm. Color piceous, very slightly aeneous; legs, antennae and apex of rostrum reddish; integument with an alutaceous sheen due to the microsculpture (see head in Fig. 222 and depression of prothorax in Fig. 219). Head and rostrum as in Diagnosis; dorso-lateral edges more rounded than they appear in Fig. 222. Side of head and rostrum with numerous, very fine, prostrate setae or scales of off-white with cupreous reflections. Prothorax 1.1x wider than long. Pronotum with lateral depressions very pronounced, sculptured as in Fig. 219; disc glabrous except for setae arising from tubercles, a few setae on basal constriction, and a few minute setae in depressions. Sides of prothorax with thin line of wider scales forming a vitta; with many long, wispy setae. Elytra across humeri 1.6x wider than prothorax; elytra 3.5x longer than prothorax; elytra 2.3x longer than width across humeri. Sides gradually divergent to just caudad of middle where they are 1.3x wider than across humeri, thence gradually rounded to apex, apical umbone not evident in dorsal outline. Apex of elytra (Fig. 221) truncate at end of intervals 1 and 2 and with a very brief tooth. Striae as in Fig. 220. Intervals with tubercles across base, on intervals 7-10, and on declivity as illustrated; in dorsal view tubercles silhouetted along entire outline of elytra. Sides of elytra

(Fig. 225) and declivity sparsely set with minute seta-like scales. Setae of intervals on disc very sparse, very inconspicuous, appressed, only a few long, wispy setae on base; setae of sutural interval at summit of declivity slightly larger, very fine, completely arched. Fore femur only 1.1x wider than hind femur, distally with vague tibial groove but no distinct flange on inner edge. Tibiae with the parallel grooves typical of group reduced. Fore tibia with approximately 10 small sharp teeth on inner edge. Metasternum sparsely set with small, round, shiny granules. Ventrite 5 across base 1.5x wider than long. Genitalia as in Figs. 341, 342; spermatheca almost in one plane; spermathecal duct 2.3 mm long.

Type Series.-Holotype, PERU, Huánuco, Huamincha, 1600 m, Mar. 1946, F. Woytkowski Coll., Donor Wm. Proctor (New York).

Remarks.-Ventrite 5 appeared almost flat before dissection. If it proves to be flat in future specimens, this would be a useful additional diagnostic character.

The bursa copulatrix appeared particularly tough and muscular; it was opaque even after treatment with KOH. Figure 342 shows a ventral view of the bursa with the inflatable flap and common oviduct pulled back to reveal the base of the spermathecal duct.

This is a very distinctive species with the integument beautifully sculptured producing a sheen like that of black silk broadcloth, hence the name "bombycinus" meaning silken.

36. Hadromeropsis (Hadrorestes) brachypterus n. sp.

Figs. 223, 224, 226-230, 344

Diagnosis.-Based on unique female. Sculpture of prothorax and elytra composed of crowded, small, uniform tubercles, each bearing a recumbent, short, white seta. Brachypterous, wing only 0.7x as long as elytron. Elytra (Fig. 229) inflated medially, 1.4x wider at about middle than across humeri, thence tapered directly to apex with no trace of apical umbone. Mesepimeron slender, protuberant (Fig. 226).

Description.-Holotype, female, length 12.5 mm, width 5.4 mm. Color castaneous; sparsely set with small white setae and seta-like scales. With no long, wispy setae on dorsal or lateral surfaces, a few on prosternum and base of femora. Head and rostrum coarsely rugose (Fig. 223); median line finely impressed between eyes, weakly carinate on rostrum. Rostrum dorsally almost flat, lateral edges abrupt. Sides of head and rostrum with scales only very slightly longer and more numerous. Prothorax (Fig. 227) 1.1x wider than long. Pronotum with lateral and median depressions replaced by slightly flattened area; sculpture as in Diagnosis and in Fig. 227. Scutellum rectangular. Elytra as in Diagnosis and Figs. 228, 229; elytra across humeri 1.2x wider than prothorax, 3.6x longer than prothorax. Base of elytra obliquely inclined anteriorly, humeral angle oblique. Only striae 1-4 discernible on disc, remaining striae irregular and confused or replaced by tubercles and rugosities. Intervals with very fine scales or setae, 1-5 abreast, each smaller or larger than the single white seta at anterior edge of each strial puncture; in addition with a few larger white scales in a rudimentary subapical

V-shape. Apex of elytron thickened, not ending in tooth (Fig. 228).
In dorsal view tubercles on sides silhouetted along entire outline of
elytra; tubercles rounded, not sharp in outline. Wing as in
Diagnosis, without folds. Fore femur (Fig. 224) 1.3x wider than hind
femur; more coarsely sculptured than in impressicollis female; flange
obsolete. Fore tibia with 10 minute teeth on inner edge. Mesepimeron
as in Diagnosis and Fig. 226; paler than surrounding integument, with
about 10 extremely small setae. Metasternum short, 1.7 mm long at
suture with metepisternum, sculptured as in impressicollis female, its
antero-lateral edge forming a raised lip over adjacent mesosternum.
Abdomen similar to that of institulus; ventrite 5 (Fig. 230) 1.6x
wider than long; apex with a few deep punctures. Genitalia as in Fig.
344. Spermatheca with cornu and ramus bent at an angle; spermathecal
duct 2 mm long; end of duct within bursa copulatrix without obvious
plate.

   Type Series.-Holotype, 282, Baly, Bowring Coll. (London).

   Remarks.-Although the castaneous color suggests a teneral
condition, the dark spermatheca indicates a mature specimen. The
slender, protuberant mesepimeron, the short metasternum, the
triangular apical half of elytra without an umbone, and the relatively
narrow and oblique base of the elytra, although not extreme, are all
attributes of a flightless species. The meso- and metathorax of
bombycinus (Fig. 225) illustrate the typical appearance of these
structures in a fully winged species. The edge of the apex of the
elytra lacks the denticles typical of Hadrorestes, but this condition
prevails in all three flightless species - see Remarks on striatus.

   The few broader scales on the elytra suggest that a pattern may
sometimes occur in this species. The absence of long, wispy setae on
the dorsal surface may be of diagnostic value. The minute teeth on
the fore tibia may also prove to be diagnostic.

   Within the impressicollis group the male of this species should be
distinguishable by the sculpture of the rostrum and prothorax,
possibly also by a protuberant mesepimeron and color.

   H. brachypterus lacks several characteristics of the
impressicollis group: the long, wispy setae of the elytra and sides
and dorsum of the prothorax are absent, the edge of the apex of the
elytra is without denticles, and the usual depressions of the pronotum
are weak. I interpret this to be a secondary loss in a highly derived
species since the sculpture of the tibiae, densely squamose scutellum,
and dense patches of elongate scales at diagnostic spots on the
ventral surface are all well-developed group characteristics.

The transandinus Group

37. transandinus n. sp.   39. apicalis n. sp.
38. nanus n. sp.

   Characteristics of Group.-Small, elongate species, 7.6-11.2 mm in
length. Epistoma wide, rostrum moderately short. Mandible without a
groove on dorsal surface, one or two dorsal setae set in a large
puncture. Eye large. Prothorax with strong basal and apical
constrictions, median line distinctly to conspicuously depressed;
sides of prothorax with sharp tubercles. Scutellum with long whitish

seta-like vestiture. Postocular vibrissae well developed. Elytra long, slender, 3.0-4.5x longer than prothorax; flat in lateral view; strial punctures very large. Fore femur not greatly enlarged, posterior face of fore femur of male smooth (male of apicalis not known). Ventrite 5 of female without lateral depression but may be flattened (female of nanus not known). Proximal end of internal sclerite of internal sac strongly reflexed (Figs. 320-323). Apex of internal sac directed dorsally or distally. Female with flat sclerite at junction of spermathecal duct within bursa copulatrix.

Remarks.-These three species, closely related to each other, occur in a transverse band across northern Ecuador. Unfortunately, both sexes are known for only transandinus and until both sexes are available for all three species, they must be considered incompletely diagnosed.

Of the two species represented by males, one has the apex of the internal sac curved and dorsally directed (transandinus, Fig. 323) and one has the apex straight and distally directed (nanus, Fig. 321). The latter is more likely a reduction from a dorsally directed apex (as in transandinus) than intermediate between the distally directed apex as in impressicollis and exilis. See Remarks, nanus. The transandinus group, therefore, apparently derived from a common ancestor with a pectinatus group species.

37. Hadromeropsis (Hadrorestes) transandinus n. sp.

Figs. 33, 231-234, 305, 322, 323, 343; Map 10

Diagnosis.-Prothorax of male (Fig. 234) with strong basal and apical constrictions and median line deeply impressed, the latter forming a "Y" with apical constriction. Elytra with vague to distinct flattened area behind scutellum. Elytra of female in profile (Figs. 231, 232) with summit of declivity distinct, apex attenuate into a conspicuous tooth distinctly ventrad of apical umbone. Spiculate apex of internal sac curved and directed dorsally (Fig. 323).

Description.-Holotype, male, length 9.0 mm, width 3.0 mm. Black, shiny; with small greenish white scales. Scales of head, rostrum, and prothorax very sparse, confined to protected areas as around eye and sides of basal constriction of prothorax. Most scales of elytra in irregular transverse depressions. Head and rostrum very smooth, polished, with only a few minute punctures. Rostrum with median line finely, deeply impressed from between posterior edge of eyes to about middle, sides rather strongly rounded and deflected to median line. Segment 2 of funicle slightly longer than segment 1. Prothorax as in Diagnosis and Fig. 234; 1.05x wider than long, sides only weakly rounded between constrictions; lateral tubercles acute in dorsal view. Pronotum with an area on side somewhat flattened and with sculpture weaker. Elytra across humeri 1.4x wider than prothorax, elytra 3.2x longer than prothorax. In dorsal view (Fig. 33) sides of elytra parallel for basal 0.5 thence very gradually convergent to apex, apical umbone not prominent, just touching outline, apices very briefly individually rounded. Dorsal outline with obsolete tubercles on basal 0.2, on apical 0.3 with small, sharp tubercles followed by denticles on edge of apex. Base of stria 5 and postscutellar area weakly depressed; apparently without long setae. Sculpture of elytra similar to Figs. 231, 232 of female; intervals with fine setae each about as long as adjacent strial puncture, setae becoming longer, erect apicad and laterad. Fore femur 1.9x wider than hind femur; 1.5x wider than rostrum; surface very smooth and polished except weakly sculptured distally; set with fine, erect white setae. Fore tibia straight; faintly punctulate; inner edge with 10 equidistant teeth, the central 3 teeth slightly larger; with numerous erect straight or curved fine dark setae. Middle and hind tibiae with bristles of inner edge obsolete. Ventrite 5 approximately 2x wider than long, slightly convex, apex truncate-emarginate. Genitalia of paratypes as in Figs. 322, 323; aedeagus 2.8 mm long.

Allotype, female (Fig. 231), length 10.6 mm, width 3.8 mm. Differs from type as follows. Antenna, mandible, epistoma and legs in part reddish yellow. Scales and some setae (especially on hind coxa and scutellum) iridescent green. Head and rostrum (Fig. 233) more strongly punctured, rostrum with some punctures coalescing into brief rugae. Prothorax 1.09x wider than long; median line not impressed, disc instead rather broadly flattened medially with confluent tubercles; lateral flattened area scarcely evident. Elytra across humeri 1.5x wider than prothorax; elytra 3.9x longer than prothorax, elytra 2.4x longer than width across humeri. In dorsal view sides slightly divergent to middle thence very gradually convergent to apex, apical umbone weakly arcuate in outline; profile as in Diagnosis, Fig. 232. Strial punctures smaller, striae more confused apically by tubercles; summit of declivity with 4 long setae. Fore femur 1.2x wider than hind femur, sculpture more pronounced. Fore tibial teeth uniform in size. Fore tarsi longer (Fig. 305). Middle and hind tibiae with 5 to 8 stout bristles on inner edge. Caudal surface of ventrites 2, 3, and 4 equal in height, slightly anteriorly directed, the edge rounded medially. Ventrite 5 across base 1.4x wider than long; entire segment slightly convex. Genitalia as in Fig. 343; spermathecal duct approximately 2.2 mm long.

Type Series.-Holotype, ECUADOR, Pichincha, 10 km E Tandapi, 2000 m, 11.VI.1976, S & J Peck, beating vegetation (Howden). Allotype, ECUADOR, Napo, 2 km S. Oritoyacu, 22 km S Baeza, 1500 m, III.4-5.1976, J M Campbell (Howden). Paratypes, 3 males. ECUADOR. Pichincha: 2 males, same data as type (Ottawa, Howden); 1 male, Quito, Coll. Kuschel (Auckland).

Note: Tandapi appears as Cornejo Astorga on recent maps.

Remarks.-The paratypes are 8.5-10.5 mm in length and 2.8-3.4 mm in width. They vary from the type as follows. Rostrum with a few brief rugae. Prothorax with sides more strongly rounded in some, 1.01-1.08x wider than long. Elytra 3.0-3.5x longer than prothorax, apical umbone arcuate in dorsal outline.

The name transandinus refers to the fact that the species is known from both the Pacific and Amazonian slopes of the Andes.

38. Hadromeropsis (Hadrorestes) nanus n. sp.

Figs. 238, 320, 321; Map 10

Diagnosis.-Based on unique male. Very small. Base of elytra evenly convex in scutellar area. Fore femur slender, approximately as wide as rostrum. Endophallic structures large, larger than in transandinus; spiculate apex short, straight (Fig. 321).

Description.-Holotype, male, length 7.6 mm, width 2.5 mm. shiny black; antenna, epistoma, mandible and legs reddish yellow. Sparsely set with round scales of slightly iridescent pale blue or whitish; scales more numerous than in transandinus, on ventral surface almost as numerous as on dorsum, occurring even on fore coxa and base of fore femur. Head and rostrum (Fig. 238) sparsely punctate, median line finely impressed from frontal fovea to anterior edge of eye, entire rostrum longitudinally concave. Segment 2 of funicle 1.4x longer than segment 1. Prothorax (Fig. 238) 1.09x wider than long, similar to that of transandinus. Elytra cylindrical, as in Diagnosis. Elytra across humeri 1.3x wider than prothorax, elytra 3.6x longer than prothorax, elytra 2.4x longer than width across humeri. In dorsal view sides of elytra parallel on basal 0.13, feebly rounded, thence very slightly convergent to apical umbone, apex broadly rounded from between striae 5, apices briefly divergent. Dorsal outline with no tubercles from base to apical 0.3, thence with small, acute tubercles similar to denticles of apical edge of elytra. Stria 5 weakly depressed at base. Base of elytra with several rows of long, wispy setae. Sculpture of elytra, strial punctures similar to those of transandinus, striae more confused by more numerous squamose depressions. Setae of lateral intervals longer, more conspicuous, especially on declivity. Fore femur 1.3x wider than hind femur, as wide as full width of rostrum; distal flange very weak; sculpture of femur obsolete; setae of femur very conspicuous, long, white, erect. Fore tibia weakly bowed distally, obsoletely punctate, inner edge with 8 or 9 very small, uniform teeth; with numerous long, curved and some long, straight erect setae. Middle tibia with row of 4 or 5, hind tibia with 2 or 3 bristles on inner edge. Ventrite 5 across base 1.7x wider than long, slightly convex, apex truncate-emarginate. Last tergite with inflexed apex (lip) short. Genitalia as in Figs. 320, 321, and Diagnosis; aedeagus 2.4 mm long.

Type Series.-Holotype, ECUADOR, Pichincha, km 27, Old Santo Domingo Rd, 3200 m, 18.II.1979, H. Howden, on alder (Howden).

Remarks.-The erect setae of the fore legs are particularly conspicuous, as are the long, wispy setae of the base of the elytra; on older specimens, they would be less conspicuous from wear.

Because of the extremely small size and delicate proportions of the type specimen, the narrow fore femur might be attributed to allometric growth. However, allometric growth could not account for the convex base of the elytra nor the large, very distinctive endophallic structure. The short, straight, distally directed apex of the internal sac is apparently a reduction from a dorsally directed form since the internal sclerite is very large and strongly reflexed proximally.

In addition to the diagnostic characters listed, nanus differs from transandinus in the setae at the base of the elytra numerous and long, teeth of the fore tibia very fine and uniform, and the elytra more cylindrical in caudal view.

Since nanus shares many characters with apicalis, including a more dorsal apex of elytra and similar habitat in Ecuador, the suspicion exists that they may be male and female of the same species. However, from the evidence on hand, I feel they are separate taxa. The base of the elytra at the scutellum is so completely convex in nanus that it is unlikely to be depressed in the opposite sex.

39. Hadromeropsis (Hadrorestes) apicalis n. sp.

Figs. 30, 235-237, 345; Map 10

Diagnosis.-Based on female only. Similar to female of transandinus, but with a pronounced aeneous lustre; elytral declivity very brief, attenuate apex of elytra in lateral view thus close to dorsal surface; elytral 4.0-4.5x longer than prothorax.

Description.-Holotype, female, length 11.2 mm, width 3.6 mm. Shiny black with aeneous lustre; femora, antenna, mandible, and epistoma piceous. Sparsely clothed with off-white or pale greenish scales of small size. Head and rostrum as in transandinus female. Prothorax as in Fig. 236; 1.1x wider than long; sides moderately rounded between constrictions; basal constriction narrow, complete dorsally; apical constriction obsolete on disc. Pronotum flattened along median line, here irregularly transversely sculptured; either side of this area punctate, almost smooth, gradually changing to distinct, discrete tubercles on vertical sides. Elytra (Fig. 30) across humeri 1.5x wider than prothorax, elytra 4.0x longer than prothorax. In dorsal view humeri very prominent, sides subparallel for basal 0.3, thence broadly rounded to apical umbone, the umbone weak; sides of apex slightly sinuate from beneath stria 4; apices briefly attenuate, slightly inwardly directed. Ventrally (Fig. 237) apical attenuation expanded beyond border of elytra which remains parallel to stria 10. Dorsal outline with small tubercles on basal 0.25, conspicuous tubercles on apical 0.5. Base of stria 5 impressed; with distinct postscutellar flattened area; with some moderately long, tapered setae. Setae of intervals becoming longer toward sides and apex. Fore femur 1.2x wider than hind femur; fore femur with moderate flange, its edge weakly serrulate; weakly sculptured with faintly impressed honeycomb lines changing to impressed longitudinal lines distally. Fore tibia straight with 11-13 small teeth; outer edge sparsely punctate. Middle and hind tibiae with 5-8 stout bristles on inner edge. Caudal surface of ventrites 2, 3, and 4 equal in height, slightly anteriorly directed. Ventrite 5 (Fig. 237) across base 1.5x wider than long. Genitalia as in Fig. 345; spermathecal duct 2.0 mm long.

Type Series.-Holotype, ECUADOR, Napo, 4 km W Papallacta, 3200 m, III.2.76, J. M. Campbell (Howden). Paratype, 1 female, ECUADOR, Chimborazo et Pichincha, Mandeville, 272, 1853 (Paris).

Remarks.-The paratype is 11 mm long and 4.2 mm wide. It differs from the type in the following respects. Sculpture of head (Fig. 235) more complex. Prothorax almost uniformly tuberculate. Elytra across humeri 1.2x wider than prothorax, elytra 4.5x longer than prothorax, elytra 2.5x longer than width across humeri. Elytra in profile slightly concave medially. Apex much more developed; sutural interval at summit of declivity with 2 or 3 long setae (these abraded in type). Metasternum and ventrites 1 and 2 with a fine microsculpture of parallel, transversely impressed lines.

The flared and attenuate apex of the elytra (Fig. 237), for which the species is named, is of a different origin than that of institulus (Fig. 218) in which the margin is simply crimped outwardly, stria 10 following the margin closely.

In addition to the diagnostic characters, apicalis differs from transandinus in the very prominent humeri and the more conspicuous tubercles of the elytra in dorsal outline. H. apicalis occurs at a much higher elevation than transandinus, its range bisecting the lower-elevation range of transandinus. H. nanus occurs in a habitat similar to that of apicalis; see Remarks of that species.

The pectinatus Group

40. dialeucus n. sp.   45. magicus (Pascoe)
41. contractus n. sp.   46. nitidus n. sp.
42. pectinatus n. sp.   47. mandibularis n. sp.
43. conquisitus n. sp.   48. spiculatus n. sp.
44. scambus n. sp.

Characteristics of Group.-Proximal end of internal sclerite within internal sac strongly reflexed (Figs. 324-328). Apex of internal sac produced in a curved, dorsally directed, short to greatly elongated cone or tube. Dorsal surface of internal sac above proximal end of heavy internal sclerite without a cupped inflatable lobe. Junction of spermathecal duct inside bursa copulatrix unmarked or with a faint small sclerite.

Remarks.-The species are arranged here according to the length of the apex of the internal sac and spermathecal duct. It then becomes evident that the species are similarly related in other features as well.

Both sexes are known for six of the nine species.

In addition to the obvious similarities in the male genitalia, the sculpture on the posterior face of the fore femur of the male shows a close relationship between impressicollis and the first few species in this group. In these species the distal tubercles are grossly enlarged into a shelf, the large seta-bearing surface flat or concave and abruptly perpendicular to the surface of the femur. A lesser development of these shelf-like tubercles is seen in the lower right corner of Fig. 257.

In the remaining species there is a trend for the overall sculpture of the beetle to gradually decrease as the genitalia becomes more extreme.

40. Hadromeropsis (Hadrorestes) dialeucus n. sp.

Figs. 28, 279, 281, 307, 324, 325, 355

Diagnosis.-Female elytra clothed (a) with small to minute scales clustered in irregular transverse depressions; (b) sometimes also patterned with basal oval ring, median fascia, and apical lunule of white scales of varying sizes (Fig. 28), or with fragments of this pattern; or, (c) with pattern of (b) but remainder of elytra densely covered with ochraceous, sometimes imbricate, larger (0.1 mm) scales. Prothorax (Fig. 281) with rounded tubercles of sides obsolete or completely absent on disc, disc almost smooth and shiny in some. Elytra of female with sides very gradually rounded to apex from about middle. Posterior half of hind coxa covered with rounded, separated, moderate-sized, whitish or ochraceous scales. Male with internal sac similar to that of pectinatus but apex lacking the pair of sclerotized dorsal lobes (Fig. 325).

Description.-Holotype, male, length 11.0 mm, width 3.8 mm. Reddish except head and base of elytra black. Appearing glabrous macroscopically; body dorsally and ventrally with minute, round, oval, or elongate white scales (visible in Figs. 279, 281); scales of posterior half of hind coxa as in Diagnosis, larger than any others but not full-sized. Head and rostrum similar to those of spiculatus in Fig. 288, with distinct scattered punctures, a few confluent punctures on rostrum forming brief rugae; median line briefly, finely impressed, with small frontal fovea. Funicle with segments 1 and 2 equal in length. Prothorax 1.1x wider than long, sides moderately rounded, widest behind middle. Pronotum with median line distinctly impressed basally, disappearing before apex; sculpture otherwise as in Diagnosis; setae as in pectinatus. Elytra across humeri 1.3x wider than prothorax; elytra 3.0x longer than prothorax. Elytra shaped as in pectinatus except apex less truncate. Elytral striae and sculpture similar to that of pectinatus but striae more distinct, intervals smoother, tubercles at base of elytra weaker. Fore femur 2.1x wider than hind femur; posterior face with about 17 shelf-like tubercles, these weaker than in pectinatus; sculpture of fore femur and tibia otherwise similar to that of pectinatus. Inner edge of fore tibia with 10 or 12 nearly uniform short teeth. Fore tarsus (Fig. 307) similar to that of pectinatus but shorter. Middle and hind tibiae without tubercles, with 1 or 2 stout bristles. Abdomen with punctures obsolete except on apical half of ventrite 5; ventrite 5 across base 2.2x wider than long. Aedeagus (Fig. 324) 3.1 mm long, aedeagal apodeme 1.5 mm, tegminal strut 1.8 mm. Internal sac as in Diagnosis, Fig. 325.

Allotype, female, length 12.3 mm, width 5.0 mm. Differs from type as follows. Most of integument black, legs and antenna reddish. Scales more numerous on both dorsal and ventral surfaces; elytra patterned as in Fig. 28 and Diagnosis, but some scales abraded, pattern recognizable by scars. Pronotum smoother (Fig. 281). Elytra across humeri 1.5x wider than prothorax; elytra 3.6x longer than prothorax. In dorsal view (Fig. 28) sides of elytra very slightly divergent to about middle where they are only 1.2x wider than across humeri, very gradually convergent to apex (Fig. 279), apical umbone weak, just touching dorsal outline. Sutural interval produced into very brief, caudally projecting tooth. In profile elytra more convex basally, gradually curved to summit of declivity which is scarcely perceptible, marked with 7-8 long, reddish setae. Fore femur 1.3x wider than hind femur; posterior face lacking shelf-like tubercles; teeth of fore tibia very small. Middle and hind tibiae with 1-3 strongly oblique bristles, no tubercles. Ventrites 3, 4, and 5 with rough sculpture at sides. Caudal surface of ventrites 2 and 3 moderate, smooth, shiny, anteriorly slanted; caudal surface of ventrite 4 almost perpendicular, moderate in height, straight in posterior view. Ventrite 5 across base 1.8x wider than long. Genitalia of paratypes as in Fig. 355; spermathecal duct 1.8 mm long.

Type Series.-Holotype, no data, green square of paper (Oxford). Allotype, VENEZUELA, Coll. C. Felsche, Kauf 20, 1918 (Dresden). Paratypes, 6 females. VENEZUELA. 1 female, same data as allotype (Dresden); 1 female, Coll. Kuschel (Auckland). No locality: 4 females, no data (Berlin, London, Howden).

Remarks.-Females vary in length from 11.3-12.5 mm and in width from 4.5-5.0 mm. In the maximum vestiture (one female) only the basal ring and median fascia are present; these markings conspicuous, glabrous (hence black) but with the center of the markings set with small, faintly opalescent scales; remainder of elytra and line on base of prothorax with larger (0.1 mm), often imbricate, pale ochraceous scales; head, rostrum, prothorax except basal line, elytral humerus, and legs with the smaller faintly opalescent scales; ventrally with ochraceous scales very sparse medially. The minute scales are definitely greenish on the elytral declivity in one female. Elytral setae are very slender, translucent brown or paler, up to 0.5 mm long on lateral intervals of elytra; slightly thicker and white on the sides of the prothorax and meso- and metasterna.

In addition to the diagnostic characters, dialeucus differs from pectinatus in the shorter aedeagus and the lack of tubercles on the middle tibia.

The name means "marked with white".

41. Hadromeropsis (Hadrorestes) contractus n. sp.

Figs. 37, 274-278, 354; Map 12

Diagnosis.-Based on female only. Elytral pattern as in Fig. 37. Ventrites 3, 4, and 5 much narrower than ventrites 1 and 2; adjacent elytral margins likewise convergent and attenuate. Caudal surface of ventrite 4 strongly produced in a high arc, strongly to moderately posteriorly slanted. Apical fourth of fore tibia abruptly bent inward.

Description.-Holotype, female, length 11.3 mm, width 4.3 mm. Integument shiny black except legs and antenna reddish. Dorsally sparsely but evenly clothed with small, round, pale greenish white scales in depressed areas. Elytra also with larger, round, white scales forming a faint basal arc, a median fascia and an apical lunule. All setae opaque white, longest on abdomen and metasternum, shortest on disc of elytra. Head and rostrum (Fig. 274) slightly concave, with punctures and some small depressions; median line finely impressed, frontal fovea obsolete. Funicle with segments 1 and 2 equal in length. Prothorax (Fig. 275) 1.1x wider than long; sides moderately rounded between constrictions; covered with low, round tubercles which become weaker medially; median line not marked; space between tubercles with scales. All setae of sides and disc of prothorax recumbent. Elytra across humeri 1.4x wider than prothorax; elytra 3.6x longer than prothorax. Elytra in dorsal view (Fig. 37) with sides slightly divergent to about middle thence convergent in straight line to apex, apical umbone interrupting this line in a conspicuous arc. Apex of elytra (Fig. 277) in dorsal view conspicuously narrow and attenuate, each elytron terminating in short tooth. Dorsal surface of elytra without long, wispy setae, without tubercles, basal portion of elytra with shallow, sinuous, transverse depressions; strial punctures deep, mostly regularly aligned except in white markings. Declivity and intervals 8, 9, and 10 with moderate to weak acute tubercles each with a short, white seta; tubercles obsolete on middle of sides. Fore femur (Fig. 276) 1.7x wider than hind femur; femur with fine netted microsculpture, distal portion of femur and all of fore tibia with deeply impressed parallel grooves; without long, wispy setae (possibly only a female characteristic), nowhere with tubercles, inner edge with very weak flange. Fore tibia as in Diagnosis and Fig. 276; with 11 small, equal equidistant teeth on inner edge. Middle tibia with 4, hind tibia with 1, strongly oblique bristles. Hind coxa with no scales, only white setae. Abdomen as in Diagnosis and Fig. 278. Ventrites 3, 4, and 5 with numerous, long, white setae; caudal surface of ventrites 2 and 3 finely rugulose; caudal surface of ventrite 4 smooth and shiny. Ventrite 5 only 1.4x wider than long, lateral concavities very deep and large. Genitalia of paratypes (including two examples of lateral sclerite) as in Fig. 354; no sclerite at junction of spermathecal duct with bursa copulatrix; spermathecal duct 1.6-1.9 mm long.

Type Series.-Holotype, female, BOLIVIA, Coroico, Ost Bolivien, 1913, 16 (Dresden). Paratypes, 2 females, BOLIVIA, Coroico, Ex. coll. Clerc (Paris).

Remarks.-The two paratypes are 11.4-12.4 mm long and 4.5-4.7 mm wide. They differ from the type in having the median line of the pronotum marked by an obsolete or partial impression, the apical umbone of the elytra less prominent, the elytra more strongly sculptured, and the striae consequently less distinct.

Superficially contractus resembles dialeucus but the conspicuously narrowed, attenuate apex of the elytra, the more strongly sculptured prothorax, and the non-squamose hind coxa readily distinguishes contractus. Series of both sexes of both species are needed to establish inter- and intraspecific variation.

This species is perhaps most closely related to pectinatus, the female of which differs in the broader, more gradually narrowed abdomen; more erect, longer, and more slender setae of intervals 8, 9, and 10 and declivity; more distinct declivity; absence of large scales dorsally; more strongly sculptured caudal surface of ventrites 2 and 3; less arcuate caudal margin of ventrite 4; and less bowed fore tibia.

The name, contractus, refers to the contracted ventrites 3, 4, and

5.

42. <u>Hadromeropsis</u> (<u>Hadrorestes</u>) <u>pectinatus</u> n. sp.

Figs. 35, 265-270, 308, 328, 352; Map 13

<u>Diagnosis</u>.-Neither sex with full-sized scales either dorsally or ventrally. Prothorax (Fig. 266) with sculpture composed of very uniform, rounded tubercles evenly distributed over all, including (in male) the narrowly depressed median line. Female with sutural edges of elytra separated apicad of interlocking flange, produced into strongly convergent teeth (Fig. 268). Fore femur of male on its posterior face distally with 6 to 24 well-developed shelf-like tubercles each bearing a long, wispy seta in its saucer-like follicle. Fore tibia of male with 3 or more teeth of inner edge extremely long (Fig. 267).

<u>Description</u>.-Holotype, male, length 11.5 mm, width 4.1 mm. Black except antenna, prothorax and base of fore coxa and fore femur reddish; with minute blue scales scattered over dorsum and very sparsely over ventral surface, scales of prothorax elongate, others rounded. Head and rostrum (Fig. 265) dorsally and laterally with deep punctures, some elongate and confluent forming short rugae; frontal fovea small; median line weakly impressed. Rostrum narrow, dorsally shallowly concave, sides flared outward apicad of antennal insertion; dorso-lateral edges prominent. Funicle with segment 2 slightly longer than segment 1. Prothorax as in Diagnosis and Fig. 266. Prothorax 1.1x wider than long, sides strongly rounded between constrictions; each tubercle with a very slender white seta, setae shorter dorsally and towards median line; no long, wispy setae visible dorsally but some setae of basal constriction on sides less curved. Elytra (Fig. 35) across humeri 1.3x wider than prothorax; elytra 3.0x longer than prothorax. Elytra in dorsal view with sides parallel to just caudad of middle, thence gradually rounded to apex, apical umbone obsolete and not entering dorsal outline, apex truncate. Base of elytra not depressed at stria 5, weakly depressed behind scutellum, base of dorsal intervals with short row of tubercles; each tubercle bearing a fine, long, wispy straight or curved seta, additional setae present as well. All lateral intervals and entire declivity with numerous acute tubercles, entire dorsal outline of elytra "dentate" from the profile of these tubercles; each tubercle bearing a conspicuous, long, straight or weakly curved seta. Striae distinct centrally only, remainder of disc with short to long, transverse squamose depressions about half as deep as strial punctures, these transverse depressions shorter caudad and changing until before declivity each interval with series of tubercles extending the width of interval and each strial puncture separated by a small tubercle. Fore femur 2.2x wider than hind femur; anterior face with fine netted microsculpture which distally becomes deeply impressed parallel grooves with scattered deep punctures superimposed, each puncture bearing a long, wispy seta; inner edge of fore femur with very weak distal flange; inner surface including tibial groove with scattered tiny granules; posterior face as in Diagnosis. Fore tibia (Fig. 267) slightly curved with 10-12

slanting teeth on inner edge, the 3 distad of middle extremely long (badly worn in type); fore tibia densely sculptured with deep rugae and punctures. Fore tarsus with segment 1 long, parallel-sided (Fig. 308). Middle and hind tibiae with row of tubercles on inner edge, 2 or 3 most distal tubercles each with a stout bristle arising from base. Abdomen with very fine scattered punctures on ventrites 1 and 2, punctures larger and denser on ventrites 3-5, apex of ventrite 5 very broadly truncate-emarginate. Genitalia similar to Fig. 324 of dialeucus, aedeagus 4.0 mm long. Apex of internal sac (Fig. 328) with pair of distinct sclerotized lobes slightly dorso-laterally directed (marked with arrow in figure).

Allotype, female, length 12.3 mm, width 5.0 mm. Differs from type as follows. Only base of pronotum reddish; scales white. Rostrum wide; only briefly concave apically, sculpture of head and rostrum finer. Prothorax with sides less strongly rounded, tubercles weaker, median line not marked except by break in tubercles. Elytra across humeri 1.5x wider than prothorax, elytra 3.6x longer than prothorax. Elytra in dorsal view with sides gradually divergent to just beyond middle where they are 1.2x wider than across humeri, thence gradually rounded to apical umbone which is small but distinct in outline, apex approximately triangular. Sutural edges of elytra as in Diagnosis, Fig. 268. Declivity in profile slightly concave, sutural edge at summit of declivity with 4 very long setae. Fore femur 1.4x wider than hind femur; posterior face lacking shelf-like tubercles. Fore tibia on inner edge with 8 or 9 more erect teeth, the 3 distad of middle shorter than in type but longer than in males of many species. Base of ventrites 3, 4, and 5 with rather coarse microsculpture (Fig. 269). Caudal surface of ventrites 2 and 3 approximately perpendicular, coarsely sculptured (Fig. 270); caudal surface of ventrite 4 smooth, concave; edge medially straight, not as high as ventrites 2 and 3, directed caudad. Ventrite 5 across base 1.6x wider than long, strongly medially convex from basal fourth. Genitalia as in Fig. 352; spermathecal duct 2.6 mm long.

Type Series.-Holotype, PERU, Cuzco, Valle de Lares, 75 km NW Calca, 2060 m, 6 Feb. 1979, W. E. Steiner (Washington, USNM Type No. 100193). Allotype, same data as type except 13 Jan 1979 (Washington). Paratypes, 5 males. BOLIVIA: 1 male, Limbo-Chapare, 11.1952, 2000 m (Howden); 1 male, Yungas, Coll. Kuschel (Auckland). COLOMBIA. 2 males, F. C. Nicholas Collection (New York). PERU. Cuzco: 1 male, same data as type (Howden).

Remarks.-The paratypes (all males) range in length from 11.2-14.6 mm and in width from 2.7-5.0 mm. The paratypes exhibit both deeper and shallower rostral sculpture than the type. The Colombian specimens differ in the elytral striae more distinct, even on the declivity. Variation in the number of shelf-like tubercles on the posterior face of the fore femur is apparently related to development of the specimen and not to geographical range. For instance, in the 11.9 mm specimen from Bolivia there are 7-12 shelf-like tubercles whereas in the 14.6 mm specimen from Bolivia there are 24 shelf-like tubercles. The hind tibia usually has a single groove on the outer edge, but it can vary from one side of the beetle to the other. In all the males there is a trace of the strong microsculpture of the female abdomen. The aedeagus ranges from 3.8-4.5 mm long in the paratypes, but the apex of the internal sac is very uniform.

When looking at mixed species, the very uniform sculpture of the pronotum and the extremely long teeth of the fore tibia are particularly conspicuous.

There are two females from Cochabamba, Bolivia, in the Kuschel collection (Auckland) that may be pectinatus. They are clothed dorsally and ventrally with small (0.06 mm) round, blue-green, slightly iridescent scales which are concentrated in depressed areas on the prothorax and elytra, absent medially on the abdomen. The thoracic sculpture is considerably weaker than in the type series and the tibial teeth are uniform. Association with males is required to establish the identity of these specimens.

The specimens from Peru were part of a small group of specimens on a legume-like small tree; the remainder of the group dropped when the collector approached and were not recovered.

The comb of long teeth of the fore tibia suggested the name "pectinatus".

43. Hadromeropsis (Hadrorestes) conquisitus n. sp.

Figs. 34, 254-259, 329, 330, 350; Map 12

Diagnosis.-Glabrous or sparsely clothed with scintillating iridescent green scales. Segment 2 of funicle much longer than segment 1. Apical umbone of elytra protuberant, set with acute, caudally directed tubercles (Figs. 34, 254). Fore femur of male (Figs. 257, 258) with well-developed flange on inner edge distally, with sharp tubercles on the flange, rounded tubercles on inner surface except tibial groove. Fore tibia both sexes almost smooth with very few scattered small punctures; with a few fine, pale, recumbent setae; fore tibia distally absolutely straight in female and less-developed in males, slightly bowed in well-developed males (Fig. 258). Internal sac of male at extreme apex without 4 inflatable lobes (Fig. 330).

Description.-Holotype, male, length 13.4 mm, width 4.2 mm. Black; sparsely clothed in depressed areas with minute green scales. All dorsal setae very fine and inconspicuous, white, curved. Head and rostrum with strong median concavity extending to between middle of eyes; median line finely impressed between apical half of eyes; dorso-lateral edges of rostrum prominent, rounded, smooth and shiny. Head and sides of rostrum weakly sculptured with punctures; confluent punctures or brief rugae on frons. Segment 2 of funicle 1.6x longer than segment 1. Prothorax (Fig. 255) as wide as long, sides almost parallel, slightly swollen over fore coxae. Prothorax with sculpture composed of rounded tubercles, medially tubercles slightly larger and some confluent; median line finely, partially impressed, crooked. In profile, pronotum almost flat, except immediately before basal constriction; basal constriction very short on disc. Elytra across humeri 1.3x wider than across prothorax, elytra 3.0x longer than prothorax. In dorsal outline sides of elytra slightly convergent from humeri, more strongly so from apical third, umbone right-angled, apex rounded. In dorsal view tubercles of lateral intervals acute on basal third, absent medially, prominent and acute on umbone; teeth of apical

edge of elytra prominent, even; apices of elytra not at all produced.
Base of elytra depressed at stria 5; elytra moderately concave behind
scutellum. Elytra irregularly sculptured with rounded tubercles
across base; tubercles reduced, becoming confluent on outer intervals
of disc, before declivity becoming low swellings strongly caudally
directed, as wide as interval; much smaller on declivity. Striae 1,
2, and 3 regularly punctate on disc; other striae confused by
sculpture and extra punctures. Fore femur as in Diagnosis and Figs.
257, 258; enormously swollen, 2.7x wider than hind femur. Inner
surface of fore femur with multiple shiny, small round tubercles
except in smooth tibial groove, each tubercle with long, wispy seta;
remaining surfaces of fore femur proximally very finely sculptured,
with only white or greenish, short recumbent setae; posterior surface
strigulate distally. Fore tibia as in Diagnosis and Fig. 258; inner
edge with multiple rounded tubercles and long, wispy setae. Fore
tarsus with segment 1 long and parallel-sided, segments 2 and 3
subequal in length, 0.8 as long as segment 1. Middle and hind tibiae
with row of 4-7 tubercles on inner surface near edge, a stout bristle
arising from base of each. Abdomen smooth and shiny, almost
impunctate; ventrite 5 across base 2.0x wider than long, apex broadly
truncate-emarginate. Genitalia as in Figs. 329, 330, and Diagnosis.

Allotype, female, length 13.1 mm, width 5.0 mm. Differs from type
as follows. Clothed on dorsal and ventral surfaces and legs with
scintillating green scales; black integument shining through naturally
(i.e., not abraded) on dorso-lateral edges of rostrum and on elytra
forming pattern (Fig. 34) of a black "S" basally and an oblique
preapical fascia. Scales sparse on disc of pronotum which
consequently appears darker. Prothorax slightly wider than long,
sculpture slightly weaker. Elytra across humeri 1.4x wider than
across prothorax, elytra 3.4x longer than prothorax. Elytra in dorsal
view widest just caudad of middle; apex beyond umbone more elongate;
apices individually attenuate into conspicuous tooth. Sutural
interval at summit of declivity with 3 and 4 long, erect setae. Fore
femur 1.4x wider than hind femur, distal flange greatly reduced but
still evident and its edge denticulate; remaining tubercles obsolete
or absent. Fore tibia with single row of teeth instead of tubercles.
Abdomen with very faint microsculpture similar to that of pectinatus.
Caudal surface of ventrites 2, 3, and 4 approximately perpendicular,
smooth and shiny. Ventrite 5 across base 1.6x wider than long.
Genitalia as in Fig. 350; spermathecal duct of allotype and paratypes
2.4-3.0 mm.

Type Series.-Holotype, ECUADOR [Zamora-Chinchipe], Sabanilla,
15.IX.-7.X.1905, Dr. Ohaus, 1, 1907 (Dresden). Allotype, same data as
type (Dresden). Paratypes, 2 males, 3 females. ECUADOR. 1 female,
52286, Buckley, Fry Coll. 1905-100 (London); 2 males, 1 female
[Morona-Santiago], Macas, Buckley (Berlin [1 female]; Fry Coll.
1905-100, London [2 males]); 1 female, same data as type (Dresden).

Remarks.-Both male paratypes are 12.1 mm long, 3.8 mm wide; the
three females are 14.0-15.3 mm long, 5.2-5.8 mm wide. The two males
are clothed with full-sized scales and are patterned with glabrous
lines as in the allotype, but the scales appear to be less dense than
in the squamose females. One female is almost glabrous, bearing
minute scales (as in the holotype) on the declivity only, the hind

coxae having larger elongate green scales. Funicular segment 2 is always much longer than segment 1, ranging from 1.2-1.7x longer in females, and 1.6x in all three males. The prothoracic sculpture is weaker in the squamose paratypes (Fig. 256). The elytral tubercles are much less developed in one male. The sutural interval on the summit of the declivity is abraded in all males, they may or may not have one or two slightly longer setae there; in females 2-8 setae are present. The fore femur is always enormously swollen in males, ranging from 2.4-2.7x wider than the hind femur, and only 1.4-1.7x wider in the females.

Three females were dissected; in one (the allotype) the spermathecal duct ends in the bursa copulatrix in a distinct spherical dark bulb. In the second a separate sclerite is seen beneath the duct near its end within the bursa. In the third a portion of the end of the duct within the bursa is enlarged and dark.

One of the male paratypes is an old, worn specimen missing its abdomen; the other is teneral with the aedeagus very callow and I was not successful in extracting the internal sac from it. This is especially unfortunate in view of the slight difference noted between this species and scambus in the apex of the internal sac.

The prominent elytral umbone and characters of the fore legs will readily distinguish this species. In addition, glabrous specimens can be distinguished from many species the same size by the absence of long free-standing setae on the elytra. The squamose specimens are especially distinctive and can be confused only with scambus. See discussion of scambus for further comparisons.

Buckley at Macas and Ohaus at Sabanilla collected the entire type series. The name conquisitus means "sought out eagerly" and refers to the efforts of these two early collectors.

44. Hadromeropsis (Hadrorestes) scambus n. sp.

Figs. 32, 260-264, 326, 351; Map 13

Diagnosis.-Similar to conquisitus but segment 2 of funicle subequal to segment 1; fore tibia with deep confluent punctures and curved dark, wiry setae; and fore tibia distally strongly bowed. Fore femur of male with distal flange very weak, without tubercles except minute ones on flange, no modification below tibial groove. Internal sac of male at extreme apex with circle of 4 equal, equidistant inflatable lobes (Fig. 326).

Description.-Holotype, male, length 9.9 mm, width 3.1 mm. Head and prothorax black, remainder of body shading to piceous. Sparsely clothed with minute, scintillating green scales except on legs. All dorsal setae very fine and inconspicuous; recumbent, less so on declivity. Head and rostrum similar to conquisitus but median concavity shallower, median line finely impressed to behind eye. Segment 2 of funicle 1.1x longer than segment 1. Prothorax 1.04x wider than long, similar to conquisitus. Elytra across humeri 1.3x wider than prothorax, elytra 2.6x longer than prothorax. In dorsal outline sides of elytra subparallel, gradually converging caudad of middle, apical umbone weak in outline, apex brief, rounded (Fig. 261).

In dorsal outline tubercles of lateral intervals obsolete or absent on basal 0.7, small but distinct on apical 0.3, those of umbone small (Fig. 261), same size as those of declivity and edge of elytra. Base of elytra at stria 5 and postscutellar area only obsoletely depressed. Base of elytra without tubercles; strial punctures completely regular, with only minimal confusion on declivity. Fore tibia and femur as in Diagnosis and similar to Fig. 262 of female; fore femur 2.2x wider than hind femur; distal half with wiry setae as on tibia. Fore tarsus as in conquisitus. Middle and hind tibiae with bristle-bearing tubercles much less developed than in conquisitus. Abdomen relatively longer and narrower than in conquisitus, ventrite 5 across base 1.8x wider than long; apex of ventrite 5 emarginate. Genitalia as in Fig. 326 and Diagnosis. Aedeagus 3.15 mm long.

Allotype, female, length 12.8 mm, width 3.0 mm. Differs from type as follows. Rostrum more robust (Fig. 263). Prothorax more robust, 1.1x wider than long, disc flat, not at all depressed; in profile basally much thicker (Fig. 264); setae larger, especially on sides. Elytra across humeri 1.4x wider than prothorax, elytra 3.4x longer than prothorax. In dorsal view (Fig. 32) sides of elytra slightly wider at middle, gradually narrowed to apex, apical umbone rounded in outline, apex almost triangular, individual apices not at all toothed. With weak tubercles on apical portion of intervals 5, 6, and 7 and on umbone (Fig. 260). Setae of intervals 9 and 10 much longer on apical half of elytra; summit of declivity with 4-6 long setae. Fore femur 1.5x wider than hind femur. Caudal surface of ventrite 2 distinctly anteriorly arcuate, of ventrite 3 less arcuate, of ventrite 4 almost perpendicular; caudal surface of all smooth and shiny. Ventrite 5 across base 1.5x wider than long. Genitalia as in Fig. 351; lateral sclerite divided into 2 separate pieces; spermathecal duct slightly thicker just before and throughout bursa; duct broken, more than 1.5 mm long.

Type Series.-Holotype, ECUADOR [Zamora-Chinchipe], Sabanilla, 15.IX.-2.X,05; 1.1907, Dr. Ohaus (Dresden). Allotype, same data as type (Dresden). Paratype, 1 female, Sarayacu*, Pascoe Coll. 93-60 (London).

*No country given. It is uncertain whether this is the Ecuadorian Sarayacu where Buckley collected (Brown, 1941:846) or the Peruvian Sarayacu from which Kirsch (1873:124) described material in the Abendroth collection.

Remarks.-The paratype is 13.0 mm long, 5.0 mm wide. It is clothed with full-sized scintillating green scales much as in conquisitus except the sinuous black vitta of the elytra is continuous from the base to the suture at the summit of the declivity, and there is an additional brief arcuate vitta on the basal third. The paratype has the rostrum intermediate in width between the type and allotype; the setae are more like those of the male.

This species and conquisitus bear the same type locality labels of Sabanilla, but given the general nature of the labels could, in fact, be from quite different habitats. Dr. Ohaus (1908:387) briefly describes his stay in the area at about 1900 m altitude, 4 hours above Zamora, 2 days' journey from Loja.

The two species are very close, and I interpret the circle of 4 apical lobes on the internal sac of scambus as indicating this to be the more derived species.

The word scambus means bow-legged and refers to the strongly bowed fore tibia, a character readily distinguishing this species from conquisitus.

45. Hadromeropsis (Hadrorestes) magicus (Pascoe)

Figs. 36, 249-253, 301, 331, 332, 347; Map 12

Naupactus magicus Pascoe, 1881:41:-42. Type, unique female, labelled "Type," "Brazil" on ellipse of pink paper, "Naupactus magicus Type Pasc," "Pascoe Coll. B.M. 1893. 60" (London).
Hadromerus magicus (Pascoe); Kuschel, 1955:278.

Diagnosis.-Black; elytra smooth and shiny without tubercles except a few on declivity. Male sharply marked with three conspicuous white spots on each elytron (Fig. 36), female sharply marked with vittate pattern. In male elytral interval 10 opposite metasternum much wider than adjacent interval 9, smooth; in female more nearly equal, squamose. Mesepisternum with dense cluster of large white scales.

Description.-Male, length 10.4-10.8 mm, width 3.7 mm. Female, length 12.3 mm, width 4.8 mm. Mostly glabrous; marked with large, shiny white scales as follows. Scattered scales around epistoma. Male (Figs. 36, 251) elytra with spots on base of interval 4, on intervals 4-7 just before middle, and on intervals 3-6 at apical umbone. Female with markings on head and prothorax as in Fig. 249, elytral markings as in Fig. 252. Female sharply vittate; a vitta on interval 2 from base to summit of declivity; a vitta from anterior edge of eye continuing along side of prothorax and continuing on elytral interval 4 to its apex; this vitta crossing to interval 6 at basal 0.3 continuing thence in broad vitta, at apical 0.3 gradually curving to stria 1 at summit of declivity, interrupted only by a round black spot from striae 5-7 on apical 0.3. Both sexes ventrally with similar large white scales aggregated on mesepisternum and scattered on sides of metasternum, and in front of fore and hind coxae. Head and rostrum punctate, more sparsely so in female; dorso-lateral edges of rostrum slightly divergent distally, median line finely impressed between eyes; rostrum moderately concave. Epistoma occupying approximately 0.4-0.5 of anterior edge of rostrum in male, 0.3 in female. Mandible with transverse groove shallow, broad, well-defined; weaker in male. Segment 1 of funicle 0.6-0.8x as long as segment 2. Prothorax of male 1.06x wider than long, sides moderately to strongly rounded between constrictions; prothorax of female as in Fig. 249, 1.1x wider than long. Disc of pronotum almost flat in profile, median line obsoletely impressed; disc of male smooth or with a sculpture of indistinct punctures and confluent low tubercles, changing to uniform, low tubercles on sides, each tubercle with slender white decumbent seta; a few setae more erect on sides of constrictions. Scutellum densely clothed with white hair-like scales; female with one scale in addition. Elytra across humeri 1.3x wider than prothorax in male, 1.5x in female; 3.0x longer than prothorax in male, 3.7x in female. In dorsal view sides of elytra of male (Fig. 36) subparallel to middle thence very gradually rounded to apex; apical umbone very weak and just touching outline; apices briefly individually rounded. Elytra of female wider just caudad of middle, apical umbone as in male; declivity (Fig. 253) slightly concave between suture and apical umbone, the transverse apical portion of interval 9 glabrous, prominent and with a few small tubercles; sutural interval terminating in tooth. Elytra with sculpture and interval 10 as in Diagnosis and Figs. 251, 252. Base of elytra not depressed behind scutellum, but flattened around scutellum in female; base obsoletely depressed at striae 4 and 5; base with a few short to long, fine, white setae. Striae regular except obliterated beneath squamose areas, strial punctures approximately the diameter of a scale; intervals smooth and polished with an occasional fine puncture. Intervals 9 and 10 with very fine, moderately long recumbent setae. Fore femur of male 1.7-1.8x, of female 1.2x wider than hind femur; distally with weak serrulate flange; with faint honeycomb microsculpture becoming rugulose distally, without tubercles; set with very fine, straight setae. Fore tibia straight to apical fourth whence slightly inwardly bowed; surface punctate; inner edge with 12 small, uniform teeth. Middle and hind tibiae cylindrical, with row of bristles and tubercles of inner edge not strongly developed. Caudal surface of ventrites 2, 3, and 4 of female equal in height; edge of ventrite 2 rounded; edge of ventrites 3 and 4 acute. Ventrites 3, 4, and 5 basally with conspicuous honeycombed microsculpture. Ventrite 5 of male with entire surface with shallow but broad punctures; apex broadly truncate-emarginate, deflected (Fig. 301). Ventrite 5 of female as in Fig. 250, across base 1.6x wider than long; apex emarginate. Male genitalia as in Figs. 331, 332; aedeagus 3.5-3.7 mm long, apex of internal sac 1.5 mm long. Female genitalia as in Fig. 347; spermathecal duct 2.0 mm long.

Distribution.-Map 12. BRAZIL: 1 female (the type), locality questionable (see Remarks) (London). COLOMBIA: 1 male, no data (Auckland); 1 male, Bogotá (London); 1 male, Fusagasugá, V.18 (Paris). No data: 1 male (Oxford).

Remarks.-I hesitated to associate these four males with the very differently marked unique type of magicus for two reasons. First, I do not believe the species could range from "Brazil" (as the type is labelled) to Bogotá and Fusagasugá; hence if the association is correct, the Brazil label is suspect. Secondly, the length of the apex of the internal sac of the males would seem to suggest a longer spermathecal duct. However, other factors seemed to outweigh these considerations. Especially significant is the squamose mesepisternum and the elytral spots of the males occurring in positions that are also squamose in the female.

The white spots of the males are very uniform in all four specimens although the scales in one are distinctly oval. The apex of the internal sac is intermediate in length within the group.

46. Hadromeropsis (Hadrorestes) nitidus n. sp.

Figs. 239-242, 300, 333, 334, 348; Map 12

Diagnosis.-Based on unique male. Shiny black, sparsely clothed with small, round, bright blue scales. Elytra convex in cross section. Last tergite narrow apically. Sclerotized apex of internal sac tubular, 2.4 mm long, stiff.

Description.-Holotype, male, length 11.4 mm, width 3.9 mm. Color as in Diagnosis. Scales of elytra set in brief, shallow transverse depressions or on the smooth surface. Head and rostrum sparsely punctate (Figs. 239, 242). Dorso-lateral edges of rostrum slightly divergent apically, high, surface deeply concave between; median line impressed between anterior half of eyes, a few brief impressed lines parallel to it. Epistoma (Fig. 242) occupying 0.4 of anterior edge of rostrum, as long as wide, sides almost parallel, apex broadly rounded. Mandible with weak transverse groove, its anterior edge being the edge of scar of cusp. Funicle with second segment 1.2x longer than segment 1. Prothorax (Fig. 239) 1.1x wider than long, sides strongly rounded between constrictions; disc gently arcuate in profile. Pronotum with median line unmarked; basal constriction smooth and polished with trace of impressed lines on right side; sculptured with uniform, low, round tubercles which are contiguous on sides, some confluent on disc; setae arising from tubercles very inconspicuous, as long as diameter of a tubercle, recumbent. Scutellum with very fine recumbent setae, no scales. Elytra across humeri 1.3x wider than prothorax, 3.2x longer than prothorax, 2.3x longer than width across humeri. In dorsal view sides of elytra parallel for basal half thence very gradually convergent to apex (Fig. 241), weak apical umbone just touching the outline; apex rounded, its edge conspicuously dentate. In dorsal outline, a few very weak tubercles on sides of elytra on basal fourth and apical half. Elytra transversely convex. Base of elytra with stria 5 weakly depressed, postscutellar depression obsolete; base of elytra with a few very fine, dark, moderately long, erect setae. Disc of elytra with no tubercles from base to apical fourth where weak ones appear in intervals and striae, those of declivity small but more acute. Striae 9 and 10 opposite metasternum (Fig. 240) not more widely separated than striae 8 and 9. Intervals 9 and 10 with no erect setae. Fore femur 2.0x wider than hind femur; distal flange weak, edge granulate; anterior and posterior face smooth and shiny proximally, distally with weak honeycomb sculpture rapidly changing to impressed lines, with no shelf-like tubercles on posterior face; set with fine, stiff, dark setae. Fore tibia straight until distal third whence moderately inwardly bowed; inner edge with 11 small teeth and distal half with multiple long, dark, stiff setae; surface punctate, more coarsely so on outer surface. Dorsal surface of tarsi clothed with prostrate pale blue seta-like scales and conspicuous, stiff, distally-directed black bristles. Middle and hind tibiae on inner edge with row of 4 or 5 stout bristles, without tubercles. Mesepisternum moderately convex and with brief row of blue scales on outer edge. Metasternum and abdomen smooth and polished with numerous very fine, erect setae. Ventrite 5 (Fig. 300) with a few very fine punctures; apex broadly, deeply emarginate. Last tergite as in Diagnosis. Genitalia as in Figs. 333, 334. Aedeagus 3.6 mm long; internal sac with conical apex elongated into tube exteriorly minutely spiculate on basal half, more than 2.4 mm long (extreme apex accidentally broken off).

Type Series.-Holotype, ECUADOR, Napo, 7 km S Baeza, 22 Feb. 1979,

H & A Howden, walking up blade of grass (Howden).

Remarks.-There is a female labelled "Columbia Baden" in the Dresden collection which I tentatively assign to this species. It differs from the type as follows. Length 12.3 mm, width 4.5 mm. All scales larger, very pale green except those forming a pattern on elytra dense, whitish; pattern consisting of a long loop from base of stria 4 to stria 2 at apical third and back to beneath humerus, a second oblique elliptical mark outlining apical umbone. Head, rostrum and prothorax similar to male except segment 1 of funicle slightly longer than segment 2. Elytra more coarsely sculptured and flatter; elytra 3.6x longer than prothorax; sutural interval at apex not longer than denticles along margin; striae 9 and 10 more widely separated opposite metasternum. Caudal surfaces of ventrites 2, 3, and 4 perpendicular, equal, moderate; ventrite 5 without lateral depressions; apex narrowly, deeply notched. Genitalia as in Fig. 348; spermathecal duct 5.0 mm, extreme in length as is the male endophallic structure.

The type of magicus resembles this female but differs in segment 2 of the funicle longer than 1; elytra with straight white vitta on interval 2 and vitta of stria 4 straight-edged until declivity, thus leaving intervals 1, 3, and most of 4 bare until declivity.

The mealy substance of Fig. 239 is incompletely dissolved wax; Fig. 240 shows large areas of untouched wax.

The type of nitidus resembles spiculatus in habitus but nitidus has the sides of the prothorax rounded instead of almost parallel-sided, disc of pronotum not flattened broadly, last tergite narrow, fore femur with setae of posterior face arising from almost smooth surface, elytra narrower apically, internal sac different and terminal tube spiculate proximally only and more slender.

All the definitions of the Latin adjective "nitidus" are appropriate for this species: shining, sleek, in good condition, elegant, etc.

47. Hadromeropsis (Hadrorestes) mandibularis n. sp.

Figs. 243-248, 346; Map 13

Diagnosis.-Mandible with two stout dorsal setae set in deep, sharply defined transverse groove (Fig. 243). Without full-sized scales dorsally or ventrally. Disc of prothorax with punctures in female or irregular pustules in male which gradually change to low, rounded tubercles on sides; median line unmarked. Apices of elytra of male broadly, individually rounded (Fig. 247). Fore femur of males sculptured as in Fig. 246. Spermathecal duct 4.5 mm long.

Description.-Holotype, female, length 16.2 mm, width 6.0 mm. Black; legs and antenna piceous. Clothed with minute greenish white scales; scales sparse on rostrum, head and prothorax, aggregated in transverse depressions on disc of elytra, more numerous on declivity. Head and rostrum (Fig. 243) uniformly punctate. Median line impressed for short distance on rostrum to between middle of eyes, with a few parallel grooves either side. Rostrum only slightly concave. Epistoma as in Fig. 243, 244; mandible as in Diagnosis and Fig. 244.

Funicle with segments 1 and 2 approximately equal in length. Head robust, eyes not very prominent. Pronotum as in Diagnosis and Fig. 245. Prothorax 1.09x wider than long, sides gently rounded between weak constrictions, widest before middle; anterior constriction unmarked on disc, basal constriction very weak; setae arising from lateral tubercles fine, recumbent, about as long as diameter of a tubercle. Scutellum with whitish seta-like scales. Elytra across humeri 1.6x wider than prothorax, 3.8x longer than prothorax. In dorsal view, sides almost parallel, very slightly convergent from about middle, apical umbone slightly arcuate in outline, sides thence almost straight to apex. In dorsal outline no tubercles present except small ones on apical umbone; margin of elytra conspicuously dentate around ventrite 5, sutural interval not modified into tooth. Elytra in profile flattened behind scutellum for basal seventh, then abruptly deflected, flat to summit of declivity, suture here set with 7 or 8 stout dark setae. Base of elytra with stria 5 scarcely depressed, without long setae. Intervals 1, 2, and 3 flat, smooth and polished, without squamose transverse depressions from near base almost to declivity in an elongate V. Elytra without tubercles except on declivity and just before apical umbone (Fig. 248); striae very regular; interval 10 wider opposite metasternum, interval 9 wider from hind coxa to apical umbone. Fore femur 1.5x wider than hind femur, with a weak serrulate flange; fore tibia with 10 small equal teeth on inner edge; outer edge punctate and set with numerous long, dark, wiry setae. Inner edge of middle tibia with row of 8 stout bristles each set at base of a sharp tubercle, hind tibia with 4 or 5 such bristles without tubercles. Caudal surface of ventrites 2 and 3 moderate, anteriorly slanted, equally high, edge not at all acute. Caudal surface of ventrite 4 less slanted, edge acute. Ventrite 5 across base 1.6x wider than long, feebly convex medially; apex narrowly truncate. Genitalia as in Fig. 346; spermathecal duct 4.5 mm long.

Allotype, male, length 11.6 mm, width 4.0 mm. Complete description impossible because of condition of specimen which is partly destroyed by dermestids, missing antennae except right scape and entire abdomen except ventrites 1 and 2. Differs from type as follows. Scales less numerous. Rostrum with median line sharply impressed between eyes only, entire rostrum concave, longitudinally convex either side of median line, the longitudinal grooves of the type represented by 2 traces. Transverse groove of mandible less developed. Pronotum more coarsely sculptured, punctures more or less replaced by irregular low pustules. Elytra across humeri 1.3x wider than prothorax, 3.0x longer than prothorax. Sides of elytra parallel, very gradually convergent from about 0.6, umbone weak, just touching dorsal outline; apices (Fig. 247) broadly, individually rounded. In dorsal outline, tubercles of lateral intervals obsolete basally, absent medially, conspicuous apically, denticles of apical edge very conspicuous. Smooth, polished area of intervals 1, 2, and 3 reduced to basal 0.5. Fore femur 1.8x wider than hind femur; sculpture as in Fig. 246; without shelf-like tubercles on posterior face. Fore tibia straight until basal 0.25 whence slightly curved inward. Middle and hind tibiae with bristles scarcely thicker than other setae, tubercles replaced by granules. Abdomen missing.

Type Series.-Holotype, COLOMBIA, S. Antonio, ex. coll. Clerc

(Paris). Allotype, COLOMBIA, St. Antonio, 2000 m, 16.6.1908, Coll. Pape, Hust. det., Hadromerus prope scabricollis Fst (Eberswalde).

The type locality could be the San Antonio on the road between Cali and Buenaventura where Fassl collected (Voss, 1953:29).

Remarks.-Bearing strong resemblance to nitidus and magicus in the head and rostrum and sharing the strong mandibular groove with cavifrons and magicus, mandibularis can be distinguished by the diagnostic characters. From cavifrons it also differs in the simpler rostrum and lack of tubercles on elytral intervals 8 and 9. A mandibular groove is also present in the impressicollis and alacer groups.

The very long spermathecal duct (second in length only to that of nitidus? female) suggests that the male will be found to have a very long endophallic structure, slightly shorter than that of nitidus, longer than that of magicus.

H. mandibularis is most closely related to nitidus, the males of these two species being most readily distinguishable by the shape of the elytra which in nitidus are more convex in cross section and profile, more nearly vertical at apex, more narrowly rounded at apex, and with weaker postscutellar depression. When more specimens are found, it is probable that additional characters will be found in the ventrite 5 and last tergite as well as in the genitalia.

The name mandibularis refers to the extreme development of the dorsal groove on the mandible.

48. Hadromeropsis (Hadrorestes) spiculatus n. sp.

Figs. 38, 280, 286-288, 327, 357; Map 12

Diagnosis.-Resembling pectinatus, differing especially in the following. Slender in habitus (Fig. 38). Pronotum flattened medially (Fig. 288). Apical edge of elytra of female briefly indented before sutural tooth (Fig. 280). Fore tibial teeth more uniform in size and anterior face of fore tibia punctate only. Internal sac as in Fig. 327, apex spiculate, very long.

Description.-Holotype, male, length 10.8 mm, width 3.6 mm. Piceous, except head, base of rostrum and most of prothorax black. Head and rostrum (Fig. 288) similar to pectinatus but sculpture finer, fovea absent and median line finely impressed to behind eyes. Funicle with segment 2 subequal to segment 1. Prothorax (Fig. 288) scarcely wider than long, sides only weakly rounded between constrictions; prothoracic sculpture consisting of low rounded tubercles of approximately uniform diameter, tubercles of broad median flattened area very shallow and tending to be arranged in rows; setae of prothorax long, more slender than in pectinatus. Elytra across humeri 1.2x wider than prothorax; elytra 3.1x longer than prothorax, 2.4x longer than width across humeri. In dorsal view (Fig. 38) sides of elytra parallel for basal 0.2, thence very slightly wider, tapering gradually to apex from apical 0.5, apical umbone very weak but touching outline. Base of elytra depressed at stria 5, more strongly depressed behind scutellum than in pectinatus. Elytral tubercles less developed than in pectinatus, absent on sides of elytra centrally.

Elytral surface smooth and polished between striae with no sculpture except that of declivity which gradually begins caudad of middle; striae very regular and distinct on disc, still recognizable on declivity. Fore femur 2.4x wider than hind femur. Entire surface of fore femur with fine microsculpture, on anterior face (Fig. 286) rugulose, on posterior face (Fig. 287) finely impressed netted or honeycomb sculpture, distal half with superimposed crowded large setiferous tubercles not strongly shelf-like; inner edge distally with gradual, weak flange, sharply denticulate on edge. Fore tibia (Fig. 286) with sculpture of punctures only, with 8-10 sharp teeth on inner edge and a few extra granules on posterior face. Middle tibia on inner face near edge with 3-5 sharp tubercles each with a stout bristle; hind tibia with 3 bristles without tubercles. Genitalia as in Fig. 327; aedeagus 3.5 mm; internal sac with apex elongated into stout exteriorly spiculate tube 2.5 mm long.

Allotype, female, length 13.1 mm, width 5.0 mm. Differs from type as follows. Reddish-brown (slightly teneral), only meso- and metasterna, scutellum, head and base of rostrum black. Normal vestiture uncertain because of teneral condition, but elytra with many minute elongate scales and with full-sized creamy white scales (mostly transparent in allotype) forming a vitta on left elytron from humerus on interval 7 to interval 5 at middle. Rostrum wider, more densely sculptured. Elytra across humeri 1.5x wider than prothorax; elytra 3.5x longer than prothorax; elytra 2.4x longer than width across humeri. Elytra slightly wider at middle, apex more elongate (Fig. 280). Sutural interval at summit of declivity with 15-16 long, erect setae. Fore femur 1.4x wider than hind femur, sculpture as in type but without shelf-like tubercles; flange of inner edge and denticles scarcely weaker than in type. Middle tibia without tubercles on inner face, with 6 or 7 stout bristles; hind tibia with 3 stout bristles. Caudal surface of ventrites 2 and 3 anteriorly inclined, edge not sharp; smooth and shiny but with trace of sculpture basally. Caudal surface of ventrite 4 perpendicular, edge straight in posterior view, smooth and polished. Ventrite 5 across base 1.5x wider than long, only feebly convex medially. Genitalia very extreme in development (Fig. 357), spermathecal duct of average diameter at nodulus only, gradually becoming larger and testaceous in color, expanded and bulbous before bursa, narrower again at entrance of bursa.

Type Series.-Holotype, PERU, Huánuco, Acomayo, 2100 m, 7 June 1946, F. Woytkowski Coll., Donor Wm Proctor (New York). Allotype, PERU, Huánuco, Carpish, 2800 m, 14 Oct. 1946, F. Woytkowski Coll., Donor Wm Proctor (New York). Paratype, 1 female, BOLIVIA, L. P., 9 mi NE Unduavi, Chuspipata Ridge, cloud forest, Apr. 9, 1979, Collectors: L & C.W.O'Brien (O'Brien).

Remarks.-The paratype is 11.3 mm in length, 4.2 mm in width. Like the allotype it also appears slightly teneral, but the vestiture is in fresh (not waxy) condition and the dorsum bears a few more scales. The paratype differs as follows. Rostrum, prothorax, elytra, appendages reddish-brown; head, meso- and metasterna and abdomen black. With minute elongate white scales scattered on dorsum of head and rostrum, sides and disc of prothorax, and elytra except medially; a glabrous area on elytra vaguely outlined by larger scales from interval 4 and 5 at base to suture at apical 0.25. Long, conspicuous, fine white setae on sides of prothorax and sides of rostrum apically. Elytra 4.1x longer than prothorax, base of elytra scarcely impressed at interval 5. Most setae abraded from sutural interval at summit of declivity. Fore femur 1.2x wider than hind femur. Ventrite 5 moderately convex on apical half. Spermathecal duct "normal", i.e., not distended as in allotype; 1.7 mm long.

The difference in the spermathecal duct between the allotype and paratype seems too great for intraspecific variation. Is the difference attributable to some physiological factor, or age, or are two species involved?

The species is named spiculatus because of the extreme spiculate apex of the internal sac.

Species Not Assigned to Group

49. picchuensis n. sp.   50. cavifrons n. sp.
Remarks.-Based on the small sample (four specimens) of females only, the relationships of these two species are not obvious, but they do belong to Hadrorestes.

49. Hadromeropsis (Hadrorestes) picchuensis n. sp.

Figs. 31, 271-273, 353; Map 12

Diagnosis.-Based on female only. Clothed with round, mostly discrete scales of cream, pale tan or rose, at most only faintly iridescent; elytra with larger white scales arranged in faint fasciae and vaguely outlining common glabrous area on basal half medially as in Fig. 31. Disc of pronotum broadly flattened (Fig. 271). Elytral intervals 8, 9, and 10 with very long setae set at a 45° angle or greater, setae very conspicuous in dorsal view. Genitalia with a large sclerite in bursa copulatrix at junction of duct (Fig. 353).

Description.-Holotype, female, length 10.5 mm, width 4.3 mm. Integument brown, head and rostrum piceous except part of mandible and epistoma reddish. Squamose as in Diagnosis, Fig. 31. Rostrum with dorso-lateral edges parallel, sides almost perpendicular, surface of rostrum almost flat (flatter than in Fig. 271 of paratype); median line briefly impressed between eyes; with a few fine, glabrous rugulae showing between scales of head and rostrum. Mandible with one long dorsal seta set in a vague shallow depression. Funicle with segments 1 and 2 subequal. Prothorax as in Diagnosis and Fig. 271; 1.2x wider than long, sides moderately rounded between weak constrictions. Sides of prothorax with scattered weak tubercles only slightly larger than a scale, each tubercle bearing a very fine seta as long as 2-3 scales, setae closely appressed. Prosternum with scales transversely elongate; with long, wispy setae. Scutellum densely clothed with imbricate scales. Elytra as in Diagnosis and Fig. 31. Elytra across humeri 1.4x wider than across prothorax; 3.7x longer than prothorax. In dorsal view sides of elytra slightly divergent from humeri to just beyond middle, thence gradually converging to apex; apical umbone weak, feeble in dorsal outline; a few weak setiferous tubercles

present in dorsal outline basally and apically. Elytra in profile
with declivity very gradual, summit unmarked except by 7 or 8 long,
wispy setae; sides of elytra strongly turned under opposite ventrites
2, 3, and 4 (Fig. 273) (partly an artifact of teneral condition?).
Base of elytra obsoletely depressed at stria 5, distinctly depressed
behind scutellum; base of elytra with many long, wispy setae, some as
long as the long setae of summit of declivity; some setae arising from
tubercles like those of prothorax; similar tubercles present on apical
third of elytra including declivity. Striae almost completely
regularly punctate. Apex of elytra as in Fig. 272; edge with
scattered denticles smaller than an adjacent scale; sutural interval
attenuate into small tooth. Fore femur 1.2x wider than hind femur,
distal flange obsolete; faint sculpture concealed by scales. Fore
tibia straight, inner edge with 8-10 acute teeth, each tooth with an
unusually conspicuous bristle exceeding the tooth in length (an
artifact of teneral condition?). Middle tibia with 5, hind tibia with
3 stout bristles on inner surface dorsally. Abdomen crumpled in type,
apparently similar to that of paratype which has caudal surface of
ventrites 2, 3, and 4 perpendicular, moderate, equal. Ventrite 5
almost flat, no lateral depressions, median convexity very weak, 1.6x
wider than long. Genitalia as in Fig. 353; spermatheca very slender,
in one plane; spermathecal duct 2.5 mm long, with a large sclerite in
bursa at junction of duct.

  Type Series.-Holotype, PERU, Cuzco, Torentoy Canyon (Base Machu
Picchu), 2000 m, VI-VII.1964, B. Malkin (São Paulo). Paratype, 1
female, PERU, Cuzco, Machu-Picchu (sobre ruinas, 2600-2800 m),
1-2.VII.1964, B. Malkin (Howden).

  Remarks.-The paratype appears to be slightly more teneral than the
type and was not dissected. It measures 9.5 mm long, 3.9 mm wide, and
differs conspicuously only in the apex of the elytra which has the
tooth almost absent.

  The pattern formed by the scales, the softer integument (although
possibly only an artifact of teneral condition), and the spermatheca
in one plane are characters more common in the nominate subgenus. The
large bursal sclerite is similar to that of the argentinensis and
nobilitatus groups, but the sclerotized streaks of the vagina of those
groups are absent. The lateral sclerites are exclusive to
Hadrorestes. Characteristics of the male genitalia are difficult to
predict.

        50. Hadromeropsis (Hadrorestes) cavifrons n. sp.

              Figs. 282-285, 356; Map 13

  Diagnosis.-Based on female only. Frons and rostrum grossly
hollowed out, the dorso-lateral edges of rostrum elevated, thus
accentuating the concavity (Figs. 282, 283). Mandible with dorsal
setae set in a deep transverse groove. Elytra with sharply defined
markings of yellowish scales forming a basal ring, median zigzag
fascia and an apical band or ellipse encircling apical umbone.
Spermathecal duct 2.5-3.0 mm long.
  Description.-Holotype, female, length 17.4 mm, width 6.8 mm.

Black; femora and tibiae pale reddish, black at ends. Integument dorsally and ventrally apparently with sparse, minute, green scales which are mostly abraded; prothorax and elytra with a pattern formed by large yellowish, dense scales; on prothorax (partly abraded) in at least a line anteriorad of basal constriction continuing ventrad beside and behind fore coxa. Elytral pattern consisting of basal ring between stria 5 and scutellum; thick median zigzag fascia; and broad, oblique ellipse just before declivity. Hind coxa covered with green and yellow scales of various shapes, medium size. Head and rostrum as in Diagnosis and Figs. 282, 283. Segment 2 of funicle slightly but distinctly longer than segment 1. Prothorax (Fig. 283) with disc almost smooth, median line obsoletely finely impressed; sides very weakly tuberculate. In profile almost flat, except for elevated basal constriction. Scutellum with a few small, round, green scales. Elytra across humeri 1.6x wider than prothorax, 3.7x longer than prothorax, 2.3x longer than width across humeri. In dorsal view sides rounded at about middle, the gentle curve continuous with apical umbone; sides of apex weakly curved from beneath stria 4 to suture, its edge strongly dentate, sutural interval elongated into a very brief tooth (Fig. 285). Base of elytra with very weak postscutellar depression, humeri suddenly produced from stria 5. Sides of elytra in dorsal view with continuous small, acute tubercles, tubercles situated on intervals 8 and 9. Interval 10 opposite metasternum wide, tuberculate; striae here and on declivity somewhat irregular, elsewhere regular. Suture at summit of declivity with 8 long, dark setae. Fore femur 1.3x wider than hind femur, with weak serrulate flange. Fore tibia with 14-16 very small teeth on inner edge; outer edge of fore tibia with long setae not stiff and wiry; outer surface punctate, some punctures coalescing into short grooves. Middle tibia on outer edge with 2 sulci, inner surface with row of 10 stout bristles. Hind tibia (Fig. 284) on outer edge with single sulcus; inner surface with row of 7 stout bristles. Caudal surface of ventrites 2, 3, and 4 slightly overhanging succeeding ventrites, equal, sharp-edged. Ventrite 5 with moderate median convexity; across base 1.4x wider than long; apex narrowly rounded. Genitalia as in Fig. 356; without lateral sclerites, no sclerite seen at junction of duct with bursa copulatrix; duct 2.5 mm long, testaceous. Spermatheca strongly twisted in several planes.

Type Series.-Holotype, PERU, Piches and Perenes Vs, 2000-3000 ft, Soc. Geog de Lima (Washington, USNM Type No. 100192). Paratype, 1 female, PERU, Huánuco, Divisoria, 1600 m, 31.7.47, Schunke leg., coll. Kuschel (Auckland).

Remarks.-The paratype is 15.8 mm long, 6.0 mm wide, and differs from the type as follows. Elytral median fascia on side of elytra at stria 7 continues anteriorly in a vitta to base of interval 9; elytra across humeri 1.5x wider than prothorax; elytra 3.8x longer than prothorax; sides of elytra in dorsal view with no tubercles in central portion; fore femur 1.5x wider than hind femur; middle and hind tibiae with only 4 stout bristles each; all tibiae without sulci on outer surface; caudal surface of ventrites 2 and 3 not overhanging but slightly anteriorly slanted; ventrite 5 across base 1.6x wider than long; spermathecal duct 3.0 mm long ending in a large disc inside bursa copulatrix. The significant differences are the presence of a bursal sclerite and the smooth, non-sulcate tibiae. Since I am certain the two females are conspecific, then the bursal sclerite and sculpture of the tibiae must be added to the long list of intraspecifically variable characters.

Because this species clearly belongs to Hadrorestes in all the characters listed for the subgenus except for the lack of paired lateral sclerites, I presume these have been secondarily lost. More material, especially males, are needed to establish the relationship of this species to others.

Incertae sedis

51. Hadromeropsis earinus n. sp.   52. Hadromeropsis striatus n. sp.

These two species, each described from a unique specimen, are not assigned to a subgenus for the reasons discussed under Remarks.

51. Hadromeropsis earinus n. sp.

Figs. 289-293, 335

Diagnosis.-Based on unique female. Eye small (Fig. 289), separated from prothorax by its length. Clothed with iridescent scintillating green scales. Prothorax (Fig. 290) with basal and apical constrictions strong laterally and dorsally, surface irregularly sculptured and subtuberculate on sides. Base of elytra between striae 5 abruptly perpendicular and slightly anteriorly produced. Margin of caudal surface of ventrites 2, 3, and 4 not acute, the edge and half the adjacent caudal surface densely covered with elongate appressed scales (Fig. 293). Spermatheca with ramus replaced by circular membranous area. With neither sclerites in the bursa copulatrix nor lateral sclerites in the vagina.

Description.-Holotype, female, length 13.0 mm, width 4.6 mm. Black. Clothed as in Diagnosis. Scales small, circular; dense, imbricate ventrally; less dense on dorsal surface, elytra appearing slightly tessellate because of varying density of scales. Dorsal setae short, broadly lanceolate to almost spatulate, curved, as long as 1-2 scales. Head and rostrum as in Fig. 289. Rostrum 1.26x longer than width at apex, strongly longitudinally concave from between eyes to about middle, dorso-lateral edges prominent, apically divergent; median line finely impressed between apical half of eyes. Surface of head and rostrum not sculptured. Mandible without dorsal groove, with only one long seta arising from a fovea. Funicle with segments 1 and 2 equal in length; antennal club very long, as long as segments 1 to 4 of funicle. Prothorax as in Fig. 290, Diagnosis; 1.08x wider than long. Postocular vibrissae very well developed, reaching 0.6 to eye. Scutellum small, squamose as elytra, broadly rounded posteriorly. Elytra across humeri 1.5x wider than prothorax, elytra 3.8x longer than prothorax; elytra 2.4x longer than width across humeri. In dorsal view base of elytra (Fig. 290) slightly arcuately emarginate between striae 5 and as in Diagnosis; humeri strongly angled posteriorly, sides of elytra gradually divergent to middle where they are 1.15x wider than across humeri, thence gradually converging to elongate apex, apical umbone briefly entering outline. Sides of apex straight caudad of umbone, apex (Fig. 291) elongate, suture attenuated into conspicuous tooth approximately 0.5 mm long. Lateral edge of elytra absolutely smooth. Elytra nowhere with tubercles; with strong, transverse postscutellar depression; interval 7 faintly keeled between middle and apical umbone; umbone slightly keeled, this keel continuing obliquely towards apical tooth (Fig. 291). Remainder of surface very faintly, irregularly, undulating and faintly depressed around punctures. Suture at summit of declivity with approximately 15 very conspicuous, long, straight, white setae. Striae regularly punctate; each puncture approximately as wide as a scale, each with a broad seta curled over it; striae 9 and 10 concealed by densely imbricate scales, but traceable by setae, clearly visible from inside elytra. Legs without sculpture. Fore femur slender, 1.08x wider than hind femur, 1.2x wider than middle femur; with no trace of flange on inner edge distally. Fore tibia long, slender, distal 0.3 bowed inward; inner edge with 13 small, equidistant teeth, the oblique bristle on distal face of each tooth longer than tooth. Middle tibia on inner face with 9 strongly oblique bristles, most set on small tubercles; hind tibia with 6 oblique bristles without tubercles. Suture between ventrites 1 and 2 strongly anteriorly arcuate on central half (Fig. 292). Caudal margin of ventrites 2, 3, and 4 as in Diagnosis and Fig. 293. Ventrite 5 across base 1.4x wider than long, with a slight convexity on either side near base, moderate median convexity, apex narrowly truncate. Genitalia as in Diagnosis and Fig. 335, spermatheca in one plane, spermathecal duct approximately 1.4 mm long.

Type Series.-Holotype, ECUADOR, - -[illegible], Coll. J. Faust, Ankauf 1900 (Dresden).

Remarks.-Some of the characteristics of this unique female are so extreme that its relationship to other species groups is obscure. the partially squamose and rounded edge of the caudal margin of ventrites 2, 3, and 4 are the most extreme in the genus, but a glabrous male of this species might have no scales and the edge acute. The suture between ventrites 1 and 2 is much more strongly arcuate than in any Hadromeropsis seen. The fore femur is slightly more slender than in any other female seen. The membranous ramus is unique and so unusual that one might suspect an aberration. I take a conservative position here and wait for more material to clarify the situation.

When a male of this species is available, subgeneric placement can be reconsidered on the condition of its vestiture, including the caudal margin of ventrites 2, 3, and 4; the edge of the elytra (males of Hadrorestes often have the denticles of the edge of the elytra stronger than in the female); and the form of the endophallic structure.

The habitus of earinus is more like that of Hadrorestes than the nominate subgenus, but it does not fit there because the edge of the elytra is smooth, striae 9 and 10 are well-separated their full length, the striae are regular beneath the apical umbone, and there are no lateral sclerites in the vagina. Of the species groups, earinus seems most closely related to the alacer group which has the eye distant from the prothorax, scarcely enlarged fore femur in the female and less acute caudal margin of ventrites 2 and 3. It differs from that group in the strong postocular vibrissae, regularly aligned strial punctures, strongly depressed postscutellar area, and segments 1 and 2 of funicle equal.

The name is derived from the Greek word "earinos" meaning the color of spring green.

52. Hadromeropsis striatus n. sp.

Figs. 358-362, 368-370

Diagnosis.-Based on unique male. Rostrum narrow in dorsal view, thin in profile (Figs. 358, 359). Elytra with prominent, straight striae; strial punctures deep; intervals convex. Brachypterous: wing 0.6x length of elytron, mesepimeron narrow, elytra extremely narrow across humeri. Fore femur enormously swollen; without shelf-like tubercles on posterior face; with strong flange on inner edge distally. Internal sac (Figs. 368, 369) without internal sclerites; with heavily sclerotized distal tube.

Description.-Holotype, male, length 11.8 mm, width 4.0 mm. Black except antennae reddish; with minute blue scales scattered over dorsum and legs, scales round, oval or elongate. Side of head below eye with patch of white to gray-blue scales of various sizes and shapes. Head and rostrum as in Diagnosis, Figs. 358, 359. Dorsum of rostrum very shallowly concave. Surface of head and rostrum including epistoma and mandible with fine microsculpture; head and rostrum also with additional moderate, irregular sculpture, not distinctly punctate; frontal fovea moderate, median line not marked except as vague carina from interantennal line to apex of epistoma. Epistoma occupying 0.4 of anterior edge of rostrum, posterior margin thickened. Scrobe more distant from eye than is usual for genus (Fig. 359). Mandible with no long dorsal setae, but one moderately long dorso-lateral seta in an elongate fovea. Antennal scape in resting position reaching anterior margin of prothorax (Fig. 359, but head twisted in photo) funicle with segment 1 slightly longer than segment 2. Eye separated from anterior margin of prothorax by 0.7x its own diameter. Prothorax as in Fig. 359. Prothorax 1.1x wider than long; sides in dorsal view almost straight, distinctly wider before middle. Surface slightly uneven; surface evenly, densely covered with slightly irregular rounded tubercles; each tubercle with a very slender seta, those on sides much longer and more nearly erect than dorsal setae. Postocular vibrissae weak. Elytra across humeri only 1.05x wider than prothorax. Elytra 2.3x longer than width across humeri; elytra 2.9x longer than prothorax. Elytra as in Diagnosis. Sides of elytra parallel for basal 0.26, thence gradually divergent, widest just caudad of middle, convergent to apex; apex (Fig. 360) truncate from approximately stria 3, sutural interval briefly produced into a tooth. Striae and intervals very regular, straight; strial punctures very deep and intervals convex (Fig. 361) from base to unmarked declivity; on declivity (Fig. 360) strial punctures rapidly becoming obsolete and intervals flat; small tubercles on basal 0.25 of elytra only. All surfaces of elytra with irregular microsculpture, more pronounced laterally and on declivity. Edge of elytra absolutely smooth. Apparently without long, wispy setae; all intervals with row of pale, slender setae, those of alternate intervals very slightly longer than seta of even-numbered intervals; setae suddenly longer, straight and erect on apical half of elytra laterally and on declivity, the longest approximately 0.5 mm. Wing as in Diagnosis. Fore femur as in Diagnosis and Fig. 359, 2.2x wider than hind femur. Surface of fore femur almost smooth proximally, becoming minutely rugulose distally, without punctures; inner edge with small scattered tubercles, tubercles pronounced on distal flange. Fore tibia (Fig. 362) straight medially, weakly bowed distally and proximally; inner edge very irregular, proximal half with small tubercles, distal half with large medial tooth followed by 4 or 5 teeth of various sizes, one bifid on right tibia of type; fore tibia with numerous granules on anterior and posterior face, outer edge with small punctures on smooth surface. Fore tarsus similar to that of pectinatus in Fig. 308. Middle and hind femora slender for proximal half then abruptly swollen; middle tibia on inner edge with 5 small indistinct tubercles, larger tubercles with a stout bristle arising from base; hind tibia without tubercles or teeth on inner edge. Ventrites 1, 2, and 3 with very fine microsculpture of transverse rugulae; ventrites 1 and 2 with setae arising from minute punctures, punctures stronger on ventrite 3. Ventrite 4 with punctures more numerous. Ventrite 5 (Fig. 370) 1.9x wider than long; with crowded punctate-rugulose sculpture; sharp lateral edge stops abruptly at middle of sides, at which point ventrite becomes convex; apex emarginate. Genitalia as in Figs. 368, 369; aedeagus 3.9 mm, aedeagal apodeme 2.5 mm, tegminal strut 2.6 mm. Internal sac without internal sclerites; with heavily sclerotized, stout distal tube.

Type Series.-Holotype, COLOMBIA, Paso Bella Vista above Duria- maina, 3760 m, 7 Dec. 1978, H. Sturm, in dead leaves of Libanothamnus, 78, coll. Kuschel (Auckland).

Remarks.-The integument of the type was cleaned of a thin film of transparent wax even though the long, wispy setae usually associated with wax were not evident.

This species is tentatively assigned to Hadromeropsis because it shares more characters with that genus than with any other genus I know. The species does not conform to either of the two subgenera of Hadromeropsis. Most important, the unique internal sac lacks the heavy internal sclerites characteristic of all Hadrorestes and of the argentinensis and nobilitatus groups of the nominate subgenus, but is not closely related to the remaining flagellate groups either. The long sclerotized tube at the apex of the internal sac is difficult to relate to any Hadromeropsis I have seen and a polyphyletic element may have been introduced by including the species. However, it is conceivable that the distal fraction of the sclerotized tube represents the sclerotized apex of Hadrorestes and the long proximal portion represents the heavy internal sclerites of Hadrorestes without the membranous case.

Other extreme expression of generic characters are: posterior margin of epistoma thickened, antennal scape reaching anterior margin of prothorax (but almost as long in pectinatus), antennal scrobe well separated from eye, prothorax with anterior constriction distinct in profile.

The species resembles a Hadrorestes on external characters but differs from that subgenus in the well-separated striae 9 and 10 and the lack of denticles or tubercles around the edge of the elytra. It is possible though, that both of these qualities are a consequence of brachyptery, especially in this species with the shortest wing of the three brachypterous species.

The species is named striatus for the very conspicuous elytral striae.

ACKNOWLEDGMENTS

Because the genus Hadromeropsis is not well represented in collections, each of the 31 institutional collections and four personal collections studied was unique and vital. The institutions and people responsible for loans of specimens are all thanked for their valuable contributions.

A seven-week study in Argentina was greatly enhanced by the many kindnesses of Antonio and Juana Martínez and by the assistance of the Instituto de Investigaciones Entomologicas Salta (INESALT) in Rosario de Lerma through M. and J. Viana and G. Williner. Patricia Hoc, Universidad Nacional de Buenos Aires, kindly identified the food plants of Curculionidae collected in Argentina.

David Maddison, University of Alberta, painted the habitus of Hadromeropsis nobilitatus for the frontispiece. Jennifer Read, Carleton University, assisted with all aspects of the art work, and her assistance is gratefully acknowledged. All scanning electron micrographs were taken from uncoated specimens by Lewis Ling, Carleton University, whose efforts are greatly appreciated. Donald R. Whitehead, Systematic Entomology Laboratory, Washington, read the manuscript and provided many helpful comments.

I especially thank my husband, Henry F. Howden, for patiently advising and discussing ideas and for assisting with the many more tedious aspects of preparing the manuscript for publication.

LITERATURE CITED

Bates, H. W.
    1910. The naturalist on the River Amazons. Popular Edition. John Murray, London. 394 pp.
Blackwelder, R. E.
    1957. Checklist of the Coleopterous insects of Mexico, Central America, the West Indies, and South America. USNM. Bull. 185: 1-1492.
Blanchard, E.
    1846. In Brullé, A., 1843, Insectes de l'Amérique méridionale recueillis par Alcide d'Orbigny, et decrits par Emile Blanchard. Paris. pp. 105-222 (= 1846 Blackwelder 1957:968).

Boheman, C. H.
    1840. In Schoenherr, Genera et species curculionidum, cum
        synonymia hujus familiae. Vol. 6, pt. 1. Roret, Paris.
        474 pp.
    1845. In Schoenherr, Genera et species curculionidum, cum
        synonymia hujus familiae. Vol. 8, pt. 2. Roret, Paris.
        504 pp.
Brown, F. M.
    1941. A gazetteer of entomological stations in Ecuador. Ann.
        Ent. Soc. Amer. 34: 809-851.
Champion, G. C.
    1911. Otiorhynchinae alatae. In Biologia Centrali-Americana,
        Coleoptera 4(3): 178-354.
Clark, W. E.
    1977. Male genitalia of some Curculionoidea (Coleoptera):
        musculature and discussion of function. Coleopt. Bull.
        31(2): 101-115.
Fassl, A. H.
    1915. Tropische Reisen. V. Das obere Caucatal und die
        Westcordillere. Ent. Rund. (1914) 31(10): 57-58.
Faust, J.
    1892. Reise von E. Simon in Venezuela. Curculionidae. Stett.
        Ent. Zeit. 53: 1-44.
Gilbert, E. E.
    1952. The homologies of the male genitalia of Rhynchophora and
        allied Coleoptera. Ann. Ent. Soc. Amer. 45(4): 633-637.
Günther, L., and F. Zumpt.
    1933. Curculionidae: Subfam. Tanymecinae. Coleopterum
        catalogus. Pars 131. Junk, Berlin. 131 pp.
Hayward, K. J.
    1960. Insectos Tucumanos Perjudiciales. Rev. Industrial y
        Agricola de Tucuman 42: 144. 1958 (1960).
Heller, K. M.
    1921. Nuevos Curculionidos de la Argentina. An. Soc.
        Cientifica Argentina 1921: 19-35.
Horn, G. H.
    1876. In Leconte, John L., and George H. Horn, The Rhynchophora
        of America North of Mexico. Proc. Amer. Phil. Soc. 15
        (96): 455 pp.
Howden, A. T.
    1959. A revision of the species of Pandeleteius Schönherr and
        Pandeleteinus Champion of America North of Mexico
        (Coleoptera: Curculionidae). Proc. Calif. Acad. Sci. 29
        (10): 361-421.
    1966. Airosimus, a new genus of neotropical Tanymecini
        (Coleoptera: Curculionidae). Trans. Amer. Ent. Soc. 92:
        173-229.
    1969. The genus Pandeleteinus Champion with the description of
        a new species from Mexico (Curculionidae, Tanymecini).
        Coleopt. Bull. 23 (3): 76-83.
    1976. Pandeleteius of Venezuela and Colombia (Curculionidae:
        Brachyderinae: Tanymecini). Mem. Amer. Ent. Inst. No. 24.
        310 pp.

Howden, H. F., and J. M. Campbell.
    1974. Observations on some Scarabaeoidea in the Colombian Sierra Nevada de Santa Marta. Coleopt. Bull. 28 (3): 109-114.

Hustache, A.
    1928 (1926). Curculionides de la République Argentine (Première note). Anales del Museo Nacional de Historia Natural "Bernardino Rivadavia" 34: 155-261.
    1938. Curculionides de l'Argentine et des regions limitrophes. Rev. Soc. Ent. Argentina 10 (1): 3-17.

International Commission on Zoological Nomenclature.
    1964. International code of zoological nomenclature adopted by the 16th International Congress of Zoology. International Trust for Zoological Nomenclature, London: 176 pp.

Kirsch, T.
    1867. Beiträge zur Käferfauna von Bogota. (Drittes Stück: Brenthiden und adelognathe Curculionen.). Berl. Ent. Zeitschr. 11: 216-243.
    1873. Beiträge zur Kenntniss der Peruanischen Käferfauna auf Dr. Abendroth's Sammlungen basirt. Berl. Ent. Zeitschr. 17 (1): 121-152.

Kuschel, G.
    1955. Nuevas sinonimias y anotaciones sobre Curculionoidea. Rev. Chilena Ent. 1955 (4): 261-312.

Lucas, P. H.
    1857. (=1859?). Entomologie. In Animaux nouveaux ou rares recueillis pendant l'expédition dans les parties centrales de l'Amérique du Sud, de Rio de Janeiro a Lima, et de Lima au Para; exécutée par ordre du gouvernement Français pendant les années 1843 a 1847, sous la direction du Comte Francis de Castelnau. Paris. 204 pp.

Marshall, G. A. K.
    1952. Taxonomic notes on Curculionidae (Col.). Ann Mag. Nat. Hist. (12) 5: 261-270.

Ohaus, F.
    1908. Die Ruteliden meiner Sammelreisen in Südamerika (Col.). Deutsch. Ent. Zeitschr. 1908: 239-262, 383-408.

Olivier, A. G.
    1807. Entomologia, ou histoire naturelle des insectes, avec leurs caractères génériques et spécifiques, leur description, leur synonymie, et leur figure enluminée. Coléoptères. Vol. 5. Desray, Paris. 612 pp.

Papavero, N.
    1971. Essays on the history of Neotropical Dipterology. Vol. 1. Museu de Zoologia, Universidade de São Paulo, São Paulo, 216 pp.

Pascoe, F. P.
    1881. New Neotropical Curculionidae.-Part IV. Ann. Mag. Nat. Hist. (5) 7: 38-45.

Pierce, W. D.
    1913. Miscellaneous contributions to the knowledge of the weevils of the families Attelabidae and Brachyrhinidae. Proc. U.S.N.M. 45 (1988): 365-426.

Schoenherr, C. J.
    1823. Curculionides. Tabula synoptica familiae curculionidum.
        Isis Oken, 1823, col.: 1133-1146.
    1826. Curculionidum dispositio methodica cum generum
        characteribus, descriptionibus atque observationibus
        variis, seu prodromus ad synonymiae insectorum. Partem 4.
        Lipsiae. 338 pp.
    1833. Synonymia insectorum, oder Versuch einer Synonymie aller
        von mir bisher bekannten Insecten. Mit Berichtigungen und
        Anmerkungen, wie auch mit Beschreibungen neuer Arten. Band
        1, Teil 4. Roret, Paris. 381 pp.
    1834. Genera et species curculionidum, cum synonymia hujus
        familiae. Vol. 2, pt. 1. Roret, Paris. 326 pp.
Selander, R. B., and P. Vaurie.
    1962. A gazetteer to accompany the "Insecta" volumes of the
        "Biologia Centrali-Americana." Amer. Mus. Nov., No. 2099.
        70 pp.
Sharp, D. E.
    1918. Studies in Rhynchophora. A preliminary note on the male
        genitalia. Trans. Entomol. Soc. London. 66:209-222.
    1920. Studies in Rhynchophora. IX. The sexes of Conotrachelus
        brevisetis Champ. J.N.Y. Entomol. Soc. 28(1): 74-78.
Simon, E.
    1889. Voyage de M. E. Simon au Venezuela (Decembre 1887 - Avril
        1888) 4e Memoire, Arachnides. Ann. Soc. ent. Fr., Aout
        1889: 169-220.
Spieth, H. T.
    1950. The David Rockefeller Mexican expedition of the American
        Museum of Natural History. Introductory Account. Amer.
        Mus. Nov. No. 1454: 1-67.
Thompson, R. T.
    1977. A revision of the New Guinea weevil genus Apirocalus
        Pascoe (Coleoptera: Curculionidae). Bull. Brit. Mus. (Nat.
        Hist.) Ent. ser. 36(5): 193-280.
Van Dyke, E. C.
    1943. Additional new species of west American Coleoptera.
        Pan-Pacific Ent. 19 (13): 101-108.
Van Emden, F. I.
    1944. A key to the genera of Brachyderinae of the world. Ann.
        Mag. Nat. Hist. (11) 11: 503-532, 559-586.
Voss, E.
    1953. Neue und bemerkenswerte Curculioniden aus Columbien und
        Bolivien. (118. Beiträg zur Kenntnis der Curculioniden).
        Entomologische Mitteilungen 1953 (2): 55-84).
    1954. Curculionidae (Col.). In Beiträge zur Fauna Perus. Bd.
        IV: 193-376.

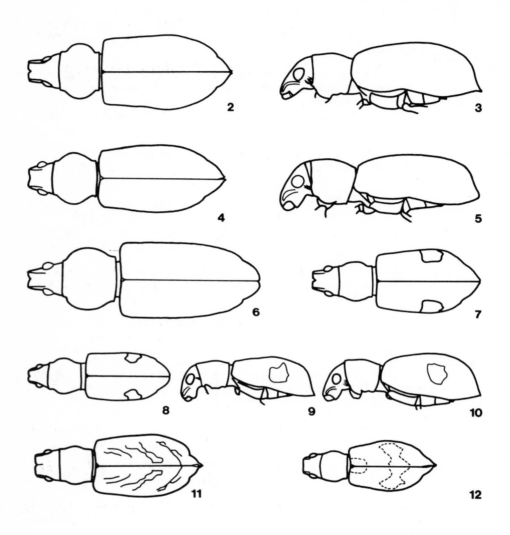

Figs. 2-12. H. (Hadromeropsis) spp. habitus. 2-3, superbus, female: 2, dorsal; 3, lateral. 4-5, meridianus, male: 4, dorsal; 5, lateral. 6, gemmifer, male, dorsal. 7-10, argentinensis: 7, female, dorsal; 8, male, dorsal; 9, male, lateral; 10, female, lateral. 11, speculifer, female, dorsal. 12, pulverulentus, female, dorsal.

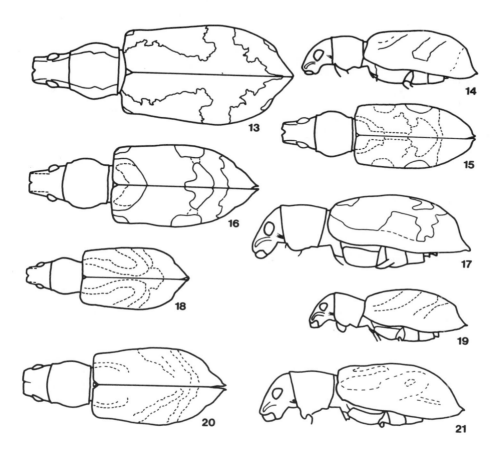

Figs. 13-21. H. (Hadromeropsis) spp. habitus. 13, togatus, female, dorsal. 14, nobilitatus, female, lateral. 15-17, plebeius: 15, male, dorsal; 16, female, dorsal; 17, female, lateral. 18-19, atomarius, female: 18, dorsal; 19, lateral. 20-21, pallidus, female: 20, dorsal; 21, lateral.

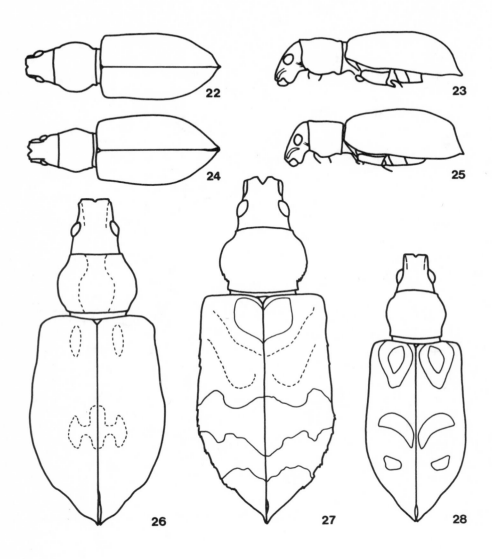

Figs. 22-25, 27. H. (Hadromeropsis) spp. habitus. 22-25, beverlyae: 22, male, dorsal; 23, male, lateral; 24, female, dorsal; 25, female, lateral. 27, fasciatus, female, dorsal.
    Figs. 26, 28. H. (Hadrorestes) spp. habitus. 26, alacer, female, dorsal. 28, dialeucus, female, dorsal.

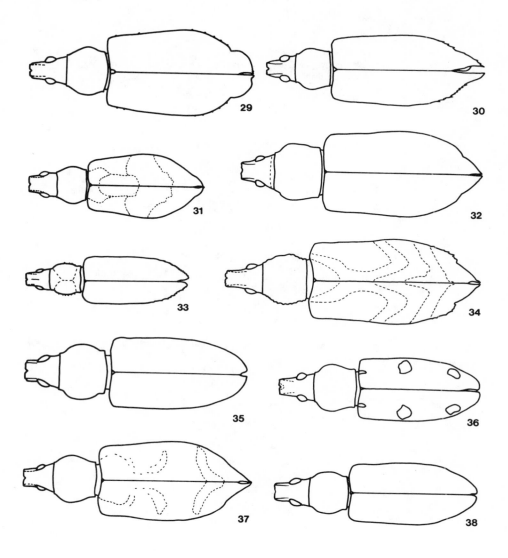

Figs. 29-38. H. (Hadrorestes) spp. habitus. 29, institulus, female. 30, apicalis, female. 31, picchuensis, female. 32, scambus, female. 33, transandinus, male. 34, conquisitus, female. 35, pectinatus, male. 36, magicus, male. 37, contractus, female. 38, spiculatus, male.

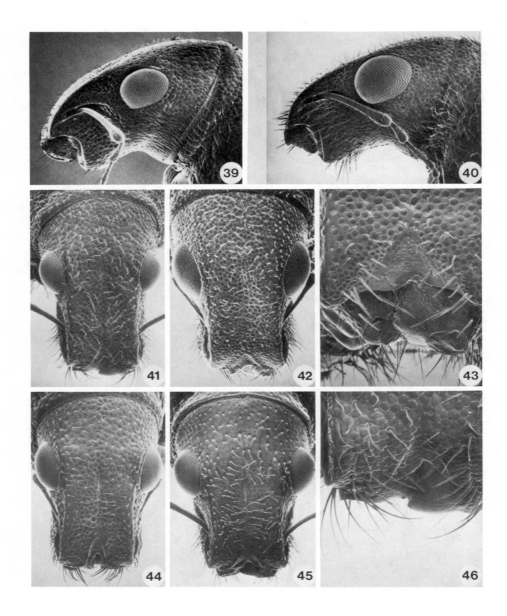

Figs. 39-46. H. (Hadromeropsis) spp. 39, opalinus, female, head, lateral view. 40, crinitus, male, head, lateral view. 41, fulgens, male, head, anterior view. 42, crinitus, male, head, anterior view. 43, crinitus, male, epistoma. 44, dejeanii, male, head, anterior view. 45, flagellatus, male, head, anterior view. 46, fulgens, male, epistoma.

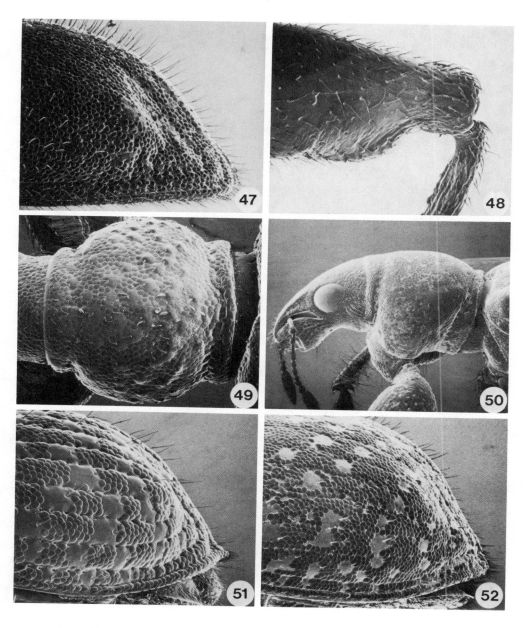

Figs. 47-52. H. (Hadromeropsis) spp. 47, crinitus, male, apex of elytra. 48, flagellatus, male, femoral flange. 49, aureus, male, prothorax, dorsal view. 50, cretatus, male, head and prothorax, lateral view. 51, aureus, male, apex of elytra, lateral view. 52, cretatus, male, apex of elytra, lateral view.

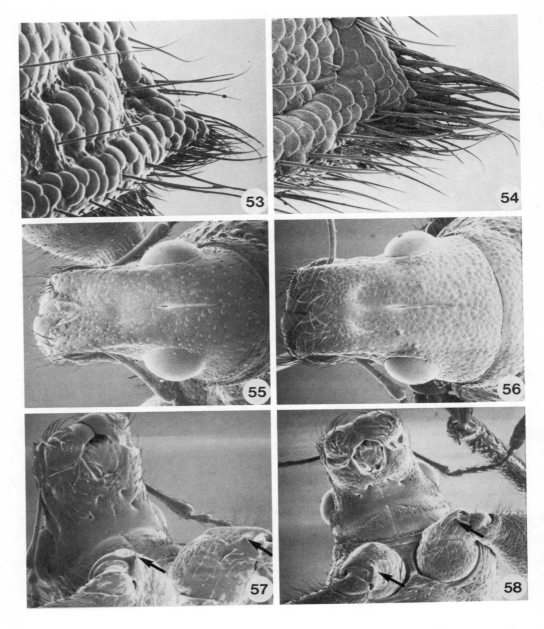

Figs. 53-58. H. (Hadromeropsis) spp. 53, aureus, male, apex of elytra, lateral view. 54, cretatus, female, apex of elytra, lateral view. 55-58, gemmifer: 55, male, head, anterior view; 56, female, head, anterior view; 57, male, coxal tubercle; 58, female, coxal tubercle.

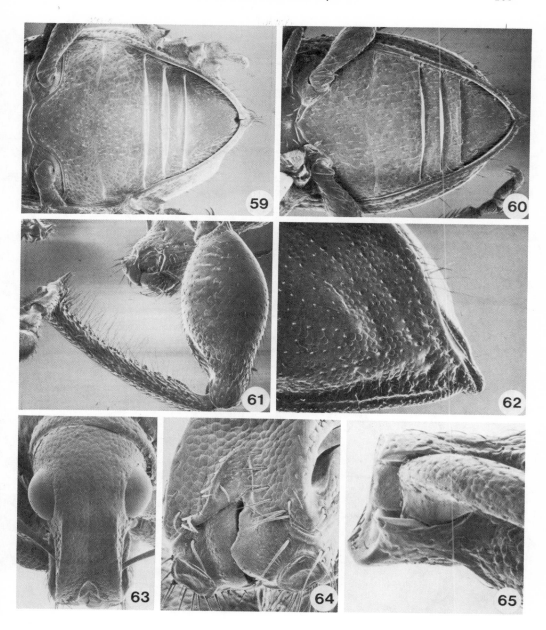

Figs. 59-65. H. (Hadromeropsis) spp. 59, gemmifer, female, abdomen; 60, meridianus, female, abdomen. 61, gemmifer, male, fore leg, ventral view. 62, rufipes, female, apex of elytra, lateral view. 63-65, batesi, male: 63, head, anterior view: 64, apex of rostrum; 65, distal end fore femur.

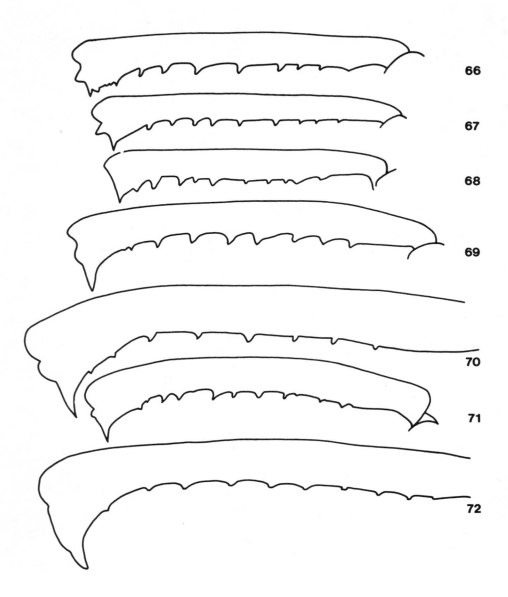

Figs. 66-72. H. (Hadromeropsis) spp. fore tibia. 66, fulgens, male, Cuernavaca. 67, opalinus, male, Durango. 68, flagellatus, male, Apizaco. 69, scintillans, male, lectotype, Quiche Mts. 70, micans, female, lectotype. 71, dejeanii, male, Playa Vicente. 72, rufipes, female, type.

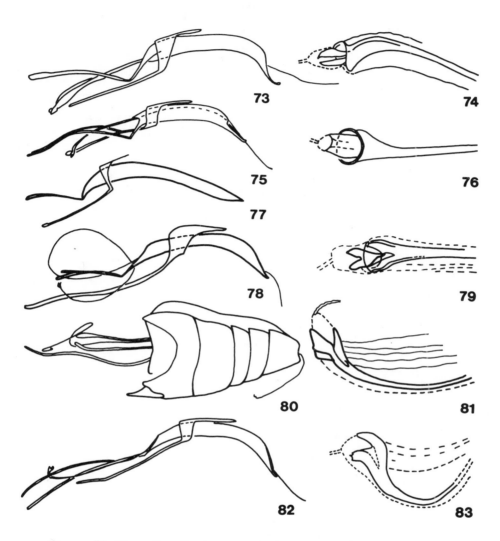

Figs. 73-83. H. (Hadromeropsis) spp. male genitalia. 73, fulgens, aedeagus with flagellum atypical paralectotype from Amula. 74-76, opalinus: 74**, proximal end of flagellum, lateral view; 75, aedeagus with flagellum; 76**, proximal end of flagellum, ventral view. 77, crinitus, aedeagus. 78-81, flagellatus: 78, aedeagus with flagellum; 79**, proximal end of flagellum, ventral view; 80, genitalia in normal position in abdomen, thorax removed; 81**, proximal end of flagellum, lateral view. 82-83, dejeanii: 82, aedeagus with flagellum; 83**, proximal end of flagellum, lateral view.

**shown 6x larger than unmarked figures

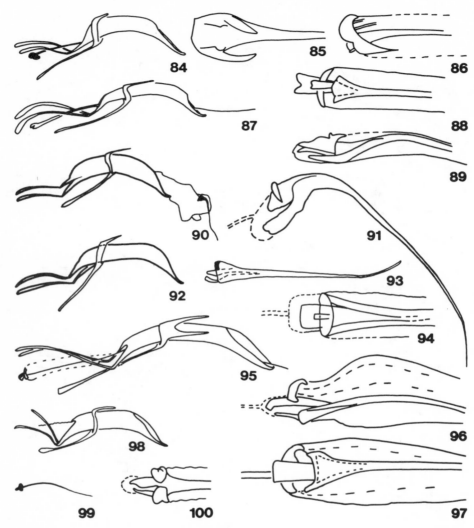

Figs. 84-100. H. (Hadromeropsis) spp. male genitalia. 84-86, amoenus: 84, aedeagus with flagellum; 85**, proximal end of flagellum, ventral view; 86**, as 85, lateral view. 87-89, scintillans: 87, aedeagus with flagellum; 88**, proximal end of flagellum, ventral view; 89**, as 88, lateral view. 90-91, brevicomus: 90, aedeagus with internal sac extruded; 91**, flagellum. 92-94, aureus: 92, aedeagus; 93*, flagellum; 94**, proximal end of flagellum ventral view. 95-97, cretatus: 95, aedeagus with flagellum; 96**, proximal end of flagellum, lateral view; 97**, as 96, ventral view. 98-100, superbus: 98, aedeagus; 99, flagellum; 100**, proximal end of flagellum, lateral view.
    *shown 3x longer than unmarked figures
    **shown 6x larger than unmarked figures

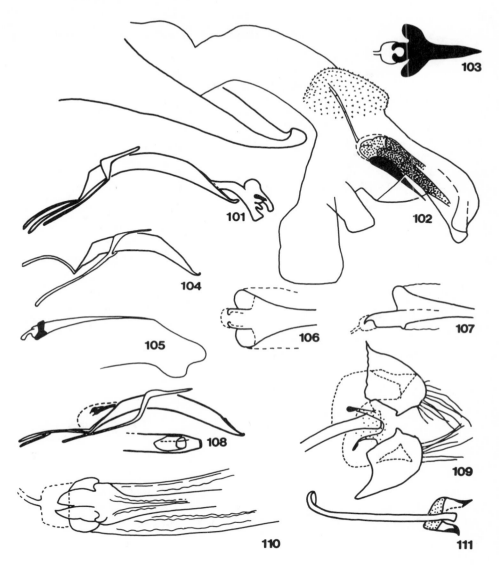

Figs. 101-111. H. (Hadromeropsis) spp. male genitalia. 101-103, gemmifer: 101, aedeagus; 102**, internal sac; 103**, "flagellum", ventral view. 104-107, meridianus: 104, aedeagus; 105*, flagellum; 106**, proximal end of flagellum, ventral view; 107**, proximal end of flagellum, lateral view. 108-111, batesi: 108, aedeagus and dorsal view of apex; 109*, sternite 8, ventral view; 110**, proximal end of flagellum, ventral view; 111, spiculum gastrale.
     *shown 3x larger than unmarked figures.
     **shown 6x larger than unmarked figures.

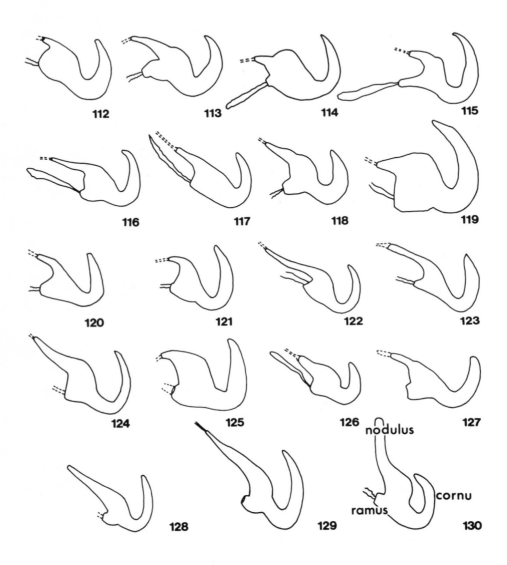

Figs. 112-130. H. (Hadromeropsis) spp. spermatheca. 112, opalinus. 113-115, flagellatus. 116, fulgens. 117, crinitus. 118, dejeanii. 119, micans. 120, scintillans. 121, amoenus. 122-123, brevicomus. 124, aureus. 125, cretatus. 126, superbus. 127, gemmifer. 128-129, meridianus. 130, batesi.

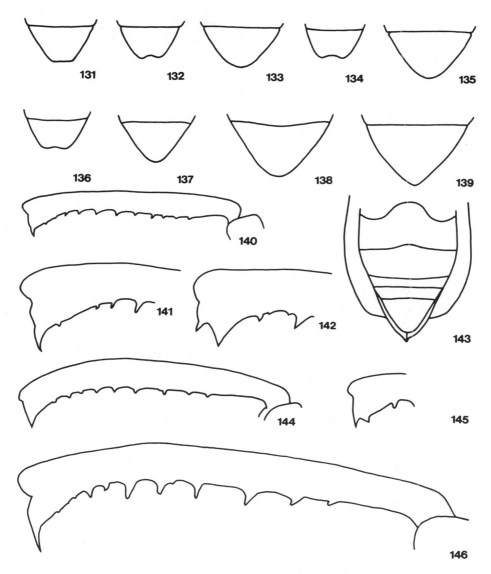

Figs. 131-146. H. (Hadromeropsis) spp. 131-139, ventrite 5: 131, crinitus, male; 132-133, flagellatus: 132, male; 133, female; 134, amoenus, male; 135, amoenus, female; 136, scintillans, male; 137, scintillans, female; 138, rufipes, female; 139, micans, female. 140, speculifer, male, fore tibia. 141-142, excubitor, apex fore tibia: 141, male; 142, female. 143, speculifer, female, abdomen. 144-145, atomarius, fore tibia: 144, male; 145, apex, female. 146, fasciatus, male, fore tibia.

Figs. 147-152. H. (Hadromeropsis) spp. 147, plebeius, female, apex of elytra, lateral view. 148-150, pulverulentus, female: 148, apex of elytra, lateral view; 149, head, anterior view; 150, base of elytra, dorsal view. 151, pallidus, male, rostrum, oblique view. 152, beverlyae, male, fore leg, posterior view.

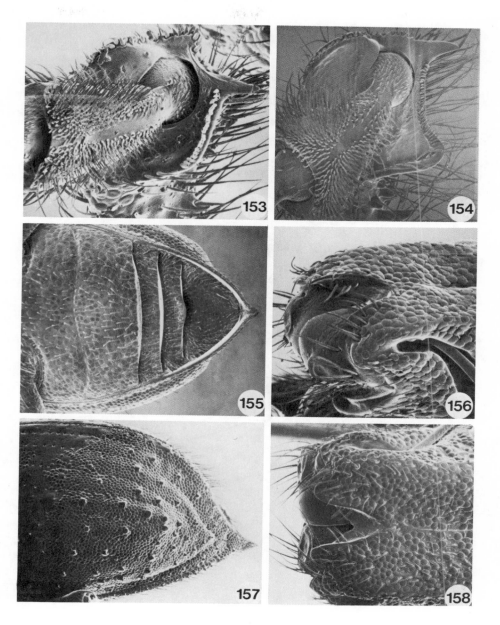

Figs. 153-158. H. (Hadromeropsis) spp. 153, opalinus, male, apex fore tibia. 154, beverlyae, male, apex fore tibia. 155, atomarius, female, abdomen. 156, beverlyae, male, rostrum. 157-158, fasciatus, female: 157, apex of elytra, lateral view; 158, apex of rostrum, dorsal view.

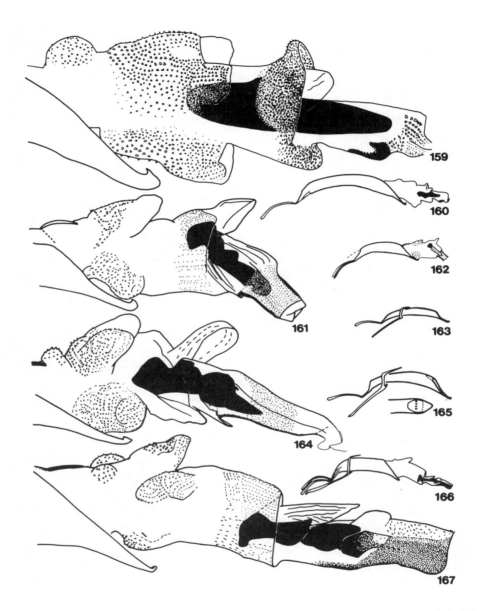

Figs. 159-167. H. (Hadromeropsis) spp. male genitalia. 159-160, togatus: 159**, internal sac; 160, aedeagus. 161-162, argentinensis: 161**, internal sac; 162, aedeagus. 163, pulverulentus, aedeagus. 164-165, plebeius: 164**, internal sac; 165, aedeagus. 166-167, pallidus: 166, aedeagus; 167**, internal sac.
**shown 6x larger than unmarked figures

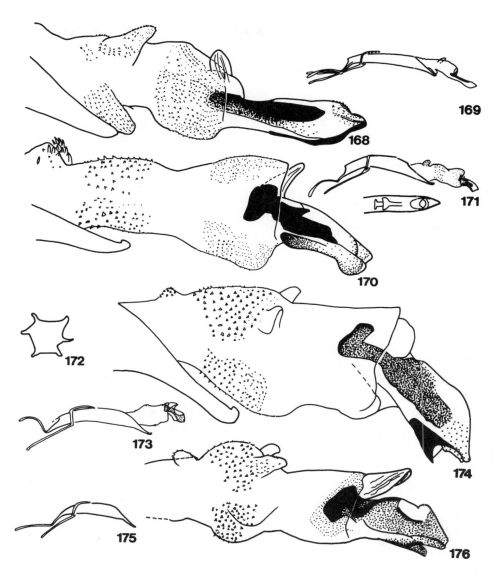

Figs. 168-176. H. (Hadromeropsis) spp. male genitalia and rectal ring. 168-169, speculifer: 168**, internal sac; 169, aedeagus. 170-171, beverlyae: 170**, internal sac; 171, aedeagus in lateral view and apex in dorsal view. 172*, flagellatus, rectal ring. 173-174, nobilitatus: 173, aedeagus; 174**, internal sac. 175-176, atomarius: 175, aedeagus; 176**, internal sac.
   *shown 3x larger than unmarked figures
   **shown 6x larger than unmarked figures

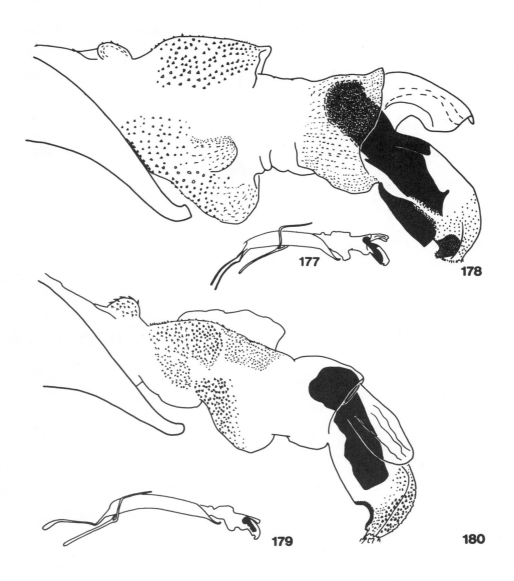

Figs. 177-180. H. (Hadromeropsis) spp. male genitalia. 177-178, excubitor: 177, aedeagus; 178**, internal sac. 179-180, fasciatus: 179, aedeagus; 180**, internal sac.
**shown 6x larger than unmarked figures

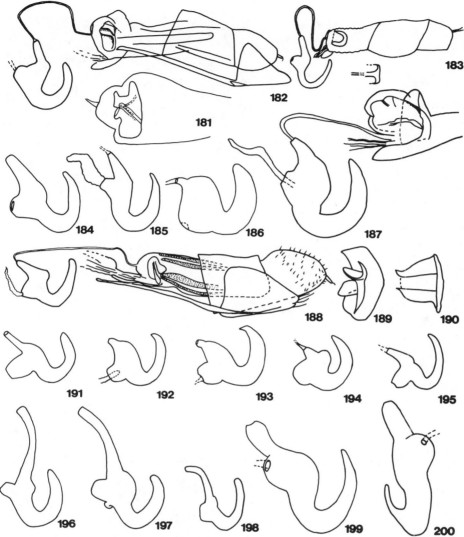

Figs. 181-200. <u>H</u>. (Hadromeropsis) spp. female genitalia. 181-182, <u>argentinensis</u>: 181*, bursal sclerite, ventral view; 182, genitalia, setae omitted. 183, <u>pulverulentus</u>, genitalia, setae omitted, including ventral view of bursal sclerite. 184, <u>togatus</u>, spermatheca. 185, <u>speculifer</u>, spermatheca. 186, <u>plebeius</u>, spermatheca. 187, <u>pallidus</u>, spermatheca. 188-190, <u>beverlyae</u>: 188, genitalia; 189*, bursal sclerite, oblique view; 190*, bursal sclerite, dorsal view. 191, <u>nobilitatus</u>, spermatheca. 192-194, "smooth" <u>nobilitatus</u>, spermatheca, three examples of variation. 195, <u>atomarius</u>, spermatheca. 196-198, <u>excubitor</u>, spermatheca, three examples of variation. 199-200, <u>fasciatus</u>, spermatheca, lateral and edge views.

*shown 3x larger than unmarked figures

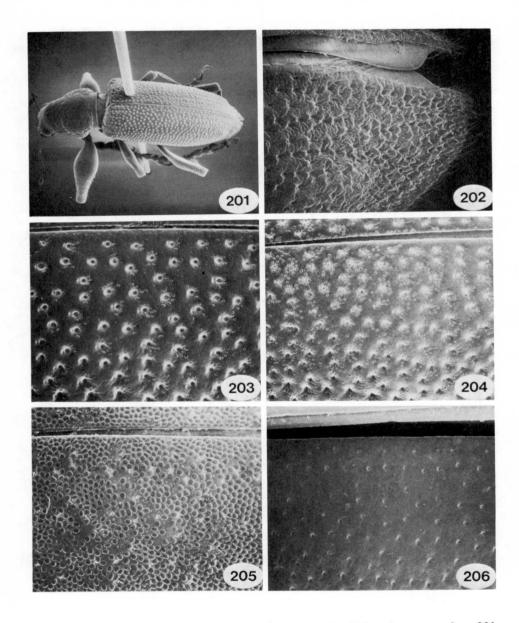

Figs. 201-206. H. (Hadrorestes) spp. 201-203, alacer, male: 201, dorsal habitus, paratype from Baeza; 202, apex of elytra, type, dorsal view; 203, elytral sculpture at suture medially, paratype from Baeza. 204, inconscriptus, female, elytral sculpture at suture medially. 205, nebulicolus, female, elytral sculpture at suture medially. 206, silaceus, male, elytral sculpture at suture medially.

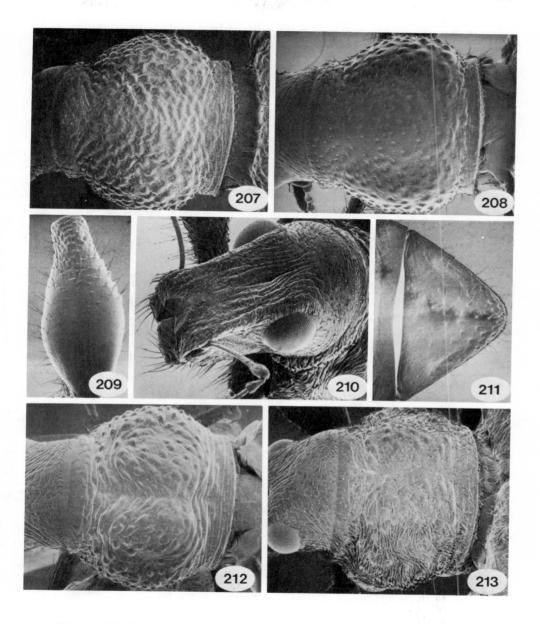

Figs. 207-213. H. (Hadrorestes) spp. 207-209, alacer: 207, prothorax, dorsal view, type; 208, prothorax, dorsal view, male from Baeza; 209, fore femur, outer edge, male from Baeza. 210-213, impressicollis: 210, female, head, antero-lateral view; 211, ventrite 5, female; 212, prothorax, dorsal view, male; 213, prothorax, dorsal view, female.

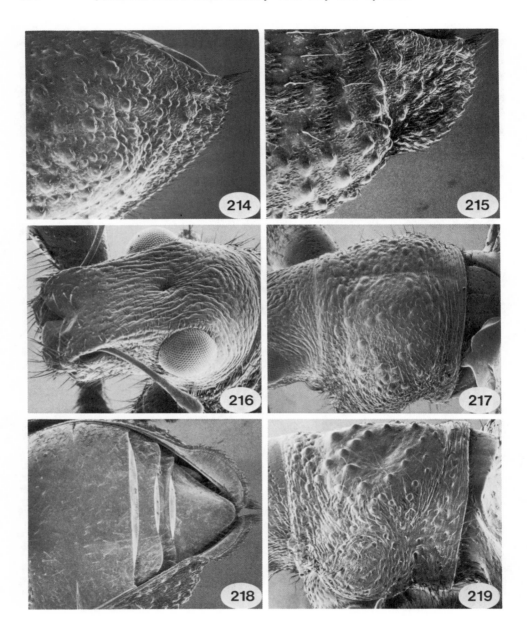

Figs. 214-219. H. (Hadrorestes) spp. 214, impressicollis, female, apex of elytron, dorsal view. 215-218, institulus, female: 215, apex of elytron, dorsal view; 216, head, antero-lateral view; 217, prothorax, dorsal view; 218, abdomen and edge of elytra. 219, bombycinus, prothorax, female, lateral view.

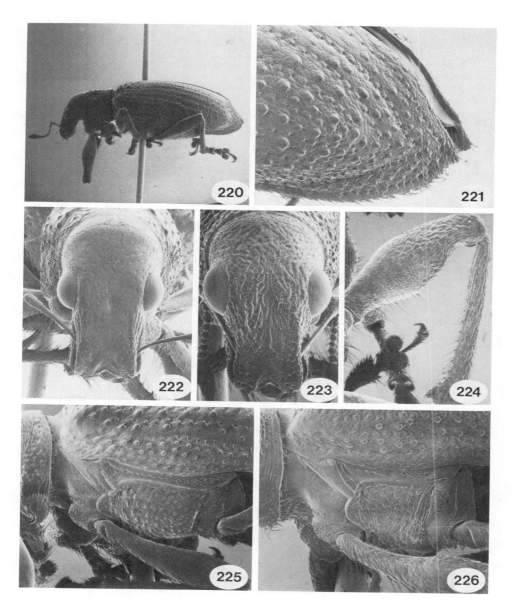

Figs. 220-226. H. (Hadrorestes) spp. 220-222, bombycinus, female: 220, habitus, lateral view; 221, apex of elytra, oblique dorsal; 222, head, anterior view. 223-224, brachypterus, female: 223, head, anterior view; 224, fore leg, anterior view. 225, bombycinus, female, metasternum, lateral view. 226, brachypterus, female, metasternum, lateral view.

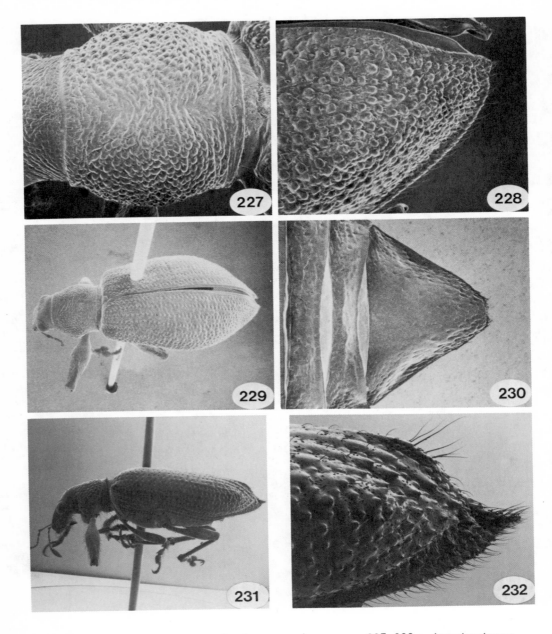

Figs. 227-232. H. (Hadrorestes) spp. 227-230, brachypterus, female: 227, prothorax, dorsal view; 228, apex of elytra, oblique dorsal; 229, dorsal habitus; 230, ventrites 3, 4, and 5. 231-232, transandinus, female: 231, habitus, lateral view; 232, apex of elytra, lateral view.

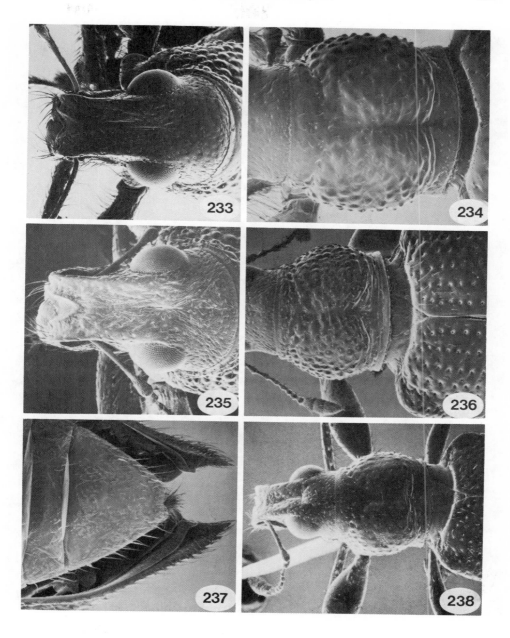

Figs. 233-238. H. (Hadrorestes) spp. 233-234, transandinus: 233, head, female, anterior view; 234, prothorax, male, dorsal view. 235-237, apicalis, female: 235, head, anterior view; 236, prothorax and base of elytra, dorsal view; 237, apex of abdomen and elytra, ventral view. 238, nanus, male, head and prothorax, dorsal view.

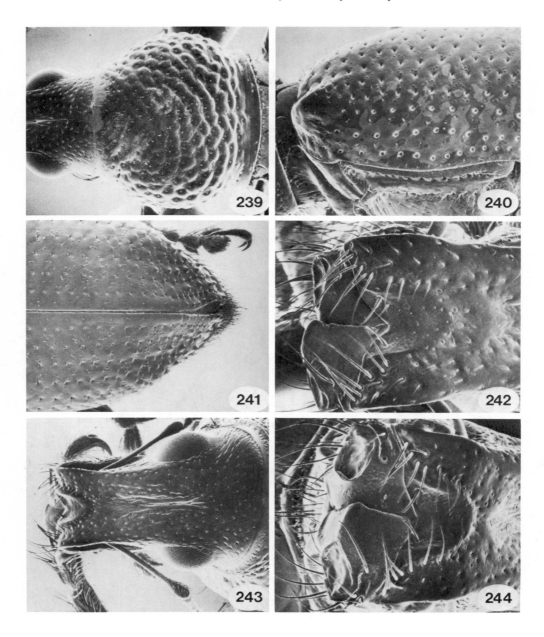

Figs. 239-244. H. (Hadrorestes) spp. 239-242, nitidus, male; 239, prothorax, dorsal view; 240, metasternum, base of elytra, lateral view; 241, apical portion of elytra, dorsal view; 242, apex of rostrum, dorsal view. 243-244, mandibularis, female: 243, head, anterior view; 244, apex of rostrum.

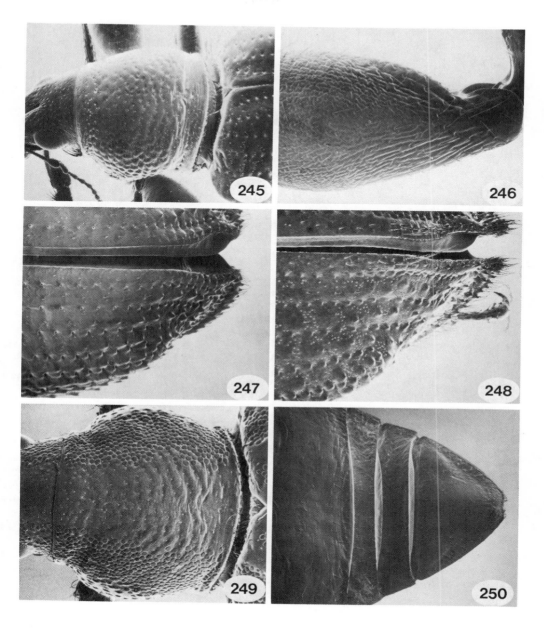

Figs. 245-250. H. (Hadrorestes) spp. 245-248, mandibularis:
245, prothorax, female, dorsal view; 246, fore femur, male, anterior
view; 247, apex of elytra, male, dorsal view; 248, apex of elytra,
female, dorsal view. 249-250, magicus, female: 249, prothorax, dorsal
view; 250, abdomen.

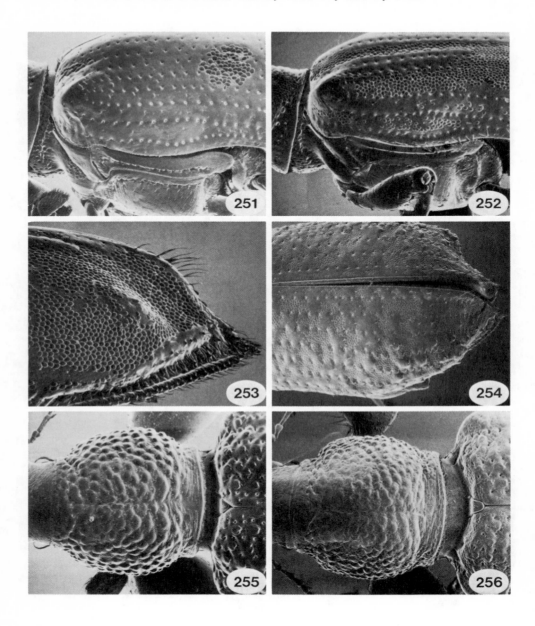

Figs. 251-256. H. (Hadrorestes) spp. 251-253, magicus: 251, base of elytra, male, lateral view; 252, base of elytra, female, lateral view; 253, apex of elytra, female, lateral view. 254-256, conquisitus: 254, apical portion of elytra, female, dorsal view; 255, prothorax, holotype, male, glabrous, dorsal view; 256, prothorax, male, squamose specimen, dorsal view.

Figs. 257-262. H. (Hadrorestes) spp. 257-259, conquisitus, male: 257, fore femur, posterior view; 258, fore leg, anterior view; 259, apex of elytron, dorsal view. 260-262, scambus: 260, apex of elytron, female, dorsal view; 261, apex of elytron, male, dorsal view; 262, fore leg, female.

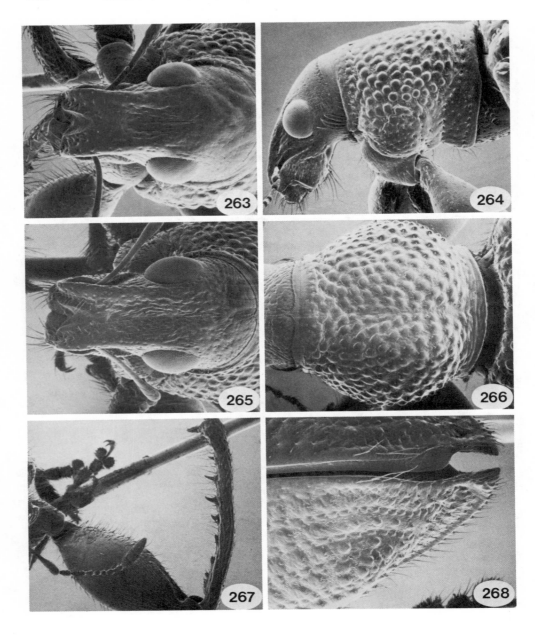

Figs. 263-268. H. (Hadrorestes) spp. 263-264, scambus, female: 263, head, anterior view; 264, head and prothorax, lateral view. 265-268, pectinatus: 265, head, type, male, anterior view; 266, prothorax, type, male, dorsal view; 267, fore leg, male from Colombia, anterior view; 268, apex of elytra, female, dorsal view.

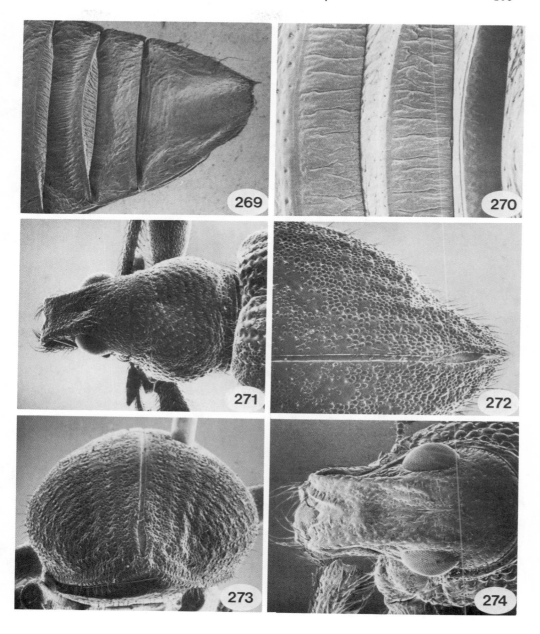

Figs. 269-274. H. (Hadrorestes) spp. 269-270, pectinatus, allotype: 269, abdomen; 270, detail of caudal surface ventrites 2, 3, and 4. 271-273, picchuensis, female: 271, head and prothorax, dorsal view; 272, apex of elytra, dorsal view; 273, elytra, caudal view. 274, contractus, head, female, anterior view.

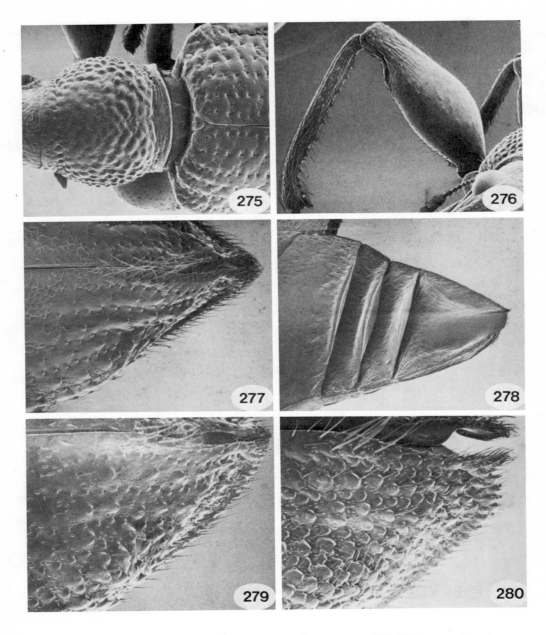

Figs. 275-280. H. (Hadrorestes) spp. 275-278, contractus, female: 275, prothorax, dorsal view; 276, fore leg, anterior view; 277, apex of elytra, dorsal view; 278, abdomen. 279, dialeucus, female, apex of elytra, dorsal view. 280, spiculatus, female, apex of elytra, dorsal view.

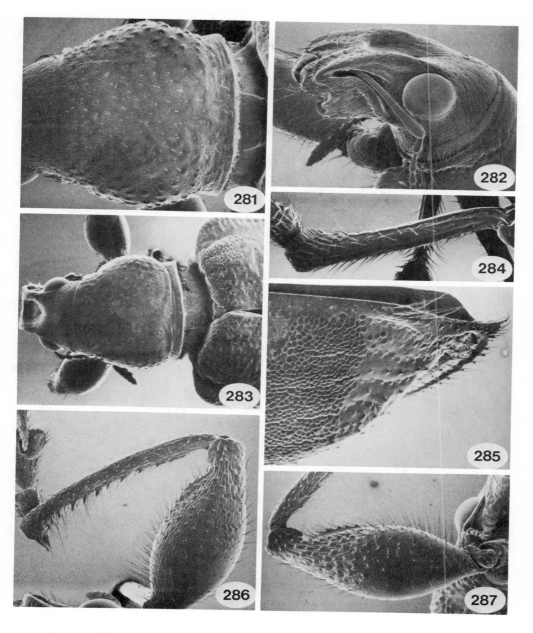

Figs. 281-287. H. (Hadrorestes) spp. 281, dialeucus, female, prothorax, dorsal view. 282-285, cavifrons, female: 282, head, oblique lateral view; 283, prothorax and head, dorsal view; 284, hind tibia, outer view; 285, apex of elytron, dorsal view. 286-287, spiculatus, male, fore leg: 286, anterior view; 287, posterior view.

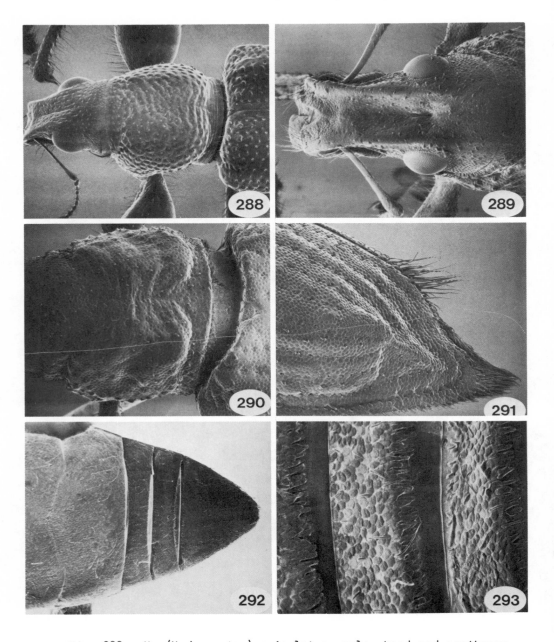

Fig. 288. <u>H</u>. (<u>Hadrorestes</u>) <u>spiculatus</u>, male, head and prothorax, dorsal view.
Figs. 289-293. <u>Hadromeropsis earinus</u>, female: 289, head, dorsal view; 290, prothorax, dorsal view; 291, apex of elytra, lateral view; 292, abdomen; 293, detail of caudal edge of ventrites 2, 3, and 4.

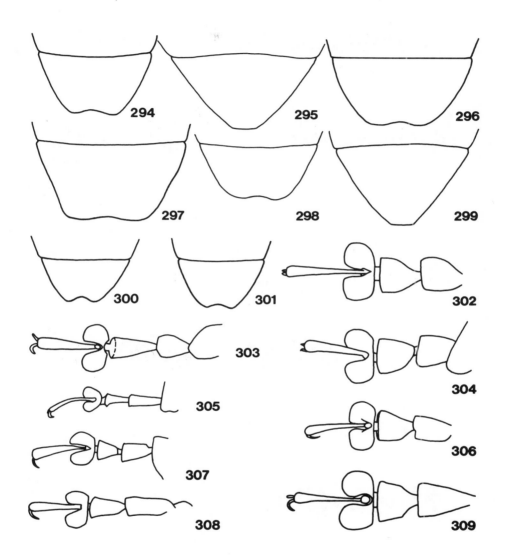

Figs. 294-309. H. (Hadrorestes) spp. 294-301, ventrite 5: 294, alacer, male; 295, alacer, female; 296, nebulicolus, male; 297, silaceus, male; 298, inconscriptus, male; 299, inconscriptus, female; 300, nitidus, male; 301, magicus, male. 302-309, fore tarsus: 302, alacer, male; 303, impressicollis, male; 304, inconscriptus, male; 305, transandinus, female; 306, nebulicolus, male; 307, dialeucus, male; 308, pectinatus, male; 309, silaceus, male.

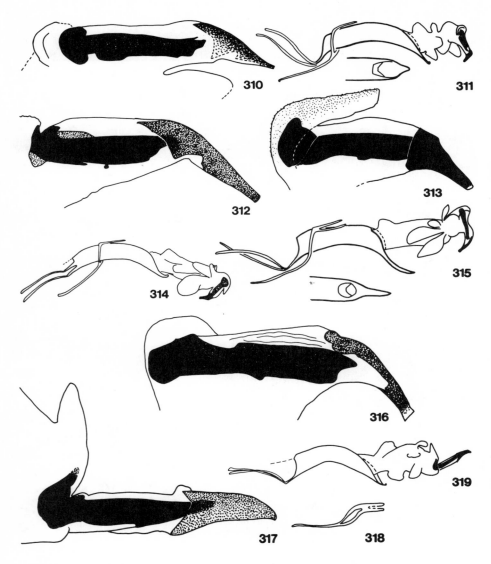

Figs. 310-319. H. (Hadrorestes) spp. male genitalia. 310-311, alacer: 310**, apex of internal sac; 311, aedeagus. 312**, inconscriptus, apex of internal sac. 313-314, nebulicolus: 313**, apex of internal sac; 314, aedeagus. 315-316, silaceus: 315, aedeagus and apex of aedeagus in dorsal view; 316**, apex of internal sac. 317-319, impressicollis: 317**, apex of internal sac; 318, tegmen; 319, aedeagus.
      **shown 6x larger than unmarked figures

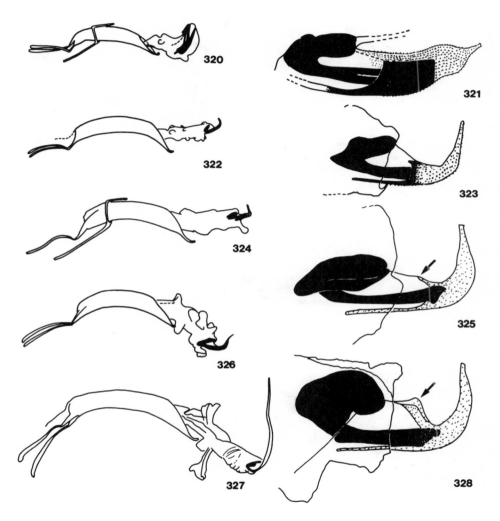

Figs. 320-328. H. (Hadrorestes) spp. male genitalia. 320-321, nanus: 320, aedeagus (sac not completely extruded); 321**, apex of internal sac. 322-323, transandinus: 322, aedeagus; 323**, apex of internal sac. 324-325, dialeucus: 324, aedeagus; 325**, apex of internal sac. 326, scambus, aedeagus. 327, spiculatus, aedeagus. 328**, pectinatus, apex of internal sac.
      **shown 6x larger than unmarked figures

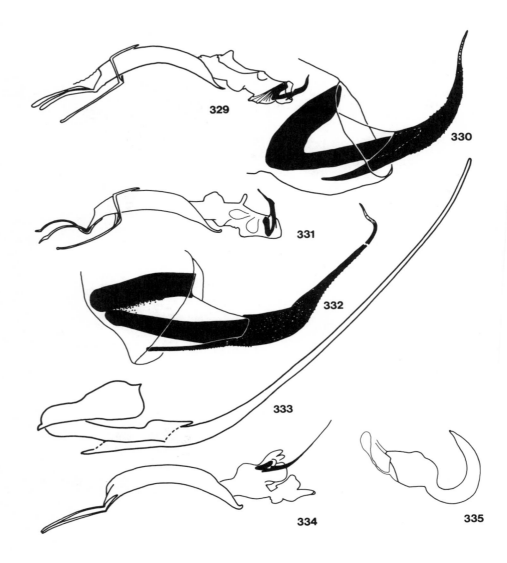

Figs. 329-334. H. (Hadrorestes) spp. male genitalia. 329-330, conquisitus: 329, aedeagus; 330**, apex of internal sac. 331-332, magicus: 331, aedeagus; 332**, apex of internal sac. 333-334, nitidus: 333**, apex of internal sac; 334, aedeagus.
    Fig. 335*. Hadromeropsis earinus, spermatheca.
   *shown 3x larger than unmarked figures
  **shown 6x larger than unmarked figures

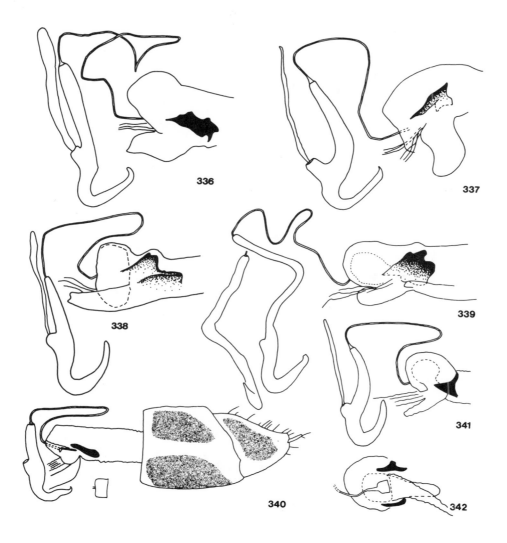

Figs. 336-340. H. (Hadrorestes) spp. female genitalia. 336,
alacer. 337, silaceus. 338, nebulicolus. 339, inconscriptus,
spermatheca in lateral and edge view. 340, institulus, plus ventral
view of bursal sclerite. 341, bombycinus. 342, bombycinus, ventral
view of bursa copulatrix with common oviduct pulled aside.

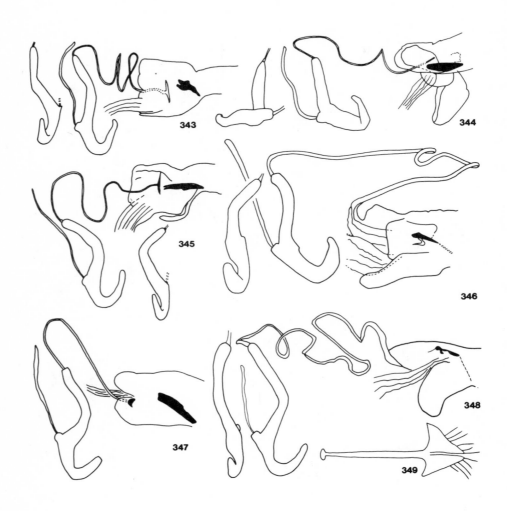

Figs. 343-348. H. (Hadrorestes) spp. female genitalia. 343, transandinus, including edge view of spermatheca. 344, brachypterus, including edge view of spermatheca. 345, apicalis, including edge view of spermatheca. 346, mandibularis, including edge view of spermatheca. 347, magicus. 348, nitidus, including edge view of spermatheca.
Fig. 349. H. (Hadromeropsis) argentinensis, spiculum ventrale.

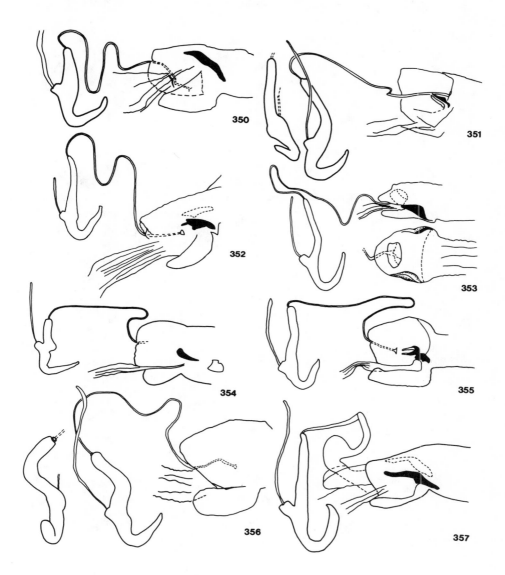

Figs. 350-357. H. (Hadrorestes) spp. female genitalia. 350, conquisitus. 351, scambus, including edge view of spermatheca. 352, pectinatus. 353, picchuensis, plus ventral view of bursa copulatrix with common oviduct pulled aside. 354, contractus, plus second example of lateral sclerite. 355, dialeucus. 356, cavifrons, including edge view of spermatheca. 357, spiculatus.

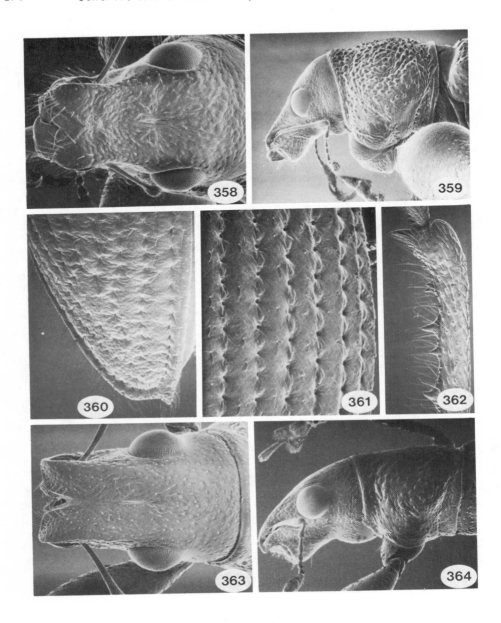

Figs. 358-362. *Hadromeropsis striatus* male. 358, head, anterior view; 359, head and prothorax, lateral view; 360, apex of elytra, dorsal view; 361, center of right elytron, dorsal view; 362, fore tibia, distal half, anterior view.

Figs. 363, 364. *H.* (*Hadrorestes*) *exilis* male. 363, head, anterior view; 364, head and prothorax, lateral view.

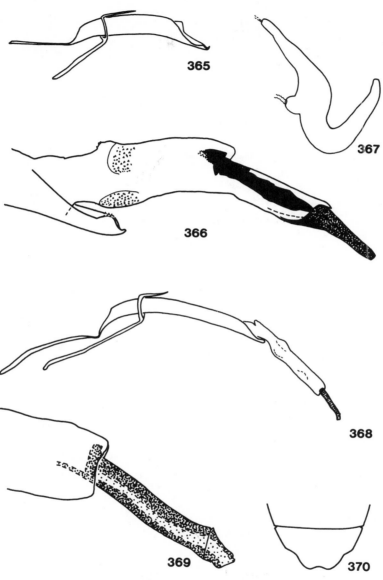

Figs. 365, 366. H. (Hadrorestes) exilis male genitalia. 365, aedeagus; 366*, internal sac.

Figs. 367*. H. (Hadromeropsis) rufipes spermatheca.

Figs. 368-370. Hadromeropsis striatus male. 368, aedeagus; 369**, internal sac; 370, ventrite 5.

*shown 3x larger than unmarked figures
**shown 6x larger than unmarked figures

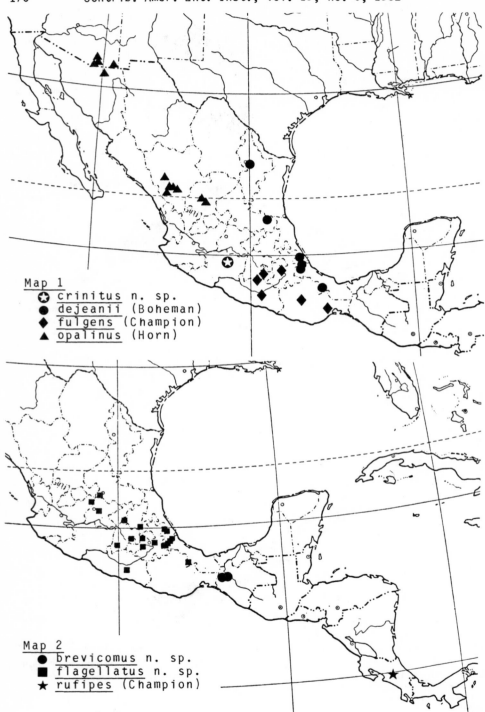

Map 1
⭐ crinitus n. sp.
● dejeanii (Boheman)
♦ fulgens (Champion)
▲ opalinus (Horn)

Map 2
● brevicomus n. sp.
■ flagellatus n. sp.
★ rufipes (Champion)

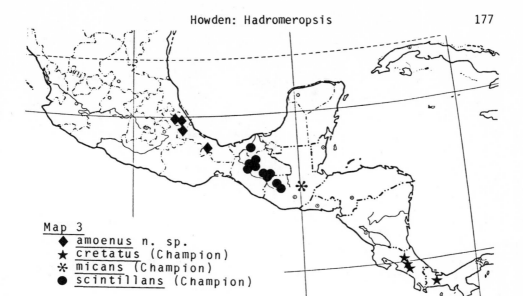

Map 3
- ◆ amoenus n. sp.
- ★ cretatus (Champion)
- ✳ micans (Champion)
- ● scintillans (Champion)

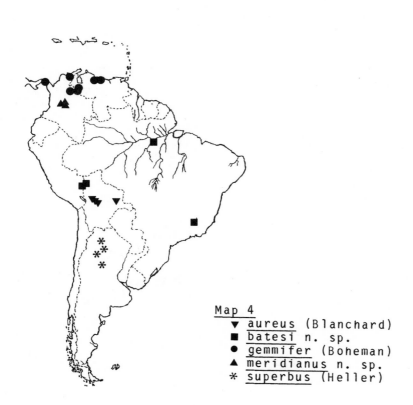

Map 4
- ▼ aureus (Blanchard)
- ■ batesi n. sp.
- ● gemmifer (Boheman)
- ▲ meridianus n. sp.
- ✳ superbus (Heller)

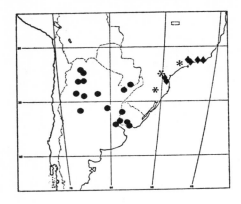

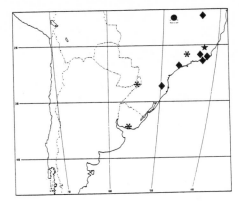

Map 5
- ● argentinensis (Hustache)
- * plebeius n. sp.
- ◆ togatus (Boheman)

Map 6
- ● beverlyae n. sp.
- ◆ fasciatus (Lucas)
- * pallidus n. sp.
- ★ pulverulentus n. sp.

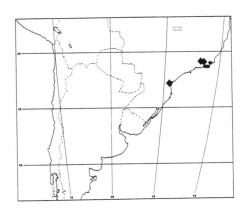

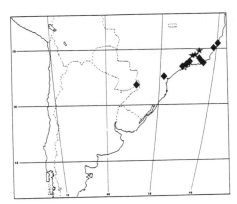

Map 7
- ● excubitor n. sp.
- ◆ speculifer n. sp.

Map 8
- ★ atomarius (Boheman)
- ◆ nobilitatus (Gyll.)

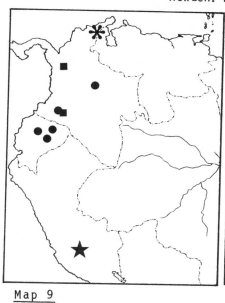

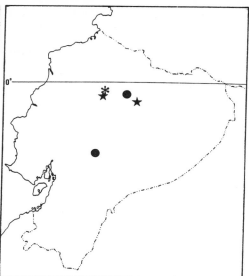

Map 9
- ● alacer n. sp.
- ★ inconscriptus n. sp.
- ✳ nebulicolus n. sp.
- ■ silaceus n. sp.

Map 10
- ● apicalis n. sp.
- ✳ nanus n. sp.
- ★ transandinus n. sp.

Map 11
- ● bombycinus n. sp.
- ■ impressicollis (Kirsch)
- ✳ institulus n. sp.

Map 12
- 3 conquisitus n. sp.
- 6 contractus n. sp.
- 1 magicus (Pascoe)
- 2 nitidus n. sp.
- 5 picchuensis n. sp.
- 4 spiculatus n. sp.

Map 13
3 cavifrons n. sp.
1 mandibularis n. sp.
4 pectinatus n. sp.
2 scambus n. sp.

THE ICHNEUMONID PARASITES ASSOCIATED WITH THE GYPSY MOTH *(LYMANTRIA DISPAR)*

By

VIRENDRA GUPTA

Center for Parasitic Hymenoptera
Institute of Food and Agricultural Sciences
University of Florida, Gainesville, Florida 32611

This research was financed by the U. S. Department of Agriculture, Systematic Entomology Laboratory, Washington. It was started at the American Entomological Institute, Ann Arbor, and completed at the Center for Parasitic Hymenoptera, Gainesville.

CONTENTS

THE ICHNEUMONID PARASITES ASSOCIATED WITH
THE GYPSY MOTH *(LYMANTRIA DISPAR)*

INTRODUCTION

The gypsy moth, *Lymantria dispar,* is an important defoliator of hardwood trees, especially oaks, over much of Eurasia and Eastern North America. Although the moth caterpillars prefer the oaks, they feed on the leaves of over 500 plant species (Forbush and Fernald, 1896, Mosher, 1915). Their enormous feeding activity in seasons of outbreaks defoliate many important ornamental trees of great aesthetic value, such as the oaks, birch, poplars, willows, maple, elms, etc. Repeated defoliations reduce the vigor of the trees and secondary infestations by insects and diseases often kill the trees. The conifers, which are also attacked, are more susceptible to defoliation, and one complete defoliation by the gypsy moth may be fatal to the trees.

Ever since the escape of the gypsy moth into the forests of the New England states before the turn of the century, intensive and extensive work has been done in the United States to contain the pest and eradicate it. However, the efforts have not been successful and the pest has continued to spread. During the past eighty years or so, the pest has spread to most of the northeastern states, extending westwards to Pennsylvania and southwards to Maryland. The pest has also spread to certain pockets in Ohio, North and South Carolina, Virginia, West Virginia, Michigan, Wisconsin, Washington and Oregon.

Since 1905, millions of parasites of various species have been imported into the United States from Europe and Japan to control the gypsy moth. Only ten of them got established, including only one ichneumonid, *Phobocampe unicincta.* They have apparently not been very effective in checking the spread of the moth. Two of them, *Anastatus disparis* and *Ooencyrtus kuvanae* are egg parasites; two, *Apanteles melanoscelus* and *Phobocampe unicincta (= disparis),* parasitize smaller larvae (first to third instars); four, *Blepharipa pratensis, Compsilura concinnata, Exorista larvarum,* and *Parasetigena silvestris* attack mature larvae; and two, *Monodontomerus aereus* and *Brachymeria intermedia* are pupal parasites. Hoy (1976) has given a list of all the species imported by then and possible reasons for their non-establishment.

In Eurasia, the original home of the gypsy moth, there appears to be some sort of a natural balance between the moth and its parasites. Yet periodical outbreaks do occur in limited areas. It appears that the aggregate effect of the introduced natural enemies and the indigenous parasites of the gypsy moth in North America is approaching that which exists in Central Europe. In many New England localities infested with the gypsy moth, the natural enemies have been important factors in preventing outbreaks, or at least responsible for prolonging the intervals between the outbreaks.

The importance of the native ichneumonid parasites has also been underestimated. The success of a parasite has often been measured in terms of the number of offspring a parasite is able to produce on the gypsy moth larvae or pupae, rather than upon the damage and "killings" a parasite accomplishes during the process of oviposition. A case in point is the work of Campbell (1963), on the four native ichneumonid parasites of the gypsy moth: *Itoplectis conquisitor*, *Coccygomimus pedalis*, *Theronia atalantae fulvescens* and *T. hilaris* in Glenville, New York State. They attack the gypsy moth pupae. They stung as many as 250 host pupae for each pupa that was parasitized successfully (leading to the development of an ichneumonid offspring). He further showed that the puncture wounds made by the ichneumonids while stinging the moth pupae, permitted the entrance of the larvae of the Sarcophagidae (which could enter the host pupae only when the integument was broken), eventually killing the host pupa. About half of the stung pupae contained the sarcophagid maggots.

By such an ichneumonid-sarcophagid relationship, and also by the mechanical injury caused to the gypsy moth pupae by stinging, many of them fail to produce adult moths. The ichneumonids, therefore, play a greater role in controlling the gypsy moth than that is generally assigned to them.

In recent years there has been a renewed activity to survey the natural enemies of the gypsy moth in different parts of the world and to introduce the promising species in the United States. Information on such activities is available in the reports of Drea (1978) in Europe, Iran and Japan; Drea and Fuester (1979) in Poland; Györfi (1963) in Hungary; Hedlund and Mihalache (1980) in Rumania; Herard and Fraval (1980) in Morocco; Herard, Mercadier and Abai (1979) in Iran; Pschorn-Walcher (1974) in Europe; Rao (1966, 1972) in India; Romanyk (1965) in Spain; Shapiro (1956) in Russia; and Vasic (1958) in Bulgaria. In a recent publication on the gypsy moth (Doane and McManus, editors, 1981), Dr. Coulson has reviewed the introductions of the parasites in the U.S.A. for the control of the gypsy moth.

In the United States, the states of New Jersey and Pennsylvania have recently been active in the release of exotic parasites. According to the publications of the Pennsylvania Bureau of Forestry, four species of *Coccygomimus* were released in Pennsylvania during 1973-79, as follows.

Coccygomimus disparis, 73,215 ex stocks from India and Japan.
C. instigator   3,250 from Yugoslavia
C. turionellae   34,134 from India
C. moraguesi   5,600 from Morocco

Of these only two specimens of *C. disparis* were recovered in July 1981 (Dr. Fusco, personal correspondence), which appears significant because releases were made in 1979 or possibly in 1980, but not in 1981.

Phobocampe unicincta, which was established during 1911-1912 in the New England states has also spread to New Jersey and Pennsylvania and it appears that parasitism by this species is also increasing gradually in Pennsylvania since 1970, particularly in post-climax gypsy moth populations. The native pupal parasites were also active there in killing the host pupae.

The Ichneumonidae are either larval or pupal parasites of the gypsy moth. The typical larval parasites are the members of the genera *Phobocampe*, *Casinaria* and *Hyposoter*, belonging to the subfamily Porizontinae. They lay their eggs within the young gypsy moth larvae and emerge from older larvae and spin their own cocoons. Other internal parasites or endoparasites of the larvae are the members of the subfamily Banchinae, Cremastinae and Ophioninae. Members of the subfamily Anomalinae are endoparasites of the larvae, but the emergence is from the host pupa. The external parasites of the larvae or ectoparasites are the members of the subfamily Tryphoninae. Members of the tribe Pimplini are external parasites of late larvae that have just spun their cocoons or the prepupae and the development is upon host larva within the cocoon. The Mesosteninae are apparently similar to the Pimplini in their host relations.

Members of the tribe Gelini and of subfamily Mesochorinae are secondary parasites of the various gypsy moth parasites, including Braconidae and Ichneumonidae. Some genera are parasitic upon Tachinidae, particularly the genera *Mesoleptus* and *Phygadeuon*. Species of *Itoplectis* and *Theronia* are also hyperparasitic upon occasions.

The typical pupal parasites belong to the subfamily Ichneumoninae and the tribes Ephialtini and Theroniini of the subfamily Pimplinae, e.g., the genera *Coccygomimus, Itoplectis, Ephialtes,* and *Theronia*. They are internal parasites of exposed or semiexposed pupae. Oviposition is into the prepupa or freshly formed pupa and the emergence of the adult parasite is from the host pupa.

MATERIAL AND METHODS

The purpose of the present study was to establish the identities of the various ichneumonid species that have been reported as parasites of the gypsy moth. Over a hundred species of the Ichneumonidae have been recorded as primary parasites and nearly 25 as secondary parasites. A near complete list of the species and the various taxonomic combinations under which they have been reported previously is given by Griffiths (1976). Previous useful compilations are of Howard and Fiske (1911), Stadler (1933), Schedl (1936) and Thompson (1946).

A literature search was made to establish the first records for each species from the gypsy moth. This revealed that several erroneous records had crept in the literature in the process of compilations. Many records have never been confirmed by subsequent rearings and in such cases it was difficult to assess the role of the parasite in the economy of the gypsy moth.

Attempts were made to gather specimens of Ichneumonidae that have been reared in the past. Unfortunately voucher specimens do not exist for most of the earlier records. The only source from which such material could be obtained was the Forest Insect Laboratory, Hamden, Connecticut, wherefrom specimens received and reared prior to 1930 were available.

In recent years several surveys have been made in Europe, Japan, India, Iran and Morocco, to gather insect parasites of the gypsy moth, and to introduce the promising species in the United States. Only a dozen or so species of Ichneumonidae have usually been reared belonging to the subfamilies Pimplinae and Porizontinae and one to the Ichneumoninae. Fortunately many of these specimens were available for study. This helped tremendously to establish the true identities of the species that are commonly encountered on the gypsy moth.

On the basis of these studies, 24 species of Ichneumonidae are confirmed as parasites of the gypsy moth in the world (List I), 31 species are shown definitely not to be associated with the gypsy moth (List III), and 38 species are listed as unconfirmed records from the gypsy moth (List II). Some of these may be occasional parasites, while some others appear to be misdeterminations. Two new species or subspecies are described: *Phobocampe lymantriae* from *Lymantria dispar* in Europe and Japan, and *Theronia atalantae himalay-*

ensis from *Lymantria obfuscata* from India. The taxonomic identity of all the species has been updated.

The hyperparasites are listed in List IV. The hyperparasites were not studied. Ichneumonid parasites reared in India from *Lymantria obfuscata*, a related species, are mentioned in List V. Many of these are being reared in the U.S.A. for possible releases against the gypsy moth.

The institutions that loaned the species for the present study are listed below, along with the names of the persons who arranged for the loans.

Bangalore, India	Commonwealth Institute of Biological Control, Indian Station, P.O. Box 2484, Bangalore, 560 024, India. (T. Sankaran).
Newark, DE	Beneficial Insects Research Laboratory, 501 S. Chapel St., Newark, Delaware, 19713. (L. R. Ertle, R. W. Fuester, R. J. Dysart, P. W. Schaefer).
Middletown, PA	Division of Forest Pest Management, Bureau of Forestry, 34 Airport Drive, Middletown, Pa. 17057. (R. A. Fusco).
Washington, D.C.	National Museum of Natural History, Smithsonian Institution, Washington, D. C. 20560. (A. S. Menke).
Trenton, NJ	N.J. Dept. of Agriculture,Beneficial Insects Laboratory, 101 Oakland St., Trenton, NJ 08618. (R. Chianese).
EPL - Paris	European Parasite Laboratory, USDA,c/o American Embassy, AGR APO New York, NY 09777. (B. D. Perkins).
Hamden, CT	Forest Insect and Disease Laboratory, Forest Service, USDA, 51 Mill Pond Road, Hamden, CT 06514 (W. E. Wallner).

In the lists and the text, that follows, the arrangement of the subfamilies is in the order of their importance, and not taxonomic. Most species are parasitic on several other lepidopterous pests besides the gypsy moth. The list of alternative hosts is by no means complete. Thompson (1957), Aubert (1967, 75), Townes et al. (1965) and Krombein et al. (1979) provide information on the hosts of the various species of the Ichneumonidae. The nomenclature of the lepidopterous pests has been updated after Lerut (1980).

ACKNOWLEDGMENTS

This research was funded by the U.S. Department of Agriculture. I am thankful to Dr. Paul Marsh for processing the grant to work on the gypsy moth parasites. I am grateful to Dr. Henry Townes for his constant help and advice during the execution of this work. Drs. R. W. Carlson and J. R. Coulson were very helpful in arranging for the loan of specimens from various sources. They also provided valuable information on the parasite rearings, etc. My thanks are also due to them, as well as to the persons listed above under loan of specimens for sending me specimens for the present study.

I record my appreciation and thanks to Mr. L. E. Ling, who took the SEM photographs for me.

LIST I

Ichneumonid species that are parasitic upon the gypsy moth and which have been commonly reared from the pest. Specimens of these species have been examined, reared from the gypsy moth.

SUBFAMILY PIMPLINAE

1. *Coccygomimus instigator* (Fabricius) Page 23.
   Pupal parasite in Eurasia, Iran and Morocco. Released in the U.S.A. during 1906-1909 and again in 1972-79, but not established.
2. *Coccygomimus pedalis* (Cresson) Page 25.
   Native North American parasite, stinging and killing a number of host pupae, without successfully parasitizing them.
3. *Coccygomimus turionellae* (Linnaeus) Page 26.
   (= *Pimpla examinator* Fabricius).
   Pupal parasite, reared in Eurasia and India. Released several times in the U.S.A., but not established.
4. *Coccygomimus moraguesi* (Schmiedeknecht) Page 28.
   (= *C. turionellae moraguesi*)
   Apparently first reared from the gypsy moth pupae by Herard and Fravel (1980) during 1973-75 in Morocco. Also occurring in Algeria and Spain. Cultured and released in Pennsylvania during 1973-79 but not established.
5. *Coccygomimus disparis* (Viereck) Page 29.
   (= *Pimpla porthetriae* Vier.)
   A Japanese parasite of the gypsy moth, subsequently collected in India, China, Mongolia and USSR. Released in Pennsylvania during 1973-79, bred from stocks from India and Japan. Recently recovered there.
6. *Coccygomimus luctuosus* (Smith) Page 31.
   (= misdetermination of *Pimpla pluto* and *porthetriae* in gypsy moth literature).
   A pupal parasite of the gypsy moth. Distributed in Japan and eastern Asia.
7. *Itoplectis conquisitor* (Say) Page 35.
   Native North American pupal parasite, stinging and killing a large number of host pupae without successfully parasitizing them.
8. *Itoplectis alternans alternans* (Gravenhorst) Page 36.
9. *Itoplectis maculator maculator* (Fabricius) Page 37.
10. *Itoplectis clavicornis* (Thomson) Page 38.
    Specimens of the first two species have been examined, labelled as reared from the gypsy moth and also from its parasite *Phobocampe unicincta*. The third species is labelled as reared from *Phobocampe unicincta* only. They occur in Europe. *I. alternans spectabilis* occurs in Japan and is reported parasitic on the gypsy moth.
11. *Ephialtes capulifera* (Kriechbaumer) Page 41.
12. *Ephialtes compunctor* (Linnaeus) Page 42.
    (= *Pimpla brassicariae* Poda)
    Reared specimens of the above two species have been examined from Europe and Japan. Vasic (1958) and Hedlund and Mihalache (1980) reported having reared the latter species in Yugoslavia and Rumania, respectively.

13. *Theronia atalantae atalantae* (Poda)   Page 45.
14. *T. atalantae gestator* (Thunberg)   Page 48.
15. *T. atalantae fulvescens* (Cresson)   Page 46.
    Various subspecies of *Theronia atalantae* parasitize pupae of the gypsy moth in Europe, Japan and North America. A related subspecies, *himalayensis*, is described here as new from *Lymantria obfuscata* in India. Sometimes *T. atalantae* is hyperparasitic.
16. *Iseropus (Gregopimpla) himalayensis*   Page 55.
    A reared specimen from Japan has been examined.

SUBFAMILY PORIZONTINAE

17. *Casinaria tenuiventris* (Gravenhorst)   Page 63.
    (= *Campoplex conicus* Ratzeburg)
    Reared several times in Europe from the gypsy moth and also in Iran in 1976. It has apparently not been seriously considered for introduction in the U.S.A.
18. *Casinaria nigripes* (Gravenhorst)   Page 64.
    (= *Casinaria anastomosis* Uchida)
    This species occurs in Europe and Japan and has been reared in small numbers only.
19. *Campoletis* sp.   Page 66.
    Two reared specimens from France seen but the specific identity is uncertain.
20. *Phobocampe unicincta* (Gravenhorst)   Page 68.
    (= *Hyposoter disparis* Viereck = *Phobocampe* sp. of Herard *et al.* from Iran).
    A widespread species in Eurasia and already established in the U.S.A.
21. *Phobocampe lymantriae* Gupta   Page 73.
    (= *Phobocampe* n. sp. of Drea and Fuester, 1979, from Poland)
    (= *Hyposoter* spp of Burgess and Crossman, 1929).
    A species sympatric and similar to *unicincta*, but biologically rather different. Occurs in Europe and Japan. The majority of the specimens mentioned under "*Hyposoter* spp." by Burgess and Crossman belongs to this species.
22. *Hyposoter vierecki* T. M. & T.   Page 77.
    (= *Campoplex (Diadegma) japonicus* Viereck)
    Japan. Rearing records are rather few.
23. *Hyposoter tricoloripes* (Viereck)   Page 78.
    Reared several times in Europe in small numbers and apparently released in North America, without success.
    A related species, *H. lymantriae* occurs in India on *Lymantria obfuscata*. This has been recently reared in large numbers at BIRL, Delaware on the gypsy moth for possible release in the U.S.A.

SUBFAMILY ICHNEUMONINAE

24. *Lymantrichneumon disparis* (Poda)   Pages 81, 87.
    (= *Ichneumon flavatorius* = *Trogus flavatorius*).
    A common but not abundant pupal parasite of the gypsy moth in Europe. Also reported from Iran by Herard *et al.* (1979). Never reared in

large numbers for release in the U.S.A.

List II

Ichneumonid parasites that have been reported from the gypsy moth, but the records of which could not be confirmed by examination of the reared material. Some of them may be occasional parasites. The taxonomic identity of quite a few species is doubtful.

SUBFAMILY PIMPLINAE

1. *Coccygomimus spurius* (Gravenhorst) Page 32.
   Yafaeva (1959) recorded it from Ukraine. Aubert (1969) and Kasparyan (1974) did not list it as a gypsy moth parasite.
2. *"Coccygomimus* sp."* of Picard (1921) from France and of Herard and Fraval (1980) from Morocco can be any one of the known species of *Coccygomimus.* Identity doubtful.
3. *Itoplectis alternans spectabilis* (Matsumura) Page 37.
   Reported by Fukaya (1936) from Japan. No subsequent confirmation.
4. *Itoplectis viduata* (Gravenhorst) Page 39.
   Reported by Meyer (1929) from Russia.
5. *Ephialtes rufatus* (Gmelin) Page 43.
   (= *rufata* Gravenhorst)
   Reported by Rudow (1911) and Meyer (1927) from Germany and Russia respectively. Not listed subsequently by Meyer (1934, 1936).
6. *Theronia hilaris hilaris* (Say) Page 50.
   (= *T. melanocephala*). North America.
   Earlier records of this species were considered to be misidentifications of *Theronia atalantae fulvescens*. Campbell (1963) mentioned having observed this species also on the gypsy moth, but did not discuss it further. It may be an occasional parasite of the gypsy moth.
7. *Acropimpla didyma* (Gravenhorst) Page 51.
   Šedivý (1963) mentioned it from Czechoslovakia as *Ephialtes didymus*.
8. *Iseropus (I.) stercorator stercorator* (Fabricius) Page 53.
   (= *Pimpla holmgreni*)
   Rudow (1911) reported it from Europe from *Lymantria dispar* and *L. monacha,* but in 1917 listed only *L. monacha* as its host. There are several such cases and his 1911 records appear rather doubtful.
9. *Iseropus (Gregopimpla) inquisitor* (Scopoli) Page 54.
   Stadler (1933) apparently is the first to list it as a gypsy moth parasite in Europe. Vasic (1958) reported it from Yugoslavia. No specimens, however, could be examined to confirm its occurrence on the gypsy moth. It may be an occasional parasite or a hyperparasite associated with the gypsy moth.

SUBFAMILY PORIZONTINAE

10. *Campoplex difformis* (Gmelin) Page 60.
    (= *?Campoplex difformis* Gravenhorst)
    Species of *Campoplex* have often been misidentified in Europe and it
    is doubtful what species, if any, is associated with the gypsy moth.
11. *Campoplex sugiharai* (Uchida) Page 61.
    Momoi (1961) reported it from Japan as a parasite of the gypsy moth.
12. *Phobocampe pulchella* Thomson Page 67.
    Shapiro (1956) reported it from Russia. Drea (in Doane and McManus,
    1981: 317) mentioned it from Yugoslavia (determination by Vasic).
    Apparently a misdetermination of either *P. unicincta* or *P. lymantriae*.
13. *Hyposoter takagii* (Matsumura) Page 76.
    Fukaya (1950) first reported it from Japan. There are no subsequent
    confirmations.

SUBFAMILY ICHNEUMONINAE

Almost all Ichneumoninae except *Lymantrichneumon disparis* (Poda)
have not been reared from the gypsy moth subsequent to their original records.
They have been merely catalogued subsequently, chiefly by Schedl (1936) and
Thompson (1946). Those cataloguers, however, missed several species
recorded from the gypsy moth by Rudow (1917, 1918).

Györfi (1963) mentioned *Protichneumon rubens* and *P. fabricator*, from
the gypsy moth, but the authenticity of these records cannot be confirmed.
Herard and Fraval (1980) mentioned having reared a species of *Melanichneumon* (= *Vulgichneumon*) from Morocco. In the list that follows, the names
used in gypsy moth literature are given together with the original records.
Taxonomic details are given in the text.

14. *Paracoelichneumon rubens* (Fonscolombe) Page 88.
    (= *Ichneumon* = *Protichneumon* = *"Ichneumon rubens* Wesm.")
    Rudow (1918). Europe.
15. *Callajoppa cirrogaster cirrogaster* (Schrank) Page 89.
    (= *Ichneumon lutorius* F. = *Trogus flavitorius lutorius* F. = *Trogus
    lutorius* F.) Howard and Fiske (1911). Europe.
16. *Ichneumon cyaniventris* Wesmael   Page 91.
    Rudow (1918). Europe.
17. *Stenaoplus pictus* (Gravenhorst) Page 94.
    (=*"Ichneumon pictus* Gmelin" of Stadler, Schedl and Thompson
    = *Stenichneumon pictus* Gmelin of Thompson)
    Mocsary (1885). Berthoumieu (1895) . Europe.
18. *Melanichneumon leucocheilus* (Wesmael) Page 95. Europe.
    (= *Ichneumon leucocheilus* of Rudow, 1918 = *"Ichneumon leucocherrus "*
    of Stadler and Schedl, 1936, and *I. leucocerus* of Thompson, 1946).
    = *?Melanichneumon (Vulgichneumon)* sp. of Herard and Fraval, 1980
    from Morocco?
19. *Cratichneumon fabricator* (Fabricius) Page 96.
    (= *Ichneumon* = *Protichneumon*)
    Cecconi (1924), Györfi (1963). Europe.

20. *Chasmias paludator* (Desvignes) Page 97.
    (= *Ichneumon* = *Chasmodes paludicola*)
    Rudow (1917). Europe.
21. *Pterocormus sarcitorius sarcitorius* (Linnaeus) Page 98.
    (= *Ichneumon sarcitorius*)
    Meyer (1929). Russia.
22. *Triptognathus amatorius* (Mueller) Page 99.
    (= *Ichneumon* = *Amblyteles* = *Diphyus*)
    Rudow (1917), Uchida (1926). Japan, Sakhalin.
23. *Spilichneumon occisor* (Fabricius) Page 100.
    (= *Ichneumon* = *Amblyteles*)
    Rudow (1917). Europe.
24. *Amblyteles armatorius* (Förster) Page 102.
    (= *Amblyteles fasciatorius*)
    Uchida (1930). Japan.
25. *Cotiheresiarches dirus* (Wesmael) Page 104.
    (= *Eurylabus dirus*)
    Fahringer (1922). Europe.

SUBFAMILY MESOSTENINAE

26. *Gambrus amoenus* (Gravenhorst) Page 106.
    (= *Cryptus, Aritranis, Spilocryptus*)
    = *Gambrus nuncius* (Say). N. America. N. syn.
    Howard and Fiske (1911). Europe.
27. *Meringopus cyanator* (Gravenhorst) Page 108.
    (= *Cryptus, Trachysphyrus*)
    Howard and Fiske (1911). Europe.
28. *Ischnus inquisitorius inquisitorius* (Mueller) Page 109.
    (= *Ichneumon assertorius* = *Ischnus assertorius*)
    Rudow, (1917). Europe.

SUBFAMILY TRYPHONINAE

29. *Netelia (Netelia) vinulae* (Scopoli) Page 111.
    (= *Panicus cephalotes*)
    Rühl (1914). Europe.
30. *Netelia (Netelia)* sp. Page 111.
    (= *?Paniscus testaceus*)
    Rühl (1914). Europe.

SUBFAMILY BANCHINAE

31. *Banchus hastator* (Fabricius) Page 113.
    (= *Banchus femoralis*)
    Kolubajiv (1934). Europe.

SUBFAMILY CREMASTINAE

32. *Pristomerus vulnerator* (Panzer) Page 115.
    Barsacq (1913), Mokrzecki (1913). Europe.

SUBFAMILY ANOMALINAE

33. *Trichomma (Trichomella) enecator* (Rossi) Page 118.
    (= *Anomalon enecator*)
    Kirchner (1856). Europe.
34. *Barylypa delictor* (Thunberg) Page 120.
    (= *B. perspicillator*)
    Kovacevic (1925). Europe.
35. *Barylypa pallida* (Gravenhorst) Page 120.
    (= *Anomalon pallidum*)
    Rudow (1911). Europe.
36. *Agrypon flaveolatum* (Gravenhorst) Page 120.
    (= *Ophion, Anomalon*)
    Rudow (1911). Europe.

SUBFAMILY OPHIONINAE

37. *"Ophion luteus* (Linnaeus)" Page 122.
    Kolomiyetz (1958). Siberia. Identity of this species doubtful, as
    O. luteus, has often been misidentified.
38. *Enicospilus merdarius* (Gravenhorst) Page 123.
    (= *Ophion*)
    Rühl (1914). Europe.
    Several other species recently collected in Morocco, Iran, etc., chiefly
by the staff of the European Parasite Laboratory, Paris, are mentioned in
reports as Ichneumonid sp. A, B, C, D, etc. A recent letter from
Dr. Lawrence R. Ertle (dated June 18, 1982), states, "The Ichneumonids as
"A", "B", and "C" were hyperparasites of *Apanteles liparidis* and *A. melanoscelus*; Ichneumonids "D" and "E" were not collected from *L. dispar* (L.),
but from *Orgyia* sp."

LIST III

Ichneumonid species that are definitely not associated with the gypsy moth.
Reasons for excluding them are given in details in the text.

SUBFAMILY PIMPLINAE

1. *Perithous septemcinctorius* (Thunberg) Page 16.
    = *Hybomischos*. Parasitic upon Sphecidae rather than on Lepidoptera.
2. *Dolichomitus tuberculatus* (Fourcroy) Page 16.
    Parasitic upon wood boring Coleoptera.

3. *Exeristes roborator* (Fabricius)   Page 17.
   (= *Iseropus roborator*)
   Parasitic upon *Ostrinia* and related lepidopterous borers.
4. *Coccygomimus aethiops* (Curtis)   Page 19.
   Wrongly listed by Thompson (1946).
5. *Coccygomimus pluto* (Ashmead)   Page 19.
   Misdetermination of *luctuosus* Smith.
6. *Coccygomimus tenuicornis* (Cresson)   Page 19.
   Misdetermination of either *C. pedalis* or *Itoplectis conquisitor.*

SUBFAMILY PORIZONTINAE

7. *Sinophorus validus* (Cresson)   Page 58.
   Fusco and Simons (1977) listed it as a native gypsy moth parasite, but correspondence with them failed to confirm its occurrence on the gypsy moth.   Not listed from the gypsy moth by Carlson (1979).
8. *Campoplex difformis* Gravenhorst   Page 60.
   Reported by Ratzeburg (1844).   Its identity is uncertain and it was almost certainly a misdetermination.
9. *"Omorgus* sp. " of Kolomietz (1958)   Page   59.
   Reported as a parasite of the pupa of the gypsy moth.   Appears to be a misidentification.
10. *Casinaria ischnogaster* (Thomson)   Page 62.
    Morley and Rait-Smith erroneously reported it as a European parasite of the gypsy moth.
11. *Hyposoter fugitivus* (Say)   Page 77.
    (= *Limneria fugitiva* , *Limnerium* sp.)
    Erroneous record (Howard and Fiske, 1911).
12. *Campoplex rapax* Gravenhorst   Pages 59, 75.
    (= *Anilastus* = *Anilasta*)
    Recorded by Rudow (1911).   Its identity is very doubtful.

SUBFAMILY SCOLOBATINAE

13. *Opheltes glaucopterous* (Linnaeus)   Page 124.
    Record of its occurrence on the gypsy moth, originating from Rühl (1914) is erroneous, as members of the genus *Opheltes* are parasitic upon *Cimbex* (saw-flies).   Györfi (1963) mentioned it and several other doubtful species from Hungary.

SUBFAMILY XORIDINAE

14. *Xylonomus irrigator* (Fabricius)   Page 124.
15. *Xorides praecatorius* (Fabricius)   Page 124.
    Rudow (1911) erroneously listed the above two species as gypsy moth parasites.   They belong to *Xorides*, which are parasites of wood boring Coleopterous larvae.   Morley (1908) also erroneously reported *Odontomerus dentipes* as a parasite of *Lymantria monacha* .

SUBFAMILY ICHNEUMONINAE

16. *Ichneumon raptorius* Gravenhorst Page 80.
17. *Ichneumon sugillatorius* Linnaeus Pages 80, 91.
18. *Ichneumon melanoceras* Ratzeburg Page 80.
    These species were mentioned by Rudow (1911) as parasites of both *Lymantria dispar* and *L. monacha*, but later (1918) only as parasites of *L. monacha*. Many such records of Rudow (1911) appear erroneous for the gypsy moth.
19. *Ichneumon* sp. Page 93.
    Picard, 1921. Identity uncertain.
20. *Ichneumon leucocerus* (Gravenhorst) Page 92.
    Thompson (1946). Misquotation of *Ichneumon leucocheilus* Wesmael.
21. *"Ichneumon leucocherrus"* Wesmael Page 92.
    Stadler (1933), Schedl (1936). *Lapsus* for *I. leucocheilus* Wesmael.
22. *"Ichneumon pictus* Gmelin" Page 92.
    Stadler (1933). Identity of this species doubtful.
23. *"Ichneumon flavus* Rd." Page 93.
    (= *Ischnus flavus* Rd. of Stadler, 1933)
    Rudow (1933) manuscript name. *Nomen nudum.*
24. *"Amblyteles varipes* Rudow" Page 103.
    Rudow (1888). Identity doubtful.
25. *"Amblyteles camelinus* Wesmael" Page 103.
    Meyer (1936). Not a gypsy moth parasite.

SUBFAMILY MESOSTENINAE

26. *"Ischnus flavus* Rd." Page 105.
    Stadler (1933). Europe.
    Nomen nudum.
27. *"Cryptus liparidis* Rd." Page 105.
    Rudow (1918). Stadler (1933). *Nomen nudum.* Subsequent authors listed the author as "Rond."

SUBFAMILY GELINAE

28. *Mesoleptus laevigatus* (Gravenhorst)
    (= *Exolytus laevigatus*)
    Shapiro (1956). Russia.
    A parasite of Diptera (mostly on Sarcophagidae) and not of Lepidoptera. A related *M. filicornis* recorded erroneously on *L. monacha*.

SUBFAMILY TRYPHONINAE

29. *"Paniscus melanurus* Thomson" Pages 111, 112.
    Meyer (1936) apparently confused it with *testaceus*. Identity uncertain.
30. *"Paniscus testaceus* Gravenhorst" Pages 110, 111.
    What species was actually involved (Rühl, 1914) is doubtful as *testaceus* has been very often misidentified in the past. The true identity of it is unknown.

SUBFAMILY BANCHINAE

31. *Banchus falcatorius* (Fabricius) Page 114.
     (= *Banchus falcator*)
     Morley (1915). Data as recorded by Morley indicate that it was never
     reared from the gypsy moth, but collected flying where gypsy moth
     larvae were also present.

LIST IV

Ichneumonid hyperparasites associated with the gypsy moth

SUBFAMILY MESOCHORINAE

Several species of *Mesochorus*, parasitic upon *Apanteles* species have
been associated with the gypsy moth, chiefly in Europe.

1. *Mesochorus ater* Ratzeburg. Europe.
2. *M. confusus* Holmgren. Europe, N. Africa.
3. *M. discitergus* (Say).
        (= *facialis* Bridgman). Holarctic, Oriental.
4. *M. dilutus* Ratzeburg. Europe.
5. *M. dorsalis* Holmgren. Europe.
6. *M. gracilis* Brischke. Europe.
7. *M. pallidus* Brischke. Europe.
8. *M. pectoralis* Ratzeburg. Europe.
9. *M. semirufus* Holmgren. Europe.
10. *M. splendidulus* Gravenhorst. Europe.
11. *M. sylvarum* Curtis. Europe.
12. *M. vitreus* Walsh. N. America.

SUBFAMILY GELINAE

1. *Acrolyta nigricapitata* (Cook & Davis). N. America.
2. *Atractodes croceicornis* Haliday (= *A. compressus* Thomson). Europe.
     Parasite of Diptera. Might be secondary on Tachinidae associated
     with the gypsy moth.
3. *Bathythrix triangularis* (Cresson) (= *Thysiotorus, Mesoleptus*). North
     America.
3a. *Dichogaster aestivalis* Gravenhorst (= *Hemiteles aestivalis*). Incorrect
     record. Parasitic upon Chrysopidae and not associated with *Lymantria*.

Genus *Gelis*
(Many species reported under *Pezomachus* and some under *Hemiteles* .)

4. *G. agilis* Fabricius. Europe.
5. *G. apantelis* Cushman. U. S. A.
6. *G. areator* Panzer (= *Hemiteles areator*). Europe.

7. *G. cinctus* Linnaeus (= *Hemiteles bicolorinus* Grav.) Europe.
8. *G. cushmani* Carlson (= *Hemiteles apanteles* Cushman) U. S. A.
9. *G. hortensis* Gravenhorst (= *Pezomachus*) Europe.
10. *G. instabilis* Foerster (= *Pezomachus*) Europe.
11. *G. intermedius* Foerster. Europe.
12. *G. inutilis* Cushman. U. S. A.
13. *G. nigriceps* Foerster. Europe.
14. *G. nigritus* Foerster. Europe.
15. *G. nocuus* Cushman. U. S. A.
16. *G. obscurus* Cresson. U. S. A.
17. *G. pulchellus* Gravenhorst. Eurasia.
18. *G. pulicarius* Fabricius. Europe.
19. *G. tenellus* Say. U. S. A.
20. *G. urbanus* Brues (= *Hemiteles cingulator* Gravenhorst). Wrong record.
21. *Lysibia nana* Gravenhorst (= *Astomaspis nana, Hemiteles nanus,*
       Hemiteles fulvipes Grav.) Europe.
22. *Mesoleptus filicornis* Thomson (= *Exolytus*) Europe.
       Parasitic on Tachinidae. May be on tachinid parasites of gypsy moth.
23. *Phygadeuon subfuscus* Cresson. U. S. A.
       Parasitic on *Sturmia scutellata.*
       Four other species of *Phygadeuon* have been reported in association
       with *Lymantria monacha,* viz., *P. flavimanus* Grav., *fumator* Grav.
       grandiceps Grav., and *variabilis* Gravenhorst. They are parasitic on
       Tachinidae, perhaps associated with the nun-moth.
    Species of the Pimpline genera *Itoplectis* and *Theronia* are also occasion-
ally hyperparasitic through other ichneumonids of the gypsy moth.

Euceros albomarginatus Cushman
    Two males examined, bearing the following data:
"Eastern Pa., Summer 1976. Reared from *Phobocampe* cocoon from *L. dis-*
par (Pennsylvania Bur. Forestry)."

LIST V

Ichneumonid parasites associated with *Lymantria obfuscata* Walker

    Lymantria obfuscata occurs in India and its habits are similar to those of
L. dispar. The two were confused for a long time. Nagaraja *et al.* (1968) and
Rao (1972) showed that the two were different species.
    The following ichneumonid parasites occur on *Lymantria obfuscata* in
India:
1. *Coccygomimus turionellae* (L.).
2. *C. disparis* (Viereck) (= *Pimpla* sp., *turionellae* Gr.)
3. *C. laothoë* (Cameron) (= *poesia*).
4. *Theronia atalantae himalayensis* Gupta (= *Theronia* sp., *T. atalantae*
       atalantae).
5. *Hyposoter lymantriae* Cushman (= *Anilastus* sp., *Hyposoter* sp.).
    Some of the above species have been reared on the gypsy moth in the
U. S. A. and also released in the field.
    The identities of other ichneumonids reported by Rao (1966, 1972) have
not been established. These are "Cryptinae gen. et sp. indet.", *Cryptus* sp.,
Metopiinae - Exochini "gen. et sp. indet." and *Goryphus* sp. nr. *inferus.*

KEY TO THE SUBFAMILIES OF ICHNEUMONIDAE ASSOCIATED WITH
THE GYPSY MOTH

1. Wingless. . . . . Some GELINAE
   Winged. 2

2. Clypeus and face forming a broad, weakly convex surface. Clypeus not
   separated from face by a distinct groove. Areolet rhombic, large.
   First tergite with a large glymma, its spiracles near middle. Male
   claspers rod-like. Female subgenital plate large, triangular.
   Brownish species, hyperparasitic within braconid cocoons.
   (Mesochorus). MESOCHORINAE
   Not as above. Clypeus separated from face by a distinct groove except
   in Porizontinae. Areolet various, never rhombic. 3

3. Spiracle of first tergite placed behind the midlength of the tergite. First
   tergite usually petiolate, slender at base. 4
   Spiracle of first tergite placed near midlength of the tergite. First
   tergite usually quadrate to narrowly trapezoid, not slender at
   base. 10

4. Abdomen compressed laterally. Third and fourth segments deeper than
   wide. 5
   Abdomen depressed or cylindric. Third and fourth segments wider than
   deep. 8

5. First intercubital vein (or the only intercubital vein present) joining
   cubitus vein distad of second recurrent vein by a distance greater than
   half its length. Epomia absent. Medium to large sized, pale brown
   colored slender species with very compressed abdomen. *(Ophion)*.
   IX. OPHIONINAE
   First intercubital vein joining cubitus basad of, opposite, or less than
   half its length distad of second recurrent vein. 6

6. Propodeum coarsely reticulated. Areolet absent. Hind tarsus often
   swollen, especially in males. Slender species with long and much
   compressed abdomen. . . . . VIII. ANOMALINAE
   Propodeum not reticulated, with distinct carinae bounding separated
   areas. Areolet present or absent. Hind tarsus not swollen. 7

7. Clypeus usually confluent with face. Face usually black. Ovipositor
   straight or decurved. Hind femur not toothed below.
   II. PORIZONTINAE
   Clypeus separated from face by a groove. Face usually pale. Ovipositor
   sinuate and hind femur toothed below in the genus associated with the
   gypsy moth. *(Pristomerus)*. . . . . VII. CREMASTINAE

8. Ovipositor very short, hardly surpassing the tip of abdomen, its sheaths
   always rigid. Notauli and sternaulus weak and short, or absent.
   Clypeus broader, with its apex flat and truncate.
   III. ICHNEUMONINAE

Ovipositor long to short, conspicuously extending beyond the tip of abdomen. Notauli and sternaulus conspicuous. Ovipositor sheaths usually flexible. 9

9. Second recurrent vein with two bullae, sloping outwards. Propodeum usually fully areolated. Face of male usually black. Hyperparasites.
                                                                GELINAE
Second recurrent vein with a single bulla, not sloping outwards. Propodeum with only transverse carinae, usually the basal transverse carina alone prominent. . . . . IV. MESOSTENINAE

10. Tarsal claws not pectinate in females, sometimes with a large basal tooth on some claws. Abdomen depressed. . . . . . I. PIMPLINAE
Tarsal claws pectinate. Abdomen compressed. 11

11. Prepectal carina present. Upper tooth of mandible a sharp point. (Netelia). . . . . V. TRYPHONINAE
Prepectal carina absent. Upper tooth of mandible obliquely chisel-shaped. (Banchus). . . . . VI. BANCHINAE

I. SUBFAMILY PIMPLINAE (= EPHIALTINAE)

Members of the subfamily Pimplinae are characterized by having a depressed abdomen, with first segment short, stout, broad, with its spiracle at or in front of the middle. Glymma nearly always present. Apical margin of clypeus usually thin and with a median notch. Mandible with two teeth. Notauli weak or absent. Sternaulus absent or short and weak. Postpectal carina incomplete. Propodeum usually not areolated (except *Xanthopimpla*, *Theronia* and related genera). Tarsal claws never pectinate. Areolet usually present and triangular. Nervellus intercepted variously. Abdominal tergites usually with paired swellings. Ovipositor long, its tip without any notch, the lower valve tip with ridges.

The larva of the Pimplinae differs from that of all other Ichneumonidae in having the hypostomal spur and the stipital sclerite both well developed and reaching each other at their apices, rather than the hypostomal spur reaching the stipital sclerite before its apex. The larval antenna is usually well developed and the mandible is with teeth, but in the tribe Ephialtini the antenna is vestigial and the mandible is without teeth (cf. figures 123-132).

The adults are rather large and slender, with body and ovipositor elongate. They are generally black in color (*Coccygomimus*, *Itoplectis*, *Ephialtes*, etc.) or yellow to yellowish-brown (*Theronia*, *Xanthopimpla*).

Several species of Pimplinae are important parasites of the gypsy moth. They are discussed below. The following species were erroneously recorded as gypsy moth parasites, or their records are doubtful.

1. *Perithous semptemcinctorius* (Thunberg)
     Griffiths (1976) reported this species as a parasite of *Lymantria dispar* from Bulgaria, quoting Stefanov and Keremidchiev (1961). This species belongs to the genus *Hybomischos*, the species of which are parasitic upon Sphecidae nesting in canes or twigs.
2. *Dolichomitus tuberculatus* (Fourcroy)
     This species has often been recorded as a parasite of *Lymantria monacha*.

Members of the genus *Dolichomitus* are parasitic on wood boring Coleoptera, and not on *Lymantria*. Dalla Torre (1902) cited *L. monacha* as a host, referring to Ratzeburg (1844), who however, mentioned *"Curculio pini"* as a host and not *Lymantria*. Also cited as *Ephialtes tuberculatus* by some authors.

3. *Iseropus roborator* (Fabricius)

Meyer (1936, pt. 6: 296-297) mentioned this species as a parasite of *Lymantria dispar* and *L. monacha*. However, in his taxonomic treatment of this species (1934, pt. 3: 22), the only host mentioned was *Ostrinia nubilalis*. There is no subsequent record confirming its association with the gypsy moth. This species properly belongs to the genus *Exeristes* and is widely distributed in southern Europe and the Mediterranean area.

KEY TO THE TRIBES AND GENERA OF PIMPLINAE ASSOCIATED WITH THE GYPSY MOTH

1. Mesopleural suture straight, without a distinct angulation, the area before it not depressed. When hind tibia banded, apical and basal dark bands and a median pale band — the extreme base of tibia dark. (Tribe Ephialtini). 2
   Mesopleural suture with a weak angulation near middle. Mesopleural pit area depressed. When hind tibia banded, there are apical and sub-basal dark bands and median and basal pale bands—the extreme base of tibia thus pale. 4

2. Inner margin of eye weakly concave above antennal socket. Face of male black. Tarsal claws of female without a basal tooth.
   1. Coccygomimus (= Pimpla of authors)
   Inner margin of eye rather strongly concave opposite antennal socket. Face of male white, yellow or black. Fore tarsal claws of female with a large basal tooth. 3

3. Ovipositor straight. Face and orbits of both sexes black. . 2. Itoplectis
   Ovipositor hooked downwards at tip. Face of male largely or entirely white or yellow. Orbits of female narrowly whitish in front.
   3. Ephialtes

4. Tarsal claws of female, or at least the front claws, with a basal tooth. Male subgenital plate wider than long. Body black or with black stripes. (Tribe Pimplini). 5
   Tarsal claws of female without a basal tooth. Male subgenital plate longer than wide. Body yellow or brownish-yellow. (Tribe Theroniini).
   4. Theronia

5. Nervellus intercepted below the middle, rarely at middle. Ovipositor sheath shorter than fore wing. First intercubitus of areolet shorter than the second. Areolet wider than high, triangular, and receiving second recurrent vein at apical corner. Propodeum with short median carinae at base. . . . . 5. Acropimpla
   Nervellus intercepted at or above the middle. Ovipositor sheath as long as or longer than fore wing. Areolet variable. . . . . . . 6. Iseropus

6. Face of male white or yellow; clypeus of female usually red. Clypeus
     weakly convex sub-basally, weakly depressed apically. Nervellus inter-
     cepted near its upper 0.33. Ovipositor 0.5 as long as fore wing. Ocellar
     triangle narrow. Interocellar distance a little less than ocellocular dis-
     tance (0.8 to 0.9:1). . . . . 6a. Iseropus (Iseropus)
  Face of male and clypeus of female black. Clypeus strongly convex sub-
     basally and depressed apically. Nervellus intercepted between its upper
     0.4 and center. Ovipositor 0.6 to 0.8 as long as fore wing. Ocellar
     triangle wider. Interocellar distance a little more than ocellocular dis-
     tance (10:8). . . . . 6b. Iseropus (Gregopimpla)

Tribe EPHIALTINI

   Members of the tribe Ephialtini have the clypeus usually a little swollen
basally and flattened apically, mesoscutum without transverse wrinklings,
prepectal carina complete, mesopleural suture without an angulation near the
middle or sometimes with a weak angulation, propodeal carinae absent to more
or less complete *(Xanthopimpla)*, propodeal spiracle elongate, areolet present,
nervellus intercepted usually above the middle, first tergite short, its lateral
carina usually strong, and ovipositor stout, long or short, with its tip a little
depressed.
   The members of this tribe are common throughout the world. They are
usually internal parasites of exposed or semi-exposed pupae of Lepidoptera.
Species of *Itoplectis* are sometimes secondary parasites. Oviposition is into
the prepupa or pupa and the emergence is from the pupa. The host range of
an individual species is wide. The males are usually smaller than the females.
There is one parasite per pupa. They overwinter in the host pupa, or species
of *Itoplectis* sometimes overwinter as adults.
   Three genera, *Coccygomimus*, *Itoplectis* and *Ephialtes*, are associated
with the gypsy moth, but only species of *Coccygomimus* appear to be of some
importance in the control of the pest.

1. Genus COCCYGOMIMUS   (Fig. 1)

   Coccygomimus Saussure, 1892. *In* Grandidier: Histoire Physique Naturelle
      et Politique de Madagascar, 20 (Hyménoptères), Part 1, Pl. 14, Fig. 12.

   This genus is often called erroneously as *Pimpla* by various authors. For
synonymical references, refer to Townes (1969). Townes and Townes (1960)
treated the Nearctic species, and Kasparyan (1974) reviewed the Palaearctic
species.
   Moderately large-sized insects with fore wing 3.2 to 17.5 mm. long.
Antenna slender, thin and long. Inner margin of eye only weakly emarginate
above antennal socket. Malar space usually long. Mandibular teeth of almost
equal length. Notauli weak or absent. Propodeum and metapleurum usually
with coarse punctures and some striation. Median longitudinal carinae of propodeum usually present basally. Tarsal claws large, simple, without a basal
tooth or an enlarged hair with a flattened tip. Abdomen usually closely punctate (sometimes impunctate). Ovipositor straight, long and stout.
   Members of the genus *Coccygomimus* are solitary, internal parasites of
prepupae and pupae of various Lepidoptera found beneath leaf litter. The
development is completed within the host pupa. They are polyphagous and

multivoltine. Some species overwinter within the host pupa. All species give off a strong pungent odor when captured.

Various species of *Coccygomimus* have been recorded as parasites of *Lymantria dispar* in Eurasia, Japan and North America. There have been many misdeterminations in the past. The following species have been confirmed by various authors as parasitic on the gypsy moth by rearings:

1. *C. instigator* (Fabricius). Eurasia.
2. *C. pedalis* (Cresson). North America.
3. *C. turionellae* (L.). Eurasia. Introduced in North America.
4. *C. moraguesi* (Schmiedeknecht). Morocco. Introduced in U.S.A.
5. *Coccygomimus disparis* (Viereck) (= *C. porthetriae* Viereck). Japan.
6. *C. luctuosus* (Smith). Japan. This species was misidentified by Howard and Fiske (1911) as *C. pluto*.

Many of the above species are being cultured in the U.S.A. for field releases. Metterhouse (in Doane and McManus, 1981: 363-365) provided some biological information on these species.

Species of *Coccygomimus* that have been reported erroneously as parasites of the gypsy moth are:

1. *Coccygomimus aethiops* (Curtis) (= *parnarae* Viereck = *aterrima* Gravenhorst). Europe.

Thompson (1946) listed this species as a parasite of the gypsy moth, citing Stadler (1933) and Schedl (1936). These authors did not list this species. It is also not listed as a parasite of the gypsy moth in the recent catalogs of Oehlke (1967) and Aubert (1969).

C. aethiops is a rather distinct species, wholly black, closely, coarsely punctate, and dull body sculpture. It belongs to the Turionellae Group and is somewhat related to *C. pluto* and *C. luctuosus*.

2. *Coccygomimus pluto* (Ashmead). Japan

This species was reported as a parasite of *Lymantria dispar* by Howard and Fiske (1911), who reared it from pupae received from Japan during 1908-1910. I have seen a female with the data, "Ex. *Porthetria dispar* Linn., Japan, July 1908, Gip Moth Lab 1650", bearing a determination label, "*Pimpla pluto* Ashm." by Viereck. This is actually a specimen of *Coccygomimus luctuosus* (Smith) which is very closely related to *pluto*. These two species have often been confused in the past. *C. pluto* has not been subsequently reared nor mentioned as a parasite of the gypsy moth, except by cataloguers like Thompson and Schedl, who based their information on Howard and Fiske. This species should therefore be deleted from the list of the gypsy moth parasites. Kasparyan (1974) provides a diagnostic key to separate the three very similar looking species: *parnarae* = *aethiops*, *pluto* and *luctuosus*.

3. *Coccygomimus tenuicornis* (Cresson). U. S. A.

Howard and Fiske (1911: 138) stated, "Recorded as a parasite [of gypsy moth] by Forbush and Fernald, but never reared at the Laboratory. Possibly *P. conquisitor* was actually the species reared."

Thompson (1946), quoting Schedl (1936), reported it from Japan and U.S.A.

Simons et al. (1979: 31) have the following entry under *C. tenuicornis* (Cresson):

"Riley and Howard (1894) reported occasional parasitism of the gypsy moth. However, this was probably a misidentification of *C. pedalis* (Cresson) or *Itoplectis conquisitor* (Say) (Carlson, 1978)."
It is not a parasite of the gypsy moth.

4. *Coccygomimus* sp.

Picard (1921) reported this species from France. Its identity is uncertain without Picard's specimens. It can be any of the known European species of *Coccygomimus* parasitizing the gypsy moth.

5. *Coccygomimus* sp.

Herard and Fraval (1980) reported this from pupae of the gypsy moth from Morocco. Its identity uncertain.
The species of *Coccygomimus* that have been reported from other *Lymantria* species are:
    1. *C. arcticus* (Zett.) from *L. monacha*. Europe.
    2. *C. instigator* (Fab.) from *L. monacha*. Europe.
    3. *C. contemplator* (Mueller) from *L. monacha*. Europe.
Prior to 1932, specimens determined as *contemplator* were called *turionellae* L. Several species have been mixed up under *contemplator* and it is not certain whether true *contemplator* is parasitic upon *Lymantria monacha*.
    4. *"Pimpla dentata* Thomson" from *L. monacha*. Europe.
Tragardh (1920) reported *"Apechthis dentata* Thomson" as a parasite of *Lymantria monacha* for the first time from Sweden. Kolubajiv (1937) indicated that *"dentata* Thomson" was a lapsus for *Pimpla quadridentata* Thomson. This species is now placed in *Ephialtes*.
    5. *C. turionellae* (L.) from *L. obfuscata*. India.
    6. *C. disparis* (Viereck) from *L. obfuscata*. India.
    7. *C. laothoë* (Cameron) from *L. obfuscata*. India.
Rao (1966) mentioned two *"Pimpla* sp." as parasites of pupae of *L. obfuscata* in Kashmir and Kotgarh. In 1972 he reported three species, *C. turionellae*, *C.* sp. nr. *turionellae*, and *C. laothoë* from *L. obfuscata*. I have checked these specimens. The specimens from Kashmir represent *C. turionellae* and *C. disparis* (= *C.* sp. nr. *turionellae* of Rao). A female from Chimla, H. P., July 20, 1964, Kotgarh substation, det. as *Pimpla poesia* Cam. by Kerrich, is *C. laothoë* (Cameron), *C. poesia* is a synonym of *C. laothoë*.

KEYS TO THE SPECIES OF *COCCYGOMIMUS* PARASITIC UPON THE
GYPSY MOTH PLUS THOSE OFTEN CONFUSED WITH THEM

Females

1. All legs, including their coxae, black. Eastern Palaearctic. Two species
   of the Turionellae Group. 2
   All legs not wholly black, either coxae or femora and tibiae red or
   orange-yellow. 3

2. Body length 12-14 mm. Body pubescence brownish. Face with scattered
   punctures, shiny in between. Mesopleurum with well separated shallow

punctures. Tergite 5 with shallow, not well formed punctures. Hind
coxa subpolished, mat. Propodeum punctate. Whole body more subpolished with shallower punctures. Japan (wrongly associated with the
gypsy moth). . . . .   pluto (Ashmead)
Body length 15-20 mm. Body pubescence whitish or golden. Face strongly
punctate. Mesopleurum strongly punctate. Tergite 5 with coarse, well
formed punctures. Hind coxa punctate. Propodeum rugose to rugoso-
reticulate centrally. Whole body dull with coarse punctures. Japan.
                                              6. luctuosus (Smith)

3. All legs including coxae, red. Scutellum red. Hind tibia with a distinct
   white band, its basal half fuscous. Hind tarsus black. Mesopleural
   punctation coarse. Morocco. . . . .   4. moraguesi (Schmiedeknecht)
   All coxae black or at least one pair of coxae black to blackish. Scutellum
   black (sometimes yellow in *laothoë*).   4

4. All coxae black. All femora reddish-brown or orange-brown.   5
   Hind coxa reddish or yellowish-brown. Fore coxa black, or blackish-
   brown. Middle coxa yellowish-brown or black. Femora reddish-
   brown.   8

5. Body pubescence brownish. Hind tibia (and femur) wholly rufous, without
   a white annulus. First tergite strongly to weakly pyramidal in profile.
   Tegula black or yellow. Epipleura narrow. Fourth epipleurum about
   3.0 as long as wide. (Instigator Group). Length 12 to 18 mm. . . .   6
   Body pubescence white. Hind tibia with a pale submedian band or wholly
   black. First tergite short, not pyramidal in profile, without strong or
   distinct humps. Epipleura wider. Fourth epipleurum the widest, about
   2.5 as wide as long. (Turionellae Group). Length 9-15 mm.   7

6. Tegula black. Dorsal humps on tergite 1 prominent. Mesopleurum not
   densely punctate, interspaces about 1.0 to 1.5 the diameter of punctures.
   Eurasia, Japan, Iran, and Morocco. . . . .   1. instigator (Fabricius)
   Tegula yellow to orange-brown. Dorsal humps on tergite 1 weaker.
   Mesopleurum densely punctate, punctures largely contiguous, or interspaces less than the diameter of the punctures   India, *ex Lymantria
   obfuscata*. . . . .   laothoë (Cameron)

7. Hind tibia with a submedian pale band, or spot, or rarely wholly reddish.
   Hind femur fuscous apically. Hind corner of pronotum yellow. Abdomen
   with coalescing punctures, a little coarser than in disparis. Length
   9-12 mm. Palaearctic and Oriental Regions.
                                           3. turionellae (Linnaeus)
   Hind tibia wholly black. Hind femur with a distinct black apical ring which
   may be as wide as 0.2-0.25 the length of femur. Hind corner of pronotum without yellow mark (only the postspiracular sclerite yellow). Abdomen punctate but not coarsely so, punctures rather well formed and
   smaller. Length 12-15 mm. Japan and India. Introduced in U. S. A.
   and possibly established in Pennsylvania. . . . . 5. disparis (Viereck)

8. Fore and middle coxae black or brownish-black. Hind coxa reddish or
   yellowish-brown. (Parasites of *Lymantria monacha*).   9

Fore coxa black or brownish-black. Middle and hind coxae orange or
yellowish-brown. (Parasites of *Lymantria dispar*).   10

9. Body 12-15 mm. long. Body pubescence brownish. Hind tibia brownish-
yellow, without any pale band. Propodeum trans-striate, with strong
basomedian carinae bounding a finely striate basal area. First tergite
with dorsal humps. Epipleura narrow. (Instigator Group). Europe.
                                    arcticus (Zetterstedt)
Body 8-10 mm. long. Body pubescence white. Hind tibia with a submedian pale band. Propodeum with a smooth basal area bounded by
weak interrupted carinae. First tergite flat, without dorsal humps.
Epipleurae, particularly 4th and 5th, wide. (Aequalis Group). Europe.
                                    contemplator (Mueller)

10. Body 12-18 mm. long. Pubescence brownish. Epipleura narrow. First
tergite conical, with dorsal humps. Hind tibia and tarsus black.
(Instigator Group). North America. . . . . . .   2. pedalis (Cresson)
Body 8-10 mm. long. Pubescence whitish. Epipleura wide. First
tergite flat dorsally, without humps. Hind tibia and tarsus brownish,
with a faint suggestion of a submedian band on tibia. Hind tibia sometimes
rufous. (Aequalis Group). Europe. . . . .   7. spurius (Gravenhorst)

Males

1. All coxae red. Legs in general red. Hind tibia with a yellow subbasal
band. Hind tarsus black. Hind tibia sometimes a little fuscous.
Scutellum red. Morocco. . . . . . .   4. moraguesi (Schmiedeknecht)
All coxae and legs never wholly red. Scutellum black or yellow marked.
                                                                    2

2. All coxae wholly black. Body pubescence various. Tyloids present or
absent.   3
Middle coxae black or yellowish-brown. Hind coxa wholly or partly
yellowish-brown. Body pubescence brown. Tyloids absent. Hind
tibia and tarsus black.   9

3. Hind femur black.   4
Hind femur red.   5

4. Flagellar segments 6-9 with tyloids. Wing bases yellow. Scutellum with
a yellow spot. Body pubescence usually whitish, sometimes brownish.
Fore and middle tibiae and femora with yellow marks. Mesopleurum
coarsely punctate. Length 10-15 mm. Japan. . . . . luctuosus (Smith)
Flagellar segments 6-7 (only) with tyloids. Wing bases black, though
tegula brown. Body pubescence white. Scutellum black. Fore femur
and tibia with brownish marks. Mesopleurum smoother. Length 8-10
mm. Japan.   6. pluto (Ashmead)

5. Tyloids absent. Body small and slender, less than 10 mm. long. Hind
tibia with a distinct submedian white band.   6

Tyloids present on flagellar segments 6-7, 6-9 or 6-11. Body medium-sized, over 10 mm. long. Hind tibia brownish, brownish-red, or black, with or without a whitish band. 7

6. Tergite 6 mat. Tegula whitish-yellow. Subapical band on hind tibia sharply demarcated from the black tibia. Europe. . . . contemplator (Mueller)
Tergite 6 coarsely punctate on front half. Tegula yellow, with a brown spot. Subapical band on hind tibia with indistinct boundries. Hind tibia brownish. Europe. . . . . 7. spurius (Gravenhorst)

7. Body pubescence brown. Tyloids present on flagellar segments 6-11. Tegula black, or with a large yellow spot. Hind femur without any apical black mark. Hind tibia brownish-red. Body 8-15 mm. long. Eurasia. . . . . 1. instigator (Fabricius)
Body pubescence whitish. Tyloids present on flagellar segments 6-7 only. Tegula yellow. Hind femur black apically. Hind tibia brownish or black, with or without a white band. 8

8. Hind corner of pronotum yellow. Hind tibia black and with a distinct yellowish-white submedian band. Mesopleural sculpture smoother. Length about 9-12 mm. Eurasia. . . . . . . 3. turionellae (Linnaeus)
Hind corner of pronotum black. Hind tibia blackish or blackish-brown, without a distinct yellow band. (Sometimes there is a faint suggestion of a band). Mesopleural sculpture coarser. Length about 12-13 mm. Japan and India. Introduced in U. S. A. and possibly established in Pennsylvania. . . . . 5. disparis (Viereck)

9. Middle and hind coxae yellowish-brown. Face comparatively smoother. North America. . . . . 2. pedalis (Cresson)
Middle coxa black. Hind coxa orange-brown dorsally and black ventrally. Face more distinctly punctate. Europe. . . . . arcticus (Zetterstedt)

1. COCCYGOMIMUS INSTIGATOR (Fabricius) (Figs. 9,15,21,27,33)

Ichneumon instigator Fabricius, 1793. Entomologia Systematica, 2: 164. Germany.
Pimpla instigator: Gravenhorst, 1818. Nov. Acta Physio Medica Acad. Caesareae Leopoldino-Carolinae Nat. Curio, 9: 291.
Apechthis flavipes Matsumura, 1912. Thousand Insects of Japan, Suppl. 4: 144. Japan.
Coccygomimus instigator: Townes, Momoi and Townes, 1965. Mem. Amer. Ent. Inst., 5: 51. China, Japan, Korea, Kuriles, Sakhalin, Russia, and Europe. Several host records from Eurasia.

The earliest record of this parasite from the gypsy moth can be traced to Rondani (1873) in Europe. Rudow (1911) mentioned it as a parasite of *Lymantria dispar*, *L. monacha*, *Euproctis similis* and *E. chrysorrhoea*. Carlson (1979) lists several synonyms.
Female: Face closely punctate, with a median ridge. Frons rather deeply excavated, shiny, with minute punctures and a few trans-striations. Meso-scutum somewhat shallowly punctate, shiny in between punctures. Scutellum shiny, with a few scattered minute punctures. Mesopleurum with deep punc-

tures, tending to be punctato-striate along mesopleural suture. Punctures not dense, separated by about 1.0 to 1.5 their diameter. Metapleurum rugoso-striate in anterior half and distinctly striate posteriorly, including the prominent submetapleural flattened ridge. Propodeum rugoso-reticulate to rugoso-striate, without median carinae, apically shiny. First tergite pyramidal in profile, with distinct dorsal humps, its apical half coarsely punctate, tending to be rugose laterally. Second tergite coarsely punctate. Tergites 3-5 punctate, punctures tending to be smaller progressively rearward, the following tergites mat, subpolished. Hind coxa punctate. Tarsal claws broad, apically strongly decurved. Fourth segment of fore tarsus strongly notched ventrally. Nervulus slightly distad of basal vein. Epipleura 2 and 3 narrow. Fourth epipleurum a little wider, rectangular, about 3.0 as long as wide.

Black. All coxae and trochanters black. All femora and tibiae orange-brown, their tarsi brownish, with apical segments fuscous. Sometimes hind tarsus wholly fuscous. Hind tibia without a pale band. Body pubescence brownish. Tegula black.

Male: Similar to the female in sculpture and color, but with fuscous hind tarsus, or hind tarsus often black. Tegula black or with a yellow spot. Flagellum with tyloids on segments 6-9, 6-10, or 6-11.

Length: Male 8-15 mm. Female 15-18 mm. Fore wing 6-13 mm.

Specimens: 8♀, 4♂, bred from gypsy moth at Melrose Highlands Lab., from pupae received from Hungary (Baja), July 1925; Austria, August 1907; France, August 1910; Germany, August 1909; and Italy, August 1911 (Forest Insect Laboratory, Hamden, Ct.). South Rumania, 1♂, 1♀, June 26-28, 1978, *ex Lymantria dispar,* R. C. Hedlund. Poland 1♂, August 1, 1975. Morocco, North Kemitra, Mamora Forest, 1♀, July 13, 1974, *ex Lymantria dispar,* A. Fravel (all B.I.R.L., Newark, Delaware). 3♂, 3♀, bred at Trenton, N. J., from stock from Yugoslavia *ex Lymantria dispar* (N.J. Dept. Agriculture, Trenton). Several European specimens seen in Townes collections (not reared).

Distribution: Widespread in Eurasia and Japan. Also occurs in North Africa. It has been reared from gypsy moth pupae in Iran and Morocco (Herard *et al.,* 1979, 1980).

Hosts: Several hosts have been recorded in Europe and Japan, for which reference may be made to Townes et al. (1965) and Aubert (1969).

Coccygomimus instigator belong to the Instigator Group as defined by Townes and Townes (1960). It is close to the Oriental *C. laothoë* (Cameron), the European *C. arcticus* (Zetterstedt), and the Russian *C. palmiricus* (Kasparyan). In *C. laothoë* the mesopleural sculpture is denser and tegula orange-brown to yellow. It is parasitic on *Lymantria obfuscata. C. arcticus,* a parasite of *Lymantria monacha,* has smoother and shiny thorax, with meso-scutum leathery, almost impunctate, and hind leg wholly orange-brown, with its tarsus sometimes a little fuscous. The male has black coxae, but the tyloids are absent. *C. palmiricus* is distinguished by its dark red hind femur, narrower tarsal claws, which are apically less decurved, and fourth segment of fore tarsus weakly notched ventrally.

Biological notes: Howard and Fiske (1911: 237) recorded that this species was received in considerable numbers from Europe in shipments of brown tail moth pupae during 1906 to 1909. During that time it was liberated in New England states for control of the gypsy moth. It was, however, not recovered. During 1973-79, 3,250 specimens of this species were released in Pennsylvania from stocks from Yugoslavia. No recoveries have so far been reported.

Györfi (1963: 51) mentioned *C. instigator* as an important parasite of

Lymantria dispar in Hungary. Romanyk (1965: 34) mentioned that it has been known as a gypsy moth parasite in Spain for a long time, but without having had any great impact on the gypsy moth population there. The only exception noted by him was its abundance during 1960-61, in a small gypsy moth infestation on Minorca Island, where about 20% of the pupae were destroyed by this parasite. Drea and Fuester (1979) reported that in Poland it was parasitizing 1.3 to 3.7% of the pupae in 1975.

Howard and Fiske (1911) mentioned that its biology was similar to that of *C. pedalis* and *C. turionellae* but that *instigator* had a tendency to hibernate within the host pupa. They also mentioned its becoming hyperparasitic on occasions on *Iseropus coelebs* (Walsh) [= *Pimpla (Epiurus) inquisitoriella* D. T.]. It is a polyphagous, multivoltine species.

2. COCCYGOMIMUS PEDALIS (Cresson) (Figs. 11, 17, 23, 29, 35, 123)

Pimpla pedalis Cresson, 1865. Proc. Ent. Soc. Philadelphia, 4: 268. Colorado, U. S. A.
Pimpla pedalis: Fernald, 1892. Bull. Hatch Exp. Sta. Mass. State Coll., 19: 116. Massachusetts. Host: *Lymantria dispar*. Howard and Fiske, 1911. U. S. Dept. Agr. Bur. Ent. Bull., 91: 137-138, 237-239. Biol.
Coccygomimus pedalis: Townes, 1944-45. Mem. Amer. Ent. Soc., 11: 62-63. Synonymical references. Host records.

This is a polyphagous parasite, first recorded from the gypsy moth in the U. S. A. by Fernald (1892). Townes (1944-45) has given a full list of hosts. Carlson (1979) provides further taxonomic and biological references. It is one of the native North American parasites of the gypsy moth.

Female: Face dull-shiny, convex, with small, scattered punctures, punctures denser close to antennal sockets. Frons excavated, subpolished, without striations. Mesoscutum and mesopleurum shiny to subpolished, with minute scattered punctures. Scutellum subpolished. Metapleurum shiny and with weak oblique striations. Submetapleural projection narrow and much less pronounced than in *instigator*. Propodeum leathery, punctate laterally and trans-striate centrally. Sculpture much weaker than in *instigator*. Median propodeal carinae faintly to distinctly visible in basal half and diverging apically. Hind coxa impunctate. Nervulus slightly distad of basal vein. First tergite somewhat conical, with dorsal humps, which are blunt. Apical half of first tergite (except for its apical margin) with coalescing punctures. Second, third and fourth tergites punctate. The following tergites progressively less punctate to mat. Abdomen shiny between punctures. Epipleura narrow, those of 4th and 5th tergites narrower than in the preceding species.

Black. Body pubescence brown. Tegula black. Fore coxa black to brownish-black. Fore leg otherwise yellowish-brown. Middle leg wholly yellowish-brown. Hind coxa, trochanters, and femur yellowish-brown to orange-brown. Hind tibia and tarsus wholly black, without any bands. Hind femur fuscous apically.

Male: Similar to the female, but face more closely punctate. Frons with scattered punctures. First tergite less conical. All legs yellowish-brown with fore coxa yellowish apically and hind tibia and tarsus wholly black. Flagellum without tyloids.

Length: 12-15 mm. Fore wing 8-12 mm.

Specimens: U.S.A.: Whiteford, Harford Co., Md., 1♀, June 30, 1981, ex pupa of *Lymantria dispar,* R. Tateman (BIRL, Newark, Delaware). Winters State Park, Pa, 1♂, *ex L. dispar,* July 17, 1975 (Delaware). Schuylkill Co., Tremont Township, Pa. 1♂, 1♀, July 4, 1974 (Bureau of Forestry, Pennsylvania). North Saugus, Mass., 1♀, *ex Lymantria dispar,* Gypsy Moth Lab. (F.I.S., Hamden).

Distribution: Common in Transition and Canadian zones in North America.

Coccygomimus pedalis belongs to the Instigator Group and is readily distinguished from C. *instigator* by its smoother thorax, and impunctate hind coxa. It is related to C. *arcticus* in having brown body pubescence, red hind coxae and by the absence of tyloids on the male flagellum. The general body sculpture of the two is also similar, but *arcticus* has a more striate propodeum. In *pedalis,* the middle coxae are yellowish-brown and face smoother, while in *arcticus,* the middle coxae are black and face comparatively more punctate. The two may be considered subspecies (Townes, 1940).

Coccygomimus pedalis is a polyphagous multivoltine species, overwintering within the prepupa in the host pupa. It is one of the native prepupal or pupal parasites of the gypsy moth, but parasitism is not common. It is therefore not considered important. There are two to three generations per year. It takes about three weeks to complete its development within the host pupa. It attacks several other hosts (Townes, 1945; Carlson, 1979). Campbell (1963) observed it attacking gypsy moth pupae and prepupae in Glenville, New York State. He found that although parasitism was scarce, it killed more pupae and prepupae through stinging than those successfully parasitized. These stung pupae and prepupae were subsequently infested by sarcophagid larvae which acted as scavengers. He observed that *pedalis* was common in the field and exerted a great influence on the population of the gypsy moth by killing the prepupae and pupae rather than by successfully parasitizing them.

3. COCCYGOMIMUS TURIONELLAE (Linnaeus) (Figs. 7, 13, 19, 25, 31, 124)

Ichneumon turionellae Linnaeus, 1758. Systema Naturae, 10: 1: 564.
    Europe.
Pimpla turionellae: Gravenhorst, 1818. Nova Acta Physico Medica Akad.
    Caesareae Leopoldino-Carolinae Nat. Curio, 9: 291.
Cryptus examinator Fabricius, 1804. Systema Piezatorum, p. 85.
    Austria.
Pimpla examinator: Ratzeburg, 1844. Die Ichneumonen der Forstinsecten,
    1: 116. Germany. Hosts: *L. monacha* and others.
Coccygomimus turionellae: Townes and Townes, 1960. U. S. Natl. Mus.
    Bull., 216(2): 323. Synonymy and other references.

This species has often been referred to as *Pimpla examinator* in gypsy moth literature. Howard and Fiske (1911) mentioned having reared it in the laboratory from both the gypsy moth *(Lymantria dispar)* and the brown tail moth *(Euproctis chrysorrhoea).* It was received in the U. S. A. in considerable numbers in shipments of brown tail moth pupae from Europe and was released in the field during 1906-1909, but not recovered.

Carlson (1979) provides further biological references. Aubert (1969) and Oehlke (1967: 32) give all synonymical references.

Male and *female:* Male flagellum with tyloids on segments 6 and 7. Face largely rugulose to ruguloso-punctate, convex, more so in the male. Frons excavated, punctato-striate. Interocellar distance 1.5-1.8 the ocellocular distance. Ocellocular distance about half the ocellar diameter. Ocellar area smoother. Mesoscutum punctate. Punctures dense, small, but not coarse, areas between them subpolished. Scutellum flat to subconvex, punctures sparse. Scutellum punctate. Mesopleurum punctate, punctures coarser than on mesoscutum, well separated with interspaces shiny, on lower half with well separated but larger punctures. Metapleurum finely punctato-striate, the striations extending over flattened and expanded submetapleural ridge. Propodeum convex, rugoso-striate on basal half, without any median carinae, except rarely at base. Extreme base of propodeum in the middle, and apical half of propodeum polished to subpolished, the basal smooth area often depressed. Hind coxa punctate. Nervulus interstitial. First tergite short and wide, convex in profile with weak humps, not conically produced medially. Postpetiole largely shiny with scattered punctures. Abdominal tergites 2 to 4 closely punctate, with shiny interspaces. Apical tergites finely punctate to mat. Apical margins of tergites shiny. Epipleura of tergites 1-3 narrow, of tergites 4-5 wider, that of 4th the widest.

Black. Body pubescence white. Tegula and hind corner of pronotum yellow. All coxae black, trochanters blackish-brown, and femora reddish-brown. Fore and middle tibiae and tarsi yellowish-brown, with tibiae fuscous dorsally and with a pale band or streak. In male tibiae and tarsi paler. Hind tibia black and with a distinct yellow submedian band. Hind tarsus black. Hind femur fuscous apically. Some males from USSR have the hind tibia wholly reddish. Sometimes apices of abdominal tergites, particularly in males, brownish.

Length: 9-12 mm. Fore wing 7.5 to 10 mm.

Specimens: Germany, 1♀, bred from *Lymantria dispar* at Gypsy Moth Lab., Sept. 30, 1907 (Hamden). S. Rumania: 2♂, *ex Lymantria dispar,* EPL-78-71, June 26-28, 1978, R. C. Hedlund (BIRL, Newark). 2♂, 9♀, reared at Trenton, N.J. *ex Lymantria dispar,* 1973-1975, presumably from stocks from India (BIRL, Delaware), 2♂ (Bureau of Forestry, PA). Imported in the U.S.A. from India and Rumania (cf. computer printout of BIRL, Delaware). India: Kashmir: Srinagar, 1♂, 1♀, July, 1964, *ex Lymantria obfuscata* (CIBC, Bangalore). Several males and females from Europe (Townes).

Coccygomimus turionellae resembles superficially *C. instigator,* from which it can be readily distinguished by its white pubescence and black hind tibia having a yellowish-white annulus. It is a comparatively smaller sized species with finer body sculpture. It has wide fourth and fifth epipleura (Turionellae Group).

Distribution: Widely distributed in the Palaearctic and Oriental Regions. Imported into North America several times during 1906 to 1979 from Europe and India. Apparently not established. Recent rearing records from the gypsy moth are by Vasic (1958), Yugoslavia; Rao (1972), India; and Hedlund and Mihalache (1980), Rumania.

Hosts: Several hosts are known in Europe, USSR and Japan (Townes, Momoi and Townes, 1965), Šedivý (1963). It is a polyphagous parasite of Lepidoptera and Coleoptera.

Biological Notes: Several thousand specimens of this species were released against the gypsy moth in Pennsylvania during 1973-79, bred in the laboratory from specimens of Indian origin. According to Pennsylvania Bureau of Forestry Report (? 1979), it was not recovered during subsequent years. Carlson (1979: 346) provides a historical record of its introductions in Eastern North America and Canada during 1906-1955. The target hosts in North America were introduced pests, like *Lymantria dispar, Rhyacionia buoliana, Cydia (Cydia) pomonella* and *Operophtera brumata*. A number of native pests were target hosts also for some of the introductions. Establishment has apparently not occurred, although a few recoveries have been reported in Ontario, Canada.

Rao (1966) provides some information on the life history of this species under *"Pimpla* sp."* reared in Kashmir, India, from pupae of *Lymantria obfuscata*. Two generations were observed in Kashmir. The adults of the first generation appeared in early May and those of the second generation in late June. Mating occurred readily between freshly emerged males and older females, lasting 35 seconds to one minute. Pre-oviposition period is 3-4 days. Oviposition is in freshly formed pupae and takes two to two-and-a half minutes. Generally a single egg is laid at a time. The egg measures 2.0 x 0.35 mm., and is round at one end and tapering at the other. Soon after oviposition the female feeds upon the oozing fluid from the puncture. The egg hatches in about 2 days. Larval period is 9-11 days. The first larval instar is 2.5 mm., and the fully grown larva is 7-13 mm. long. Pupation occurs inside the host pupa. The pupal period is 7-8 days. The adult males and females lived in the laboratory for about 25 and 40 days, respectively. The ratio of males to females was 1:3. Only one parasite develops per host.

Metterhouse (in Doane and McManus, 1981) has also provided biological information on this species apparently based upon his observations during culturing it in the laboratory. His data differ only slightly from what has been given by Rao from field studies. The biology of *C. disparis, instigator* and *moraguesi* is similar to that of *turionellae*.

In the field collected pupae of the gypsy moth, the larvae and pupae of *Monodontomerus* sp. were seen with the remains of *"Pimpla* sp" larvae.

Rao (1972) reported three species of *Coccygomimus* from *Lymantria obfuscata* in Kashmir and Kotgarh, India. These three species, confirmed after examining the specimens, are *C. turionellae, C. disparis* and *C. laothoë*. He reported that the aggregate parasitism by the three species was generally low in all the four localities studied in Kashmir. The percentage parasitism was 0.45 to 12.58% in 1968-69 and 1.36% in 1971. However, in 1970, the parasitism was reported to be 17.35 to 67.41 percent.

4. COCCYGOMIMUS MORAGUESI (Schmiedeknecht) (Figs. 8,14,20,26,32)

Pimpla Moraguesi Schmiedeknecht, 1888. Zool. Jahrb., 3: 479. Morocco.
Coccygomimus turionellae moraguesi: Oehlke, 1967. Hymenopterorum
    Catalogus, 2(1): 32.

Coccygomimus moraguesi has usually been referred to as *C. turionellae moraguesi* in gypsy moth literature in the U. S. A. It is mentioned in the reports of the Pennsylvania Bureau of Forestry, among the 15 exotic parasites that were released against the gypsy moth in Pennsylvania since 1973. Carlson (1979) states that *Pimple freyi* Héllen from the Canary Islands is probably a synonym of it. Both *moraguesi* and *freyi* were treated by Aubert (1969) as synonyms of *turionellae*.

Simons *et al*. (1979) differentiated *C. turionellae moraguesi* from *C. turionellae turionellae* by stating that the former species has hind coxa and disc of scutellum reddish, while the latter species has both these structures black. Both taxa have a yellow or yellowish-white sub-basal band on hind tibia. I find that the body sculpture of *moraguesi* is coarser than that of *turionellae*, justifying separation as a species.

Male and *female:* Essentially similar and related to *C. turionellae* and differing as follows: Face distinctly punctate. Punctures coarse, tending to run into striations near antennal sockets. Face with a median smooth raised ridge in female and a convex median punctate area in male. Frons excavated, largely striate and punctate laterally. Ocelli larger than in *turionellae,* closer to the eye. Interocellar distance 2.0 to 2.3 the ocellocular distance. Ocellocular distance 0.3 the ocellar diameter. Thorax closely and coarsely punctate, with scutellum convex and smoother, sometimes with minute scattered punctures. Mesopleural punctures rather coarse and close together. Metapleurum coarsely punctato-striate, sometimes tending to be rugoso-striate in female. Propodeum rugoso-striate with short median carinae at base. Abdomen coarsely punctate. Postpetiole densely punctate.

Black. Tegula and hind corner of pronotum yellow. Tegula of female often black on apical half. Scutellum, metascutellum and all legs reddish. Fore and middle tibiae with light fuscous marks. Hind tibia with fuscous basal and apical marks and a sub-basal yellow band. Hind tibia fuscous. Often middle tibia also with a yellow subbasal band. Fore tibia and tarsus sometimes with yellow lines, or tibia largely yellow in male. Abdominal tergites with reddish-brown margins.

Length: 9-13 mm. Fore wing 6-9 mm.

Specimens: Algeria, 1♀, bred from *Lymantria dispar* at Gypsy Moth Laboratory, No. 13029 (Hamden, Ct.) [det. as *Ephialtes* sp. by Muesebeck]. Morocco: North Kemitra, Mamora Forest, 1♀, July 13, 1974 A. Fraval, ex live shipment of *Lymantria dispar.* 1♂ 5♀, "Lab culture, Trenton, N. J." (BIRL, Delaware). 2♂ "Parasite Lab, Middletown, Pa." (Pennsylvania) and 5♂, ♀ (Trenton, N. J.). A computer printout received from BIRL, Delaware, indicates importation of this parasite in the U. S. A. from Morocco during 1976-77.

Biological notes: Coccygomimus moraguesi is a polyphagous multivoltine parasite of pupae of various Lepidoptera (as are other *Coccygomimus*). It is known from North Africa: Algeria and Morocco. Seyrig (1927) recorded it from Spain from *Malacosoma neustria.* Its biology is similar to that of *C. turionellae.*

A specimen from Algeria bred from *Lymantria dispar* at the Gypsy Moth Laboratory in Massachusetts prior to 1930, has been seen. It had remained unnoticed due to lack of specific identification. Only recently this species was introduced in the U. S. A. from Morocco and bred in the laboratory for field releases. According to Pennsylvania Bureau of Forestry Report, 5,600 specimens of it were released against the gypsy moth in Pennsylvania during 1973-79, but it has not established itself there.

5. COCCYGOMIMUS DISPARIS (Viereck)   (Figs. 10,16,22,28,34)

Pimpla (Pimpla) disparis Viereck, 1911. Proc. U. S. Natl. Mus., 40: 480.
     Japan. Host: *Lymantria dispar.*
Pimpla (Pimpla) porthetriae Viereck, 1911. Proc. U. S. Natl. Mus.,
     40: 480. Japan. Host: *Lymantria dispar.*
Coccygomimus disparis: Townes, Momoi and Townes, 1965. Mem. Amer.
     Ent. Inst., 5: 48. Japan, Korea, China, Sakhalin.

Several taxonomical and biological references to this species are listed in Townes *et al.* (1965). Kasparyan (1974) records it from Mongolia, China, and USSR.

Male and *female:* Face closely punctate, punctures of moderate size and more regularly disposed. In female a smooth median raised area below antennal sockets. Male flagellum with tyloids on flagellar segments 6-7. Tyloid on 6th segment smaller. Frons excavated, faintly rugoso-striate. Mesoscutum with well separated punctures, interspaces shiny. Mesopleurum also with well separated punctures, but deeper than on mesoscutum, interspaces shiny. Scutellum a little elongate, laterally margined and punctate. Metapleurum rugoso-striate. Submetapleural ridge thinner, narrower and weakly striate. Propodeum rugose, with a few irregular striations centrally and smoother apically. Median carinae present basally or obliterated. When carinae present, the area in between them smooth. Hind coxa shallowly punctate. Nervulus interstitial. First tergite wide and flat apically, without dorsal humps, though medially convex in the female. Postpetiole densely punctate, smoother medially. Tergites 2-4 densely punctate, punctures coalescing. Tergite 5 onwards basally punctate and smoother apically, the punctures progressively becoming smaller to mat. Epipleura of tergites 4-5 wider than those of tergites 1-3, the fourth widest.

Black. Tegula yellow, often apically black marked. Postspiracular sclerite alone yellow near hind corner of pronotum. All coxae and trochanters black. All femora reddish-yellow. Hind femur black in apical 0.2 to 0.25. Fore and middle tibiae and tarsi yellowish-brown to orange, with faint fuscous marks on tibiae. Hind tibia and tarsus wholly black or blackish. Body pubescence white.

Length: 12-15 mm. Fore wing 10-12 mm.

Specimens: Japan, 1♂, 2♀ (paratypes, Nos. 13078), "Gipsy Moth Lab." (Washington). 1♀, Gypsy Moth Lab, No. 3309B. 1♀, Gifu, Japan, Y. Nawa Coll. (Washington). Japan, 1♂, *ex L. dispar*, Gipsy Moth Lab, No. 1650 (Hamden). Nishigahara, Japan, bred from *L. dispar* (labelled Paratype of *P. disparis* Vier.) (Hamden). Japan: Nara, Honshu, 1♀, *ex* Pupa, *Lymantria dispar*, June 23, 1977, P. Schaefer, emerged July 7, 1977 BIRL, (Newark). 1♂, 2♀, Lab reared in New Jersey Dept. of Agriculture from stocks from India (Pennsylvania) and 1♂ (Trenton). India: Srinagar, Kashmir, *ex Lymantria obfuscata* on willow, July 1963 (CIBC, Bangalore).

Distribution: Japan, Korea, China, Sakhalin, Mongolia, and India.

Hosts: Several host records, besides *Lymantria dispar,* are listed in references given in Townes *et al.* (1965) and Kasparyan (1974).

Coccygomimus disparis belongs to the Turionellae Group and is close to *C. turionellae* in the nature of pubescence, epipleura and general body sculpture. It differs from *turionellae* and *moraguesi* in having the hind tibia wholly black, without a white or yellow subbasal band, hind corner of pronotum not yellow (except for the postspiracular sclerite), tegula yellow, hind femur black in apical 0.3± 0.05, and abdominal punctures well formed but not coarse. It is comparatively larger in size. The tyloid on the 6th segment of the flagellum of the male is smaller. The scutellum also appears distinctive.

Coccygomimus indra (Cameron) from India, belonging to the Instigator Group, has similar coloration of the legs, but the abdomen has sparse punctures, particularly on the second tergite.

Coccygomimus nigricoxata Oehlke from Europe and USSR is also close to *disparis*, but differs in having coarser punctation on abdominal tergites with margins of punctures indistinct, especially on the apical half of fourth and whole of the fifth tergite. The interspaces between the punctures are mat, with indistinct fine sculpture. The hind femur is brownish or red with darker apex.

Biological notes: Although this parasite was received in the U.S.A. during the first decade of the century, it has never been given serious attention because of the scanty material that was received. During 1973-79, this species was bred at the New Jersey Department of Agriculture, Trenton, from stocks from India and also bred and released in Pennsylvania by the Pennsylvania Bureau of Forestry. In one of their reports, it is stated that 73, 215 specimens were released from stocks from Japan and India, but that it was never recovered except in the year of release. Metterhouse (in Doane and McManus, 1981), mentions that 10 specimens of this species were recovered in N. J. during 1978, but that the species is not yet known to be established in North America.

A recent letter from Dr. Fusco (Jan. 19, 1982) indicates that it was recovered in Bradford County, Pennsylvania, near Towanda on July 13, 1981. Two males of this species emerged from *Lymantria dispar* pupae collected there. No releases of *C. disparis* were made in 1979 or 1980 in that county. The closest release site was about 30 miles away in an adjacent county. The New York state border is rather close, where *C. disparis* was released in 1979 and possibly 1980, but not in 1981.

Rao (1966, 1972) gave some biological information on three *Coccygomimus* species, including the present species (see under *turionellae*).

6. COCCYGOMIMUS LUCTUOSUS (Smith)

Pimpla luctuosa Smith, 1874. Trans. Ent. Soc. London, 1874: 394. Japan.

Coccygomimus luctuosus: Townes, Momoi and Townes, 1965. Mem. Amer. Ent. Inst., 5: 53.

For full synonymical references, distributional records and host records (up to 1964) refer to Townes *et al.* (1965). Kasparyan (1974) provides a key to distinguish it from the Eurasian species.

This species is widespread in Japan, and Asian USSR and has also been reported from Korea, Taiwan, Sakhalin, etc. It has often been confused with *C. pluto* and *C. porthetriae* in the Japanese literature. Uchida (1930) and Matsumura (1931) reported it from *Lymantria dispar* (as *porthetriae*) in Japan. Yasumatsu and Watanabe (1964) include it in their list of parasites of the gypsy moth in Japan.

A female bred by Howard and Fiske (1911) in 1908 from pupae sent from Japan and determined as *pluto* by Viereck, actually belongs to this species, confirming the occurrence of this parasite on the gypsy moth.

Male and *female:* Face punctate. Punctures coarser and running into striations below antennal sockets. Face with a median smooth area. Face of male more regularly punctate and convex. Flagellar segments 6 to 11 in the male with tyloids. Mesoscutum punctate, punctures a little smaller and shallower than on mesopleurum. Scutellum flat to subconvex and shiny, with scattered punctures. Mesopleurum strongly punctate, but punctures not coarse or alveolar (as in *aethiops*), interspaces shiny and equal to the diameter of punctures. Punctures sometimes closer or contiguous at places. Metapleurum rugoso-striate anterodorsally and striate posteroventrally. Striations extending on the expanded and rather wide submetapleural ridge. Mesopleural suture below the mesopleural pit with well separated 10-13 transverse ridges. Propodeum rugose to rugoso-reticulate, its extreme base with two small median carinae, its apical region not smooth. First tergite sharply angled medially, with two small humps. Postpetiole and second and third tergites rather coarsely punctate. Central part of fifth tergite with well formed punctures. Apices of all tergites smooth and shiny. Hind coxa punctate. Fourth segment of fore tarsus deeply notched apically. Nervulus interstitial or a little distad of basal vein. Male with first tergite flatter and thoracic sculpture a little weaker. Smaller sized specimens with weaker sculpture.

Black. Body pubescence whitish or light brown. Wing bases and their sclerites yellow. Tegula black or brownish-black. Legs black. Fore tibia yellowish anteriorly. Sometimes fore and middle femora yellow apically and their tibiae and tarsi brownish to brownish black. Hind tibia sometimes with a faint suggestion of sub-basal light colored spot or band. In male scutellum with a yellow spot. Tegula yellow with a black apical spot. Fore and middle femora, tibiae and tarsi often partly to largely reddish-brown to yellowish-brown.

Length: 15-20 mm. Fore wing 11-14 mm. Sometimes the male smaller.

Specimens: Japan: 1♀, July 1908, Gypsy Moth Lab., No. 1650, *ex Lymantria dispar*, det. by Viereck as *"Pimpla pluto* Ashm."* (Hamden). Japan: Sapporo, 1♀ (homotype of *Pimpla luctuosa* by Townes), T. Uchida (Townes). Japan: Yokohama, 1♂ (homotype of *Pimpla aethiops neustriae* Uchida by Townes), May 20, 1933 (Townes). In addition, a few males and females from Fukien, China and Japan seen in the Townes collection.

Distribution: Japan, Korea, China, Sakhalin, and Eastern USSR.

This species is related to *Coccygomimus pluto* (Ashmead) and *C. aethiops* (Curtis). *C. aethiops* has densely punctate abdominal tergites and mesopleurum, the punctures on mesopleurum touching each other and becoming alveolar. The body is dull. The first tergite is without dorsal humps. The fourth tarsal segment of fore leg is bilobed, but not deeply notched. *C. pluto* differs from *luctuosus* in having the wing bases and tegula black. The scutellum of male is black. Only 6-7th flagellar segments in the male have tyloids. The body sculpture is a little sparser, with punctures on face less dense, on mesopleurum well separated, shallower, and on tergite 5 not well formed and shallow. It is also smaller in size, about 12-14 mm. long and with brownish-black pubescence, though in male the pubescence is yellowish.

7. COCCYGOMIMUS SPURIUS (Gravenhorst) (Fig. 125)

Pimpla spuria Gravenhorst, 1829. Ichneumonologia Europaea, 3: 179. Europe.
Pimpla spuria: Yafaeva, 1959. Ukrainskaya Akad. Selsk. Nauk, p. 227. Ukranian SSR. Host: *Lymantria dispar*.
Coccygomimus spurius: Townes and Townes, 1960. Bull. U. S. Natl. Mus., 216(2): 338.

Coccygomimus spurius has only recently been recorded as a parasite of the gypsy moth by Yafaeva (1959) from Ukraine, Russia. Kasparyan (1974), however, does not list that reference, nor that host. He records this species as a usual parasite of *Cydia (Cydia) pomonella*, distributed in Iran, Central and Southern Europe, Soviet Central Asia, and Caucasus, etc. Aubert (1969) lists several common hosts belonging to various families of Lepidoptera and also Coleoptera: Tenebrionidae. No reared specimen has been seen. The following description is based upon specimens seen in the Townes Collection.

Male and *female:* Male flagellum without tyloids. Face a little convex, shiny, with scattered but distinctly formed punctures. Frons shallowly excavated, minutely punctate, shiny and trans-striate centrally. Ocellar diameter about equal to ocellocular distance. Interocellar distance 1.2 the ocellocular distance. Mesoscutum shiny, with minute scattered and shallow punctures. Mesopleurum shiny, with minute scattered punctures. Metapleurum shallowly punctate anteriorly and striato-rugose posteriorly. Propodeum punctato-rugose, its median carinae distinct in basal 0.3, its basal area smooth. Coxae not punctate, mat. Fourth tarsal segment of fore leg not deeply notched. Nervulus interstitial or a little distad of basal vein. First tergite flat in apical half, without humps. Abdominal tergites punctate, punctures well formed and interspaces shiny. Postpetiole punctate. Apical two tergites smoother. Apices of all tergites smooth and shiny. All epipleura wide.

Black. Body pubescence white or a little yellowish. Tegula black. Middle and hind coxae reddish or orange-brown. Fore coxa and trochanters blackish-brown. Fore and middle legs otherwise and hind trochanters and femur orange-brown. Hind femur without any fuscous mark. Hind tibia brownish, without a pale band or with an indistinct pale band, the margins of which are indistinct. Hind tibial color variable, sometimes entirely red or reddish-brown. Hind tarsus blackish.

In male the coloration is somewhat different. All coxae are black, first trochanters blackish, hind tibia black with a pale sub-basal band, and hind tarsus black. Tegula brownish-yellow. Apical margins of abdominal tergites subpolished.

Length: Male, 5-8 mm, female, 8-10 mm. Fore wing 4 to 7 mm.

Specimens: Several males and females from Ireland, Scotland, Germany, Russia and Rumania in the Townes Collection. 1♀, Debreczen, Hungary, 1923, R. T. Webber, Gypsy Moth Lab., but without any host label (Hamden).

This species is very close to *C. contemplator* (Mueller) in sculpture and general coloration. Both belong to the Aequalis species Group of Townes and Townes. The latter species, however, can be differentiated by having the fourth tarsal segment of fore leg apically deeply notched, first two flagellar segments shorter, less than or just equal to the longitudinal diameter of eye, and ovipositor sheath equal to or a little shorter than the hind tibia. The male of *C. spurius* has the basal half of 6th tergite coarsely punctate, while in the male of *contemplator* the 6th tergite is almost impunctate, mat.

Coccygomimus spurius also is close to *C. melanacrias* (Perkins), the latter having a distinct white annulus on hind tibia, and apex of hind tibia fuscous. *C. nipponicus* (Uchida) and *C. confinis* (Kasparyan) from Eastern Palaearctic are also related to *spurius* and *contemplator*, but they have black coxae.

2. Genus ITOPLECTIS (Fig. 2, 126, 127)

Itoptectis Foerster, 1869. Verh. Naturh. Ver. Rheinlande, 25: 164.

For synonymy and relationships with other Ephialtini, refer to Townes (1969).

Small to medium sized ichneumonids with fore wing 2.5 to 12.5 mm. long. Clypeus without a transverse suture. Mandible not narrowed apically, teeth equal to subequal. Occipital and prepectal carinae present. Inner margin of eye rather strongly concave at antennal socket. Mesopleural suture without a distinct angulation in the middle. Front tarsal claws of female with a basal tooth, the middle and hind tarsal claws of female and all tarsal claws of male simple. Nervellus intercepted far above the middle. Ovipositor straight.

Face and orbits of both sexes entirely black (cf. *Ephialtes*). Hind tibia banded (except in *viduata*) with apical and basal dark bands and a median pale band.

Members of the genus *Itoplectis* are widely distributed. They usually
are internal parasites of lepidopterous pupae, but some (e. g. *I. conquisitor*
and others) are often secondary parasites of Ichneumonidae, while others may
be normally secondary parasites.

Unlike *Coccygomimus* and *Ephialtes*, most species of *Itoplectis* do not give
off a strong odor when captured or disturbed.

Five taxa of *Itoplectis* have been reported as parasitic on the gypsy moth,
viz., *I. alternans alternans* (Gravenhorst), *alternans spectabilis* (Matsumura),
conquisitor (Say), *maculator maculator* (Fabricius), and *viduata* (Gravenhorst).
Another species, *I. clavicornis* is here recorded from the gypsy moth. Some
of them, including *I. clavicornis*, are also hyperparasites of Ichneumonidae
associated with the gypsy moth.

Itoplectis alternans spectabilis (Matsumura) (= *Exeristesoides spectabilis*)
was first reported from *Lymantria dispar* in Japan by Fukaya (1936). It is also
a hyperparasite of *Hyposoter takagii*, and some other ichneumonids and
braconids (Townes *et al.* 1965). *I. alternans alternans* (Gravenhorst) is a
European parasite of *Lymantria monacha*, which Oehlke (1967) reported from
L. dispar rather than from *L. monacha*.

Itoplectis conquisitor (Say) (= *Pimpla conquisitor*) was observed by Howard
and Fiske (1911) to frequently attack gypsy moth pupae in the U.S.A., but the
development of the parasite was rather infrequent. *"Pimpla tenuicornis*
Cresson"*, recorded as a parasite of gypsy moth pupae, by Forbush and
Fernald, but never reared at the laboratory, was believed by Howard and Fiske
to be *Itoplectis conquisitor*.

Itoplectis maculator maculator (Fabricius) was reported as a parasite of
Lymantria dispar by Oehlke (1967) from Europe. However, I could not trace
it back to its original record and did not see it mentioned in other publications.

Itoplectis viduata (Gravenhorst) (= *Pimpla viduata*) was reported as a parasite of *Lymantria dispar* by Meyer (1929).

Useful taxonomic and biological information on *Itoplectis* can be found in
Townes and Townes (1960), Šedivý (1963), Townes, Momoi and Townes (1965),
Oehlke (1967), Aubert (1969), Kasparyan (1973) and Carlson (1979).

KEY TO THE SPECIES OF ITOPLECTIS ASSOCIATED WITH
THE GYPSY MOTH

1. Hind tibia uniformly reddish-brown (sometimes with a faint suggestion of a
    yellow subbasal band, not sharply demarcated). Coxae and basal tro-
    chanteral segments black. Tarsal segments generally pale brownish, not
    clearly black marked. Europe, North America, USSR, and China.
                                         5. viduata (Gravenhorst)
    Hind tibia with a white or yellow subbasal band. Coxae reddish brown to
    black. 2

2. Antennal flagellum thickened apically, the subapical segments transverse.
    Ovipositor sheath short, about as long as the first tergite. Fore and
    middle legs largely yellow in male and yellowish-brown in female.
    Eurasia. . . . . 4. clavicornis (Thomson)
    Antennal flagellum not thickened apically, sometimes only slightly wider,
    but segments elongate rather than transverse. Ovipositor sheath con-
    siderably longer than tergite 1. 3

3. Mesopleurum with dense punctures. In male punctures smaller. All
   coxae and basal trochanteral segments black in male and female.
   Abdominal tergites usually reddish-brown laterally, sometimes apically
   also. Eurasia. . . . . 3. maculator (Fabricius)
   Mesopleurum finely punctate. Coxae and trochanters reddish-brown except
   in male of alternans where hind coxa black. Abdominal tergites narrowly
   yellow apically, or not so. Sides of tergites not reddish or yellow. . 4

4. Ovipositor sheath long, about 2.0-2.2 times the length of first tergite and
   1.5 times the length of hind femur. Hind femur with a fuscous black
   apical mark. Punctures on abdominal tergites uniform, regular but on
   tergites 4-5 of male rather sparse. Pale bands on tergites rather conspicuous. North America. . . . . 1. conquisitor (Say)
   Ovipositor sheaths shorter, about 1.5-1.7 as long as the first tergite
   and 1.1-1.2 as long as the hind femur. Hind femur without a fuscous
   apical mark. Pale bands on tergites faint to inconspicuous. Tergites
   coarsely punctate. . . . . 2. alternans. . 5

5. Trochanters reddish-brown in female and yellow in male. Fore and
   middle legs reddish-brown in female and yellow in male. Basal black
   mark on hind tibia wider so that the yellow band is narrow. Apical half
   of hind tibia often reddish-brown rather than black. Eurasia.
                            2a. alternans alternans (Gravenhorst)
   Trochanters reddish with yellow markings. Fore and middle legs largely
   yellow or yellowish-brown. Basal black mark on hind tibia very small,
   so that the yellow band is wide. Apical half of hind tibia black. Japan
   and Far east. . . . . 2b. alternans spectabilis (Matsumura)

1. ITOPLECTIS CONQUISITOR (Say) (Figs. 38, 40, 42, 127)

Cryptus conquisitor Say, 1936. Boston J. Nat. Hist., 1: 232.
Pimpla conquisitor: Howard and Fiske, 1911. U. S. Dept. Agr. Bur.
      Ent. Bull., 91: 138, 237.
Itoplectis conquisitor: Townes and Townes, 1960, U. S. Natl. Mus. Bull.,
      216(2): 287.
Itoplectis conquisitor: Campbell, 1963. Canad. Ent., 95: 337.
Itoplectis conquisitor: Carlson, 1979. *In* Krombein, *et al.* Catalog of
      Hymenoptera in America north of Mexico, 1: 340.

Useful taxonomical and biological references are given in Townes and
Townes (1960) and Carlson (1979). Information pertaining to the association of
this parasite with the gypsy moth is given by Howard and Fiske (1911) and
Campbell (1963).
    Male and *female:* Flagellum weakly widened apically. Temple receding
from eye. Face moderately wide, punctate. Mesoscutum and mesopleurum
with scattered small punctures. Notauli absent. Propodeum short, its medium
carinae extending up to 0.3 its length. Fore tarsal claw of female with a large,
broad tooth. Third tergite about as long as wide in male and 0.5 to 0.6 as long
as wide in female. Fourth and fifth tergites of male polished and with widely
separated punctures. In females punctures dense. Tergites 2-7 with weak
depressions and elevations. Ovipositor sheath about 2.2 as long as the first
tergite.

Black. Palpi of male white. Maxillary palpus of female white. Labial palpus of female light brown. Scape and pedicel of male antenna white in front. Flagellum reddish-brown beneath, more extensively so in male. Tegula and hind corner of pronotum white. Legs of female largely reddish-brown. Fore coxa blackish basally. Fore tibia and tarsus with faint yellow patches. Middle tibia with a pale subbasal band. Middle tarsal segment yellowish basally and fuscous apically. Apex of hind femur, base and apical 0.4 of hind tibia and apical half of hind tarsal segments black. Fourth tarsal segment wholly black. Hind tibia with a broad yellowish-white subbasal band. Hind tarsal segments whitish in basal half. In male, fore and middle legs largely yellow with tibiae yellowish-brown. Hind leg color same as in the female. Apices of abdominal tergites yellow.

Length: 9-14 mm. Fore wing 6.0-2.5 mm. Some specimens are rather short, 5-6 mm. long.

Specimens: Several males and females from different states of U.S.A., in Townes Collections. No reared specimens from the gypsy moth seen.

This species is readily distinguished by the color pattern of its legs and abdomen. The hind leg and abdomen gives a banded appearance. The sculpture of the fourth and fifth tergites of male is characteristic.

Itoplectis conquisitor is a polyphagous parasite and is hyperparasitic upon occasions. It is widely distributed in the Nearctic Region. Howard and Fiske (1911) observed that this parasite frequently attacked the gypsy moth pupae, but that the young larvae of it rarely completed their development upon it. Campbell (1963) substantiated those findings. He observed that *Itoplectis conquisitor* stung a number of prepupae and pupae of gypsy moth in the field, but hardly ever developed upon them. The species preferred to hunt in the open and exhibited a striking positive response to defoliated areas. In the Glenville area, it was the dominant species attacking gypsy moth pupae and prepupae. Adults occurred from early spring to late fall. A wide variety of exposed or weakly protected lepidopterous pupae or prepupae served as hosts. Development from egg to adult took about 20 days.

Itoplectis conquisitor apparently attacks the gypsy moth pupae and prepupae for getting food, rather than oviposition. However, in this process they puncture the pupal skin and kill the host. Sarcophagids oviposit in such pupae. They have a scavenger relationship with the host.

This species   therefore kills pupae of the gypsy moth in large numbers. Three other native Ichneumonids, *Coccygomimus pedalis*, *Theronia atalantae* and *Theronia hilaris* have also been observed to kill more pupae than they parasitize, thus exerting a greater influence on the populations of gypsy moth than is usually credited to the Ichneumonidae.

Further information on its biology and biological references are given by Townes (1940) and Townes and Townes (1960).

2a. ITOPLECTIS ALTERNANS ALTERNANS (Gravenhorst) (Figs. 12, 30, 36,
37, 39, 126)

Pimpla alternans Gravenhorst, 1829. Ichneumonologia Europaea, 3: 201.
Europe.

Itoplectis alternans: Townes, Momoi and Townes, 1965. Mem. Amer.
Ent. Inst., 5: 38. Hosts and distributional records in Eastern
Palaearctic Region.

Itoplectis alternans: Oehlke, 1967. Hymenopterorum Catalogus (nova
editio), 2(1): 26. Palaearctic. Various hosts including *Lymantria
dispar*

Oehlke (1967) provides some biological references. Kasparyan (1973) provides a key to the Eurasian species of *Itoplectis*.

This subspecies is extremely close to *Itoplectis conquisitor* from North America, differing mainly as follows:

Mesopleural punctures sparser and minute. Propodeal carinae short and widely diverging. Propodeum smoother. Abdominal punctures coarse and contiguous in both the sexes. Ovipositor short, its sheaths 1.5 to 1.7 as long as the first tergite and about 1.1 to 1.2 as long as the hind femur.

Hind corner of pronotum very narrowly yellow, mainly in the region of postspiracular sclerite. Palpi yellowish-brown. Fore coxa usually reddish-brown. Hind femur without an apical fuscous mark. Apex of hind tibia reddish-brown. Abdominal tergites with faint to rather narrow brownish margins.

Length: 5-7 mm. Fore wing 4-6 mm.

Specimens: Italy: Portici, 1♂, 4♀, 1912, *ex Hyposoter disparis* Vier. [= *Phobocampe unicincta*], Gypsy Moth Lab., No. 7429F and 7429 (Hamden). Hungary: Olasziszka, 1♂, June 25, 1927, *ex Lymantria dispar*, labelled *Ephialtes (Itoplectis)* sp. (Hamden). Several unreared ♂, ♀ from Europe (Townes Coll.).

The above specimens reared at the Gypsy Moth Laboratory, indicate that it is a primary parasite of the gypsy moth, as well as a hyperparasite of a gypsy moth parasite, *Phobocampe unicincta*. In literature (Townes *et al.*, 1965, Aubert, 1969, etc.) it is reported as a parasite of several other Lepidoptera including *L. monacha*, and as a hyperparasite of *Rhogas*, *Casinaria* and *Hyposoter* species.

2b.   ITOPLECTIS ALTERNANS SPECTABILIS (Matsumura)

Pimpla (Pimpla) spectabilis Matsumura, 1926. J. Coll. Agr. Hokkaido
        Imp. Univ., 18: 30. Japan. Host: *Dendrolimus spectabilis*.
Exeristesoides spectabilis: Fukaya, 1936. Oyo Dobutsugaku Zasshi,
        8: 232, 335. Japan. Host: *Lymantria dispar*.
Itoplectis alternans spectabilis: Townes, Momoi and Townes, 1965. Mem.
        Amer. Ent. Inst., 5: 38. China, Japan, Korea. Several hosts in
        Japan.

This subspecies is close to *Itoplectis conquisitor* and differs from the latter as given in the key and under *I. alternans alternans*. It differs from the subspecies *alternans* in having a rather broad yellowish-white band on hind tibia, with the basal black mark small to almost obliterated and apical black mark prominent. The fore and middle legs are lighter in color. In males they are largely yellow.

Length: 5-7 mm. Fore wing 4-5 mm.

Specimens examined: Japan: 2♂, 4♀, reared from *Grapholitha molesta* (Townes Coll.). No specimens reared from gypsy moth seen. It has been reported as a hyperparasite of *Hyposoter takagii*, *Eriborus molestae*, and *Apanteles* sp. in Japan.

3. ITOPLECTIS MACULATOR MACULATOR (Fabricius) (Fig. 2)

Ichneumon maculator Fabricius, 1775. Systema Entomologiae, p. 337.
        Germany.
Itoplectis maculator: Townes, Momoi and Townes, 1965. Mem. Amer.

Ent. Inst., 5: 40. Eurasia, North America, North Africa.
Synonymical references and host records.
Itoplectis maculator maculator: Oehlke, 1967. Hymenopterorum
     Catalogus (nova editio), 2 (1): 28. Host: *Lymantria dispar,* besides
     several other hosts.

This species conforms in general to the description of *I. conquisitor* and is
differentiated as follows:
     Thorax rather closely punctate. Mesoscutum with small close punctures.
Mesopleurum with dense well formed punctures, which are of moderate size
and depth. In males punctures smaller. Propodeum densely punctate laterally,
smoother medially, its median carinae extending in basal 0.25 and weakly
diverging apically. Punctures on abdominal tergites dense and coalescing.
     Coxae and first trochanteral segments of all legs black. Fore and middle
legs yellowish-brown, with yellow patches. Hind leg brownish-yellow or
orange colored, with tibia having a yellow sub-basal band and tarsal segments
basally yellow. Abdominal tergites laterally brownish. Their apices also
brownish. Extent of lateral brownish marks on tergites variable, sometimes
tergites more extensively brownish-yellow, particularly in males.
     Length: 5-10 mm. Fore wing 4 to 7.5 mm.
     Specimens: Spain: Madrid, 1♂, June 1925, *Lymantria dispar,* Gypsy
Moth Lab. Italy: Portici, 1♀, 1912, *ex Hyposoter disparis* Vier. [= *Phobocampe unicincta*], Gypsy Moth Lab., No. 7429e. 1♀, "bred from *P. dispar*",
Gypsy Moth Lab., No. 1094, July 15, labelled "*Scambus maculatus* F. (All
Hamden).
     Distribution: Eurasia, N. Africa, and N. America. Walkley (1958)
recorded its establishment in North America in Oregon. See Townes and
Townes (1960) and Carlson (1979) for further information about its establishment in the U.S.A., and for biological references.
     Šedivý (1963) provided a list of hosts in Europe and Meyer (1934) in USSR.
     The above mentioned specimens indicate its occurrence at times as para-
sitic on *Lymantria dispar,* and that it is hyperparasitic on occasion. Vasic
(1958) reported that it parasitized 25% of *Casinaria tenuiventris* cocoons in
Yugoslavia.

4. ITOPLECTIS CLAVICORNIS (Thomson)

Pimpla clavicornis Thomson, 1889. Opusc. Ent., 13: 1409. Europe.
Itoplectis clavicornis: Oehlke, 1967. Hymenopterorum Catalogus (nova
     editio), 2(1): 27. Europe. England.

According to Townes and Townes (1960) *I. clavicornis* is usually a
secondary parasite on Ichneumonidae. It is a rather distinct species with
flagellum thickened apically, the subapical segments being wider than long.
Ovipositor short. Ovipositor sheaths about as long as the first tergite. Mesopleurum smoother, polished. Median propodeal carinae short. Abdomen
slender, punctate.
     Legs in general yellowish-brown with hind femur orange colored and with-
out any apical fuscous mark. Hind tibia with a broad yellow sub-basal band.
Fore and middle legs in male more yellow. Hind coxa of male black and that
of female brown to partly blackish. Abdominal tergites without conspicuous
pale bands.
     Length: 8-9 mm. Fore wing 5-6 mm.

Specimens: Italy: Portici, 1912, 1♂, 1♀, *ex Hyposoter disparis* Vier. [=Phobocampe unicincta], Gypsy Moth Lab., No. 5427 and 7435 DA. Hungary: Vees, 3♂, May 13, 1929, *ex Hyposoter disparis*, Gypsy Moth Lab., No. 13039 D (Hamden).

Distribution: Europe.

The above specimens attest to the fact that *I. clavicornis* is a hyperparasite of *Phobocampe unicincta* and not useful in the control of the gypsy moth.

5. ITOPLECTIS VIDUATA (Gravenhorst)

Pimpla viduata Gravenhorst, 1829. Ichneumonologia Europaea, 3: 214. Europe.

Pimpla viduata: Meyer, 1929. Izv. Otd. Prikl. Ent. GIOA (Repts. Bur. Appl. Ent.), 4: 235, 240. USSR. Hosts: *Lymantria dispar* and *Cosmia subtilis.*

Itoplectis viduata: Townes and Townes, 1960. U. S. Natl. Mus. Bull., 216(2): 293. China: Manchuria, Europe, N. America. Various hosts in North America.

This species is readily distinguished from all other species of *Itoplectis* by its hind tibia being uniformly reddish-brown and all coxae and first trochanteral segments black. Middle and hind coxae are not mentioned to be black in European specimens by Kasparyan (1973).

It is somewhat robust species with deeper punctation on mesoscutum and mesopleurum, propodeal spiracles elongate-oval (usually short oval), propodeum punctate laterally, its median carinae extending up to the middle, abdominal tergites strongly punctate, and ovipositor sheath about 2.5 as long as the first tergite.

All legs uniformly orange to reddish-brown, with their coxae and trochanters black (Nearctic specimens seen). Abdomen without yellow bands.

Length: 9-12 mm. Fore wing 7.5 to 11.5 mm.

Specimens: 3♂, 3♀, from California, Colorado and Arizona in Townes Collection. No reared specimens seen.

Distribution: Widely distributed in Europe, USSR and western North America.

Hosts: Recorded as a parasite of *Lymantria dispar* by Meyer (1929). Townes and Townes (1960) list several hosts in North America. Aubert (1969) lists European hosts.

3. Genus EPHIALTES (Figs. 3, 128, 129)

Ephialtes Schrank, 1802. Fauna Boica, 2: 316.

This genus has often been called *Apechthis*. Refer to Townes (1969) for synonymy and its relationships with other members of the tribe Ephialtini, subfamily Pimplinae.

Ephialtes can easily be distinguished from other pimpline ichneumonids by its sharply decurved ovipositor tip. It is further characterized by having the face of the male largely to entirely white or yellow. Orbit of female narrowly whitish in front. Inner margin of eye rather strongly concave opposite antennal sockets. Malar space 0.15 to 0.2 the basal width of mandible. Clypeus normally formed, basally a little concave, without any transverse suture. Mandible broad, teeth subequal. Notauli weak or absent. Propodeum with only median carinae extending in basal 0.5 to 0.7. Fore, middle and sometimes the hind tarsal claws of female with a large tooth. Tarsal claws without an enlarged flat-tipped hair. Nervellus intercepted well above the middle. Abdomen with coarse punctures. Epipleurum of fifth tergite narrow, at least 3.0 as long as wide in female and 4.0 in male.

Members of the genus *Ephialtes* are medium-sized to moderately large ichneumon-flies. They are internal parasites of lepidopterous prepupae or pupae.

Three species have been reported in literature as being parasitic upon *Lymantria dispar*. These are *Ephialtes compunctor* (L.) (= *brassicariae*), *E. rufatus* (Gmelin) (= *rufata* Gravenhorst), and *E. capulifera* (Kriechbaumer). They have not been commonly encountered, however.

Howard and Fiske (1911) reported rearing *Ephialtes compunctor* (L.) (as *Pimpla brassicariae* Poda) at the Gypsy Moth Laboratory (Massachusetts) infrequently from the European collections of the gypsy moth and the brown tail moth. Rudow (1911) mentioned *Pimpla varicornis* Gr. = *compunctor* (L.) as a parasite of gypsy moth and the nun moth. Stadler (1933) mentioned it (= *P. brassicariae*) as a parasite of *Lymantria dispar* in Europe.

Rudow (1911) reported *Ephialtes rufatus* (Gmelin) (= *Pimpla rufata* Grav.) as a parasite of *Lymantria dispar* and *L. monacha* in Europe. Meyer (1927) reported it from the gypsy moth in Russia. However, he did not mention this host in his subsequent publications on USSR Ichneumonidae (Meyer, 1934, 1936). Thompson's source was Meyer (1927) (RAE, A, 16: 200).

Ephialtes capulifera (Kriechbaumer) was first reported as a parasite of *Lymantria dispar* by Uchida (1958) (as *Apechthis capulifera* var. *nigriabdominalis*) in Japan. Kamijo (1962) gave some biological notes.

All these three species have been frequently mentioned as parasites of *Lymantria monacha* as well as several other hosts in Europe and Japan. Aubert (1969) and Townes *et al.* (1965) may be consulted for listings of hosts, taxonomic references and distributional records.

"*Pimpla dentator* Thomson" mentioned in literature as a parasite of *Lymantria monacha* is a lapsus for *Pimpla quadridentata* Thomson, which is a species of *Ephialtes*.

KEY TO THE SPECIES OF <u>EPHIALTES</u> PARASITIC UPON THE GYPSY MOTH

1. Hind coxa and apex of hind femur black. Usually all coxae black and hind femur black. Tibia black, with a wide pale submedian band. Body pubescence white. Eurasia and Far East.
                    1. <u>capulifera</u> (Kriechbaumer)
   Hind coxa and hind femur usually wholly reddish. 2

2. Body pubescence brownish. Mesoscutum without yellow markings, except sometimes in males. Scutellum narrowly yellow apically. Metascutellum black. First tergite with two conical projections in the middle. Hind tibia without any pale band, rust-red colored (except faintly so in a subspecies, orientalis, from Eastern Siberia). . . . 2. <u>compunctor</u> (L.)
   Body pubescence whitish. Mesoscutum with two stripes beyond its middle and usually also along its basolateral corners. Scutellum largely yellow. Metascutellum yellow. First tergite without conical projections, only convexly angled medially. Hind tibia with a faint to somewhat distinct

white band.   3. rufatus (Gmelin)

1. EPHIALTES CAPULIFERA (Kriechbaumer)

Pimpla destructor Smith, 1874. Trans. Ent. Soc. London, 1874: 394.
    Japan. Name preoccupied by Smith, 1863.
Pimpla capulifera Kriechbaumer, 1887. Ent. Nachr., 13: 119. Germany.
Apechthis capulifera var. *nigriorbitalis* Uchida, 1958. Shin Konchu,
    8(5): 8. Japan. Hosts: *Lymantria dispar* and others.
Apechthis capulifera: Kamijo, 1962. Kôshunai Rinboku Ikushujo Hokoku,
    1: 87. biol. Japan. Host: *Lymantria dispar*.

Male and *female:* Face a little longer than wide, with scattered punctures, wrinkled in a median area below antennal sockets. Clypeus basally a little convex and with scattered punctures, its apical 0.75 flat and somewhat mat. Malar space 0.2 the basal width of mandible. Frons and vertex smooth. Interocellar distance 1.7 to 1.8 the ocellocular distance. Vertex narrow, sharply receding from behind ocelli. Mesoscutum dull mat, leathery in texture. Scutellum subpolished with scattered minute punctures. Pronotum polished, with scattered minute punctures along its upper margin. Mesopleurum polished and with scattered punctures, speculum glabrous. Metapleurum polished with minute punctures in its upper 0.3. In larger specimens the punctures a little spread out on side of thorax. Propodeal spiracle oval, small. Propodeum finely rugoso-punctate except the petiolar area which is smooth. Median propodeal carinae a little diverging apically and ending in the middle of propodeum. Area between the carinae smooth. Areolet more trapezoidal with first intercubitus shorter than the second and second recurrent in the apical 0.3. Abdomen closely punctate. First tergite with coarser shallow punctures, tending to be shallowly rugoso-punctate. Its median dorsal carinae forming moderately convex humps, broadly angulate in the female and evenly rounded in the male. Ovipositor sharply decurved, though not quite angled or L-shaped, more sigmoid.
    Black. Flagellum black or yellowish-brown ventrally and darker dorsally. Face narrowly yellow along inner orbits and sometimes on vertical orbits. Palpi and tegula brown. Scutellum and metascutellum apically yellow. Coxae wholly to largely black. Legs otherwise yellowish-brown with yellow patches on fore femur, tibia and tarsus, middle tibia and tarsus, and apex of middle femur. Hind tibia with a broad yellow sub-basal band. Apex of hind femur, tibia except for the yellow band, and hind tarsus black to blackish-brown. Stigma and veins brownish-black. Body pubescence white.
    In many specimens from Japan (var. *sapporoensis*) the legs are more extensively black, with hind femur wholly black and black marks on trochanters and middle femur. The tegula is also blackish. In some other specimens the coxae are partly black and partly yellow, and yellow on hind tibia more extensive. The face is more extensively yellow on sides. The scutellum and metascutellum are also yellow. A few specimens, usually the males have the legs extensively yellow. A female from Hungary from gypsy moth has all femora orange colored and all coxae black.
    Length: 12-18 mm. Fore wing 10-13 mm.
    Specimens: Hungary: Baja, 1♀, *ex Lymantria dispar*, "Gipsy Moth Lab., No. 3483C". (labelled as *Apechthis* sp 1) (Hamden). Japan: Hokkaido, Higashimokoto, 1♀, July 24, 1978, *ex* pupa of *Lymantria dispar*, P. Schaefer

(BIRL, Delaware). Hokkaido, Jyozonkei, 1♀, July 21, 1975, emerged from pupa of *Lymantria dispar*, August 4, 1975, P. Schaefer (BIRL, Delaware). Several ♂♀ from Japan (Townes Coll.).

Distribution: Eurasia, Japan, Korea, Taiwan and China. According to Kasparyan (1973) it is a transpalaearctic forest species, which is of rather rare occurrence in the European part of USSR.

Hosts: Lymantria dispar, *L. monacha* and several other hosts belonging to the following families (cf. Aubert, 1969): Tortricidae, Geometridae, Lasiocampidae, Arctiidae, Lymantriidae, Noctuidae, Hesperidae, Papilionidae, Pierididae, Nymphalidae, and Cerambycidae.

2. EPHIALTES COMPUNCTOR (Linnaeus) (Figs. 3, 128)

> *Ichneumon compunctor* Linnaeus, 1758. Systema Naturae,(Ed. 10) 1: 564.
> Europe.
> *Ichneumon brassicariae* Poda, 1761. Insecta Musei Graecensis, p. 105.
> Yugoslavia. Host: Pieri brassicariae.
> *Pimpla (Apechthis) brassicariae*: Howard and Fiske, 1911. U. S. Dept.
> Agri. Bur. Ent. Bull., 91: 85, 238. Host: *Lymantria dispar*.
> *Ephialtes compunctor*: Townes, Momoi and Townes, 1965. Mem. Amer.
> Ent. Inst., 5: 44. References.

Male and *female:* Essentially similar in sculpture to *E. capulifera* and differing as follows:

Metapleurum largely smooth and polished. First tergite convex and dorsally humped, with its median carinae angled in the middle, the area between them depressed. Fifth tergite shallowly punctate with interspaces shiny. Side of thorax more polished. Propodeum basally smoother. Median propodeal carinae longer, extending up to basal 0.4 to 0.5 and widely diverging. Areolet more triangular, with second recurrent vein near its middle. Ovipositor tip strongly decurved, almost L-shaped.

Black. Flagellum yellowish-brown. Palpi and tegula blackish-brown. Scutellum yellow near metascutellum. Metascutellum black or only faintly yellow-marked. Mesoscutum as a rule without yellow markings, yellow markings only sometimes present in males. Legs largely reddish-brown with fore coxa and base of middle coxa sometimes black. Hind tibia rust-red, usually without banding (except faintly so in *compunctor orientalis* Kasparyan). Hind tarsal segments fuscous apically. Body pubescence brownish.

Length: 12-16 mm. Fore wing 10-12 mm.

Specimens: France: Charroux, 1♀, bred from *Lymantria dispar*, Gypsy Moth Lab, No. 1683, det. as *Apechthis brassicariae* Poda, by Muesebeck. Germany, 1♂, bred from *Lymantria dispar*, Gypsy Moth Lab, No. 829 Ag 5 (Hamden). Several males and females from Europe in Townes Collection.

Distribution: Europe, Russia, Japan and Korea. Kasparyan (1973) described an Eastern Siberian subspecies, *E. compunctor orientalis*, with banded hind tibia.

Hosts: A polyphagous parasite developing within the pupae of several families of Lepidoptera and some Coleoptera (cf. Aubert, 1969), including *Lymantria dispar* and *L. monacha.* Townes, *et al.* (1965) give host records from Japan and Russia. Hedlund and Mihalache (1980) reared two specimens of *Ephialtes compunctor* from the gypsy moth from Site B in Southern Rumania, which they considered to be a new host record. According to them the parasitism was less than 1%. Vasic (1958) reported it from Yugoslavia from the gypsy moth.

3. EPHIALTES RUFATUS (Gmelin) (Fig. 129)

Ichneumon rufatus Gmelin, 1790. *In* Linnaeus: Systema Naturae, Ed
13 (1) 5: 2684. Europe.
Pimpla rufata: Gravenhorst, 1829. Ichneumonologia Europaea, 3: 164.
Europe.
Pimpla rufata: Grav.: Rudow, 1911. Internatl. Ent. Ztschr., 5: 99.
Europe. Hosts: *Lymantria monacha*, *L. dispar.*
Pimpla rufata: Meyer, 1927, Izv. Otd. Prikl. Ent. GIOA (Repts. Bur.
Appl. Ent.), 3 (1): 75-91. Russia. Hosts: *Lymantria dispar,
Aporia crataegi, Malacosoma neustria, Diprion pini.*
Ephialtes rufatus: Townes, Momoi and Townes, 1965. Mem. Amer. Ent.
Inst., 5: 45. Germany, Japan, Kamchatka, Korea, Kuriles, Russia,
Sakhalin. Full synonymical references and hosts.

Further taxonomical and biological references are given by Oehlke (1967)
and Aubert (1969), which may also be consulted for the various hosts.
Essentially similar in sculpture to *Ephialtes capulifera*, and differing as
follows:
Malar space 0.25 the basal width of mandible. Metapleurum with
scattered punctures. Propodeum smoother basally. Propodeal sculpture
less coarse and somewhat shiny. Median propodeal carinae generally shorter,
almost parallel-sided and ending in a central rugose area. Propodeal
spiracles bean-shaped. Areolet more triangular in outline, with the two intercubiti almost equal and second recurrent vein meeting areolet close to its
middle. First tergite not conspicuously humped medially. Abdominal tergites
less closely punctate. Punctures on fourth and fifth tergites well separated
with interspaces shiny. Ovipositor tip only slightly decurved.
Black. Face, frons and vertex with narrow yellow orbital lines. Flagellum
yellowish-brown ventrally and blackish dorsally. Palpi and tegula orange-
yellow. Mesoscutum with two yellow lines behind its middle. Sometimes
its antero-lateral corners also yellow. Scutellum broadly and metascutellum
almost wholly yellow. Legs, including coxae, reddish-brown, with yellow
patches on fore trochanters, apex of femur, tibia and tarsus. Apex of middle
femur, tibia and tarsus also with yellow marks. Hind tibia fuscous at extreme
base and with a faint to rather distinct sub-basal yellow band. Hind basitarsus
yellow except apically. Coxae partly black in Japanese populations *(rufatus
geometriae)*. Body pubescence whitish.
Length: 10-15 mm. Fore wing 8-12 mm.
Specimens: Several males and females, from Europe and Japan in Townes
Collection. No reared specimens seen. The Japanese populations have the
coxae black marked and yellow band on hind tibia more pronounced. Perhaps
they do represent a distinct subspecies, *Ephialtes rufatus geometriae* Uchida,
1928. This has, however, been synonymized under the nominate subspecies
(Townes *et al.*, 1965).
Distribution: Europe, Russia, Japan and Korea.
Hosts: Lymantria dispar, L. monacha, and several others belonging to
various families of Lepidoptera (Townes, Momoi and Townes, 1965). Aubert
(1969) records the hosts from Europe, including a few cases of parasitism of
Diprionidae. Vasic (1958) reported it from the gypsy moth in Yugoslavia.

Tribe THERONIINI

Members of the tribe Theroniini are characterized by having the meso-pleural suture angulate near middle, propodeum more or less areolated, areola often distinctly formed, tarsal claws simple, without a basal lobe, areolet present, subgenital plate of male often longer than wide, and ovipositor short to long, of uniform width.

Only one genus *Theronia* is associated with the gypsy moth. They are usually yellowish-brown in color (although some species are black) and have an enlarged hair on each tarsal claw which is flattened at the tip.

4. Genus THERONIA   (Figs. 4, 132)

Theronia Holmgren, 1859. Öfvers. Svenska Vetensk. Akad. Forh.,
      16: 123.

Townes (1969) may be consulted for generic synonymy and key to distin-guish it from other Theroniini.

Mandible not strongly tapered, its teeth equal. Clypeus truncate or with a median notch. Prepectal carina present. Mesopleural suture with a weak angulation near middle. Propodeal carinae strong. Tarsal claws very large, simple, but with an enlarged bristle arising sub-basally below and extending to apex of claw, the tip of the bristle with a spatulate enlargement. Nervellus intercepted far above the middle. Areolet present. Abdomen polished, with minute setiferous punctures. Ovipositor short to moderately long, cylindric, its sheath usually about 0.45 as long as the forewing.

Species of *Theronia* are medium-sized ichneumonids inhabiting mostly dense forests. They are pale colored and generally fly low. They are primary or secondary parasites within lepidopterous pupae. Some species parasitize pupae of aculeate wasps in their nests. The egg is deposited within the pre-pupa or a freshly formed pupa and emergence is from the pupa. As a primary parasite the insect lives within its host. If the host attacked has already been attacked by another ichneumonid, it becomes a secondary parasite external to the latter. The larvae of *Theronia* have a large internal tooth on the mandible.

Two species, *Theronia atalantae* (and its subspecies) and *T. hilaris* (=*melanocephala*) have been associated with gypsy moth. *T. atalantae* has been commonly reared from the gypsy moth in Europe, USSR, North America, and Japan. A subspecies of it occurs on *Lymantria obfuscata* in India. *T. zebra diluta* was reported from *Lymantria serva* in Taiwan by Gupta (1962).

T. atalantae and *T. hilaris* have been reported as hyperparasitic on other ichneumonids associated with the gypsy moth on occasions.

KEY TO THE SPECIES AND SUBSPECIES OF THERONIA
ASSOCIATED WITH THE GYPSY MOTH

1. Hind femur sharp beneath for part of its length, and usually with a weak to sharply irregularly serrate ridge. Head rufous-brown.
                                        1. atalantae 2
   Hind femur rounded beneath, without a ridge. Head black. (In other subspecies from Europe and Japan whole body black). North America.
                                        2. hilaris hilaris (Say)

2. Abdominal tergites with dark brown to black basal bands or spots. Flagellum brown. Thorax with extensive fuscous markings. Japan, Korea, Manchuria, and Siberia. . . . . .   2c. atalantae gestator (Thunberg)
   Abdominal tergites uniformly rufous brown, or rarely with a few spots. Thorax rufous brown or a few spots and stripes on mesopleurum, base of propodeum or around scutellum.   3

3. Wings dark brown. Flagellum blackish-brown to black. Ridge or crest along lower margin of hind femur weak. Mesopleurum and base of scutellum with fuscous markings. India: Kashmir and Himachal Pradesh. . . . .   1d. atalantae himalayensis, n. subsp.
   Wings clear hyaline, or only lightly yellowish-tinged. Flagellum black to brown. Lower margin or hind femur usually with a serrated ridge. Fuscous marks on thorax absent or less extensive (except sometimes in *atalantae atalantae*).   4

4. Flagellum light brown, of the same color as the thorax. Mesopleurum often with fuscous markings. Sometimes mesoscutum also with fuscous markings. Europe. . . . .   1a. atalantae atalantae (Poda)
   Flagellum dark brown, darker in color than the thorax. Mesopleurum partly to largely yellowish, usually without fuscous markings. North America. . . . .   1b. atalantae fulvescens (Cresson)

1. THERONIA ATALANTAE (Poda) (Figs. 4, 132)

Clypeus apically thin and only slightly concave. Prepectal carina roundly curved forward and almost reaching margin of pronotum. Notauli distinct at front end of mesoscutum. Scutellum not carinate laterally. Thorax shiny, with minute scattered setiferous punctures. Propodeal spiracle elongate, linear. Median and lateral propodeal carinae present. Median carinae parallel-sided and extending to middle of propodeum, where they meet the strongly curved apical transverse carina. Lateral carinae obsolete basally, opposite spiracles. Petiolar area hexagonal. Median section of apical transverse carina bounding the combined areola and basal area often dipped and notched in the middle in the female. Hind femur with a ventral crest or ridge, which is often serrate in larger specimens. First tergite about 1.5 as long as wide and with strong dorsal carinae. Abdomen polished. Ovipositor about 0.7-0.75 as long as the abdomen, its sheaths about 1.5 as long as the hind femur.
Rufous-brown. Wings hyaline or brown tinged. Thorax and abdomen with or without fuscous marks depending upon the subspecies.
Size variable, depending upon the host. Those reared from ichneumonid hosts much smaller than those reared from the gypsy moth.

1a. THERONIA ATALANTAE ATALANTAE (Poda) (Figs. 4, 24, 41, 132)

Ichneumon atalantae Poda, 1761. Insecta Musei Graecensis, p. 106.
     Europe. Host: *Vanessa atalanta*.
Ichneumon flavicans Fabricius, 1793. Ent. Syst., 2: 182. Germany.
Pimpla flavicans Fabr.: Ratzeburg, 1844: 118. Germany. Hosts: pupae
     of *Aporia crataegi*, *Dendrolimus pini*, and *Lymantria dispar*.

Theronia atalantae: Howard and Fiske, 1911. U. S. Dept. Agr. Bur.
    Ent. Bull., 91: 85, 236. Lab. reared from gypsy moth.
Theronia atalantae atalantae: Townes and Townes, 1960. U. S. Natl.
    Mus. Bull., 216(2): 354.

This subspecies has several synonyms in Europe. Refer to Townes,
Momoi and Townes (1965) and Oehlke (1967) for synonymical references, and
host records.

Ratzeburg (1844) first reported it from *Lymantria dispar* in Europe.
Howard and Fiske reared it in the Gypsy Moth Laboratory from shipments of
gypsy moth pupae from Europe. Several authors have recorded it as a parasite of the gypsy moth in Europe (Thompson, 1946). Some of the recent
studies are of Györfi (1963), and Hedlund and Mihalache (1980).

Theronia atalantae atalantae is characterized by having the flagellum light
brown in color (resembling the coloration of the thorax), mesopleurum brownish, mesoscutum and mesopleurum usually with only small fuscous marks,
which at times may be extensive, or reduced, wings light yellowish-hyaline,
and legs and abdomen pale yellowish-brown, without fuscous marks. The
scutellum is yellow. Two females in Townes Collection from Germany and
Poland have extensive fuscous marks on thorax, and hind coxae partly blackish.

Length: 8-13 mm. Fore wing 7-11 mm.

Specimens: Southern Rumania: Site B, 1♂, 1♀, July 1978, *ex Lymantria
dispar* pupae, R. C. Hedlund (E.P.L., Paris). Hungary: Vees, 2♂, July 12,
1928, *ex Hyposoter* sp., Gypsy Moth Lab., Nos. 13039D and 13039B2.
Hungary: Olaszliszka, 2♂, bred from *L. dispar*, Gypsy Moth Lab., No.
13028B. Italy, 1♂, July 22, 1911, bred from *L. dispar*, Gypsy Moth Lab.,
No. 3416. Italy, 1♂, June 1911, *ex Hyposoter disparis*, Gypsy Moth Lab.,
No. 3410B (all FIS, Hamden). Several males and females without host data,
from Europe in Townes Collection.

Distribution: Widely distributed in Europe.

Hosts: Several hosts are recorded from Europe and USSR, including
Lymantris dispar and *L. monacha*. It is also recorded as a hyperparasite of
genera like *Casinaria, Iseropus, Rhogas,* etc. It is here recorded as a
hyperparasite of *Phobocampe unicincta* (= *Hyposoter disparis*).

It is likely that many USSR records may pertain to *T. atalantae gestator,*
which occurs in the Asian part of USSR.

It is a casual parasite of the pupae of *Lymantria dispar* and has never
been reared in abundance on that host, although it is likely that it may be
killing more host pupae by puncturing the pupae than by parasitizing, as has
been seen in the U.S.A. by Campbell (1963) in *T. atalantae fulvescens.*

1b. THERONIA ATALANTAE FULVESCENS (Cresson) (Fig. 18)

Pimpla fulvescens Cresson, 1865. Proc. Ent. Soc. Philadelphia, 4: 268.
    U.S.A.: Colorado.
Theronia fulvescens: Howard and Fiske, 1911. U. S. Dept. Agr. Bur.
    Ent. Bull., 91: 137, 141, 142, 236-237. Hosts: *Lymantria dispar,
    Euproctis chrysorrhoea.*
Theronia atalantae fulvescens: Townes and Townes, 1960. U. S. Natl.
    Mus. Bull., 216(2): 354. Syn., distribution and biological refer-
    ences.

This subspecies was first mentioned by Howard and Fiske (1911) as a native North American parasite of the gypsy moth. According to them the record of *Theronia melanocephala* (Brullé) = *hilaris hilaris* (Say) of Forbush and Fernald (1896) pertained to this subspecies. They considered it a primary parasite of the gypsy moth. It has been also reared from *Itoplectis conquisitor* (Say), another primary parasite of the gypsy moth, which often fails to develop on that host. They believed it to be a case of superparasitism: the host pupa by chance containing larva of *Itoplectis* and *Theronia* parasitizing that pupa. *T. atalantae fulvescens* is parasitized by a chalcid, *Dibrachys cavus*.

Theronia atalantae fulvescens differs from the nominate subspecies by having the flagellum black to brownish-black, mesopleurum largely yellow, usually without fuscous markings, though in larger specimens sometimes present on prepectus near lower corner of pronotum, and wings hyaline, lightly tinged with yellow. Specimens from southwestern U.S.A. tend to have wings a little darker.

This subspecies is sometimes difficult to separate from *atalantae atalantae*, as coloration of the two often approach each other.

Length: 6-13 mm. Fore wing 4-11 mm.

Specimens: U.S.A.: Providence, R.I., 1♂, July 24, 1916, "ex tray of b.t. pupae", Gypsy Moth Lab., No. 12099. 2♂, August 6, 1907, Gypsy Moth Lab., No. 826-07 (Hamden). Perry Co., Pa., Millers Gap, *ex* pupa, 1♂, June 27, 1978, E. M. Blumenthal. Dauphin Co., Pa., Jackson Twp., July 6, 1977 (both these apparently reared from *Lymantria dispar*) (Pennsylvania).

Distribution: Widely distributed in North America. Schedl (1936) mentioned it as occurring in Japan, but that record should pertain to *T. atalantae gestator*.

Hosts: Several host records are given by Townes (1940) and Carlson (1979). Townes and Townes (1960) give several biological references.

Biological Notes: Howard and Fiske (1911) considered *Theronia atalantae fulvescens* as "the most common American parasite completing its transformation upon the gypsy moth" pupae. Parasitism at times amounted to 2 per cent. It appears that the adult female overwinters and that there is a single generation per year.

According to the observations of Campbell (1963), this species seeks its host under shady conditions, in forests and seldom is seen in the open, though at times it may be seen along the border of a defoliated area. About 25% of the pupae stung by the parasite produced ichneumonid offspring. None of the pupae stung were fed upon, but all prepupae stung were fed upon. The effectiveness of this parasite is both by parasitization as well as by killing the host pupae and prepupae through stinging.

Townes (1940) summarized the biological information on this species as follows:

"*Theronia atalantae* may be either a primary or a secondary parasite. If the host attacked has already been parasitized by another ichneumonid it becomes a secondary parasite, if not it lives as a primary. There is no clear evidence that it is obligated to be a secondary parasite on any host, although it has been shown usually to live thus in the case of *Malacosoma*. As a primary parasite the insect lives inside its host. When it is a secondary parasite, it lives inside its primary host and outside of its secondary host. The egg is deposited only in a prepupa or a freshly formed pupa. Fiske believes the egg stage to last about a day. He has observed that on the average the first three larval stages last two days each and the fourth stage feeds for two days and then spins its cocoon for one day and rests two days before forming the pupa. After two to four days the pupa hatches. Development from the egg to the adult thus consumes fourteen to eighteen days. When a secondary parasite on *Malacoma americana* (through *Itoplectis conquisitor*) in New Hampshire, most of the adults emerged July 10-20, the females about four days behind the males. *Itoplectis conquisitor* emerged from the same cocoons on an average of seven days earlier than its parasite."

1c. THERONIA ATALANTAE GESTATOR (Thunberg)

Ichneumon gestator Thunberg, 1822. Mem. Acad. Imp. Sci. St.
    Pétersbourg, 8: 262; 1824, 9: 312.
Theronia japonica Ashmead, 1906. Proc. U. S. Natl. Mus., 30: 181.
    Japan.
Theronia japonica: Howard and Fiske, 1911. U. S. Dept. Agr. Bur.
    Ent. Bull., 91: 121. Japan. Host: *Lymantria dispar*.
Theronia atalantae gestator: Townes, Momoi and Townes, 1965. Mem.
    Amer. Ent. Inst., 5: 64. Synonymical and biological references.
    Japan, Korea, China, USSR.

This subspecies is readily distinguished from the other subspecies of *atalantae* by having dark brown bands or spots on its abdomen, and the thorax also extensively dark marked. The wings are pale-hyaline, flagellum brown, hind coxa with a black or dark brown mark, and the hind femur is often fuscous ventrally.

The Japanese and the Korean populations of this subspecies are uniformly distinctive by the blackish markings, but populations from USSR and China tend to intergrade with *atalantae atalantae*. Rarely the American specimens approach *gestator* by having black markings on basal abdominal tergites or somewhat extensively on the thorax.

Length: 8-13 mm. Fore wing 5.5-11.5 mm.

Specimens: Japan: 4♂, 3♀, bred from *Lymantria dispar*, 1908, 1910, some without date, Gypsy Moth Lab., Nos. 1600, 1825, & 3399A (Hamden). 1♂, labelled "Europe ex *L. dispar*, Istria Austria thru Ruhl, Gypsy Moth Lab. No. 853-07, Au. 12. 07." (Hamden). Japan: Sekigahara, Honshu, 2♂, 1♀. ex pupa of *Lymantria dispar*, June 20, 1977, emerged July 6, 7 and 10, 1977, P. Schaefer (B.I.R.L., Delaware).

Distribution: Eastern Palaearctic Region, typically in Japan, Korea, North China, Siberia and adjacent areas. Occurrence in Europe (specimen from Austria above) is rather unusual, though it is reported to intergrade with *atalantae atalantae* in Europe.

This subspecies was first reported as a parasite of *Lymantria dispar* by Howard and Fiske (1911). Further references are given by Townes, Momoi and Townes (1965).

1d. THERONIA ATALANTAE HIMALAYENSIS, n. subsp.

(= *Theronia* sp., or *Theronia atalantae atalantae* from *Lymantria
    obfuscata* from Kashmir)

Male and *female:* This subspecies is characterized by its brownish wings, black flagellum, and rufous brown body, with fuscous marks on antennal scrobes, prepectus, below subtegular ridge, base of scutellum, wing bases, lower half of mesopleural groove, and along groove separating metapleurum from mesopleurum. The anterior margin of the middle lobe of mesoscutum rather vertically and conspicuously raised and prominently separated from the rest of mesoscutum. In profile view the front edge of mesoscutum appears concave and raised. Ventral margin of hind femur with a shorter ridge, inconspicuously serrated.

Length: 10-13 mm. Fore wing 9-11.5 mm.

Holotype: ♀, India: Manali, 1828 m., in Northwest Himalaya (Himachal Pradesh), May 29, 1970, Dauli Ram, Coll. No. K252 (Gupta). *Paratypes:* 25♀, same locality as the holotype, collected between May 17 to June 3, 1970 by various collectors (Gupta). Kashmir: Srinagar, 1♂, 1♀, July 1966, *ex* pupa of *Lymantria*, Gupta, No. 277; 1♀, June 1970, Gupta (Gupta). Srinagar, 4♂, 4♀, July 1963, C. I. B. C., Indian Station, *ex* pupae of *Lymantria obfuscata* on willow (CIBC, Bangalore).

This subspecies has earlier been determined as *Theronia atalantae atalantae* (Poda), or simply as *Theronia* sp. (Rao, 1966, 1972). The following information on the biology of this subspecies is summarized from those reports:

Adults of *Theronia atalantae himalayensis* appeared in the middle of June in Srinagar. The females were seen flying over the congregations of pupae of *Lymantria obfuscata* in search of suitable hosts. Freshly formed pupae were oviposited upon. In the laboratory freshly emerged adults paired immediately and mating lasted 8-15 minutes. The ratio of males to females was 1:2.

The egg is cream colored, elongate with one end narrower than the other, and measures about 2 mm. in length. It is laid in the body cavity of the pupa. The larva hatches in three days and starts feeding soon after. Once the parasite larva inside the host pupa is active, the latter stops all movements after about two days and becomes hard and stiff. The full grown parasite larva is yellowish in color and measures about 12-14 mm. in length and 3 mm. in width. It is tapering at the caudal end. Because of the developing parasite, the host turns slightly blackish between the fourth to ninth abdominal segments. The full grown larva spins a cocoon which forms a lining to the abdominal wall of the host. The parasite pupa is also yellowish in color. The adult parasite gnaws an irregular hole and comes out. Males emerge earlier than the females and are generally smaller in size.

The total life-cycle from the egg to the adult takes about 21-24 days.

This subspecies does not appear to show any preference for any particular ecological condition or host plant. In general the percentage parasitism varied from 2% to 20%, although Rao (1972) recorded as high as 40.82% parasitism in a *Salix* plantation and stated that "It was a constant mortality factor operating against the pupae of *L. obfuscata* in all the four experimental localities."

Rao (1972) made some observations on the emergence of the moths from pupae that were pricked by *T. atalantae himalayensis* (as well as other pupal parasites). It was observed that the adult parasite fed on the host haemolymph after every prick. When freshly formed pupae were pricked either once or thrice, more parasites emerged than from 4-5 day old pupae pricked. The mortality of the pupae increased with increasing number of pricks. The moth emergence was greater when 4-5 day old pupae were pricked by the parasite, particularly after a single prick.

Rao (1966) also reported that, "on one occasion, 4 parasite larvae were found feeding externally on a fourth instar larva of *L. obfuscata* and one of them developed to adult stage in June 1963."

2. THERONIA HILARIS HILARIS (Say)

Ichneumon hilaris Say, 1829. Contrib. Maclurean Lyceum Arts Sci.,
    1: 71. U.S.A.
Pimpla melanocephala Brullé, 1846. Hist. Nat. Insectes Hymén., 4: 99.
    North America. Name preoccupied.
Theronia melanocephala: Fernald, 1892. Bull. Hatch. Exp. Sta. Mass.
    State Coll., 19: 116. Massachusetts. Host: *Lymantria dispar*.
Theronia melanocephala: Viereck, 1917. Bull. Conn. Geol. Nat. Hist.
    Survey, 22: 323. Hosts: *Lymantria dispar* and others.
Theronia hilaris: Townes and Townes, 1960. U. S. Natl. Bull., 216(2):
    356.
Theronia hilaris hilaris: Carlson, 1979. *In* Krombein *et al.:* Catalog of
    Hymenoptera in America. . ., 1: 347.

This subspecies was first mentioned (as *T. melanocephala*) by Fernald
(1892) as a parasite of the gypsy moth in Massachusetts, U.S.A. Forbush
and Fernald (1896) called it "the most abundant of the hymenopterous parasites" in 1895. Howard and Fiske (1911), however, considered that the record
of *T. melanocephala* of Forbush and Fernald was actually of *T. fulvescens*
[=*atalantae*] and that "the true *T. melanocephala* appears not to have been
reared from this host." Campbell (1963) observed both *Theronia hilaris*
(=*melanocephala*) and *T. atalantae,* attacking the gypsy moth in Glenville,
N.Y. He, however, did not discuss *hilaris* further in that paper. It, therefore, appears that it is just an occasional parasite of the gypsy moth.

This subspecies is readily distinguished from *atalantae* by not having a
ridge along the ventral margin of the hind femur and by its black head. It
differs from the Eurasian subspecies of *hilaris* by its rufous thorax and abdomen. *T. hilaris laevigata* (Europe) and *T. hilaris nigra* (Japan and Siberia)
are wholly black. The legs of *laevigata* are largely reddish, while those of
nigra are wholly black.

No reared specimens from the gypsy moth have been seen. It is widely
distributed in eastern and midwestern North America. Townes and Townes
(1960) provides a detailed description and distributional data and Carlson
(1979) records the hosts of *T. hilaris hilaris*.

Tribe PIMPLINI

Members of the tribe Pimplini are distinguished from the other tribes of
Pimplinae by having a uniformly convex mesoscutum, without wrinkles, prepectal carina present, mesopleural suture with an angulation just above middle,
propodeum without transverse carinae, often median carinae present at base
and pleural carinae complete, tarsal claws of female, at least the front claws,
nearly always with a basal tooth, areolet present (except rarely), nervellus
intercepted variously, first tergite with a glymma, first sternite separate
from first tergite, with dorsolateral carinae, subgenital plate of male transverse, with its apex truncate or retuse, and ovipositor long to very long,
slender and of a uniform diameter.

The coloration is generally black. The pattern on hind tibia is rather
characteristic, with apical and subbasal dark bands and median area and
extreme base of tibia pale yellow.

Members of the Pimplini genera treated below are external parasites of late larvae of various Lepidoptera that have just spun their cocoons, or those living in leaf rolls, etc. The development is completed upon the host larva within the cocoon. Two genera, *Acropimpla* and *Iseropus* are associated with the gypsy moth. *Gregopimpla* is treated here as a subgenus of *Iseropus*.

5. Genus ACROPIMPLA

Acropimpla Townes, 1960. U. S. Natl. Mus. Bull., 216(2): 159.

Townes (1969: 84) and Gupta and Tikar, (1976: 128) may be consulted for generic synonymy and affinities of the genus.

Some of the salient features of the genus are:

Body moderately long and slender, clypeus of male, and often the face of male white or yellow. Malar space very short to almost wanting. Face usually with its upper margin raised and often narrowly and strongly, or broadly cleft. Occipital carina complete, dipped a little on the mid-line above. Propodeum rather short and convex, with or without median longitudinal carinae. Areolet triangular, oblique, usually short petiolate, receiving second recurrent vein at its outer corner (sometimes the areolet lacking). Nervellus intercepted below middle. First tergite short and wide, its median dorsal and dorsolateral carinae rather strong. Second tergite with short, moderately strong oblique grooves cutting off its basolateral corners. Third and fourth tergites with distinct tubercles. Ovipositor straight, compressed, usually nearly as long as fore wing, but shorter in some species, its apex usually slender, seen in profile, concave above and the ridges on the tip of its lower valve very oblique.

Acropimpla is predominantly Oriental; but it also occurs in the Holarctic and Ethiopian regions. Their known hosts are several species of Microlepidoptera.

1. ACROPIMPLA DIDYMA (Gravenhorst)

Pimpla didyma Gravenhorst, 1829. Ichneumonologia Europaea, 3: 178. Europe.

Acropimpla didyma: Oehlke, 1967. Hymenopterorum Catalogus (novo editio),2(1): 15. Host: *Lymantria dispar*.

This species was reported as a parasite of *Lymantria dispar* for the first time by Šedivý (1963) from Czechoslovakia, as *Ephialtes didymus* (Gravenhorst).

Female: Face polished, centrally raised, its upper margin straight. Mesoscutum hairy, subpolished. Mesopleurum and metapleurum polished and shiny, with scattered setiferous punctures. Propodeum with faint median carinae at base, which are somewhat divergent apically. Propodeum smooth dorsally, its pleural area with scattered punctures. Abdomen with coarse and scattered punctures. Tubercles on tergites punctate. Ovipositor long, as long as the fore wing.

Black. Face with an inverted crescentic yellow mark just below antennal sockets. Hind corner of pronotum, tegula, and wing bases, yellow. Legs yellowish-brown, with apical 0.3 of hind tibia, and tarsus except at extreme base, blackish.

Male: Generally similar to the female, but face more convex and with scattered punctures. Face and clypeus wholly yellow. Scape yellow ventrally. Hind margin of pronotum also yellow. Fore and middle legs more yellowish, lighter in coloration than the hind leg.

Length: 8 mm. Fore wing 6 mm.

Specimens and *distribution:* Several males and females from Europe examined in the Townes Collection. No reared specimens were seen. It is widely distributed in Europe.

Hosts: Lymantria dispar (*vide* Šedivý, 1963), and various other hosts belonging to the families Lasiocampidae and Noctuidae as mentioned by Šedivý (1963) and Aubert (1969).

This species runs to Group B and to *Acropimpla varuna* Gupta and Tikar from Java in the Gupta and Tikar (1976) Monograph on Oriental Pimplini. The two species are rather close. *A. varuna*, however, is distinguished by having a yellow scape in the female and the mesopleurum with a red mark in the dorsal half. The other European species of the genus, *A. pictipes* (Gravenhorst) is black, without yellow marks on face, and the propodeum and abdomen are coarsely sculptured.

6. Genus ISEROPUS   (Figs. 5, 6, 130, 131)

Iseropus Foerster, 1868. Verh. Naturh. Ver. Rheinlande, 25: 164.
Gregopimpla Momoi, 1965. Mem. Amer. Ent. Inst., 5: 601. [Subgenus of *Iseropus*].

Body moderately slender and long. Clypeus of female black to yellow. Clypeus convex, its apical margin thin and with a median notch. Malar space short. Occipital carina complete, with a moderate dip medially above. Propodeum moderately convex, with median longitudinal carinae. Areolet present. Nervellus intercepted at or above the middle. First tergite short and wide, its dorsal and lateral carinae strong. Basolateral oblique grooves on second tergite moderately strong to obsolete. Third and fourth tergites with tubercles, their impunctate bands occupying about 0.2 their length. Ovipositor straight, compressed, its apex slender, concave above, the ridges of basal teeth very oblique. Ovipositor 0.5 to 0.8 as long as the fore wing.

Two subgenera are recognized: *Iseropus* (*Iseropus*) and *Iseropus* (*Gregopimpla*). Diagnostic characters are mentioned in the key that follows. Members of the genus are gregarious parasites of lepidopterous prepupae in thin cocoons, particularly of the families Lymantriidae, Lasiocampidae and Notodontidae.

Further taxonomic information may be obtained from Townes (1969) and Gupta and Tikar (1976). Information on larval morphology and affinities is provided by Short (1978).

Three species have been associated with the gypsy moth. They are *Iseropus* (*Iseropus*) *stercorator* Fabricius (= *Pimpla holmgreni* Schmiedeknecht), *Iseropus* (*Gregopimpla*) *himalayensis* (Cameron) (= *Iseropus hakonensis* Ashmead) and *Iseropus* (*Gregopimpla*) *inquisitor* (Scopoli).

KEY TO THE SUBGENERA AND SPECIES OF ISEROPUS ASSOCIATED WITH THE GYPSY MOTH

1. Interocellar distance 0.8 as great as ocellocular distance. Second tergite with conspicuous basolateral grooves. Abdominal tergites only moderately punctate, with shiny areas in between the punctures. Areolet as high as wide, more triangular in outline with second recurrent vein close to its middle. Face of male white. (Subgenus *Iseropus*).
                                         1. stercorator Fabricius

Interocellar distance 1.25 as great as ocellocular distance. Second tergite with faint to indistinct basolateral grooves. Abdominal tergites wholly rather densely punctate. Areolet wider than high, more rectangular, with second recurrent vein close to its outer corner. Face of male black. (Subgenus *Gregopimpla*). 2

2. Face and metapleurum impunctate and shiny. Side of thorax shiny. Female flagellum 24-27 segmented. Median propodeal carinae extending up to 0.7-0.8 the length of propodeum. Propodeum evenly convex. Tergite 2 without oblique grooves. Legs slender, reddish, with hind tibia banded with fuscous and hind tarsus black except basally. Stigma pale brown. . . . . 2. inquisitor (Scopoli)

Face and metapleurum with coarse scattered punctures. Side of thorax with scattered punctures. Female flagellum 29-30 segmented. Median propodeal carinae confined to basal 0.5-0.6 of propodeum. Propodeum more convex in profile. Tergite 2 with faint oblique grooves. Legs a little thicker with bands on hind tibia rather conspicuously black and coxae and trochanters yellow marked. Stigma blackish-brown.

3. himalayensis (Cameron)

1. ISEROPUS (I.) STERCORATOR STERCORATOR (Fabricius) (Figs. 5, 131)

Ichneumon stercorator Fabricius, 1793. Entomologia Systematica, 2: 172. Germany.
Ichneumon graminellae Schrank, 1802. Fauna Boica, 2(2): 301. Czechoslovakia.
Pimpla Mussii Hartig, 1838. Jahresber. Forstschr. Forstwiss. Forstl. Naturk., 1: 253. Germany.
Pimpla Holmgreni Schmiedeknecht, 1888. Zool. Jahrb. System., 3: 448, 502. Germany.
Iseropus stercorator Roman, 1912. Zool. Bidr. Från Uppsala, 1:280.
Iseropus stercorator Townes, Momoi and Townes, 1965. Mem. Amer. Ent. Inst., 5: 29. syn., ref., distr., hosts, etc. Eurasia, Japan.

Iseropus stercorator Fab. in older literature was used as a senior synonym of *flavipes* Gravenhorst or a junior synonym of *inquisitor* Scopoli. It has also been cited as *stercorator* Gravenhorst. It appears that the host records are mixed up because of this confused nomenclature.

Rudow (1911) reported it as a parasite of *Lymantria dispar* and *L. monacha*, but in 1917 recorded only *monacha* as its host. Stadler (1933), Schedl (1936) and Thompson (1946) report it as a parasite of the gypsy moth under the name *Pimpla holmgreni*. No *Lymantria* hosts are mentioned by Oehlke (1967) or Aubert (1969).

The association of *Iseropus stercorator* with *Lymantria dispar* has not been confirmed by recent rearings.

Iseropus stercorator orgyiae (Ashmead) is the Nearctic subspecies.

Female and *male*: Face punctate, smoother along eye orbits and near base of clypeus. Clypeus subpolished, convex in basal half. Malar space short, 0.2 to 0.25 the basal width of mandible. Frons and vertex smooth, subpolished. Interocellar distance a little less than ocellocular distance (8:10). Mesoscutum hairy, subpolished with shallow indistinct punctures. Scutellum convex subpolished. Pronotum minutely punctate and with a smooth central area. Mesopleurum convex, polished, with minute punctures in the prepectal area. Metapleurum polished and with a few scattered minute punctures. Propodeum largely smooth but the scattered punctures and rugosities at places, not convexly sloping, its median carinae divergent and extending only in the basal 0.5 of propodeum. Areolet triangular. First recurrent vein equal to the second. Nervellus intercepted at its upper 0.3. First tergite thinner, smoother laterally. Abdominal tergites with irregularly formed punctures with shiny interspaces. Swellings on tergites tending to be smoother. Oblique grooves on second tergite sharper. Ovipositor compressed, about 0.6-0.7 as long as the fore wing and the abdomen.

Female: Color black. Flagellum, clypeus and palpi brownish. Legs reddish-brown. Hind tibia with broad black subbasal and apical bands. Hind tarsal segments black in apical half (or more). Stigma and veins brown. Hind femur without fuscous mark. Tegula brown with a yellow anterior spot.

Male: Face, clypeus and scape ventrally, yellow. Fore and middle coxae with yellow spots. Otherwise like the female.

No reared specimens seen. Several European specimens examined in the Townes Collection.

Distribution and *hosts:* This subspecies is widespread in Europe and also occurs in Japan and eastern Russia. Several hosts have been reported by Aubert (1969: 42) belonging to the families Curculionidae, Tortricidae, Ypnomeutidae, Oecophoridae, Gelechidae, Pyralidae, Zygaenidae, Lasiocampidae, Notodontidae, Lymantridae, and Noctuidae. He does not mention any *Lymantria* species as host.

Iseropus (I.) stercorator stercorator is rather close to the Nearctic *stercorator orgyiae* (Ashmead), and *coelebs* (Walsh) and also to the Japanese *orientalis* Uchida. *I. stercorator orgyiae* differs from the nominate subspecies in having a black apical mark on the hind femur, white tegula, and rougher propodeum with longer median carinae. *I. coelebs* has denser punctures on the pronotum, mesopleurum and metapleurum, and elongate male claspers. Townes and Townes (1960) provide additional distinguishing characters. *I. orientalis* has pale labial palpi, pale tegula, black hind coxa and hind femur lightly fuscous apically. The side of thorax is shiny, with a few scattered punctures. The propodeum is more like that of *orgyiae* and *coelebs*.

2. ISEROPUS (GREGOPIMPLA) INQUISITOR (Scopoli)   (Figs. 6, 130)

Ichneumon inquisitor Scopoli, 1763. Entomologia Carniolica Exhibens
    Insecta Carnioliae Indigena, p. 286.
Pimpla inquisitor Stadler, 1933. Ent. Anz., 13: 30.
Gregopimpla inquisitor Townes, Momoi and Townes, 1965. Mem. Amer.
    Ent. Inst., 5: 27. Eurasia. Host records after Meyer, 1934.

For further synonymical references refer to Oehlke (1967) and Aubert (1967).

The first report of this species (as *Pimpla inquisitor*) from gypsy moth could be traced only to Stadler (1933). The only other record is by Vasic (1958) from Yugoslavia. As mentioned under *I. stercorator*, the present species has been confused in the past with *stercorator* and it is not sure which of the two, if *either*, is really parasitic on the gypsy moth.

Aubert (1969) gave an elaborate list of hosts belonging to the following families of Coleoptera, Lepidoptera and Hymenoptera: Anobiidae, Curculionidae, Tortricidae, Cochylidae, Yponomeutidae, Pyralidae, Geometridae, Lasiocampidae, Lymantriidae, Noctuidae, Cephidae, and Tenthredinidae. He listed *Lymantria monacha* as a host after Brischke, 1878

(Brischke cited *Pimpla stercorator* 'Gravenhorst' ♂ = *flavipes*). He did not mention *L. dispar* as a host of this species. Oehlke (1967) did not mention any *Lymantria* hosts.

Male and *female:* Flagellum 24-27 segmented. Face smooth and shiny. Clypeus subpolished, not strongly convex. Interocellar distance a little more than the ocellocular distance (8:7). Frons and vertex smooth. Temple a little more convex than in *himalayensis*. Head in dorsal view quadrate. Mesoscutum subpolished, shiny, and with scattered punctures. Scutellum with scattered punctures and shiny. Pronotum polished. Mesopleurum and metapleurum polished and with scattered punctures. Propodeum evenly convex, its median carinae more parallel-sided and extending up to 0.7-0.8 the length of propodeum. Propodeum with shallow rugosities across middle. Nervellus intercepted at the middle or at upper 0.45. Areolet 1.5 as wide as high. Legs slender. Hind femur about 5.4 as long as wide. First tergite slender, its median carinae making an angle of 30⁰ with the horizontal. Whole of abdomen densely punctate. Apical smooth bands on tergites a little wider than in *himalayensis*. Second tergite without grooves. Ovipositor about 0.85 as long as fore wing.

Black. Pedicel and basal flagellar segments ventrally yellowish. Palpi, hind corner of pronotum, and tegula yellow. Legs reddish-brown, with tibiae and tarsi a little paler. Hind tibia with conspicuous black bands. Hind tarsus black with basal 0.3 to 0.4 of first segment yellow. Stigma pale brown. In male, the fore and middle legs largely pale yellow and scape and pedicel yellow ventrally.

Specimens: Several males and females from Europe examined in the Townes Collection. No reared specimens seen.

Distribution and *relationships: Iseropus (G.) inquisitor* is distributed in Eurasia. It is close to *I. kuwanae* Viereck from Japan and is distinguished from the latter by having the hind tarsal segments largely black, first tergite slender, abdomen coarsely punctate, and propodeum being evenly convex. It differs from *I. himalayensis* by having a smoother face and metapleurum, slender hind femur, and median propodeal carinae extending up to 0.7 the length of propodeum.

3. ISEROPUS (GREGOPIMPLA) HIMALAYENSIS (Cameron)

Pimpla himalayensis Cameron, 1899. Mem. & Proc. Manchester Lit.
   Phil. Soc., 43(3): 178. India.
Epiurus hakonensis Ashmead, 1906. Proc. U. S. Natl. Mus., 30: 179.
   Japan.
Pimpla japonica Ulbricht, 1911. Soc. Ent. Stuttgart, 26: 54. Japan.
   Preoccupied. Host: *Samia cynthia pryeri*.
Epiurus satanus Morley, 1913. Fauna British India, Hymenoptera,
   3(1): 173. India.
Itoplectis attaci Habermehl, 1917. Ztschr. f. Wiss. Insektenbiol., 13: 117.
   New name for *Pimpla japonica* Ulbricht.
Epiurus quersifoliae Uchida, 1928. J. Fac. Agr. Hokkaido Imp. Univ.,
   25: 59. Japan. Host: *Gastropacha quercifolia*.
Itoplectis attaci: Kamiya, 1934. Bull. Forst Exptl. Sta. Govt.-Gen.
   Chosen, 18: 66. Japan. Hosts: *Dendrolimus spectabilis,*
   Lymantria dispar (first record).
Gregopimpla himalayensis: Townes, Momoi and Townes, 1965. Mem.
   Amer. Ent. Inst., 5: 26. China, India, Japan, Korea. Host records,

synonymical references.

Male and *female:* Flagellum 29-30 segmented. Face punctate below antennal sockets, smoother along inner orbits and near clypeus. Clypeus subpolished. Malar space 0.2 the basal width of mandible. Interocellar distance a little more than ocellocular distance (10:8). Frons and vertex polished, shiny. Mesoscutum elongate, leathery in texture. Scutellum with shallow punctures apically. Side of thorax shiny, polished with scattered shallow punctures on pronotum and mesopleurum. Metapleurum with more definite punctures than mesopleurum. Punctation on metapleurum similar to that on face. Propodeum roundly convex, its median carinae confined to its basal 0.5, diverging apically and ending in a rugose band across propodeum. Basolateral areas of propodeum with scattered but definite punctures. Legs short, not slender. Hind femur 4.5 as long as wide. Nervellus intercepted at upper 0.4. Areolet elongate 1.5 to 2.0 as wide as long. First tergite short, convex, with sharp evenly arched median carinae, which may be a little angled medially. Second tergite without conspicuous basolateral grooves. Whole of abdomen densely punctate, only apices of tergites smoother. Ovipositor about 0.8 the length of fore wing.

Black. Antenna yellowish-brown ventrally. Maxillary palpi, tegula, and hind corner of pronotum, yellowish-brown. Legs brownish with yellowish-brown patches on fore and middle coxae, femora, and tibiae. Hind tibia with conspicuous fuscous sub-basal and apical bands, clearly separated by a wide yellow area. Hind tarsus blackish but first tarsal segment yellow on basal 0.5-0.6. Bases of other tarsal segments yellow.

Specimens examined: Several males and females from India and Japan in Townes and Gupta Collections. In the Townes Collection, a specimen bears the following data: "Emerged from pupa of *Lymantria dispar* on 5-VI-1930", "Kyushu, Fukuoka (Chikuzen)."

Distribution: Widespread throughout the northern belt of the Oriental Region (Northern India, China, Japan, and Korea).

Hosts: Lymantria dispar, and several other hosts in Japan and India [Townes, Townes & Gupta (1961) and Townes, Momoi & Townes (1965).] Kamiya (1934) first reported it from the gypsy moth (as *Itoplectis attaci*) in Japan.

Iseropus (G.) himalayensis is close to the Palaearctic *I. (G.) bernuthii* (Hartig) in having similar punctuation of face, metapleurum, shorter hind femur and hind tibia distinctly banded (which characters separate these two from the other species of the subgenus *Gregopimpla*). *I. (G.) bernuthii*, however, is different from *himalayensis* in having parallel median propodeal carinae, propodeum rugose only in central area between carinae, propodeum dorsally flatter, and the fuscous bands on hind tibia not separated by a clear pale band.

II. SUBFAMILY PORIZONTINAE

Members of the subfamily Porizontinae are characterized by having an apically compressed abdomen, with spiracle placed behind the mid-length of first tergite. Clypeus usually not distinctly separated from face. Mandibular teeth usually equal. Notauli absent. Sternalus absent or short. Propodeum partly to completely areolated. Tarsal claws usually pectinate. Areolet present or absent—when areolet absent, intercubitus basad of second recurrent vein (except in Hellwigiini). Epipleurum of tergites 2-3 separated by a crease.

Subgenital plate transverse, not enlarged. Male clasper usually rounded apically. Ovipositor long or short, with a subapical dorsal notch and lower valve without teeth.

They are usually small sized species. They are usually primary internal parasites of lepidopterous larvae, except for some genera which parasitize larvae of Coleoptera, Rhaphidiidae or Tenthredinidae.

Six genera have been associated with the gypsy moth. They are keyed below:

KEY TO THE TRIBES AND GENERA OF PORIZONTINAE ASSOCIATED WITH THE GYPSY MOTH

1. Cross-section of petiole near its basal third depressed oval or circular. Suture separating its sternite from tergite at or above the mid-height, the suture always present. First tergite without a pit (glymma) before spiracle, sometimes a lateral groove may be present. (Porizontini). 2
   Cross-section of petiole near its basal third quadrangular or prismatic; suture separating its tergite from sternite below the mid-height, the suture present or obsolete. First tergite with or without a pit before its spiracle. (Macrini). 4

2. Pronotum narrow with a deeper groove. Epomia long and strong, almost reaching upper margin of pronotum. Propodeum narrowed towards apex, without carinae (except at base). Eyes distinctly emarginate just above antennal sockets. Ovipositor hardly exerted, as long as the apical depth on abdomen. 3. Casinaria
   Pronotum broad with a shallow groove in the middle. Epomia short, not extending to upper margin of pronotum. Propodeum short, not narrowed, apically, more convex in profile, with distinct transverse and longitudinal carinae. Ovipositor long, 0.5 to 1.0 as long as the abdomen. Eyes weakly emarginate. 3

3. Postpetiole broad, parallel-sided, flatter dorsally and with a lateral carina extending from spiracle to its apex. Basal part of petiole prismatic and flattened above. Frons with a weak median vertical carina. Combined areola and petiolar area of propodeum forming a broad, deep concave trough .
   . . . . 1. Sinophorus
   Postpetiole narrower, more globular in profile, without a lateral carina between spiracle and its apex. Petiole basally more cylindrical or depressed-oval. Frons with or without a median carina. Areola usually constricted posteriorly and the median trough not very deep.
   . . . . 2. Campoplex

4. Nervellus intercepted, though discoidella unpigmented. Areolet receiving second recurrent vein basad of its middle. Apical margin of clypeus with a short median tooth, which is sometimes rather weak. Glymma present. . . . . 4. Campoletis
   Nervellus not intercepted. Areolet receiving second recurrent vein at or distad of the middle. Clypeus without any median projection. Glymma present or absent. 5

5. Nervulus distad of basal vein by about 0.3 its length, strongly sloping to
    make an angle of about 70° with the discoidal vein. Clypeus rather wide,
    flat and with its apical margin truncate. Petiole without a conspicuous
    lateral pit or glymma. . . . .   5. <u>Phobocampe</u>
   Nervulus interstitial or only slightly distad of basal vein, almost vertical
    or slightly arched. Clypeus short and convex. Petiole with a distinct
    lateral pit or glymma. . . . .   6. <u>Hyposoter</u>

1. Genus SINOPHORUS (Fig. 43)

Sinophorus Foerster, 1869. Verh. Naturh. Ver. Rheinlande, 25: 153.

    For taxonomical and synonymical references refer to Townes (1970) and
Gupta and Maheshwary (1977).
    Small sized insects, about 6-11 mm. long. Body stout. Fore wing 3 to 8
mm. long. Clypeus large and flat with apical margin weakly rounded and
truncate. Mandible with a ventral lamella. Malar space 0.43 to 0.56 as long
as the basal width of mandible. Frons with a weak median vertical carina.
Eye weakly emarginate. Pronotum broad, with a shallow groove. Epomia
short, not extending to upper margin of pronotum. Propodeum short, with
areola and petiolar area completely confluent and forming a broad deep concave trough. Median longitudinal carinae of propodeum widely separated.
Areolet present. Second recurrent vein inclivous. Nervellus usually not
intercepted, the base of discoidella usually detached from nervellus. Petiole
prismatic basally, the suture separating the tergite from sternite distinct and
a little below the mid-height. First tergite without a lateral pit or glymma,
but often with a shallow long groove. Postpetiole broad, parallel-sided, flatter dorsally and with a lateral carina extending from spiracle to its apex.
Apex of male clasper rounded. Ovipositor about 1.7 to 2.1 as long as the
hind femur.
    Members of the genus *Sinophorus* are internal parasites of lepidopterous
larvae or sometimes sawflies. This genus has not yet been certainly associated
with the gypsy moth and the records in the literature appear erroneous.

1. SINOPHORUS VALIDUS (Cresson) (Fig. 43)

    Rühl (1914) reported *Limnerium validum* Cresson (type-species of
Sinophorus) as a parasite of *Lymantria dispar* in the U.S.A. on the authority
of Timberlake (1912). Timberlake was working on the biology of *L. validum*,
which is a parasite of *Euproctis chrysorrhoea*, and attempted to rear it on
Lymantria dispar, but failed. Subsequent cataloguers like Stadler, Schedl,
and Thompson, mentioned it as a parasite of the gypsy moth in Europe and
North Africa—on whose authority is not clear.
    Fusco and Simons (1973) mentioned *S. validus* as an unimportant native
parasite of the gypsy moth found in association with the tent caterpillar and the
webworm. Simons *et al.* (1979), quoting Carlson (1973, personnal correspondence) mentioned that this species had been reared from the gypsy moth,
but only rarely. However, Carlson (1979: 625) does not list *L. dispar* as a
host of this species, and states, "I have listed only hosts from labels of
specimens which I have identified as *validus*. Some of the host records in
literature pertain to misidentifications in USNM collections."

It is thus evident that all records of this parasite ex gypsy moth are erroneous.

2. Genus CAMPOPLEX (Fig. 44, 134)

Campoplex Gravenhorst, 1829. Ichneumonologia Europaea, 3: 453.

For full synonymical and other references, refer to Townes (1970) and Gupta and Maheshwary (1977).

Body slender. Small-sized insects about 3-11 mm. long. Fore wing 2-8 mm. long. Clypeus weakly convex, its apical margin rounded, sometimes truncate, usually weakly depressed in the middle. Mandible with a ventral lamella, which is often weak. Malar space 0.4 to 0.8 as long as the basal width of mandible. Frons without a median carina, except rarely. Eye not or only weakly emarginate. Pronotum broad, with a shallow groove. Epomia short. Propodeum with its median longitudinal carinae somewhat closer *(cf. Sinophorus)* and angulate at the junction of areola and petiolar area—these two areas confluent but the junction between them discernible. Areola and petiolar area forming a flat, weakly depressed or a little excavated area, and not in the form of a broad deep trough. Areolet present, stalked or weakly sessile. Sometimes areolet absent. Second recurrent vein slanting outward. Nervellus usually intercepted. Petiole usually cylindrical in cross-section in basal 0.3, sometimes a little squarish. Postpetiole more globular in shape, usually wider in the middle, and somewhat raised in profile, without any lateral carina. Suture separating first tergite from sternite distinct and at middle or a little below middle. Glymma absent, but in its place usually a narrow, shallow groove present. Apex of male clasper broad and round, or sometimes with a shallow apico-dorsal emargination. Ovipositor 1.5 to 4.1 as long as the hind femur.

Ratzeburg (1844) recorded two species of *Campoplex* as parasites of the gypsy moth, *viz.*, *Campoplex difformis* Gravenhorst and *C. conicus* Ratzeburg. The latter species is a synonym of *Casinaria tenuiventris* (Gravenhorst). The identity of the species reported as *C. difformis* is uncertain, as according to several authors it has been a mixed and often misidentified species. Perhaps two different species, *C. difformis* and *Venturia deficiens* were involved, but the true identity of the species reared from the gypsy moth can only be decided when voucher specimens or freshly reared material turns up.

Campoplex deficiens Gravenhorst (= *algerica* Habermehl) is a species of *Venturia* (Horstmann, 1974). It is not discussed further.

Campoplex rapax Gravenhorst, as reported in the gypsy moth literature (Rudow, 1911) appears to be a species of *Hyposoter*. It has been referred to subsequently under *Anilasta* or *Anilastus*.

Momoi (1961) reported *Campoplex sugiharai* (Uchida) from *Lymantria dispar* in Japan.

Campoplex sp. of Burgess and Crossman (1929) is actually *Casinaria tenuiventris* (Gravenhorst).

Kolemietz (1958) reported *Omorgus* sp. as a parasite of the pupa of the gypsy moth, reared in August 1955 in Siberia. *Omorgus* is a synonym of *Campoplex,* the species of which are internal parasites of lepidopterous larvae, spinning their own cocoon after killing the host larva or prepupa and emerging from their own cocoons.

KEY TO THE SPECIES OF CAMPOPLEX ASSOCIATED
WITH THE GYPSY MOTH

1. Face rugulose. Abdomen wholly black. Hind femur, tibia and tarsus
    yellowish-brown. Occipital carina joining hypostomal carina at the
    base of mandible. Propodeal areola well formed in both the sexes,
    though widely open behind. Europe. . . . . . .   1. difformis (Gmelin)
    Face granulose. Abdomen reddish-brown laterally. Hind femur, tibia
    and tarsus blackish-brown to light brown. Occipital carina joining
    hypostomal carina just above the base of the mandible. Propodeal areola
    not well formed in the female, its median longitudinal carinae absent
    below costulae and weakly represented apically. Japan, Korea, Ryukyus,
    Thailand and southern India. . . . .   2. sugiharai (Uchida)

1. CAMPOPLEX DIFFORMIS (Gmelin) (Fig. 134)

>   *Ichneumon difformis* Gmelin, 1790. *In* Linnaeus: Systema Naturae,
>   Ed. 13 1 (5): 2,720. des. Type destroyed. (Neotype ♀, designated
>   by Horstman (1969), the same specimen as the lectotype of
>   *mutabilis* Holmgren).
>   *Campoplex difformis:* Gravenhorst, 1829. Ichneumonologia Europaea,
>   3: 458. Europe. (A mixed series).
>   *Limneria mutabilis* Holmgren, 1860. Svenska Vetensk. Akad. Handl.
>   (N.F.) 2(8): 55. Lectotype ♀, labelled and designated by Hinz (1964,
>   Entomophaga, 9: 70), Sweden: Småland (Stockholm).
>   *Campoplex difformis* Gravenhorst: Ratzeburg, 1844: 92. Host: *Lymantria
>   dispar* (vide Bouché, Garten Ins., p. 154).

Horstman (1969) has shown that the true *Campoplex difformis* Gravenhorst,
which is actually *Campoplex difformis* (Gmelin), is a senior synonym of
Campoplex mutabilis (Holmgren), and that what most authors have called
"*Campoplex difformis* Gravenhorst" is the same as *Venturia deficiens* Gravenhorst. Aubert (1975) rejects the neotype fixation of *difformis* by Horstman and
considers *difformis* and *mutabilis* as different species. According to him the
present species should be called *mutabilis* Holmgren, and *deficiens* Gravenhorst = *algerica* Habermehl, should be called *difformis* (Gmelin).

Whether true *difformis* (Gmelin) as described below is a parasite of
Lymantria dispar or not, can only be ascertained by fresh rearings.

Male and *female:* Face rugulose. Clypeus, frons, and vertex granular.
Malar space 0. 6 the basal width of mandible. Interocellar distance 1. 4 the
ocellocular distance. Temple and vertex granuloso-mat. Occipital carina
joining hypostomal carina at base of mandible. Mesoscutum granulose, tending
to be rugose at places. Mesopleurum and metapleurum granulose, with a few
scattered punctures. Scutellum granulose. Propodeum granulose in basal
areas, transversely striate in apical half, particularly in petiolar area. Propodeal carinae rather strong, including the lateral longitudinal carina (which
is weaker in other species). Areola wide, as wide or wider than the length of
costula, rather widely open behind. Areola and petiolar area forming a rather
wide trough as is usually seen in *Sinophorus*. Postpetiole finely granulose.
Abdomen mat. Ovipositor as long as or a little shorter than the fore wing and
longer than the abdomen.

Black. Mandible often partly brownish. Tegula yellow. Coxae black.
Legs otherwise reddish-brown. Hind trochanters blackish-brown. Hind tibia
and apex of hind femur fuscous. Tegula brown in the male.
Length: 6-7 mm. Fore wing 4. 5 to 5 mm. Ovipositor 4. 0 mm.
Specimens from Europe examined in the Townes Collection. No reared
specimens seen.

2. CAMPOPLEX SUGIHARAI (Uchida)

Omorgus sugiharai Uchida, 1932. Trans. Sapporo Nat. Hist. Soc.,
      12: 74. Japan.
Campoplex sugiharai: Momoi, 1961. Kontyü, 29: 272. Japan, Korea.
      Host: *Lymantria dispar*.
Campoplex sugiharai: Townes, Momoi and Townes, 1965. Mem. Amer.
      Ent. Inst., 5: 276. Japan, Korea. Momoi, 1970. Pacific Insects,
      12: 383. (Key and description of two subspecies from Thailand
      and Ryukyus). Gupta and Maheshwary, 1977. Ichneumonologia
      Orientalis, 4: 79.

This parasite has not been previously cited in the gypsy moth literature.
It has the abdomen reddish-brown laterally and hind femur, tibia and tarsus
blackish-brown to light brown. It occurs in the Orient and Japan.
Male and *female:* Face, frons, and vertex granulose. Clypeus finely
granulose. Malar space 0. 5 the basal width of mandible. Interocellar distance
1. 0 to 1. 2 as long as the ocellocular distance. Temple and occiput finely
granuloso-mat. Occipital carina joining hypostomal carina a little above the
base of mandible. Mesoscutum and scutellum strongly granulose. Pronotum
granuloso-striate. Mesopleurum and metapleurum granulose, metapleurum
more finely and densely so. Propodeum granulose. Petiolar area apically
trans-striate. Basal transverse carina and the short carinae bounding basal
area, the strongest. Areola not formed in female. Median longitudinal carinae
erased, visible only in the apical half and enclosing a moderately deep trough-
like petiolar area. Apical transverse carina visible laterally. In male median
carinae joining basal transverse carina so that the areola is formed laterally.
Areola a little constricted apically and widely confluent with petiolar area.
Second lateral area and petiolar area of propodeum trans-rugulose. Hind
coxa, and first and second tergites granulose. Petiole without a lateral pit or
groove before spiracle, but with a fine lateral carina from base to the spiracle.
Thyridium on second tergite small, oval and separated from base of second
tergite by about 2. 5 its length. Abdomen from third tergite onwards mat.
Ovipositor long, about as long as the abdomen, 0. 8 to 0. 9 as long as the fore
wing, evenly arched upwards.
Black. Mandible, palpi, and tegula partly or wholly, yellow. Fore and
middle legs largely yellowish-brown, with their coxae often brownish to
blackish. Hind coxa black. Hind leg otherwise dark brown with tibia centrally
light brown. Abdomen brownish-black to black, with third and the following
tergites yellowish-brown. Antenna brownish black to black.
Length: 7-9 mm. Fore wing 4 to 5. 5 mm. Ovipositor 3. 5 to 4. 5 mm.
Momoi (1970) recognized three subspecies: *C. sugiharai sugiharai* (Uchida)
from Japan and Korea, *C. s. okinawensis* Momoi from Ryukyus (Okinawa) and
C. s. australis Momoi from Ryukyus (Omotodake) and Thailand. The three
subspecies differ in the coloration of scape and pedicel (light brown to black

beneath), hind first trochanter (light brown to black), middle coxa (brown to black), tergites 3-8 (reddish-brown to black laterally), and hind femur (light reddish-brown to blackish-brown).

Specimens from Japan and Korea examined in the Townes Collection. No reared specimen seen.

A specimen from Naduvattom, S. India in the Townes Collection, agrees with this species, but comes somewhat in between *sugiharai sugiharai* and *sugiharai australis*.

Distribution and *hosts:* Momoi (1961) reported it as a widespread species in Japan and Korea and recorded *Lymantria dispar* as a host in Kyushu, Japan.

This species comes close to *Campoplex burmensis* Gupta and Maheshwary (1977) from Burma and *C. oriens* Gupta and Maheshwary from the Orient. It differs from the former in having a long ovipositor and areola not formed in the female. It differs from *oriens* in having blackish hind femur, and short, almost parallel-sided basal area.

3. Genus CASINARIA   (Fig. 45, 135)

   Casinaria Holmgren, 1859. Öfvers. Svenska Vetensk. Akad. Forh.,
        15: 325.

For full synonymical references, refer to Townes (1970). The Oriental species were treated by Gupta and Maheshwary (1977), and the Nearctic species by Walley (1947).

Body slender, with fore wing 4 to 9 mm. long. Eye margin strongly indented opposite antennal socket. Cheek short. Temple moderately short to very short and flat. Mesopleural suture, or at least its median 0.3, impressed as a sharp groove. Pronotum narrow, with epomia complete. Propodeum moderately long to very long, its apex between basal 0.3 and the apex of hind coxa. Propodeum usually with a median longitudinal trough, with incomplete areolation, or without carinae. Propodeal spiracle short, elongate-oval. Fore wing with areolet always present. Second recurrent vein usually inclivous. Nervellus not intercepted. Petiole cylindric or weakly depressed, moderately long to very long, the suture separating its sternite from tergite at or a little above the mid-height. Postpetiole a little bulbous. Glymma absent. Apex of male clasper rounded or a little elongate, without a subapical dorsal notch. Ovipositor short, 0.8 to 1.4 as long as the apical depth of abdomen, notched subapically.

This is a large genus of world-wide distribution. They are parasitic within lepidopterous caterpillars.

Two species of *Casinaria* are generally reported as parasites of the larvae of *Lymantria dispar, viz., Casinaria tenuiventris* (Gravenhorst) (= *Campoplex tenuiventris* Gravenhorst = *Campoplex conicus* Ratzeburg) and *C. ischnogaster* (Thomson). The record of the latter species is erroneous. Momoi (1963) added *Casinaria anastomosis* Uchida to the list of gypsy moth parasites in Japan. It is a junior synonym of *Casinaria nigripes* (Gravenhorst).

Morley and Rait-Smith (1933) recorded *Casinaria ischnogaster* as a parasite of *Lymantria dispar* on the authority of Morley (1914, Ichn. Britain, 5: 112). In the latter reference, Morley clearly states that records of the "synonymous *Campoplex conicus* from *Bombyx dispar* (Ratz., 1: 95), should be referred to Gravenhorst's species [*tenuiventris*], which Thomson also calls

C. latifrons Holmgr." He, in fact, did not record *C. ischnogaster* as a parasite of *Lymantria dispar* and was only clarifying the identity of *ischnogaster* vs. *tenuiventris* in the British fauna. *C. ischnogaster* is therefore to be removed from the list of gypsy moth parasites.

In literature two species of *Casinaria*, viz., *C. claviventris* Holmgren and *C. scutellaris* Tschek are reported as parasites of *Lymantria monacha*, the nun moth in Europe.

KEY TO THE CASINARIA SPECIES PARASITIC UPON THE GYPSY MOTH

1. Abdomen wholly black. All coxae and trochanters black. All femora and tibiae reddish-brown (except rarely the hind femur blackish). Europe, USSR. . . . . 1. tenuiventris (Gravenhorst)
   Abdomen with at least tergites 3 and 4 red. Middle and hind legs almost wholly blackish, their tibiae lighter in basal half. Japan, Europe, USSR. . . . . 2. nigripes (Gravenhorst)

1. CASINARIA TENUIVENTRIS (Gravenhorst) (Figs. 45, 135)

Campoplex tenuiventris Gravenhorst, 1829. Ichneumonologia Europaea, 3: 482. Poland.
Campoplex conicus Ratzeburg, 1844. Die Ichneumonen der Forstinsecten, 1: 95. Germany. Host: *Lymantria dispar*. (Syn. by Dalla Torre, 1901-02: 126).
Casinaria latifrons Holmgren, 1858. Svensk. Vetensk. Akad. Handl., (2) 2(8): 50. (cf. Dalla Torre, 1901-02).
Casinaria tenuiventris: Brischke, 1880. Schrif. Naturf. Ges. Danzig, N.F. 4 (4): 147.
Campoplex sp. Burgess and Crossman, 1929. U. S. Dept. Agr. Tech. Bull., 86: 104. Larval parasite. France, Czechoslovakia.

For fuller synonymical references, see Dalla Torre (1901-02), Morley (1914), Meyer (1935), and Townes, Momoi and Townes (1965). Dalla Torre misspelled *Campoplex conicus* as *C. canonicus* Ratz. Howard and Fiske (1911) reported *C. tenuiventris* as a parasite of gypsy moth larvae, though the earliest record would be that of Ratzeburg (1844), as *Campoplex conicus*. I have examined specimens from France and Czechoslovakia that were mentioned by Burgess and Crossman as *Campoplex* sp.; these belong to the present species. In recent years, it has been mentioned as having been reared from the gypsy moth in Yugoslavia (Vasic, 1958), Russia (Shapiro, 1956) and Iran (Herard *et al.*, 1979)

Male and *female:* Face almost squarish, rugulose. Clypeus narrow, slightly convex, granulose, its apical margin convex and impressed. Mandible short and broad, with a wide lower lamella. Malar space 0.75 to 0.85 as long as the basal width of mandible. Temple strongly receding behind eye. Head lenticular. Frons granulose, without a median carina. Interocellar distance 1.4 as long as ocellocular distance. Lateral ocellar diameter about equal to ocellocular distance. Occipital carina strong, sinuate below lower level of eye and meeting hypostomal carina at the base of mandible. Mesoscutum convex, finely rugose. Scutellum rugulose, slightly convex with its lateral carinae confined to its front 0.4. Mesopleurum and metapleurum granuloso-rugulose.

Pleural area of propodeum more strongly sculptured than metapleurum, generally finely rugose. Dorsal face of propodeum rugose with a shallow and broad median groove along its length, which is beset with short transverse carinae. Propodeum coarser in male than in female, particularly in the median trough and areas laterad to it. Propodeum narrowed apically and extending to the basal 0.33 of hind coxa. Propodeal spiracle elongate-oval. Hind coxa granulose. Areolet moderate-sized, short petiolate, receiving second recurrent in the middle. Nervellus not intercepted, reclivous. First tergite 1.2 as long as the second, shorter than the hind femur. Abdomen not strongly compressed, clavate or rounded apically. Male claspers broader and rounded apically. Ovipositor short, straight, hardly exerted beyond the tip of abdomen. Ovipositor sheath a little clavate.

Black. Tegula black to yellowish-brown. Coxae and trochanters black. Legs otherwise reddish-brown with tarsi and apical 0.3-0.5 of hind tibia lightly fuscous. Sometimes hind femur and tibia dark, brownish or black, with tibia faintly banded. Abdomen wholly black. Middle femur may have basal black marks.

Length: 6-8 mm. Fore wing 3-4 mm.

Specimens: Germany: Würzburg, 1♂, June 1974, *ex Lymantria dispar.* France, 1♂, June 1981, *ex Lymantria dispar* (BIRL, Newark). France [Hyeres], 1♂, *ex Lymantria dispar*, V-28-1922, Gypsy Moth Lab No. 3438 (FIS. Hamden, Ct.) Czechoslovakia, Bilky, 1925, *ex Lymantria dispar*, No. 3475, 1♂ 1♀ (Hamden, Ct.), 1♂ 1♀ (USNM), 1♂ 1♀ (Townes). Bulgaria, 1924, *ex Lymantria dispar* (Hamden, Ct.) France: Foret des,May-June 1972-73, 3♀, *ex* gypsy moth larvae (EPL, Paris). Poland: Skiern., 1♀, June 1975, *ex* gypsy moth larva (EPL, Paris). China: Heilongjiang Province. Several males and females, reared from larvae of *Lymantria dispar* and *L. mathura* by Schaefer, *et al.* in May-June 1982.

Distribution: Europe, China, Iran (Herard *et al.*, 1979).

This species belongs to the Atrata Group of Gupta and Maheshwary, 1977. The specimens from China have the hind femur wholly black. It comes close to *C. natashae* Maheshwary & Gupta from the Himalaya.

Kolemietz (1958) mentioned rearing *Casinaria tenuiventris* (Gravenhorst) from gypsy moth larvae in Siberia during August 1954. It was considered rare. Pschorn-Walcher (1974) reported it (plus another species "red" = *nigripes*) from Southern France, Austria and Bavaria as an uncommon or rare parasite of young as well as of older larvae of the gypsy moth. The larval head of this species is figured by Short (1978).

Casinaria ischnogaster, wrongly reported as a gypsy moth parasite is close to *C. claviventris* in leg color and general body sculpture, but the head is not so lenticular, propodeum broad and short, rugose dorsally, and with its median groove shallow. The postpetiole is dilated at the level of the spiracle, then narrowed apically. In *C. claviventris* the outer posterior angle of discoidal cell is slightly obtuse, and the radial cell shorter and broader. *C. scutellaris* is unknown to me. The latter two species have been reported parasitic upon *Lymantria monacha*.

2. CASINARIA NIGRIPES (Gravenhorst)

Campoplex nigripes Gravenhorst, 1829. Ichneumonologia Europaea,
    3: 598. Poland.
Casinaria nigripes: Thomson, 1887. Opuscula Entomologica, 11: 1102.

Casinaria anastomosis Uchida, 1930. Insecta Matsumurana, 4: 130.
Japan: Sapporo. <u>New synonym</u>. Host: *Ichthyura anastomosis*.

Fuller synonymical references to the two species can be found in Townes, Momoi and Townes (1965: 279). Momoi (1963: 54) first reported *Lymantria dispar* as a host of *anastomosis* in Japan. A recently bred specimen of this species from the gypsy moth in France is at hand.

Other host records are: *Ichthyura anastomosis*, *Orgyia gonostigma*, *O. thyellina* and *Epicnaptera ilicifolia* in Japan; and *Dasychira pudibunda*, *Orgyia antiqua*, *Dendrolimus pini*, and *D. sibiricus* in USSR. Its hyperparasites are *Itoplectis alternans* in Japan and USSR, and *Gelis areator*, *Gelis* sp., *Pteromalus* sp., and *Theronia atalantae* in USSR.

This species is readily distinguished by the red color of central abdominal tergites and blackish hind leg.

Male and *female:* Essentially similar in sculpture to *C. tenuiventris* and differing as follows:

Lateral ocellus separated from eye by about 0.7 its diameter. Interocellar distance 1.8 the ocellocular distance in female and about 2.0 in male. Malar space 0.5 to 0.7 the basal width of mandible. Areolet large, with a short petiole. First tergite 1.3 as long as propodeum, thinner than in *tenuiventris*, less widened apically. Propodeum somewhat coarser dorsally, particularly in male. In male median groove trans-carinate and lateral areas reticulate. Both petiolar area and lateral areas with irregular longitudinal incomplete carinae. Body sculpture coarser, tending to be finely rugose, particularly face, meso- and metapleurum.

Abdomen with apex of tergite 2 and tergites 3 and 4 wholly red. (Tergite 5 also red in one specimen before me.) Legs black with fore femur reddish and fore tibia and tarsus yellowish. Middle tibia and tarsus often yellowish-brown, particularly in male, and middle femur may also be brownish-yellow apically.

Length: 8-10 mm. Fore wing 5-7 mm. Larger than *C. tenuiventris*.

Specimens: S. France, 1♀, May-June 1972, *ex* gypsy moth larvae (EPL, Paris). Japan: Sapporo, 1♂, compared with type of *C. anastomosis* (Townes). Germany: 3♂, 3♀ det. as *C. nigripes* by Heinrich and by Teunissen (Townes). China: Heilonjiang Province, several ♂, ♀, *ex Lymantria mathura*, May-June 1982, P. Schaefer *et al.* (BIRL, Delaware).

Distribution: Europe, USSR, China, Japan.

Hosts: Lymantria dispar, *L. mathura*, and others, as mentioned above.

Pschorn-Walcher (1974) reported this as *"Casinaria* sp. red" from France, Austria and Bavaria as an uncommon parasite of the larvae of the gypsy moth, together with *C. tenuiventris*. He also referred to both of them collectively as *"Casinaria* spp."

4. Genus CAMPOLETIS (Fig. 46)

Campoletis Foerster, 1869. Verh. Naturh. Ver. Rheinlande, 25: 157.

For full synonymical references, refer to Townes (1970). *Anilastus* and *Anilasta* are junior synonyms of *Campoletis*, although in older literature *Anilastus* has been incorrectly but rather consistently used for *Hyposoter*. This was because of the misidentification of the type-species of *Anilastus*.

Small sized species with fore wing 3.3 to 7.5 mm. long. Body moderately slender. Eyes weakly emarginate. Malar space small. Clypeus moderately wide, weakly to moderately convex, its apical margin with a median tooth of varying shape and size. Sometimes the tooth indistinct. Propodeal areola usually elongate, hexagonal, distinct from petiolar area or more or less fused with it. Areolet short petiolate, or pointed above, receiving second recurrent vein basad of its middle. Nervellus intercepted, discoidella unpigmented. First abdominal segment moderately decurved, with a moderately slender petiole and moderately stout postpetiole. Suture separating tergite from sternite below the mid-height of petiole. Glymma present. Thyridium subcircular, separated from base of tergite by 0.7 to 1.5 its diameter. Ovipositor moderately stout, upcurved or almost straight, 1.6 to 3.5 as long as apical depth of abdomen.

Species of *Campoletis* have been occasionally collected from the larvae of the gypsy moth, recently in China (P. Schaefer) and France (EPL, Paris). Their specific identities are uncertain.

Species reported in the literature under *Anilastus* or *Anilasta*, like *Anilastus rapax* (Gravenhorst), or *Anilastus* n. sp. Stadler (1933) appear to belong to *Hyposoter*. Their specific identity is uncertain in the absence of the reared specimens. The figure of the larval remains in Stadler (1933) corresponds to that of *H. tricoloripes* (Viereck), which is known from Europe.

Two specimens of *Campoletis* bred from the gypsy moth in France are at hand. They are described below as *Campoletis* sp., because it is probable that the species has a name in Europe, but the name is unknown to me.

CAMPOLETIS sp.

Male: Face granuloso-punctate. Face and clypeus forming a convex surface, without a clear demarcation. Apical margin of clypeus convex and with a median acute tooth, which is rather well developed. Malar space about as wide as the basal width of mandible. Ocellar area punctate. Interocellar distance 1.4 the ocellocular distance. Ocellocular distance equal to ocellar diameter. Temple subconvex, not receding from eye. Thorax largely granulose. Areola granulose, open behind, a little constricted at its junction with petiolar area. Costula distinct. Petiolar area striate, subconcave and widened medially. Propodeal carinae more or less complete. Petiole and postpetiole finely granulose. Glymma in the form of a deep pit. Tergite 2 more than twice its basal width, mat. The following tergites subpolished.

Black. Mandible, palpi, fore and middle legs, pale yellowish-brown. Tegula yellow. Hind femur orange brown. Tibia whitish-yellow medially and base of hind tarsus yellowish-white. Second and following tergites reddish-brown laterally. Second and third also with faint reddish apical irregular bands.

Length: 5.5-6 mm. Fore wing 4 mm.
Specimens: South France, 2♂, May-June 1973 (EPL, Paris).
Cocoon: Silken white, slender, cylindrical, about 3.0 as long as its medial diameter.

5. Genus **PHOBOCAMPE**   (Figs. 48, 136)

Phobocampe Foerster, 1869. Verh. Naturh. Ver. Rheinlande, 25: 156.

For full synonymical references, etc. refer to Townes (1970: 175).
Body short, with fore wing 4 to 6 mm. long. Eye margin weakly or not at

all indented opposite antennal sockets. Cheek short. Clypeal foveae not distinctly impressed, open. Clypeus weakly convex, with a subapical groove, the apex sharp, truncate or subtruncate. Mandible short, with a fringe on its lower margin, its teeth equal. Temple rather short. Occipital carina joining hypostomal carina. Lower corner of pronotum translucent. Mesopleurum mat to granulose. Posterior mesosternal carina complete, though often weak. Propodeum short, its areola and petiolar areas confluent or with a small constriction at their junction. Costula present. Propodeal spiracles circular or oval. Hind basitarsus without a ventral row of closely spaced hairs. Tarsal claws short and pectinate. Fore wing with areolet petiolate above, receiving second recurrent vein distad of its middle. Nervulus distad of basal vein by about 0.3 its length, strongly slanted, forming an angle of about 70° with the discoidal vein. Discoidella unpigmented, usually not reaching nervellus. Nervellus vertical and usually not intercepted. Petiole slender. Postpetiole rather broad. Suture separating sternite from tergite a little below the mid-height of petiole. Glymma small to obsolescent. Abdomen short and stout. Thyridium circular, separated from base of second tergite by about its diameter. Ovipositor about as long as the apical depth of abdomen, with a submedian notch.

The hosts are small or early instar lepidopterous caterpillars.

Viereck (1911) described *Hyposoter disparis* (now *Phobocampe unicincta*) as a parasite of gypsy moth larvae received in the U.S.A. from Russia. Howard and Fiske (1911) mentioned that they were first received in 1907 in a shipment of small gypsy moth caterpillars from Kiev, Russia. This parasite was subsequently released in Northeastern United States and has become established. It has, however, proved to be of little value in the control of the gypsy moth.

Burgess and Crossman (1929) mentioned having received specimens of two distinct species of *Hyposoter* in addition to *disparis* from Europe (Spain, Czechoslovakia, Hungary and Yugoslavia) during 1924, 1925 and 1927. The adults of these emerged from the cocoons in the same season in which they were formed, rather than hibernating in cocoons and issuing the following spring, as was the case in *disparis*. They also pointed out that such adults did not mate and oviposit and that "it seemed that they might hibernate. They were placed in several types of containers for hibernation, but the last one died after living 92 days." About the same specimens Muesebeck (1933) remarked, "*Hyposoter disparis* is easily confused with an unidentified species of the same genus which is occasionally reared as a parasite of the gypsy-moth larvae in Europe. The latter differs, however, in having the antennae 28-30 segmented; in the ocellocular line being slightly shorter than the diameter of an ocellus; in the less erect areolet of the anterior wing, with the second recurrent joining the cubitus very near the second intercubitus; and in having the petiolar area wider and uniformly closely granular and opaque."

The majority of these specimens represent a new species, *Phobocampe lymantriae*, while some from Hungary are *P. unicincta*. There are some variations in the two, which occur sympatrically in many areas in Europe. Both of them also occur in Japan.

Shapiro (1956) listed *Phobocampe pulchella* Thomson as one of the ichneumonid parasites of *Lymantria dispar* in Russia. Drea (in Doane and McManus, 1981: 31) also mentions *P. "pulchella* Thomson" as a parasite of the gypsy moth from Yugoslavia. This species is rather close to *P. unicincta*, but is a different species (lectotype examined). I suspect that the species involved might be either of the two European species of *Phobocampe* discussed here.

The record of Herard *et al.* (1979) of *Phobocampe* sp. from the gypsy moth in Iran concerns *P. unicincta*. A specimen of this material, now in Washington, has been examined. The record of Drea and Fuester (1979) of *Phobocampe*, n. sp. from Poland apparently pertains to *P. lymantriae*, which is sympatric with *unicincta*. Specimens collected by Fuester, located in Washington, have also been examined.

KEY TO THE PHOBOCAMPE SPECIES ASSOCIATED WITH THE GYPSY MOTH

1. Tergite 2 yellow or orange in apical half, sometimes this may be narrow (0.5 to 0.33); in males much narrower. Tergite I, often apically yellow in female. Postpetiole parallel-sided, laterally not distinctly margined, as long as or longer than wide. Propodeal areola usually constricted below costula, and median longitudinal carinae usually distinct. Areola and petiolar areas with transverse striations. Malar space 0.5 to 0.7 the basal width of mandible. Interocellar distance 1.3 to 1.5 the ocellocular distance. Ocellocular distance longer than ocellar diameter (rarely equal). Europe, USSR, Japan, and USA. . . . .   1. unicincta (Gravenhorst)
   Tergite 2 narrowly yellow apically (0.25-0.3); in male often without yellow band or only apicolaterally faintly yellow. Tergite I hardly yellow. Postpetiole rather sharply constricted from petiole, distinctly margined and convex laterally, wider than long, widest submedially. Propodeal areola not fully formed, not constricted, widely open below. Areola and petiolar areas granulose. Malar space 0.25-0.4 the basal width of mandible. Interocellar distance 1.7 to 1.9 the ocellocular distance. Ocellocular distance slightly shorter than ocellar diameter. Europe, Japan, U.S.A. . . . .   2. lymantriae, new species

1. PHOBOCAMPE UNICINCTA (Gravenhorst) (Figs. 49-51, 57-62, 136)

Campoplex unicinctus Gravenhorst, 1829. Ichneumonologia Europaea, 3: 529. ♀. (♂ misdet.).
Hyposoter disparis Viereck, 1911. Proc. U. S. Natl. Mus., 40: 478. ♂, ♀ [Kiev, USSR] "Gypsy moth Lab." Synonymized by Carlson (1979).
Limnerium disparis: Howard and Fiske, 1911. U. S. Dept. Agri. Bur. Ent. Bull., 91: 121, 191. Japan. Russia. Host: *Lymantria dispar*.
Phobocampe disparis: Townes, 1945. Mem. Amer. Ent. Soc., 11: 646.
Phobocampe unicincta: Carlson, 1979. *In* Krombein *et al.*: Catalog of Hymenoptera in America North of Mexico, 1: 661. Syn. Introduced in U. S. A.
Biological references: Howard and Fiske (1911), Burgess and Crossman (1929), Muesebeck and Parker (1933), Schedl (1936).

This species has so far been referred to as *Phobocampe disparis* or *Hyposoter disparis* and has only recently been synonymized under *P. unicincta* by Carlson (1979), after examining the types of *unicincta* and *disparis*.

This species was first discovered in the gypsy moth laboratory in Massachusetts in 1907 from a small collection of gypsy moth caterpillars imported from Kiev, Russia. In 1911 it was found in great abundance at Gioia Tauro,

Italy. The cocoons of it gathered from Italy were shipped to Massachusetts and adults that emerged the following spring (1912) were liberated around Melrose Highland. It was recovered from the field in 1913. It has apparently been recovered each year but only in small numbers. It has established itself in the U.S.A. but has not been able to exert much influence on the population of the gypsy moth.

Male and *female:* Antennal flagellum usually 28-32 segmented. Face strongly granulose. Clypeus finely granulose, its apical margin impressed and straight. Malar space 0.5 to 0.7 the basal width of mandible, a little wider in male than in female. Frons granulose. Vertex finely so. Interocellar distance smaller, 1.3 to 1.5 the ocellocular distance. Ocellocular distance longer than ocellar diameter. Temple and occiput subpolished and receding from the eye. Mesoscutum coarsely granulose. Scutellum finely granulose. Propleurum, mesopleurum and metapleurum granulose, but granulations finer than that of mesoscutum. Granulations on propleurum and speculum a little sparser and these areas somewhat shiny. Pronotal groove and prespecular area with carinations. Granulations on metapleurum somewhat similar to that of meso-scutum. Propodeum granulose in basolateral areas. Areas apicad of basal transverse carina with coarse granulations. Petiolar area and areola trans-carinate, more so in female than in male. Sometimes males with granular areola. Propodeal carinae in general strong. Basal transverse carina usually strongly angled medially. Areola a little constricted apically (though open), its lateral carinae a little convergent or parallelsided and then widely diverging. Sometimes these carinae weak or obliterated just below costulae. Areola more flat and petiolar area a little concave. Propodeal spiracle oval, connected to pleural carina by a strong carina equal in length to the spiracular opening. Areolet small, petiolate with second recurrent vein emitted from its middle. Nervulus inclivous, distad of basal vein by about 0.3 to 0.33 its length, making an angle of 60° with the submedian vein. Discocubitus strongly arched. Petiole quadrate basally, a little flattened apically, gradually merging with the postpeti-ole. Postpetiole finely to coarsely granulose, parallel-sided, longer then wide or almost squarish. Postpetiole laterally not margined, the lateral carina thin or weak. Lateral groove of petiole weak. Tergite 2 narrower, elongate, more so in male, about 2.0 as long as its basal width in female and more than 2.0 in male. Thyridium irregularly round, separated from base of second ter-gite by a distance a little less than its maximum diameter. Spiracle of second tergite at its middle. Abdomen dull mat. Ovipositor as long as the apical depth of abdomen, or a little longer, finely tapered and slightly upcurved. Ovipositor sheath a little clavate apically. Female subgenital plate appears more hairy. Epipleurum of tergite IV with uniformly distributed hairs, not clustered along its margin.

Black. Mandible, palpi, fore and middle trochanters, and tegula, yellow. Scape and pedicel ventrally, hind corner of pronotum, fore and middle legs largely, and hind trochanters, yellowish. Sometimes fore and middle coxae, tibiae and trochanters more yellow than brown. Hind coxa and femur reddish-brown, tibia yellow to yellowish-brown and tarsus brownish. Apex of hind coxa, apex of femur, base and apical 0.25 of tibia, and tarsal segments api-cally infuscate. Apex of postpetiole narrowly, thyridia and apical 0.33 to 0.5 of second tergite yellow to reddish-brown. In males yellow on first tergite absent. Second tergite apically usually amber colored or only narrowly or laterally yellowish-brown.

The propodeum exhibits some variations. Usually the areola and petiolar area are demarcated by a constriction and both have irregular carinations in addition to granulations. Sometimes lateral carinae of areola are weak and areola may be more granular. But the nature of postpetiole, malar space and ocellocular distance in conjunction with the propodeal sculpture and the width of the yellow band on second tergite will distinguish it from the related *Phobocampe lymantriae*, n. sp. The males often have granular areola but they usually have faint or narrow apical band on second tergite and the malar space is larger than in the female. Other distinguishing characters also hold good for the male.

Length: 4 to 7 mm. Fore wing 3.5 to 5 mm.

Specimens: 30♂, 50♀, from: Italy [Gioia Tauro] April-May 1912. Vees, Hungary, May 1929. Var and Bouches du Rhone, France, May-June 1973. S. France, June 1972, May 1973. France, June 1980. Prilep, Yugoslavia, May-June 1973. Jovljak, Yugoslavia, May-June 1973. Trenton, Mass., U.S.A., May 1948. Lancaster Co., Pa., U.S.A., May 1977. Hawk Mts., Pa., May 1974. Mohonk Lake, N.Y., May, 1974, all reared from *Lymantria dispar*. Glenville, N.Y., June 1960. High Point State Park, N.J., May 1973. Starkoc. Boh., Czechoslovakia, July 1961 (not reared).

In addition, the following specimens also belong to this species: North Iran, Location A, 1♀, May 5-14, 1976, *ex Lymantria dispar*, Herard & Mercadier, EPL-Iran-76-2 (Washington). This has the postpetiole a little wider apically. Japan: Hokkaido: Hobetsu, 1♂, 1♀, June 15, 1978. *ex Lymantria dispar*, P. Schaefer, emerged Nov. 1978 (Washington). 3♀, Japan "APL-78-18C" and APL-77-26" (BIRL, Newark). 1♂, Eniwa, Hokkaido, June 3, 1977, reared, Herard. Honshu, Utsunomiya, 1♂, June 3, 1978, *ex Lymantria dispar* (BIRL, Newark).

The specimens from Japan labelled "APL-78-18C" and "APL-77-26" are typical *unicincta*, while the other specimens from Japan have somewhat less striate propodeum, hind femur is blackish, and the band on second tergite is narrow. They match with *unicincta* rather than with *lymantriae* in most of the characters. These have been previously labelled or mentioned as '*Phobocampe* n. sp.''

Cocoon: Cocoon uniformly dark brown in color, 6 x 4 mm., oval. Sometimes blackish-brown.

Distribution: Europe, USSR, Iran, Japan, and U.S.A. (Introduced).

This species is widespread in Europe and has definitely been established in northeastern United States. Pschorn-Walcher (1974) reported *Phobocampe unicincta* (= *disparis*) to be a dominant parasite of the gypsy moth in Würzburg, Germany, subdominant in eastern Austria, and of moderate abundance in southern France. Earlier, Burgess and Crossman (1929: 49) reported that it was scarce in Russia during their search in 1909-1910, but was found abundantly in Gioia Tauro, Italy in 1911. Muesebeck and Parker (1933) reported that specimens of this species have been received from various localities in Austria, Czechoslovakia, Poland, Hungary, Yugoslavia, Bulgaria and Italy. It appears to be most abundant in south-central Europe. In the United States it was recovered by them from Northeastern Massachusetts and Eastern United States. Since then it has been recovered from Pennsylvania, New Jersey and New York.

BIOLOGY

Phobocampe unicincta is a specific univoltine, internal parasite of *Lymantria dispar*. Parasitism is generally low, although on occasions heavy parasitism has been observed in different parts of Europe. Apparently the parasitism is heavier in dense woodland than in open growth or on the outer edges of wooded areas. Muesebeck and Parker (1933) published on the biology of this species (as *Hyposoter disparis*). The following information is summarized from their paper:

Phobocampe unicincta, which was released in the areas of the New England states infested with the gypsy moth in 1912 onwards, has definitely become established, but has remained of little value as a control factor. Some of the factors that contribute to its ineffectiveness, are heavy hyperparasitization, much overwintering mortality, and much loss of the egg and first-instar larvae due to phagocytosis.

About 12,500 adults were liberated for the first time in four localities in eastern Massachusetts and one point in southeastern New Hampshire, in the spring of 1912 that were reared out of cocoons received in 1911 from Gioia Tauro, Italy. Importations ceased between 1912 and 1920. During 1924-1931 small numbers of the parasite were received from Hungary and Yugoslavia, and three small colonies were liberated in the field in Massachusetts during that time.

Collections of gypsy moth larvae were made at all the five points of 1912 release and the parasite recovered from each point.

Hibernation: *Phobocampe unicincta* hibernates as an adult within the cocoon. The posterior end of the body remains immersed in the moist meconial discharge. If the meconium dries, the parasite dies within the cocoon. In the field the cocoons remain on the surface of the ground throughout the winter, usually covered by leaf litter.

Emergence and *mating:* The adult *unicincta* emerges at about the time when eggs of the gypsy moth begin to hatch, usually in late April and early May. Temperatures of 65°F. or higher stimulate mating and sunlight is essential. Most satisfactory mating was obtained in the laboratory when freshly emerged females were mated with 3-4 day old males.

Oviposition: Females oviposit readily into the first and second instar larvae of the gypsy moth. The ovipositing parasite prefers to attack moving caterpillars and tends to prod caterpillars to move, when the ovipositor is quickly inserted and the egg deposited--the whole act taking just a second. The eggs are usually deposited in the posterior part of the body cavity. Several eggs may be deposited in one host but only one parasite matures.

Fecundity: The total number of eggs deposited by a female ranged between 182 to 1,228, in experiments of Muesebeck and Parker. The average number was 561. The longevity of the ovipositing female varied between 12 days to 54 days. In most cases the female continued to oviposit until the last day of her life, but the female that lived the longest deposited no egg after the 34th day. On a single day a female can deposit from 50 to 84 eggs.

Egg: The egg (Fig. 60) is 0.40 to 0.45 mm. in length and 0.11 to 0.14 mm. in maximum width. It is slightly curved on one side, smooth and pearly white. After deposition in the host, the egg gradually increases in size and before hatching attains a length of 0.7 to 0.8 mm. and a width of 0.25 to 0.28 mm. The duration of the egg stage is usually 7 days, which may extend up to 10 days when temperatures are low.

Larva: Five larval instars were distinguished by Muesebeck and Parker, although in other related parasites only three have been observed by several authors. Fig. 62 depicts larval mandibles.

The first larval instar is elongate more or less cylindrical, and smooth, with a strongly sclerotized brown head, and a long caudal appendage, which is a prolongation of the last, or thirteenth, body segment. On hatching, the larva measures about 1.2 mm. in length, including the oval appendage, which itself is 0.30 to 0.35 mm. long. The mandibles (fig. A) are small, but heavily sclerotized and strongly hooked. The second larval instar is larger in size, with less sharply defined head, differently shaped and less sclerotized mandibles, and somewhat shorter caudal appendage. The larvae in the third and fourth instars are generally similar to the second larval instar, but larger in size and with differently shaped mandibles (fig. C, D). The fifth larval instar has more heavily sclerotized mandibles, the labial ring and the sclerotic framework in the mouth region are brown in color and conspicuous, body integument covered with minute tubercles, antennal leg and wing pads visible, and has nine pairs of open spiracles. The first pair of spiracles is situated near the posterior margin of the first thoracic segment, followed by a pair on each of the first eight abdominal segments. The caudal appendage is greatly reduced, resembling a short, thick, evenly tapering spine. The mature larva measures 8-10 mm.

The duration of each larval instar varies considerably, depending upon the temperature. The following averages are reported by Muesebeck and Parker after a large number of dissections:

I Instar	5 to 10 days
II Instar	2 to 5 days
III Instar	2 to 4 days
IV Instar	2 to 4 days
V Instar	1 to 2 days (within host)

The mature parasitic larva emerges from the fourth larval instar of the gypsy moth host. The host is killed several hours before the mature parasitic larva emerges. After emergence, practically nothing remains of the host larva.

Cocoon: The mature larva spins a cocoon on the underside of leaves or branches beside the dead host larva. The cocoon is ovoid in shape, measuring 6-7 mm. in length and 4 to 4.5 mm. in diameter (fig. 61). It is dark brown in color with a broad grey band around its middle. An outer layer of comparatively loose silk covers the dense tough more or less parchment-like envelope. The attachment of the cocoon to underside of leaf or host remains is rather weak so that the cocoon drops to the ground within 48 hours of its formation.

Pupa: Twenty-four hours after cocoon formation, the short caudal appendage retracts and becomes shrunken and dark in color. After four days within the cocoon, the larva exhibits a slight constriction at the posterior margin of the thoracic region and the developing eyes are weakly discernible. The meconium is usually cast about 6 days after the formation of the cocoon, and actual pupation occurs on the 9th or 10th day. The pupa darkens gradually, until at the end of about 5 days the head and thorax become black and the base of abdominal petiole begins to darken. Transformation into a adult usually occurs 20-21 days after cocoon formation. The pupal stage thus covers 11 days.

The adult within the cocoon is fully formed during the first half of July, although emergence does not occur until the following spring. There is thus only one generation per year. Rarely a male may emerge in the same season in which the cocoon was formed.

Muesebeck and Parker noted that the parasite dies within the cocoon if the meconium dries out. Consequently it is important that the cocoons be stored in sufficiently moist atmosphere to maintain the semiliquid condition of the

meconial discharge. Proper moisture conditions are important during shipment of cocoons and during hibernation in the laboratory.

Duration of various developmental stages

Egg	7-10 days
Larva	12-25 days
Prepupa	9-10 days
Pupa	10-11 days
Adult	12-54 days

The total life cycle thus covers from 50 days to 110 days and is temperature dependent.

Hyperparasites

Several hyperparasites were reared from shipments received from Italy
during 1911 and 1912. These were:

Gelis areator (= *Hemiteles areator* Grav.).
Gelis sp. (3 unidentified species)
Theroscopus sp.
Spilocryptus pumilus Kriechbaumer
Bathythrix sp (= *Thysiotorus* sp.)
Theronia atalantae (Poda)
Itoplectis clavicornis (Thomson)
Itoplectis alternans (Gravenhorst)
Monodontomerus aereus Walker
Monodontomerus sp.
Haltichella maculipennis De Stefani
Eurytoma appendigaster (Swederus)

2. PHOBOCAMPE LYMANTRIAE, new species (Figs. 52-56)

Male and *female*: Similar to *P. unicincta* in general sculpture and color,
and differing as follows:
Flagellum 26 to 28 (30) segments. Face somewhat rugulose. Malar space
shorter, 0.33±0.1 the basal width of mandible; in female usually 0.25 and in
male 0.35-0.4. Interocellar distance longer, 1.8±0.1 the ocellocular distance.
Ocellocular distance slightly less than the ocellar diameter. Propodeum granulose, including petiolar area and areola, both of which are depressed and concave and without striations. Sometimes a few striations seen at the junction of
areola and petiolar area only. Areola broadly open below costulae, not constricted apically, with median longitudinal carinae weak in this area. Propodeal carinae generally weaker. Basal transverse carina usually more
roundly arched medially. Areolet larger, more oblique, with a short petiole
and second recurrent vein at its outer corner. Nervulus usually distad of basal
vein by 0.25 its length. Discocubitus roundly arched, not very strongly so.
Postpetiole more abruptly widened from petiole. Junction between petiole and
postpetiole appears constricted. Lateral groove on petiole deeper and more
conspicuous than in *unicincta*. Postpetiole wider than long, its sides roundly

arched and margined by a distinct and sharp carina. Postpetiole widest at its middle. Tergite 2 wider, less than 2.0 as long as its basal width; in male a little narrower. Postpetiole finely granulose to mat, in males often granulose. Epipleurum of tergite 4 with denser hairs, clustered along its apical margin. Subgenital plate appears less hairy.

Black. Color similar to that of *P. unicincta*, but the black marks on hind coxa, apex of hind femur and base of hind tibia usually faint, obsolescent or even. Tergite 2 of female with a narrower yellow band, occupying its apical 0.25 only. Apex of tergite 1 hardly yellow. In males tergites 1 and 2 without yellow or orange bands, often extreme apex of tergite 2 brown.

Variations: A few specimens from Madrid, Spain; Vees, Hungary; and Bilky, Czechoslovakia; and a specimen each from Simantornya, Hungary; and Bibai, Hokkaido, K. Kamijo, July 11, 1962; are like *Phobocampe lymantriae* in the nature of malar space, interocellar distance, postpetiole and second tergite, but the propodeum is more like that of *P. unicincta* in that the basal transverse carina is angulate medially and areola and petiolar areas have more carinations. The hind coxa also has faint to somewhat distinct black apical marks. The malar space is rather small (0.2-0.25).

I believe these are *P. lymantriae,* as the areola is widely open behind, not constricted and the longitudinal carinae are weak.

Length: 4-7 mm. Fore wing 3.5 to 5 mm.

Holotype: ♀, SPAIN: Madrid, June 1925, *ex Lymantria dispar*, Gypsy moth Lab. (FIS, Hamden, Ct.)

Allotype: ♂: Same data as the holotype.

Paratypes: Spain: Same data as the holotype, 20♂, 80♀, June-July 1925 (Hamden). Czechoslovakia: Bilky, 4♂, 8♀, 1925, *ex Lymantria dispar,* Gypsy Moth Lab., No. 3475 (Hamden). Hungary: Olaszliszka, 3♀, June 25, 1927, *ex Lymantria dispar* (Hamden & Washington). Vees, Hungary, 1♂, 1♀, July 12, 1928, and July 6, 1929, Gypsy Moth Lab., No. 13039 D and 13044 C 3 (Hamden). Debreczen, Hungary, Gypsy Moth Lab., 2♂, No. 3469. Jugoslavia: Moscenica, 1♂, June 18, 1927, Gypsy Moth Lab., No. 13019 B (Hamden). Vees, Hungary, 7♂, 6♀, June 1928, *ex* gypsy moth (Washington). Nieborow, Poland, 1♂, 2♀, May 1975, Fuester & Mura, emer. June 1975. Skierniewice, Poland, 2♀, May-June 1975 Fuester & Drea. Burgenland, Austria, 1♂, April-June, 1974, Hoyer. South France, 2♂, 3♀, June 1972, J. Drea (Washington & EPL, Paris). Forêt d'Orleans (Loiret), France, 1♀, May 1974 (EPL, Paris). Bouches du Rhone, France, 1♂, June 1973, Fuester and Gruber (Washington). Westwood, Mass., 2♀, June 1929, Gypsy moth lab. No. 11532 (FIS, Hamden). Japan: Bibai, Hokkaido, 1♀, K. Kamijo, July 11, 1962, *ex Lymantria dispar* (Washington). Many specimens in Hamden collections were identified as *"Hyposoter* species."

Cocoon: Cocoon slender, oblong, light brown in color, often with black marks encircling the end opposite the exit hole and with a whitish silken central band, 3 x 6 mm. or smaller, but about 2.0 as long as wide. Cocoons generally smaller than those of *P. unicincta*.

Distribution: Europe (Spain, Czechoslovakia, Hungary, Yugoslavia, Poland, Austria and France). China. Japan. ?U.S.A. (Introduced). It is sympatric with *P. unicincta* over much of its range.

Biological notes: This species appears multivoltine, the adults emerge in the same season in which the cocoons are formed. According to Dysart (personal communication), Dr. Paul Schaefer also observed the multivoltine nature in *Phobocampe* species collected in southern Japan from the gypsy moth. The specimens examined from his collections, however, fit better with *unicincta* rather than with *lymantriae*.

The specimen collected by Kamijo in 1962 is the only Japanese specimen fitting better under this species. The two specimens from Massachusetts, collected as early as 1929 are interesting. There is no evidence whether they were reared from the gypsy moth stocks in Europe or from Massachusetts. Most specimens from Europe were mixed and it is likely that both these species were released in the field. There is no evidence of its establishment in the U.S. A.

A series of specimens from China: Heilongjiang Province reared from *Lymantria dispar* in May-June 1982 by Schaefer *et al.* have been examined. They are *Ph. lymantriae*. A few have blackish hind femur, and propodeum a little coarser. Three males and a female were also reared on *Lymantria mathura* feeding on *Salix*. They also have darker hind femur.

6. Genus HYPOSOTER (Fig. 47)

Hyposoter Foerster, 1869. Verh. Naturh. Ver. Rheinlande, 25: 152.

For generic synonymy, refer to Townes, 1970: 181. Species of this genus associated with the gypsy moth have often been referred to under the genus *Anilastus*.

Moderately stout to slender species with fore wing 3.2 to 9 mm. long. Eye margin weakly to strongly indented opposite antennal sockets. Malar space 0.4 to 0.85 the basal width of mandible. Clypeus small, convex, its apex also convex. Lower edge of mandible with a basal lamella that is rather abruptly narrowed beyond the middle. Lower tooth of mandible a little smaller than the upper tooth. Temple short to very short. Occipital carina joining hypostomal carina. Thorax largely granulose to rugulose. Posterior mesosternal carina complete. Propodeal areola usually distinct and, in the species treated here, closed behind. Petiolar area concave and its bounding carinae often erased so that the combined petiolar area and third lateral area form a shallow trough. Propodeal spiracle circular to short elliptic. Hind basitarsus without a midventral row of closely spaced short hairs. Tarsal claws small, pectinate. Areolet present, with second recurrent vein near apex. Nervulus opposite basal vein or a little distad. Nervellus not intercepted, vertical to weakly reclivous. Glymma present in the form of a deep pit. Abdomen compressed apically. Thyridium circular or elliptic. Ovipositor short, 1.0 to 1.5 the apical depth of abdomen.

Members of the genus are parasitic within lepidopterous larvae.

Four species of *Hyposoter* have been mentioned in literature as parasitic on *Lymantria dispar*: *H. tricoloripes* (Viereck) in Europe, *H. fugitivus* (Say) in North America, and *H. vierecki* T. M. & T. [= *Campoplex (Diadegma) japonicus* Viereck] and *H. takagii* Matsumura in Japan. *H. lymantriae* (Cushman) occurs in India on the related *Lymantria obfuscata*.

Campoplex rapax Gravenhorst, often referred to under the genera *Anilastus* and *Anilasta*, appears to be a species of *Hyposoter*. It is unknown to me. *Anilastus* n. sp. (Stadler, 1933) is also a *Hyposoter*, most probably *H. tricoloripes*, judging from the figures of the larval remains.

"*Hyposoter* spp." of Burgess and Crossman (1929) represents *Phobocampe lymantriae* from Spain, and Central Europe.

Hyposoter fugitivus (Say) (= *Limnerium* sp = *Limneria fugitiva*) was wrongly associated with the gypsy moth on guess-work. Its record in literature stems

from Howard and Fiske (1911: 138) who stated:
    "A single cocoon, which was directly associated with the remains of the host
caterpillars [*Lymantria dispar*] was collected by Mr. R. L. Webster in 1906
during his association with the laboratory. It was very likely that of *L. fugitiva*
Say, but the fact will never be known, because a specimen of *Hemiteles utilis*
Norton, a hyperparasite, actually emerged."

KEY TO THE SPECIES OF HYPOSOTER FROM THE GYPSY MOTH

1. Larger-sized species, 8-11 mm. long. Propodeum largely rugose, with
    basal areas somewhat rugoso-punctate. All coxae, trochanters and
    femora black, the anterior ones may be brown. Abdominal tergites
    marked with reddish-brown. Areola bounded by strong carinae. Japan,
    Korea, and China. . . . . 1. takagii (Matsumura)
    Smaller-sized species, about 6-7 mm. long. Propodeum largely granu-
    lose. Petiolar area often rugose. Fore and middle trochanters yellow
    to yellowish-brown. All femora yellowish-brown. Coxae black or
    anterior ones yellow. Abdomen black. Areola bounded by weak to
    moderately strong carinae. (In *tricoloripes* hind femur often brownish,
    rarely black). 2

2. Fore and middle coxae and trochanters yellow. Propodeum in basal and
    lateral areas and inside areola granulose. Petiolar area granuloso-
    rugulose. In male sculpture coarser, with petiolar area rugulose.
    Slender species. Japan. . . . . 2. vierecki T.M.&T.
    Fore and middle coxae black, as is hind coxa, or fore coxa partly yellowish-
    brown. Slender to moderately robust in build. 3

3. Hind femur and tibia often brownish though at times lighter in color. Areola
    crescentic, about 2.0 as wide as long, its apical closing carina complete
    and weakly arched. Propodeum largely granulose basolaterally. Pro-
    podeal sculpture finer than in *lymantriae*. Scutellum finely granulose.
    Second tergite granulose. Gastrocoeli usually deep and conspicuous.
    Europe. . . . . 3. tricoloripes (Viereck)
    Hind femur reddish-brown. Hind tibia yellowish-brown with fuscous basal
    and apical marks. Areola horse-shoe shaped, its apical closing carina
    usually strongly concave, and carinae bounding areola often irregular or
    incomplete. Propodeum in general coarser than in *tricoloripes*, with
    lateral areas rugulose. Scutellum granuloso-punctate. Second tergite
    mat to weakly granulose. Gastrocoeli shallow. India.
    4. lymantriae Cushman

1. HYPOSOTER TAKAGII (Matsumura)

Casinaria takagii Matsumura, 1926. J. College Agr. Hokkaido Imp. Univ.,
    18: 28. Korea. Host: *Dendrolimus spectabilis*.
Hyposoter takagii: Townes, Townes and Gupta, 1961. Mem. Amer. Ent.
    Inst., 1: 242. China, Japan, Korea.
Hyposoter takagii: Yasumatsu and Watanabe, 1964. Catalogue of Insect
    Natural Enemies of Injurious Insects in Japan, Pt. 1: 43. Hosts:
    Lymantria dispar (cf. Fukaya, 1950), *Dendrolimus spectabilis,*

Malacosoma neustria testacea (cf. Hayashi, 1933).
This species has often been misidentified in Japan as *Casinaria atrata*
Morley (*cf.* Townes, Momoi and Townes, 1965: 300). It was first reported
from the gypsy moth by Fukaya (1950). It is distinguished from other *Hyposoter*
species by its larger size (8-11 mm.) and by the reddish marks on tergites.

Male and *female:* Face rugulose. Malar space 0.8 the basal width of mandible. Interocellar distance 1.7 to 1.8 the ocellocular distance. Vertex finely
granuloso-mat. Mesoscutum granuloso-rugulose. Scutellum rugose. Mesopleurum rugulose to finely rugose. Metapleurum rugulose. Propodeum largely
rugose, with basal areas somewhat rugoso-punctate to rugulose. Areola
bounded with strong carinae. Areola pentagonal or hexagonal with its apical
closing carina assuming various shapes—straight, arched, or angled. Median
longitudinal carinae separating petiolar area from third lateral area weak to
distinct and convergent apically. Postpetiole shiny or finely mat. Second
and the following tergites mat to subpolished. Thyridium larger, wide,
separated from base of second tergite by about half its width.

Black. Mandible partly to wholly, palpi, tegula partly (or not so), and
fore and middle tibiae and tarsi, yellow. All coxae, trochanters and femora
black to blackish-brown, particularly the anterior ones. Hind tibia and tarsus
brown to blackish-brown. Abdomen black with some tergites often marked with
reddish-brown. Abdomen never appearing wholly black. Color of legs and
abdomen variable, with more or less of reddish color.

Specimens: Several males and females from Japan, China, and Korea
examined, reared from *Dendrolimus spectabilis* and *Malacosoma neustria.*
No specimen from the gypsy moth available.

Distribution: China, Japan and Korea.

2. HYPOSOTER VIERECKI T.M.&T.

Campoplex (Diadegma) japonicus Viereck, 1912. Proc. U. S. Natl. Mus.,
    42: 636. Name preoccupied by Cameron, 1906. Japan. "Gypsy Moth
    Lab. No. 1071".
Hyposoter vierecki Townes, Momoi & Townes, 1965. Mem. Amer. Ent.
    Inst., 5: 302. New name.

This species is readily distinguished from the others treated here by its
yellowish-white fore and middle coxae and trochanters and yellowish-brown
femora.

Male and *female:* Face and clypeus granulose. Malar space 0.65 the basal
width of mandible. Interocellar distance 2.0 the ocellocular distance. Frons
and vertex finely granulose. Thorax granulose. Pronotum and mesopleurum
with a few striations interposed amongst granulations. Sculpture of scutellum
a little coarser. Propodeum largely granulose. Combined petiolar and third
lateral area ruguloso-granulose. Areola rectangular to squarish, its
bounding carinae sharp to weak, apical closing carina angled medially. Areola
rugulose to granulose. Costula incomplete. Sculpture of male propodeum
coarser, particularly within areola and in petiolar area. Postpetiole and
second tergite finely granulose. Other tergites progressively mat to subpolished. Thyridium rather small and narrow, separated from base or second
tergite by about its width.

Black. Mandible, tegula and fore and middle coxae and trochanters, light
yellow. Fore and middle legs otherwise yellowish-brown. Hind coxa black,

trochanters blackish, trochantellus in female yellowish-brown, femur brownish-yellow, tibia yellowish-brown with fuscous marks apically, and tarsus fuscous. Sometimes apex of hind femur and base of hind tibia also fuscous.

Length 6-7.5 mm.   Fore wing 4-5.5 mm.

Specimens: Japan: 1♂ (type, No. 7258 of *Campoplex japonicus* Viereck), "Gypsy Moth Lab. No. 1071, June 21" (Washington). Japan: Kamifurano, Hokkaido, 1♂, 1♀, July 1, 1975, P. Schaefer, *ex Lymantria dispar* (Washington); 2♀, June-July 1975 (BIRL, Newark). Hobetsu, Hokkaido, 1♂ (broken), June 15, 1978, emerged June 25, 78, P. Schaefer (BIRL, Delaware). 1 ex (without abdomen), "Asagawa", June 16, 1923 (Washington).

According to Carlson (1979: 661) some of the records *Phobocampe unicincta* in Japan may pertain instead to the present species.

3. HYPOSOTER TRICOLORIPES (Viereck) (Figs. 63-67)

Anilastus tricoloripes Viereck, 1911. Proc. U. S. Natl. Mus., 40: 478.
    Europe. "Gypsy Moth Lab. Nos. 1079 and 1065". Type is a female,
    not a male (no. 1079). Paratype is a male (no. 1065).
Limnerium (Anilastus) tricoloripes: Howard and Fiske, 1911, U. S. Dept.
    Agr. Bur. Ent. Bull., 91: 192.

This species has often been reared in Europe at different places from the gypsy moth and imported into the U.S.A. for release. In 1911, Howard and Fiske reported, "From time to time several specimens of *Limnerium* cocoons, all of them oblong in shape, and most of them partly concealed by the skin of the host caterpillar, have been received from Europe. In no instance they have been in sufficiently large numbers to make the species appear promising as a parasite."

Pschorn-Walcher (1974) reported it to be a larval parasite of low incidence in southern France, eastern Austria and Bavaria.

Male and *female:* Head granulose. Face a little coarser in male. Temple and occiput subpolished. Malar space 0.5 to 0.6 the basal width of mandible. Ocelli comparatively large, so that they appear somewhat closer to eye. Interocellar distance 2.2 to 2.3 the ocellocular distance. Mesoscutum, scutellum and mesopleurum granulose, almost of the same intensity. Speculum distinctly granulose, dull. Metapleurum a little finely granulose. Propodeum granulose. Petiolar area granuloso-rugose or rugulose in female and rugose in male. Areola granulose, crescentic, bounded by sharp carinae, about 2.0 as wide as long, closed apically. Costula complete to incomplete. Postpetiole and second tergite granulose. Rest of abdomen granuloso-mat to mat apically. Gastrocoeli deeply impressed, oval to crescentic, separated from base of second tergite by 0.5-0.7 its width. Ovipositor sheath appears a little widened apically. Ovipositor a little shorter than apical depth of abdomen.

Black. Mandible, palpi, tegula, and fore and middle trochanters, yellow. Fore and middle femora yellowish-brown to reddish-brown, their tibia and tarsi yellowish and with fuscous marks. All coxae black. Fore coxa sometimes apically brownish or yellow. Hind leg brownish, though at times lighter in color, or seldom hind femur blackish. Apex of tibia often darker, or hind tibia brown with its base pale yellow.

Length: 5.5-7.5 mm.   Fore wing 4-5 mm.

Specimens: Europe: Type ♀ and paratype ♂, "Gypsy moth Lab., Nos. 1079 and 1065", Type No. 13799 (Washington). Europe: Several males and females reared from the gypsy moth at Oberpullendorf, Austria, June 18, 1931; Austria, July 1932; Burgenland, Austria, May-July 1974; Orleans, France, June 1972 and May 1976; Neuf Brisach, France, May 1976; France, June 1980; and Nieborow, Poland, June 1975 (Hamden, Washington and Paris). The specimens in Hamden were identified as *Hyposoter* sp.

Distribution: Europe. The female type is rather abnormal. It is a small specimen with a very different areola—which is rather oblong and open apically. Viereck (1911) noticed this difference between the type and the paratype. I believe that the areola is rather abnormal in the type-specimen, as this condition is not seen in other specimens reared from the gypsy moth.

It appears that this species has been released in the U.S.A. in recent years but the results have not been encouraging. Drea and Fuester (1979) observed, "Information on the biology of *H. tricoloripes* indicates that an alternative host is probably required for the species to complete its annual cycle. This may be the reason why this parasite has never become established in North America despite repeated releases.

4. HYPOSOTER LYMANTRIAE Cushman

Hyposoter lymantriae Cushman, 1927. Rec. Indian Mus., 29: 244. India: Kangra in Himachel Pradesh. Host: *Lymantria concolor*.
Hyposoter lymantriae: Beeson and Chatterjee, 1935. Indian Forest Rec. (N.S.) Ent., 1: 161. biol.

This species occurs in Northern India on *Lymantria obfuscata* and *Lymantria concolor*, which are related to the gypsy moth. It is rather close to the European *tricoloripes*, but is coarser in sculpture, with areola differently shaped and malar space and interocellar ratios a little different. The hind leg color is lighter.

Male and *female:* Head granulose. Temple and occiput subpolished. Malar space 0.75 the basal width of mandible. Ocelli comparatively small, appearing farther from the eye than in *tricoloripes*. Interocellar distance 1.8 to 2.0 the ocellocular distance. Mesoscutum granulose. Scutellum and mesopleurum granuloso-punctate. Speculum a little more convex and subpolished than in *tricoloripes* and finely weakly striate. Metapleurum and pleural area of propodeum weakly granulose, subpolished. Propodeum granulose on front half. Areola granulose to rugulose, more arched in front, its lateral carina and apical closing carina often weak or incomplete. Costula usually distinct and complete. Areola often irregular in outline. Second lateral area rugulose. Combined petiolar and third lateral area rugose. Postpetiole finely granulose. Second tergite mat to weakly granulose, and the following tergites mat to subpolished. Ovipositor sheaths appear slender. Ovipositor equal to apical depth of abdomen.

Black. Mandible, palpi, tegula, and fore and middle trochanters, yellow. Fore and middle femora yellowish-brown, their tibiae and tarsi yellow with pale brown marks. Hind femur reddish-brown, tibia yellowish-brown with apical and subbasal fuscous marks or bands. Hind tarsus brownish to blackish, with a pale basal band on first segment. Coxae black with fore coxae often yellow apically.

Length: 6-7 mm. Fore wing 4.5-5 mm.

Specimens: India: Kulu in Himachal Pradesh, 2♂, 2♀, and 2 broken sp., June 1964, ex larva of *Lymantria* [*obfuscata*] (Gupta). Kulu Dist.: Katrain, *ex* larva of *Lymantria [obfuscata]*, 1♂ (CIBC, Bangalore) and 1♀ (Washington). Katrain, Kulu, 1♀, June 26, 1965; Shambi, Kulu, 1♂, May 19, 1965, *ex Lymantria [obfuscata]* on *Alnus nitida*, and determined as *Anilastus* sp. (CIBC, Bangalore). Srinagar, 1♂, June 1969, *ex Lymantria obfuscata*, det. as *Hyposoter* sp. (CIBC, Bangalore). India: Kangra Forest, August 15, 1917, O. H. Walters Coll., 1♀ (type of *H. lymantriae*), parasite of *Lymantria concolor* (Washington).

Beeson and Chatterjee (1935) gave the following biological information: "Host: *Lymantria concolor* Walk. (Lymantriidae) defoliating *Quercus incana*." "The host is attacked in the early larval instars during June-July; the parasite matures in August and the host in September-October."

Dr. Roger Fuester (BIRL, Delaware) recently sent me specimens of this species, which were reared at Newark on the gypsy moth from stocks from Kulu District, India. It is intended to be released in the Northeastern U.S.A. against the gypsy moth.

III. SUBFAMILY ICHNEUMONINAE

Members of the subfamily Ichneumoninae are characterized by having the clypeus relatively flat, separated from the face by a weak groove, its apical margin weakly arcuate, or truncate, with or without a blunt median point. Upper tooth generally longer than the lower, notauli and sternaulus absent, or short and shallow, except rarely, propodeum steeply sloping in the petiolar area, with longitudinal carinae, areola present, variously shaped and often raised, propodeal spiracles linear (circular in some tribes that are not treated here), areolet pentagonal, the intercubiti convergent towards radius, abdomen flattened, usually spindle-shaped, first segment quadrate basally in cross section, with its spiracles placed far beyond the middle, postpetiole flattened and wide or pyramidally raised, gastrocoeli usually wide and distinctly impressed, and ovipositor generally short, hardly surpassing the tip of the abdomen. The female flagellum is usually widened preapically and the male flagellum is slender and tapering.

The species of Ichneumoninae are internal parasites of a variety of lepidopterous pupae. The oviposition is usually in the pupa but sometimes into the larva and the emergence from the pupa.

Several species of Ichneumoninae have been listed as parasites of the gypsy moth, chiefly in Europe. Most of them, however, have never been recorded subsequently and reared specimens are not available to confirm their occurrence on the gypsy moth. The only exception is *Lymantrichneumon disparis* (Poda), which was perhaps the first ichneumonid recorded from the gypsy moth and which has been collected subsequently in small numbers from the pupae of the gypsy moth.

There are some *lapsi* in the literature regarding the association of the ichneumonine species with the gypsy moth. Rudow (1911) listed several species as parasites of both *Lymantria monacha* and *L. dispar, viz., Ichneumon raptorius, I. sugillatorius, I. melanoceras, I. fabricator*, and *Trogus flavatorius*. In his subsequent lists (1917-1919) he was more specific as to the host records. In 1918, *Ichneumon raptorius, sugillatorius*, and *melanoceras* were listed as parasites of *Lymantria monacha* and not of *dispar*, while *I. fabricator* was not mentioned from either of them. These species are omitted here, except *I. fabricator*, which has been subsequently recorded as a gypsy moth parasite by Cecconi (1924) and Györfi (1963) and *Trogus flavatorius*, which is the same as *Lymantrichneumon disparis*. There are a few other species, apparently first reported by Rudow (1917-19) from the gypsy moth, but missed by Schedl (1936) and Thompson (1946). These are included, although, as is usually the case, subsequent rearing records are lacking.

The identification of the genera and the species of the Ichneumoninae is somewhat difficult. In the keys that follow, attempts have been made to simplify them and a field key based largely on color is also given. For more information on the taxonomy of genera and species, the following works may be consulted: Heinrich (1960-62), Perkins (1953, 1959-60), Townes, Momoi and Townes (1965), and Kasparyan (1981).

KEY TO THE ICHNEUMONINAE ASSOCIATED WITH THE GYPSY MOTH, INCLUDING A FEW RELATED GENERA

1. Petiole wider than deep, flat above. Clypeus projecting forward, triangular in outline. Mandible strong, with one tooth. Body black, femora yellow. (Tribe Pristocerotini). . . . . 13. Cotiheresiarches
    [*Cotiheresiarches dirus* (Wesmael)]
    Petiole as deep as wide, squarish in cross-section. Otherwise not as above. 2

2. Clypeus wide, its apical margin thin and convex. Face and clypeus in an even plane. Temple strongly widened behind eye. Occipital carina reaching hind corner of mandible, not joining hypostomal carina. Mandible wide, not or very little tapered apically, both teeth prominent. (Tribe Geodartiini). 3
    Clypeus narrower, its apical margin thicker and usually arcuate or truncate. Clypeus and face not forming a flat surface. Either clypeus and face with median convexities or their junction depressed. Temple moderately swollen. Occipital carina joining hypostomal carina. Mandible moderately narrow and tapered apically, with lower tooth often weak or small. 4

3. Tarsal claws not pectinate. Metapleurum separated from propodeum by a carina. (Recorded from *Lymantria* sp. and *L. monacha*. . . Geodartia (*G. cyanea*: Large size, 22 mm. Abdomen apically compressed, obtuse at apex and metallic bluish.
    Tarsal claws pectinate to apex. Metapleurum separated from propodeum by a faint groove. . . . . Pseudomaraces
    (*P. melli* Heinrich reported from *Lymantria* sp.)

4. Propodeum convex on front half or less (up to areola). Propodeum in profile view raised basally and then abruptly sloping. Areola usually constricted, horse-shoe shaped, or in the form of a raised polished "boss", its bounding carinae often flat and polished. Incisures between tergites usually deep. Tribe Ichneumonini (Protichneumonini). . . . 5
    Propodeum flat on basal half and then sloping. In profile view propodeum making an inclined slope. Areola normally formed, squarish or elongate, or sometimes arched basally. Tribe Joppini. 8

5. Areola reduced to a small polished boss that is strongly elevated, its bounding carinae obsolete. Apex of female abdomen blunt (amblypygous).

Gastrocoeli shallow and wide. . . . .   3. Callajoppa
[*Callajoppa cirrogaster cirrogaster* (Schrank)].
Areola normal or somewhat normal, bounded by distinct carinae. Apex of
female abdomen acute (oxypygous). Gastrocoeli usually deep. 6

6. Abdomen black or bluish-black, with or without white spots, spindle-shaped,
a little narrower than thorax. Areola within the general convex surface
of propodeum, not distinctly raised, usually horse-shoe shaped or
hexagonal. . . . .   4. Ichneumon
(Several species).
Abdomen more parallel-sided, reddish or brownish, as wide as thorax.
Areola raised above the general surface of propodeum, propodeum
sloped away from areola on all sides. *Protichneumon* and related
genera.   7

7. Scutellum flat, polished. Areola rugose, longer than wide. Postpetiole
not pyramidally raised. Head and thorax black. Abdomen reddish.
Wings clear hyaline. . . . .   2. Paracoelichneumon
[*Paracoelichneumon rubens* (Fonscolombe)].
Scutellum subconvex, punctate. Areola horse-shoe shaped, smooth.
Postpetiole somewhat pyramidally raised medially. Body yellowish-
brown with wings yellow tinged. . . . .   1. Lymantrichneumon
[*Lymantrichneumon disparis* (Poda)].

8. Thyridia very wide, the space between them less than 0.7 the width of
each.   9
Thyridia of moderate width or narrower, the space between them more
than 0.7 the width of each. Thyridia usually weakly impressed. . .   10

9. Lateral carina of scutellum sharp and strong up to apex of scutellum.
Postpetiole granulose to punctate, usually without a defined median
field. . . . .   5. Stenaoplus
[*Stenaoplus pictus* (Gravenhorst)].
Lateral carina of scutellum ending at basal 0.2 to 0.3. Postpetiole with a
distinct median field, which is striate. . . . .   Stenichneumon
(not associated with the gypsy moth).

10. Extreme base of propodeum with a weak median tubercle. Median field of
postpetiole with moderately dense punctures. Apical half of female
flagellum strongly flattened below, tapering apically.
                                        6. Melanichneumon
[*Melanichneumon leucocheilus* (Wesmael)].
Base of propodeum without any median tubercle. Median part of postpetiole
with fine to distinct longitudinal striations or aciculations. Apical half of
female flagellum cylindric, blunt or tapering apically.   11

11. Median field of postpetiole not sharply demarcated, with fine weak striations
and punctures. Apical half of female flagellum cylindric or weakly
flattened below. . . . .   7. Cratichneumon
[*Cratichneumon fabricator* (Fabricius)].
Median field of postpetiole demarcated and usually aciculate or strongly
striate. Apical half of female flagellum with a short to long taper, or
somewhat blunt apically (In *Spilichneumon occisor*, postpetiole finely

striate with smoother apex, and median field rather weakly demarcated). Scutellum yellow. 12

12. Tip of abdomen acutely pointed (oxypygous). Ovipositor not unusually short, distinctly exerted beyond abdominal tip. Subgenital plate not elongate. Sternite 4 membranous medially (usually medially folded).

13

Tip of abdomen rounded (amblypygous). Ovipositor unusually short, hardly exserted. Subgenital plate sometimes with a median prolongation. Sternite 4 often not membranous medially. 14

13. Apical margin of clypeus broadly, weakly concave, with a weak, broad, median tooth. Apical part of female flagellum blunt, not distinctly tapered before the last segment. . . . . 8. Chasmias
[*Chasmias paludator* (Desvignes)].
Apical margin of clypeus truncate or weakly convex, without a median tooth. Apical part of female flagellum weakly tapered.

9. Pterocormus

[*Pterocormus sarcitorius sarcitorius* (Linnaeus)].

14. Flagellum short, its second segment 0.8 to 1.4 as long as wide, its apex with a short taper. Sternite 4 not membranous medially. 15
Flagellum long, slender, its second segment 1.4 to 3.0 as long as wide, its apex with a long taper. Sternite 4 often membranous medially. . 16

15. Areola about 1.0 as long as wide. Sides of median field of postpetiole sharply defined. 10. Triptognathus
[*Triptognathus amatorius* (Mueller)].
Areola about 1.8 as long as wide. Side of median field of postpetiole indistinctly defined. . . . . 11. Spilichneumon
[*Spilichneumon occisor* (Fabricius)].

16. Subgenital plate with a median apical tuft of suberect hairs.   Eutanyacra
[Not associated with the gypsy moth].
Subgenital plate without a median apical tuft of hairs, often with a few longer hairs on its apex but these sparse and either decumbent or only weakly raised. 17

17. Propodeum with an acute tooth at apex of its second lateral area on either side. . . . . 12. Amblyteles
[*Amblyteles armatorius* (Foerster)].
Propodeum without teeth. . . . . Diphyus
[Not associated with the gypsy moth].

FIELD KEY FOR THE IDENTIFICATION OF ICHNEUMONINE PARASITES ASSOCIATED WITH THE GYPSY MOTH

1. Body wholly black. Legs may have yellow or brown marks. Wings clear hyaline or a little infuscate. 2
Body wholly to largely rufous brown, or black with abdomen striped, or thorax black and abdomen reddish or rufous. 8

2. Size large, 15 to 18 mm. long. 3
    Size small, 12 mm. or less. 7

3. Fore and middle legs rufous. Scutellum black. 4
    All legs black. Scutellum yellow (or black in *dirus*). 5

4. Antenna black. Wings infuscate (according to original description). Hind
        leg wholly black. Not seen. . . . . ♂, ♀. "Amblyteles varipes Rudow."
    Antenna brownish-black. Wings clear hyaline. Hind femur and tibia
        reddish. . . . . ♂. Chasmias paludator (Desvignes)

5. Abdomen with white spots on tergites 2-3 in ♀ and 2-4 in ♂. Areola flat.
                        ♂, ♀. Ichneumon cyaniventris Wesmael
    (*Ichneumon sugillatorius, ex Lymantria monacha*, very much like the above
        but the areola more convex).
    Abdomen wholly black, without spots. Areola convex, a little raised. . 6

6. All femora yellowish-brown. Clypeal margin triangular, its apex project-
        ing forward. Abdomen robust. Petiole flat above. Scutellum black,
        convex. . . . . ♀. Cotiheresiarches dirus Wesmael
    All femora black. Legs wholly black, except in males where fore and
        middle legs with brownish patches. Clypeal margin truncate or arcuate,
        its apex not projecting forward. Abdomen spindle-shaped, slender.
        Petiole quadrate in cross section. Scutellum yellow and flat.
                        ♂, ♀. Ichneumon leucocerus Gravenhorst
    (Incorrect record. Occurrence on the gypsy moth not confirmed).

7. Face strongly punctate. Body conspicuously marked with red and yellow.
        Inner orbits yellow. Scutellum yellow at apex. All femora reddish.
        Hind tibia without a yellow median band. Size 6-10 mm.
                        ♂. Stenaoplus pictus (Gravenhorst)
    Face with scattered punctures. Body largely black, not conspicuously
        marked. Face wholly black (♀), or largely yellow (♂). Scutellum black
        (♀) or apically yellow (♂). All femora reddish (♀) with tibiae yellow
        medially, or hind leg wholly brownish-black (♂). Size 8-11 mm.
                        ♂, ♀. Cratichneumon fabricator (Fabricius)

8. Body largely to wholly rufous-brown. Thorax may have black patches.
        Abdominal tip may be black. 9
    Body black with abdomen banded or wholly reddish. Thorax black. Head
        black or partly yellow. 10

9. Head, thorax, abdomen, and legs largely rufous brown. Abdomen black
        tipped. Gastrocoeli deep and wide. Postpetiole coarsely striato-
        punctate. Wings yellowish-brown. Size 17 mm. or more, large.
                        ♂, ♀. Lymantrichneumon disparis (Poda)
    (Some Japanese specimens have less or more extensive black marks on
        thorax, abdomen and legs).
    Head blackish, except for face. Propodeum blackish. Rest of body
        including tip of abdomen rufous. Gastrocoeli shallow. Postpetiole
        granulose. Size small, 7-10 mm. . ♀. Stenaoplus pictus (Gravenhorst)

10. Thorax black. Head wholly black or dorsally black. Abdomen reddish or yellowish-brown, with or without black on its apex. Abdominal tergites not banded. Wings tinged or clear hyaline. 11
    Thorax black. Head black. Abdomen banded (yellow and black, or red and black). Wings clear hyaline. 15

11. Size large, 20 mm. and over. Abdomen yellowish-brown, with apex black. Wings lightly to strongly yellow-tinged. Flagellum not banded. Scutellum subconvex to pyramidal. Areola quadrate or in the form of a polished boss. 12
    Size medium, less than 20 mm. long. Wings clear hyaline or a little tinged. Abdomen reddish. Flagellum with a white band in ♀. Areola elongate and rugose. Scutellum flat. 13

12. Scutellum pyramidal. Areola in the form of an elevated polished boss. Wings strongly tinged with yellow. Basal four tergites yellow.
    ♂, ♀. Callajoppa cirrogaster cirrogaster (Schrank)
    Scutellum subconvex. Areola quadrate, well formed. Wings lightly tinged with yellow. Basal segment of abdomen and fourth and the following segments black, only 2-3 segments yellow.
    ♂. Triptognathus amatorius (Mueller)

13. Tip of abdomen white marked. Abdominal tergites 2-4 reddish (♀) or with reddish bands (♂). Gastrocoeli and thyridia weak, indistinct.
    ♂, ♀. Melanichneumon leucocheilus (Wesmael)
    Tip of abdomen black or red. Scutellum yellow. Areola elongate and rugose. 14

14. Abdomen beyond tergite 3 black. Tergite 7 may have a longitudinal yellowish mark. Gastrocoeli weak and short. 4th sternite membranous medially. Tip of flagellum blunt. Propodeum flat basally. Size 14-15 mm. ♀. Chasmias paludator (Desvignes)
    Abdomen red except for tergite 1. Gastrocoeli deep and wide. 4th sternite wholly sclerotized. Flagellar tip pointed. Propodeum convex basally. . . . . ♀. Paracoelichneumon rubens (Fonscolombe)

15. Females. Flagellum with or without a white band. 16
    Males. Flagellum without a white band. 19

16. Second tergite black, only narrowly yellow basally. (Abdominal tergites black with yellow bands). Propodeum with a sublateral acute tooth on either side. Flagellum slender, finely tapered, without a white band.
    ♀. Amblyteles armatorius (Förster)
    Second tergite red. Propodeum without sublateral teeth. Flagellum short, curled apically and short tapered, with an indistinct white band. . . 17

17. Third tergite also red. Tergites 4-6 with median apical yellow incomplete bands. Median field of postpetiole less strongly demarcated and finely striate. Mandible not narrowed apically. Flagellum narrowed apically.
    ♀. Spilichneumon occisor (Fabricius)
    Third tergite largely black, with yellow apical margin or red with base broadly black. Median field of postpetiole clearly defined and distinctly striate. Mandible slightly to strongly narrowed apically. 18

18. Third and the following tergites black with narrow apical yellow stripes.
    Abdominal tip amblypygous. Tip of flagellum tapered apically. Femora
    black. . . . .   ♀. Triptognathus amatorius (Mueller)
    Third tergite black in basal 0.4 and red in apical 0.6. Tergites 4, 5, and 7
    wholly black. Tergite 6 largely yellow. Abdomen oxypygous. Flagellum
    rather blunt apically. Short tapered. Femora red.
                 ♀. Pterocormus sarcitorius sarcitorius (Linnaeus)

19. Second tergite yellow on basal half or more. Third tergite wholly to
    largely yellow. Hind femur largely to wholly black. Face without a
    black mark. 20
    Second and third tergites largely black, with yellow apical concave stripes.
    (Abdomen banded by yellow and black). Face with a black mark. Hind
    femur yellow on basal 0.75.
                 ♂. Pterocormus sarcitorius sarcitorius (Linnaeus)

20. Propodeum with a sharp tooth on either side. Flagellum brown. Hind
    femur yellow on basal 0.25, otherwise black. Tergites 4-5 black.
                 ♂. Amblyteles armatorius (Förster)
    Propodeum without teeth. Flagellum black. Hind femur wholly black.
    Tergites 4-5 with yellow apical bands.
                 ♂. Spilichneumon occisor (Fabricius)

TRIBE ICHNEUMONINI (= PROTICHNEUMONINI)

1. Genus LYMANTRICHNEUMON

Lymantrichneumon Heinrich, 1968. Ent. Tidskr., 89(1-2): 104.

Mandible apically narrowed and twisted so that the small lower tooth lies
behind the long and pointed upper tooth. Flagellum of female a little widened
subapically. Scutellum convex, raised from metascutellum and punctate. Its
lateral carina confined in its basal 0.3. Mesopleurum coarsely punctate.
Mesepisternum without a knob-like protuberance near middle coxa at its junction with mesosternum, so that in profile view this area does not appear concave (*cf. Protichneumon*). Propodeum with a horse-shoe shaped high areola.
Petiolar area concave and abruptly sloping. Postpetiole wide, knob-like.
Gastrocoeli deep and wide, the interspace between them about 0.7 as long as
the width of gastrocoeli. Abdomen punctate and acute apically in the female
(oxypygous).
    Color yellowish-brown with wings tinged with yellow. The Japanese speci-
mens often with black parts.
    Heinrich (1968) segregated *Lymantrichneumon* from *Protichneumon* for the
reception of those species which are parasitic upon Lymantriidae rather than
Sphingidae. He distinguished *Lymantrichneumon* from *Protichneumon* by the
absence of a raised knob-like projection on mesepisternum near middle coxa
in the region of sternaulus, which is angled broadly with mesosternum, convex
propodeum at a level higher than that of metascutellum, apical margin of mesosternum not concave and raised, and mesoscutum densely punctate. He further
separated it on its body color: yellowish to orange brown with yellow-tinged
wings.
    Heinrich (1960) stated that the females of the tribe Protichneumonini do not

hibernate as adults and probably produce only one generation per year, while the adult of *Lymantrichneumon* does hibernate.

1. LYMANTRICHNEUMON DISPARIS (Poda)

Sphex disparis Poda, 1761. Insecta Musei Graecensis, p. 107, no. 3.
Europe. Host: *Lymantria dispar*.
Protichneumon disparis: Morley, 1903. Ichneumonologia Britannica,
1: 20.
Ichneumon disparis: Howard and Fiske, 1911. U. S. Dept. Agr. Bur. Ent.
Bull., 91: 85, 239. "Reared in Lab.".
Lymantrichneumon disparis: Heinrich, 1968. Ent. Tidskr., 89: 104-105.
Europe.

This species was the first ichneumonid parasite described from the gypsy moth. It has been reared several times in Europe and there are several synonyms of it, which are given by Dalla Torre (1901-02) and Uchida (1941). Those synonyms that have appeared in the gypsy moth literature are: *Ichneumon flavatorius* Fabricius, *Trogus flavatorius* Panzer, *Ichneumon ventralis* Matsumura, *Protichneumon disparis orientalis* Heinrich, *Protichneumon disparis matsumurai* Uchida, and *P. disparis segmentalia* Uchida. This species has also been put under various genera, like *Amblyteles*, *Ichneumon*, *Coelichneumon*, *Protichneumon*, and lately under *Lymantrichneumon*.

Face with scattered punctures, extending to base of clypeus. Clypeus smooth and shiny on apical half, with only a few sparse punctures. In female central area of face more densely punctate. Apical margin of clypeus weakly indented medially. Mandible apically tapering, its lower tooth small, less than half as long as the upper tooth. Vertex, upper areas of frons and upper temples with shallow scattered punctures. Antennal scrobes smooth and shiny, frons just in front of median ocelli trans-striate. Area between lateral ocelli and occipital carina closely irregularly punctate. Ocellocular space granulose and a little longer (10: 9) than the interocellar space, which is smoother. Ocelli in male a little raised. Temple wide medially, a little wider than the diameter of eye in side view. Mesoscutum closely ruguloso-punctate. Scutellum subconvex, punctate, carinate laterally only at base, the carina blunt. Apex of scutellum elevated from metascutellum and not in the same plane. Metascutellum subpolished, with a few scattered punctures. Mesopleurum and metapleurum with larger punctures, which are often aciculate at places, particularly in female. Propodeum rugose, areola and basal area subpolished. Areola semicircular or oval, with its apical closing carina arched inwards. Petiolar area concave and abruptly sloping from areola. Median longitudinal carinae bounding petiolar area more or less parallel-sided. Lateral longitudinal carina indistinct so that second lateral area and second pleural area confluent. Spiracular area not separated from second pleural area. Postpetiole flat, widened, its median area raised, like a low pyramid, and striato-punctate. Its sides with irregularly coalescent punctures. Abdomen longitudinally striato-punctate. Gastrocoeli deep and wide, space between them less than their width (0.7) and aciculate. Abdomen of female acute at apex.

Yellowish-brown. Orbits yellow. Flagellum dark brown basally and black apically. Tip of abdomen (tergites 5 and beyond) black. Wings pale yellowish-brown.

Variations: Body often narrow with blackish marks on mesoscutum, mesopleurum, metapleurum, apex of propodeum, and hind femur. In the Japanese forms described as *segmentalia*, the thorax is more extensively black and in the form *matsumurai*, the thorax, abdomen, hind femur and hind coxa are largely black.

Length: 15-17 mm. Fore wing 11-12 mm.

Specimens: Several males and females reared from the gypsy moth seen from Italy, Hungary, and Austria.

Distribution: Europe.

Hosts: *Lymantria dispar*, *L. monacha*, *L. dissoluta*, *Leucoma salicis*, *Orgyia antiqua*, *Mimas tiliae*, and *Smerinthus ocellatus*.

2. Genus PARACOELICHNEUMON

Paracoelichneumon Heinrich, 1978. Eastern Palaearctic Ichneumoninae
    (in Russian). Acad. Sci. USSR., p. 13.

Mandible narrowed apically, but not twisted so that both the teeth are in the same plane. Lower tooth blunt and short. Scutellum flat, in line with metascutellum, shiny and with scattered punctures. Mesopleurum with well formed, often evenly spaced punctures, not coarsely punctate. Areola elongate, rugose, extending up to basal 0.3 of propodeum. Petiolar area abruptly sloping, not concave. Mesoepisternum without a knob-like elevation near hind coxa and in profile view not appearing concave at its margin with mesosternum *(cf. Protichneumon)*. Gastrocoeli deep and wide, space between them less than 0.7 their width, and rugoso-striate. Postpetiole rugose. Abdomen closely punctate, in female acute apically.

Paracoelichneumon is rather close to *Lymantrichneumon* and is distinguished by its flat and more polished scutellum, areola, rugose, postpetiole not pyramidally raised, and by its clear hyaline wings. The body is black with the legs and abdomen reddish.

1. PARACOELICHNEUMON RUBENS (Fonscolombe)

Ichneumon rubens Fonscolombe, 1847. Ann. Soc. Ent. France, (2) 5: 407.
    Europe.
"*Ichneumon rubens* Wesmael": Rudow, 1918. Ent. Ztschr. Frankfurt,
    32: 72. Hosts: *Cerura vinula*, *Lymantria dispar*.
Protichneumon rubens: Györfi, 1963. Külonl Allattani Kozlem, p. 50-53.
    Host: *Lymantria dispar* (of minor importance).
Paracoelichneumon rubens: Heinrich, 1978. Eastern Palaearctic Ichneumoninae (in Russian), Acad. Sci. USSR, p. 14.

This species has been referred to as *Ichneumon rubens* Wesmael in the gypsy moth literature (Stadler, 1933, Schedl, 1936). It was apparently first associated with the gypsy moth by Rudow (1918). Györfi (1963) mentioned having reared it in Hungary from the same host. Other host records in literature are of *Cerura vinula* and *Catocala elocata* in Europe.

Flagellum of the female a little widened and ventrally flattened preapically. Face punctate, punctures denser in the median area. Clypeus with scattered punctures at base, smoother apically, its apical margin arcuate. Malar space 0.85 as long as basal width of mandible. Mesoscutum subpolished, with scattered but definite punctures. Scutellum flat, polished and with scattered punctures. Apex of scutellum in level with the metascutellum. Mesopleurum with distinct large punctures, well separated from each other, by at least their diameter. Metapleurum striato-punctate. Propodeum rugose. Petiolar area and third lateral area trans-rugose. Postpetiole rugulose, its median area weakly defined. Abdomen regularly punctate. Gastrocoeli wide and deep, interspace between gastrocoeli 0.7 their width. Apex of female abdomen acute.

Black with reddish abdomen. Vertical orbits, scutellum, subtegular ridge and a median band on flagellum yellow. Frontal orbits finely marked with brown. Hind corner of pronotum and tegula brown to brownish-black. All coxae and trochanters black. Hind tarsus brownish-black. Legs otherwise reddish (brick-red) with apices of hind femur and tibia blackish. Abdomen from second segment onwards reddish. Wings hyaline, very slightly infuscated.

Length: 20 mm. Fore wing 14 mm.

Specimens: One female from Rumania in the Townes Collection examined. No reared specimens seen.

Distribution: Europe.

3. Genus CALLAJOPPA

Callajoppa Cameron, 1903. Entomologist, 36: 236.

The genus *Callajoppa* is characterized by having the mandible not tapered apically, its lower tooth about 0.5 as long as the upper, occipital carina joining hypostomal carina before the base of mandible, female flagellum widened preapically, scutellum pyramidal, without sharp angles, areola in the form of a polished raised "boss", somewhat pyramidal in profile, postpetiole flat and with scattered punctures, gastrocoeli shallow, with the interspace between them as wide or wider than the width of a gastrocoelus, and apex of female abdomen obtuse.

Only one species, *Callajoppa cirrogaster cirrogaster* (Schrank) is associated with the gypsy moth. This has generally been referred to as *Ichneumon lutorius* or *Trogus lutorius*.

1. CALLAJOPPA CIRROGASTER CIRROGASTER (Schrank) (Figs. 91-94)

Ichneumon cirrogaster Schrank, 1781. Eunmeratio Insectorum Austriae. Indigenorum, p. 348. Europe.

Ichneumon lutorius Fabricius, 1787. Mantissa Insectorum. . ., 1: 262. Italy.

Trogus lutorius: Mocsary, 1787. Tijdschr. Ent., 21: 198. Russia.

"*Trogus flavitorius* [sic] *lutorius* (Fab.)?": Howard and Fiske, 1911. U.S.D.A. Bur. Ent. Bull., 91: 85. Host: *Lymantria dispar*. Europe.

Trogus lutorius: Meyer, 1933. Tables systematiques des hyménoptères parasites (fam. Ichneumonidae) de l'URSS et des payes limitrophes, 1: 345. Russia. Europe. Hosts: *Papilio machaon, Dendrolimus pini, Lymantria dispar*.

Callajoppa cirrogaster cirrogaster: Townes, Momoi and Townes, 1965. Mem. Amer. Ent. Inst., 5: 539. Europe. Russia.

This species was first mentioned as a parasite of the gypsy moth by Howard and Fiske (1911, reference cited above), based upon the information they had on their card file as well as from Dalla Torre (1901-02). No such host record is, however, seen in Dalla Torre under *"lutorius"*. Other persons who reported this species as a parasite of the gypsy moth are Meyer (1933) and subsequent catalogers like Stadler, Schedl and Thompson. Townes, Momoi and Townes considered *Trogus lutorius* as a synonym of *Callajoppa cirrogaster cirrogaster*.

A subspecies, *C. cirrogaster bilineata* Cameron occurs in China and Japan. Its hosts are species of *Laothoe*, *Smerinthus*, and *Dendrolimus*. *C. cirrogaster caspica* Heinrich occurs in Iran and Transcaucasia.

Face and clypeus punctate. Punctures coalescing in the median area of face. Malar space ruguloso-mat, 1.15 as long as basal width of mandible. Antennal scrobes concave and polished. Vertex concave behind. Temple widened medially. Female flagellum a little indented preapically and then tapering. Mesoscutum finely punctate on a mat surface. Scutellum subpolished pyramidal, without lateral margins or angles. Side of thorax largely punctato-striate. Upper half of pronotum and mesopleurum punctate. Propodeum with areola in the form of a polished elevated "boss", whence with a straight slope to the abdominal attachment. Costula, and median and lateral longitudinal carinae strong and raised. Propodeum strongly transcarinate, especially in the female. Petiole long, narrow and squarish in cross section. Postpetiole wide and flat, weakly punctate, its median field not sharply demarcated. Abdomen finely punctate, obtuse apically. Gastrocoeli shallow and broad, the interspace between them as wide as the width of gastrocoeli and aciculate.

Black with yellowish-brown abdomen. Face, clypeus, mandible except teeth, lower temple, inner and outer orbital borders, and ocellocular area yellowish-brown. Frons, ocellar patch, vertex and temple posteriorly, and occiput, black. The extent of yellowish-brown on outer orbits variable and in some males may not be connected to inner orbits dorsally. Antenna yellowish-brown. Thorax black with yellow marks on upper part of pronotal collar, upper margin of pronotum, hind corner of pronotum, tegula, subtegular ridge, scutellum, metascutellum, and a mark along lateral carina of scutellum. Mesoscutum sometimes with a reddish-brown mark. Legs brownish, with all coxae black, and fore coxa yellowish apically. Hind femur with blackish marks. Tarsi lighter in color. Wings yellowish-brown. Petiole blackish basally. Apex of abdomen black or brown.

Length: 22-24 mm. Fore wing 16-18 mm.

Specimens from Europe seen in the Townes Collection, ex pupa of "Amorpha tremulae".

Distribution: Eurasia.

Hosts: Lymantria dispar, Papilio machaon, Dendrolimus pini and *Amorpha tremulae.* Thompson (1957) lists several others belonging to the family Sphingidae.

4. Genus ICHNEUMON   (Fig. 114)

Ichneumon Linnaeus, 1758. Systema Naturae, (Ed. 10), 1: 560.
     = *Coelichneumon* of authors.

This genus is characterized by having a bristle-shaped flagellum, more or less widened preapically in the female, apical margin of clypeus arcuate or truncate, a little impressed, occipital carina joining hypostomal carina, mandible moderately narrowed apically, propodeum convex on basal half or less, areola usually horse-shoe shaped or hexagonal, often confluent with basal area, not raised above the general surface of propodeum, abdomen spindle-shaped, a little narrower than the thorax, black or bluish-black, with or without white spots. Postpetiole usually with a distinct aciculate median area, and gastrocoeli large and deep, sometimes wider than the space between them. The tip of abdomen in the female is strongly acute, pointed (oxypygous).

Several species of *Ichneumon* have been reported in literature as parasites of the gypsy moth. There appear to be many erroneous records. The identities of some of the species are also not clear. It appears that only *Ichneumon cyaniventris* may be a parasite of the gypsy moth.

1. ICHNEUMON CYANIVENTRIS Wesmael

Ichneumon cyaniventris Wesmael, 1859. Mém. Couronnés Acad. Sci. Belgique, 8: 58. Europe.

Ichneumon cyaniventris Wesm.: Rudow, 1918. Ent. Ztschr. Frankfurt, 32: 59. Europe. Host: *Lymantria dispar*.

Ichneumon cyaniventris Wesmael: Townes, Momoi and Townes, 1965. Mem. Amer. Ent. Inst., 5: 523. Eurasia, Korea, Japan. Host: *Clostera anastomosis* (in Japan).

This species was listed by Rudow (1918) as a parasite of the gypsy moth, *Lymantria dispar*. He did not list it in 1911 in his list of parasites of various "Bombycidae". Subsequent catalogers have somehow missed this and some other species listed by Rudow.

This species is rather closely related to *Ichneumon sugillatorius* L. from *Lymantria monacha*. Both have similar color patterns with white spots on abdominal tergites. The face of male is white with a wide black vertical line, which is wider in *cyaniventris* than in *sugillatorius*. The areola of *cyaniventris* is flat and propodeum a little sloping, while in *sugillatorius*, the areola is a little raised and propodeum convex in this area.

Ichneumon leucocerus Gravenhorst is different in having no white spots on the abdomen and the propodeum convex in the region of areola. The areola is crescentic, narrower than in *cyaniventris* or *sugillatorius*.

Some of the salient features of this species are:

Face and clypeus punctate. Clypeus flat and a little arcuate apically. Sub-apical segments of female flagellum a little widened and flattened ventrally. Flagellum apically tapered. Mandible apically not strongly tapered. Thorax punctate. Scutellum flat, shiny, with scattered punctures. Areola horse-shoe shaped, a little wider than long, its apical closing carina concave. Propodeum rugose or rugoso-punctate, with areola smoother. Propodeum in the region of areola flatter and sloping or in male a little convex. Postpetiole aciculate. Petiole laterally margined and trans-striate. Second tergite medially aciculate, rest punctate to striato-punctate. Gastrocoeli deep. Thyridia wide, their interspace 0.7 the width of each thyridium. Abdomen apically acute, punctate, punctures progressively smoother apically.

Black. Face of male white laterally, rather broadly so. Flagellum with about five medial segments wholly or partly white. Scutellum yellow. Tergites 2-3 in ♀ and 2-4 in male with apicolateral white spots. Fore tibia with a yellowish-brown line. In male fore and middle legs from apex of femur onwards brownish, or with yellowish-brown patches. Thorax of male with yellow marks along upper margin of pronotum, on tegula and on subtegular ridge.

Length: 15-16 mm. Fore wing 11 to 12 mm.

Specimens from Germany seen in the Townes Collection, but no reared specimens were available.

Distribution: Eurasia, Japan and Korea.

2. ICHNEUMON LEUCOCERUS Gravenhorst

Thompson (1946) listed this species as a parasite of *Lymantria dispar* quoting Stadler (1933) and Schedl (1936) from Europe and North Africa. Stadler (1933: 30) however mentioned "*Ichneumon leucocherrus* Wesmael" as a parasite of *Lymantria dispar*, and not *Ichneumon leucocerus* Gravenhorst, which was quoted as such by Schedl (1936). How the record got converted into *leucocerus* is not clear.

"*Ichneumon leucocherrus* Wesm." is not listed in Dalla Torre (1901-02), nor could I find this name in any other publication. It was obviously a *lapsus*, not for *Ichneumon leucocerus* Gravenhorst as Thompson apparently thought, but for *Ichneumon leucocheilus* Wesmael as this latter species was reported parasitic upon *Lymantria dispar* by Rudow (1918).

Ichneumon leucocheilus Wesmael properly belongs to the genus *Melanichneumon* (authentic specimens not available).

Ichneumon leucocerus Gravenhorst is to be removed from the list of parasites of the gypsy moth.

3. "ICHNEUMON PICTUS Gmelin"

Ichneumon pictus Gmelin, 1790. *In* Linnaeus: Syst. Nat., Ed 13,1(5): 2721.
Ichneumon pictus Gmelin: Dalla Torre, 1901-02. Catalogus Hymenopterorum, 3: 968. Eurasia (in part).

The identity of this species is uncertain. In fact Gmelin (1790. *Systema Naturae*) described two "*Ichneumon pictus*", --one on page 2702 and another on p. 2721. The one described on page 2702 was synonymized under *Microcryptus sericans* (Gravenhorst) (Gelinae) by Dalla Torre (1901-02: 708). The other, described on page 2721, was considered by Dalla Torre (1901-02: 968) to be a senior synonym as well as a homonym of *Ichneumon pictus* Gravenhorst (1829: 418) [which was actually described as *Hoplismenus pictus* Gravenhorst].

A perusal of the original descriptions of the above two species indicates that they are not synonymous.

Hoplismenus pictus Gravenhorst is *Stenoblus pictus* (Gravenhorst) *vide* Rasnitsyn (1981: 132) and Kaspayran (1981: 580). The identity of "*Ichneumon pictus* Gmelin, 1790: 2721" can not now be established. Subsequent workers have overlooked this species. It is incidentally a junior primary homonym of *Ichneumon pictus* Schrank, 1776, as well as a homonym of *Ichneumon pictus* Gmelin 1790: 2702.

Berthoumieu (1895), in his treatment of the European Ichneumonidae did not mention *I. pictus* of Gmelin. He mentioned *I. pictus* Gravenhorst (= *rufescens* Stephens = *ratzeburgi* Hartig) and mentioned *Lymantria dispar* as a host of it upon the authority of Mocsáry (1885). Mocsáry's references could not be located.

Ichneumon pictus Gmelin, should therefore be removed from the list of the parasites of the gypsy moth.

Ichneumon pictus (Gravenhorst) is treated here under *Stenaoplus*. Records of *Ichneumon pictus* Gmelin as a parasite of *Lymantria dispar*, as given by Stadler (1933), Schedl (1936) and by Thompson (1946, p. 495), should be construed to belong to *Stenaoplus pictus* (Gravenhorst).

The entry in Thompson (1946, p. 499) as *Stenichneumon pictus* Gmelin also pertains to *Stenaoplus pictus* (Gravenhorst). Thompson cited Morley and Rait-Smith (1933) as his source of information. Morley and Rait-Smith referred to *Stenichneumon pictus* Gravenhorst and not Gmelin, taking the host record from Morley (1903, 1: 49).

4. "ICHNEUMON FLAVUS Rd. " Nomen nudum.

"*Ichneumon flavus* Rd. ": Rudow, 1918. Ent. Ztschr. Frankfurt, 32: 64. Europe. Hosts: *Lymantria dispar*, *Cerura vinula*.
"*Ischnus flavus* Rd. ": Stadler, 1933. Ent. Anz., 13(2-4): 30.

I could not trace the original description of this species in Rudow's publications, nor in Zoological Record. How the generic name changed from *Ichneumon* to *Ischnus* in subsequent publications of Stadler (1933) and Schedl (1936) is unclear. Thompson (1946) mentioned it as "*Ischnus flavus* Rond. "

Rudow apparently intended to describe it as a new species. It is a *nomen nudum*.

5. ICHNEUMON sp.

"*Ichneumon* sp. ": Picard, 1921. Progres Agric. Vitic., 76(33): 160-165. France. Schedl, 1936: 188. Thompson, 1946: 495. Host: *Lymantria dispar*.

The identity of this species will remain uncertain until the voucher specimen is seen, if still available. It could be one of the other Ichneumonini treated here.

TRIBE JOPPINI

5. Genus STENAOPLUS

Stenaoplus Heinrich, 1938. Mem. Acad. Malgache, 25: 116.

The genus *Stenaoplus* is characterized by having a moderately to strongly raised scutellum which is laterally sharply margined by carinae up to its apex, flagellum of female long, tapering and slender, not flattened or much widened preapically, propodeum flat on basal half and then sloping, the horizontal portion of propodeum a little shorter than the apical sloping part, apophyses absent, areola normally formed, squarish or elongate, postpetiole granulose to punctate, without a median field, gastrocoeli shallow, weakly impressed, very wide, and narrow, the space between them less than 0.7 their width.

One species, *Stenaoplus pictus* (Gravenhorst) has been mentioned in literature as a parasite of the gypsy moth.

1. STENAOPLUS PICTUS (Gravenhorst)

Hoplismenus pictus Gravenhorst, 1829. Ichneumonologia Europea, 2: 418.
Ichneumon pictus Gravenhorst: Berthoumieu, 1895. Ann. Soc. Ent.
        France, 1895, p. 567. (synonymy and host records).
Ichneumon pictus: Dalla Torre, 1901-02, Catalogus Hymenopterorum,
        3: 968. (In part). Hosts: *Lymantria dispar* (after Mocsáry),
        Thera juniperata, Semiothisa liturata, Cidaria fulvata.
Ichneumon rufescens Stephens, 1835. Illustrations to British Entomology,
        Mandibulata, 7: 207. Name preoccupied by Rossi, 1794 and by
        Cervier, 1833.
Cryptus Ratzebergii Hartig, 1838. Jahresber. Forstschr. Forstw.,
        1 (2): 263. Host: *Dendrolimus pini.*
Ichneumon rufescens Stephens: Perkins, 1953. Bull. British Mus. (N.H.)
        Ent., 3: 112. [= *Aoplus ratzeburgi* (Hartig), syn. *Stenichneumon
        pictus* (Gravenhorst) Morley.]

   Hoplismenus pictus Gravenhorst and *Ichneumon pictus* Gmelin are not primary homonyms and do not belong to the same genus now. Therefore the
name *pictus* Gravenhorst is reinstated for *rufescens* Stephens or *ratzeburgi*
Hartig, as used in the publications of Perkins (1953, 1960) and Fitton (1978).
   Stenaoplus pictus (Gravenhorst) was first reported as a parasite of the
gypsy moth by Mocsáry (1885). Berthoumieu (1895) mentioned it in his treatment of the European Ichneumoninae. All the reports of Stadler, (1933), Schedl
(1936) and Thompson (1946, p. 495 and 499), pertain to this species. The host
records as given by Morley (1903) and others are *Lymantria dispar, Cidaria
fulvata, Semiothisa liturata* and *Thera juniperata.* Kasparyan (1981) mentions
these hosts except *Lymantria dispar*—the reason for which is not stated! Perkins (1960) stated that this species is parasitic on geometrids on *Pinus.*
   Female: Small sized species. Subapical flagellar segments not conspicuously widened though a little flattened ventrally. Face and clypeus with distinct
well separated punctures. Clypeus a little convex basally, its apical margin
truncate. Mandible narrowed apically, its lower tooth short and pointed.
Malar space about 0.8 as long as the basal width of mandible. Temple not
widened medially. Mesoscutum mat, with shallow punctures. Scutellum subconvex, laterally carinate and with scattered minute punctures. Mesopleurum
and metapleurum punctate. Areola horse-shoe shaped, as long as wide, subpolished, mat. Petiolar area concave. Postpetiole mat. Tergites 2 and 3
with definite regular punctures. Thyridia wide, about 2.0 as wide as the space
between them. Ovipositor rather long for the group, as long as or longer than
the maximum width of abdomen (apical width of tergite 2).
   Brownish-yellow. Frontal and vertical orbits yellow marked. Frons,
vertex, occiput and antenna blackish-brown. Apex of scutellum, metascutellum, and subtegular ridge yellow-marked. Propodeum blackish. Coxae brownish with dorso-apical blackish marks. Hind tibia and tarsus with fuscous spots.
Abdomen brownish-yellow, without any white apical marks.
   Male: Face strongly punctate. Body conspicuously marked with red and
yellow. Inner orbits yellow. All femora reddish. Hind tibia without a yellow
median band. Coxae usually marked with yellow and red.
   Length: 6-10 mm. Fore wing 4.5-6.5 mm.

Specimens: England, 1♀, det. Perkins, 1953 (Townes). No reared specimens seen.

Perkins (1960) puts the species under *Aoplus* as *ratzeburgii* (Hartig) and provides a key to distinguish it from the other British species.

6. Genus MELANICHNEUMON (Fig. 115)

Melanichneumon Thomson, 1893. Opuscula Entomologica, 18: 1954.

The genus *Melanichneumon* is characterized by having the subapical flagellar segments of the female very wide and flat ventrally, the widest segments about 2.0 as wide as long. Male flagellum with transverse ridges ventrolaterally which are beset with short bristles. Base of propodeum with a small median tubercle. Areola more or less horse-shoe shaped, narrowed towards base. Propodeal carination sharp and of a generalized type. Abdominal tergites 2-3 strongly neatly punctured and convex. Postpetiole also punctate, not striate. Thyridia weakly impressed, the space between them more than 1.5 the width of a thyridium. Apex of female abdomen oxypygous, and usually yellow or white.

One species of this genus has been mentioned as a parasite of the gypsy moth, under the name *Ichneumon leucocheilus* Wesmael.

1. MALANICHNEUMON LEUCOCHEILUS (Wesmael) (Figs. 95, 96)

Ichneumon leucocheilus Wesmael, 1844. Nouv. Mem. Acad. Sci. Bouxelles, 18: 89. Europe.
Ichneumon leucocheilus Wesm.: Rudow, 1918. Ent. Ztschr., 32: 64. Europe. Host: *Lymantria dispar*.
Melanichneumon leucocheilus: Perkins, 1960. Handbook for the identification of British Insects, 7(2): 150. England.

The earliest reference to this species being parasitic upon *Lymantria dispar* was traced to Rudow (1918).

"*Ichneumon leucocherrus* Wesm." as quoted by Stadler (1933) and Schedl (1936) is definitely a lapsus for *I. leucocheilus* Wesm. No species as "*Ichneumon leucocherrus* Wesm." exists. Thompson (1946) erroneously reported it as "*Ichneumon leucocerus* Gravenhorst."

Specimens of this species were not available for study. The following description is taken from Berthoumieu (1895) and Perkins (1960):

Clypeus with a weak, rather broad, central projection. Areola subhexagonal, elongate. Legs robust. Postpetiole punctate. Gastrocoelus small and superficial. Thyridium indistinct. Occipital carina meeting hypostomal carina near base of mandible.

Female: Antenna with a white annulus. Head and thorax wholly black. Wings hyaline. Stigma black. Tegula reddish-brown. Legs reddish-brown with coxae black. Hind femur and tibia brown. Abdominal segments 2-4 and base of 5, reddish-brown. 6-7 segments white. Tip of abdomen white.

Male: Facial orbits white. Antenna black, ferruginous basally. Wings a little infuscate. Coxae and trochanters black. Rest of fore and middle legs red, with white or fuscous patches. Hind femur infuscate, tibia and tarsus black. Abdomen black with apex of tergites 1-3 red. Tergite 7 with a large

dorsal white spot.
Length: 10-13 mm.
Distribution: Europe.

7. Genus CRATICHNEUMON   (Fig. 117)

Cratichneumon Thomson, 1893. Opuscula Entomologica, 18: 1945.

Female flagellum cylindrical and blunt apically, only a little widened preapically. Occipital carina joining hypostomal carina at a distance equal to
0.35 the basal width of mandible. Scutellum flat, not carinate laterally.
Median part of propodeum without a weak tubercle in the middle, its dorsal
surface flat, and petiolar area excavated. Postpetiole with shallow punctures,
or its median area with only fine striations and not sharply demarcated from
the lateral areas. Gastrocoeli weak, superficial. Thyridia linear, smaller
than the distance between them. Abdomen oxypygous.
    Cratichneumon is close to *Melanichneumon.* Species of *Cratichneumon* are
more smooth and shiny, with less densely punctate mesoscutum and abdomen,
shallower gastrocoeli, which are often obsolete, and abdominal tip is rarely
white marked.
    Members of this genus are generally parasitic upon geometrid pupae.
The females do not hibernate and many species have two generations per year,
adults appearing in May-June and again in August. One species, *Ichneumon
fabricator,* is reported in the literature as a parasite of the gypsy moth.

1. CRATICHNEUMON FABRICATOR (Fabricius) (Fig. 97, 98)

Ichneumon fabricator Fabricius, 1793. Entomologica Systematica, 2: 166.
    Germany.
Ichneumon fabricator F.: Schedl, 1936. Monogr. Angew. Ent., 12: 188.
    Europe. Hosts (after Cecconi, 1924): *Lymantria dispar, Elkneria
    pudibunda, Panolis flammea, Bupalus piniaria, Operophtera brumata.*
Protichneumon fabricator: Györfi, 1963. Különenyomet az állattani
    Közleményeh L. Kötet, 1-4: 51. Hungary. Not common. Host:
    Lymantria dispar.
Cratichneumon fabricator: Townes, Momoi and Townes, 1965. Mem.
    Amer. Ent. Inst., 5: 442. Eurasia.
Cratichneumon fabricator: Kasparyan, 1981. Fauna USSR, Vol 3 (Ichneu-
    monidae): 573. Eurasia. Hosts: *Semiothisa liturata, Bupalus
    piniaria, Rhyacionia pinicolana, Notodonta dromedarius, Elkneria
    pudibunda, Panolis flammea,* and *Tethea or.*

This species was first reported from *Lymantria dispar* by Cecconi (1924).
Kasparyan (1981) listed several hosts in European USSR, but did not mention
Lymantria dispar.
    Female: Flagellum cylindrical and blunt apically, only a little widened pre-
apically. Face a little convex below antennal sockets. With well separated
punctures. Clypeus polished, with rows of punctures at base and with a few
scattered punctures in the middle. Temple wider ventrally. Malar space
equal to the basal width of mandible. Frons and vertex mat, with scattered
punctures. Vertex widened behind eyes. Thorax punctate, the mesoscutum

shallowly so and propleurum more densely so. Scutellum flat, subpolished and with scattered punctures. Propodeum flat dorsally, its petiolar area excavated. Lateral carinae of areola indistinct basally. Median field of postpetiole not sharply demarcated, with fine weak striations interposed with punctures. Tergite 2 punctate. Width of thyridia less than the distance between them. Tergite 3 with shallow, indistinct punctures.

Black. Flagellum with a white annulus. Tegula brownish-black. All coxae and trochanters black, femora brown, tibiae brown with white marks and tarsi brown. Hind tibia apically and tarsus darker, brownish-black. Wings hyaline, a little brown-tinged.

Male: Flagellum tapered, somewhat serrate. Propodeal carinae stronger. Postpetiole without striations. Gastrocoeli deeper.

Face and clypeus largely white. Face with a black central mark. Flagellum brownish below, without a white annulus. Hind corner of pronotum and subtegular ridge yellow. Fore and middle femora, tibiae and tarsi orange colored. Hind leg wholly brownish-black.

Length: 8-11 mm. Fore wing 7-8 mm.

Specimens from Europe examined in the Townes Collections. No reared specimen seen. Several host records are mentioned by Thompson (1957).

Distribution: Europe.

8. Genus CHASMIAS (Fig. 118)

Chasmias Ashmead, 1900. Proc. U. S. Natl. Mus., 23: 17.

Apical margin of clypeus broadly weakly concave, with a weak broad median tooth. Mandible not much narrowed apically, its lower tooth shorter than the upper. Occipital carina joining hypostomal carina above the base of mandible. Antennal flagellum cylindric in the female and somewhat serrate in the male. Preapical segments not flattened. Tip of female flagellum blunt and apical segments curled. Areola horse-shoe shaped, open behind, elongate in female. Abdomen oxypygous. Postpetiole with a weak median field, which is weakly striate. Gastrocoeli small to large, shallow. Ovipositor distinctly exserted beyond the tip of abdomen.

One species, *Chasmodes paludicola,* was reported by Rudow (1917) as a parasite of the gypsy moth in Europe.

1. CHASMIAS PALUDATOR (Desvignes)

Ichneumon paludator Desvignes, 1854. Trans. Roy. Ent. Soc. London,
    (N.S.) 3: 44. England.
Chasmodes paludicola Wesmael, 1857. Bull. Acad. Sci. Belgique,
    (2)2: 356. Europe.
Chasmias paludicola: Dalla Torre, 1901-02. Catalogus Hymenopterorum,
    3: 1024.
Chasmodes paludicola: Rudow, 1917, Ent. Ztschr. Frankfurt, 31: 31.
    Host: *Lymantria dispar.*

Heinrich (1937) synonymized *Chasmodes paludicola* with *paludator.* Rudow's record was missed by Schedl, Thompson and others.

Female: Face with distinct well separated punctures, interspaces 1.0 to

1.5 the diameter of the punctures and shiny. Thorax punctate. Punctures on mesoscutum small and shallow. Scutellum flat, subpolished. Areola horse-shoe shaped, open behind, elongate in the female. Median field of postpetiole very finely striate. Gastrocoeli small. Tergites 2 and 3 finely punctate. Ovipositor extending conspicuously beyond the tip of abdomen.

Black. Flagellum medially and scutellum yellow. Coxae black. Legs otherwise reddish-brown, with tarsi infuscate. Tergites 1, 2, and 3 reddish-brown, the rest black. Tergite 7 with an elongate yellowish-brown longitudinal stripe.

Male: Punctures stronger over body. Flagellum tapered and weakly serrate. Areola more squarish and widely open behind.

Black. Facial orbits narrowly yellow. Scutellum black. All femora, and tibiae reddish-brown, tarsi infuscate. Tergite 2 largely reddish-brown, with black patches, rest of abdomen black.

Length: 15-17 mm. Fore wing 12-14 mm.

Specimens from Europe examined in the Townes Collections. No reared specimens seen.

Distribution: Europe.

9. Genus **PTEROCORMUS** (Fig. 120)

Pterocormus Foerster, 1850. Arch. f. Naturgesch., 16: 71.
    = *Ichneumon* of authors.

A genus close to *Chasmias*, but apical margin of clypeus truncate, or weakly convex, without a median tooth, apical part of female flagellum more or less tapered, thyridia moderately large and distinctly impressed and median fold on sternite 4 rather distinct.

One species, *Ichneumon sarcitorius* was recorded by Meyer (1929) as a parasite of the gypsy moth in Russia.

1. PTEROCORMUS SARCITORIUS SARCITORIUS (Linnaeus) (Figs. 99, 113, 120)

Ichneumon sarcitorius Linnaeus, 1758. Systema Naturae, (Ed 10) 1: 561.
    Europe.
Meyer, 1929. Rpts. Sppl. Ent., Bur. Appl. Ent., State Inst. Exper.
    Agron, Leningrad, 4: 240. Host: *Lymantria dispar.* USSR.
Pterocormus sarcitorius sarcitorius: Townes, Momoi and Townes, 1965.
    Mem. Amer. Ent. Inst., 5: 479. Syn. references, host records.
    Eurasia. North Africa.

Female: Face punctate, interspaces shiny. Flagellum cylindrical, apical segments tapering and weakly curled. Thorax punctate. Scutellum flat and subpolished. Punctures on meso-and metapleurum becoming rugose at places. Propodeum ruguloso-punctate. Areola squarish, open behind. Median field of postpetiole clearly demarcated and finely longitudinally striate. Gastrocoeli small but deep. Tergite 2 and 3 moderately deeply punctate. Sternites 2-5 with a midventral fold. Ovipositor slightly exserted beyond abdominal tip.

Black. Frontal orbits, and tegula brown. Median flagellar segments, subtegular ridge and scutellum yellow. All coxae, trochanters and apices of hind femur and tibia black. Legs otherwise reddish-brown. Tergite 2 wholly and tergite 3 in apical 0.6 reddish-brown. Sternites 2 and 3 yellowish-brown. Tergite 6 broadly yellow dorsally.

Male: Face, clypeus and scape ventrally yellow. Clypeus and face triangularly above clypeus may be black. Antenna brown, darker dorsally. Tegula, subtegular ridge and scutellum, yellow. Hind corner of pronotum also often yellow. Apices of abdominal tergites yellow, these yellow marks widened laterally. Those on tergites 1 and 4 usually incomplete dorsally, and absent on tergite 5.

Length: 12-16 mm. Fore wing 9-12 mm.

Specimens from Europe examined in the Townes Collections. No reared specimens seen.

Distribution: Eurasia, China, North Africa.

Hosts: Lymantria dispar, Diloba caeruleocephala, Gortyna borelli, and *Agrotis segetum.*

10. Genus TRIPTOGNATHUS (Fig. 119)

Triptognathus Berthoumieu, 1904. Genera Insectorum, 18: 49.

Female flagellum short, curled, its second segment 0.8 to 1.4 as long as wide, its apex with a short taper. Flagellum of male slender. Mandible not tapered apically, only slightly narrowed apically. Occipital carina joining hypostomal carina above base of mandible. Areola about 1.0 as long as wide. Abdomen amblypygous, its tip rounded. Side of median field of postpetiole sharply defined. Gastrocoeli small to large, shallow to moderately deep, with a distinct thyridium. Sternite 3 with its median 0.3 or more membranous, visible as a longitudinal fold in dried specimens. Sternite 4 not membranous medially. Ovipositor unusually short. Female subgenital plate conspicuous. Male subgenital plate with a rather long median apical lobe. Genital claspers unusually large, in profile view the lower apical corner more produced and pointed than the upper.

One species, *Triptognathus amatorius* (Mueller), is associated with the gypsy moth.

1. TRIPTOGNATHUS AMATORIUS (Mueller) (Figs. 100-103)

Ichneumon amatorius Mueller, 1776. Zoologicae Danicae Prodromus, p. 151. Denmark.

Amblyteles amatorius: Rudow, 1917. Ent. Ztschr. Frankfurt, 31: 26. Europe. Hosts: *Lymantria dispar, L. monacha.*

Amblyteles amatorius: Uchida, 1926. J. Fac. Agr. Hokkaido Imp. Univ., 18: 118. Sakhalin. Hosts: *Lymantria dispar, L. monacha, Anaplectoides virens, Dendrolimus albolineatus.*

Triptognathus amatoria: Townes, Momoi and Townes, 1965. Mem. Amer. Ent. Inst., 5: 496. Europe, Japan, Russia, Sakhalin.

Diphyus amatorius: Kasparyan, 1981. Fauna USSR, Ichneumonidae, 3: 614, 616, 617, 618. Russia.

This species was not listed as a gypsy moth parasite by Schedl (1936) or Thompson (1946). The earliest record of its occurrence on the gypsy moth is that of Rudow (1917) in Europe and Uchida (1926) in Sakhalin.

Female: Face densely punctate. Mesoscutum closely finely punctate. Scutellum flat, subpolished with shallow indistinct punctures. Mesopleurum and metapleurum densely punctate, punctures on metapleurum finer than those on mesopleurum. Propodeum densely finely punctate. Areola rectangular, open behind. Costula indistinct. Apophyses indistinct. Median field of postpetiole longitudinally striate. Postpetiole punctate apicolaterally. Tergites 2 and 3 very finely and closely punctate. Gastrocoeli weakly impressed, rather indistinct, their width less than the space between them. Ovipositor hardly exerted.

Black. Inner orbits, flagellum medially, hind corner of pronotum, subtegular ridge and scutellum yellow. Apices of abdominal tergites yellow. Tergite 2 orange-red. Sternites 2 and 3 orange-red. Legs black with their femora and tibiae yellowish-brown.

Male: Generally similar to the female but flagellum long and slender, black. Punctures on face more regular and a little larger. Thoracic punctures comparatively larger. Areola squarish, closed behind and gastrocoeli more distinctly impressed. Sternites 2 and 3 with a midventral fold.

Face, clypeus and scape ventrally, yellow. Tegula, subtegular ridge, hind corner of pronotum and scutellum, yellow. Legs as in the female but fore femur wholly, and middle femur largely also yellow. Tergites 2 and 3 yellow. Yellow margins on tergites narrow.

Length: 15-18 mm. Fore wing 12-15 mm.

Specimens from Europe seen in the Townes Collection. No reared specimens seen.

Distribution: Eurasia.

11. Genus SPILICHNEUMON   (Fig. 116)

Spilichneumon Thomson, 1894. Opuscula Entomologica, 19: 2087.

Female flagellum short, curled, its second segment about 1.4 to 1.8 as long as wide, its apex with a short taper. Mandible wide, apically, in male gradually tapered. Male flagellum with a long taper. Areola about 1.4 (male) or 1.8 (female) as long as wide. Apical half of postpetiole with the median field indistinctly bounded laterally. Abdomen amblypygous. Ovipositor unusually short. Sternite 4 often not membranous medially. Subgenital plate of male and female and male claspers as in *Triptognathus*.

One species, *Ichneumon occisor* Fabricius, was listed by Rudow (1917) as a parasite of the gypsy moth in Europe.

1. SPILICHNEUMON OCCISOR (Fabricius) (Figs. 104-110, 116)

Ichneumon occisor Fabricius, 1793. Ent. System., 2: 142. Europe.
Ichneumon occisorius Fabricius, 1804. System. Piez., p. 61.
     Emendation.
Amblyteles occisorius Wesm: Rudow, 1917. Ent. Ztschr. Frankfurt,
     31: 26. Hosts: *Lymantria dispar, Leucoma salicis*.
Spilichneumon occisor: Townes, Momoi and Townes, 1965: 504.
     (= *occisorius* Grav. = *occisor* Fabr.)

Rudow (1917) apparently first recorded it as a parasite of *Lymantria dispar*. Subsequent cataloguers have missed this species in their lists.

This species resembles *Triptognathus amatorius* and *Pterocormus sarcitorius* in general appearance, but it can be readily separated from them by the characters given in the keys above.

Female: Face with minute well formed punctures, interspaces shiny. Clypeus rather flat. Mandible a little wider preapically. Thorax punctate with mesoscutum finely so and largely shiny. Scutellum polished. Propodeum finely ruguloso-punctate. Areola elongate and narrow, rounded basally. Postpetiole finely aciculate in the median field, polished otherwise. Tergites 2 and 3 very finely and shallowly punctate and shiny. Sternites 2 and 3 with a median fold.

Black. Frontal orbits and mandible reddish-brown. Flagellum yellowish-brown subapically. Subtegular ridge yellow. Scutellum yellow. Tergites and sternites 2 and 3 orange-red. Other tergites yellow medially along their apical margins, these stripes incomplete. Legs blackish-brown with tibiae and tarsi lighter in color.

Male: Face, clypeus, mandible medially, and underside of scape, yellow. Tegula, subtegular ridge, hind corner of pronotum and scutellum, yellow. Legs yellow with all coxae and trochanters and hind femur black. Apex of hind tibia black. Abdomen black with tergites 2 and 3 largely and apices of the following tergites, yellow. Last tergite largely yellow. Only sternite 2 with a median fold.

Length: 12-15 mm. Fore wing 8-10 mm.

Specimens from Europe in the Townes Collection examined. No reared specimens seen.

Distribution: Europe.

12. Genus AMBLYTELES

Amblyteles Wesmael, 1844. Nouveaux Mém. Acad. Roy Sci. Lett. Beaux-Arts Belgique, 18: 112.

Apical margin of clypeus truncate, without any tubercle. Lower tooth of mandible small. Occipital carina complete to hypostomal carina. Flagellum long, its second segment 1.4 to 3.0 as long as wide, its apex tapered, median segments cylindric. Propodeum with distinct sublateral triangular projections, one on either side. Abdomen of female amblypygous, its tip blunt. Gastrocoeli small to large. Thyridium small, weakly impressed. Ovipositor unusually short. In the female sternite 4 often membranous medially, and subgenital plate conspicuous and without an apical tuft of hairs. In the male the subgenital plate medially rounded or without a long lobe, and genital claspers not enlarged. Apex of first tergite and propodeum often black.

Amblyteles as defined here contains only one European species, *A. armatorius* (Förster).

Three species, *Amblyteles armatorius*, *A. camelinus* and *A. varipes* have been mentioned in the literature as parasitic upon the gypsy moth in Europe. The record of *A. camelinus* appears erroneous. Neither *camelinus* nor *varipes* belong to *Amblyteles*. Neither of these species have been subsequently reared from the gypsy moth. Reared material even of *A. armatorius* was not available to confirm its occurrence on the gypsy moth.

1. AMBLYTELES ARMATORIUS (Foerster) (Figs. 111, 112)

Ichneumon armatorius Förster, 1771. Novae Species Insectorum.
        Centuria, 1: 82. England.
Ichneumon fasciatorius Fabricius, 1775. Systema Entomologiae, 330.
        England.
Amblyteles fasciatorius: Uchida, 1930. J. Fac. Agr. Hokkaido Imp.
        Univ., 25: 354. Japan. Hosts: *Dendrolimus albolineatus,*
        Lymantria dispar, L. monacha.
Amblyteles armatorius: Townes, Momoi and Townes, 1965: 502.
        Eurasia, Iran, Algeria.

This species has been reported in literature from several hosts in Japan
and Europe. Uchida (1930) apparently reported it for the first time from
Lymantria dispar in Japan. Schedl (1936) and Thompson (1946) do not list it.
    This species can be readily distinguished by its black and yellow banded
abdomen and propodeum having spine-like projections on either side.
    Female: Face closely punctate, punctures coalescing. Clypeus smoother
apically, its apical margin truncate or a little arcuate medially. Mandible
narrowed apically, lower tooth blunt and much shorter than the upper. Malar
space as long as the basal width of mandible. Frons and vertex closely
punctato-rugose. Occipital carina meeting hypostomal carina away from
mandibular base by a distance equal to the basal width of mandible. Thorax
rugosely sculptured. Sculpture of mesoscutum finer than that of rest of
thorax. Propodeal areola squarish to a little rectangular, almost touching
base of propodeum. Petiolar area depressed and bordered by longitudinal
carinae. Lateral projections on propodeum sharp. First tergite, particularly
postpetiole, rugose, its median field distinct and with striations. Second and
third tergites punctato-rugulose. The following tergites smoother. Gastrocoeli short, carinate, its length about 0.5 the distance between the two. Tip
of abdomen blunt, obtuse (amblypygous).
    Black, with yellow marks on body and abdomen.
    Face with lateral yellow stripes along inner orbits. Flagellum brown,
tapering. Pronotal collar dorsally, hind corner of pronotum, tegula, subtegular ridge, and scutellum, yellow. Legs black with extensive yellow marks
on trochanters, base, apex and a dorsal line on fore and middle femora,
whole of fore and middle tibiae, hind femur in basal 0.25 and hind tibia in
basal 0.5. All tarsi brownish, but hind tarsus darker. Abdomen black with
yellow stripes on base of second and third tergites, extending laterally, and
apex of fourth to sixth tergites. Seventh tergite broadly yellow. Sometimes
hind tibia brownish in apical half, rather than black.
    Male: Sculpture similar to that of female, but color different as below.
Flagellum yellow ventrally. Scape largely yellow. Face and clypeus yellow.
Fore and middle legs almost wholly yellow except for dorsal black marks on
femora. Hind tibia yellow in basal 0.75. Second tergite yellow in basal 0.6.
Third tergite yellow except narrowly apically. Fourth and fifth wholly, and
basal half of sixth tergite black. Tip of abdomen yellow.
    Length: 13-16 mm. Fore wing 11-13 mm.
    Specimens from Europe examined in the Townes Collections. No reared
specimens seen. Host records are given by Thompson (1957).
    Distribution: Europe. Japan.

2. "AMBLYTELES VARIPES" Rudow.

Amblyteles varipes Rudow, 1888. Entom. Nachr., 14: 86. ♂, ♀.
   Germany. Host: *Lymantria dispar*.
Amblyteles variipes: Dalla Torre, 1901-02. Catalogus Hymenopterorum,
   3: 843.
Amblyteles varipes: Howard and Fiske, 1911. U. S. Dept. Agr. Bur.
   Ent. Bull., 91: 85. Host: *Lymantria dispar*.

This species, somehow, has not been mentioned by any of the subsequent
cataloguers of gypsy moth parasites. It is also unknown to me, as no specimens of it were available in the Townes Collections. Berthoumeiu (1894-96)
did not include it in his treatment of the European species. The identity of
this species is, therefore, uncertain.

A perusal of the original description reveals that it is not a true
Amblyteles and may belong to *Ichneumon*, *Melanichneumon* or *Ctenichneumon*.
It is characterized by having a black body, with antenna, scutellum and hind
leg wholly black. The fore and middle legs are rufous and the wings are
infumate.

The antenna is said to be serrate underneath *(cf. ♂ Ctenichneumon)*, scutellum
flat, areola regularly 5-sided *(cf. Ichneumon)* thorax and abdomen ruguloso-
punctate and only sternites 1 and 2 with ventral folds *(cf. Ctenichneumon)*.

Length: 18 mm.
Host: Lymantria dispar.
Distribution: Germany.

3. "AMBLYTELES CAMELINUS Wesmael"

"Amblyteles camelinus Wesmael, 1844. Nouveaux Mem. Acad. Sci. Lett.
   Beaux Arts Belgique, 18: 129. Belgium.

Meyer (1936) mentioned *Lymantria dispar* as one of the hosts of this
species. Earlier (1933, Pt 1: 304) he listed *"Pergesa dispar"* as one of the
hosts in his treatment of this species. No subsequent worker has reported it
as parasitic upon the gypsy moth. This species is a parasite of the pupae of
nymphalid butterflies and the record from the gypsy moth is undoubtedly
erroneous. Perkins (1953) erected a new genus *Thyrateles* for the reception
of this species. Townes *et al.* (1965: 459) synonymized *Thyrateles* under
Pterocormus Foerster (= *Ichneumon* of authors). Kasparyan (1981: 608)
treats this species as *Thyrateles camelinus* Wesmael.

TRIBE PRISTICEROTINI

13. Genus COTIHERESIARCHES   (Fig. 121)

Cotiheresiarches Talenga, 1929. Zool. Anz., 83: 185.

This genus is readily distinguished from the others treated here by its
petiole being flat dorsally, and wider than deep. The clypeus projects forward and is triangular in outline. Mandible is strong, with only one tooth.

One species *Cotiheresiarches dirus* has been associated with the gypsy
moth, which is a robust looking species with a black body and yellow femora.

1. COTIHERESIARCHES DIRUS (Wesmael) (Fig. 121)

> *Eurylabus dirus* Wesmael, 1853. Bull. Acad. Sci. Belgique, 20: 307.
> *Eurylabus dirus:* Dalla Torre, 1901-02. Catalogus Hymenopterorum,
>   3: 792. Europe. Algeria. Hosts (after several authors):
>   *Trichiura crataegi, Eriogaster lanestris, Orthosia opima.*
> *Eurylabus dirus:* Fahringer, 1922. Z. Angew. Ent., 8: 325-388. Host:
>   *Lymantria dispar.*
> *Cotiheresiarches dirus:* Townes, Momoi and Townes, 1965. Mem. Amer.
>   Ent. Inst., 5: 513. synonymy.

This species was first reported as a parasite of the gypsy moth by Fahringer (1922). Kolubayiv (in Komarek, 1937) subsequently reported it as a moderately common parasite of the nun moth *(Lymantria monacha).* There is apparently no subsequent rearing record from the gypsy moth.

It is readily distinguished from other Ichneumoninae by its petiole flattened dorsally. This flattened area is carinate laterally. The body is somewhat robustly built and is 14-16 mm. long. The abdomen is obtuse apically.

Face and clypeus closely punctate. Clypeus triangular, with its apical margin curved anteriorly and thus quite distant from the mandibles. Clypeus and face making a concave curve. Mandible tapered apically, with its lower tooth virtually absent. Antennal flagellum not widened medially, apically finely long-tapered. Thorax closely punctate, punctures rough at places. Scutellum convex and punctate. Propodeum rugose. Areola convex, combined first and second lateral areas concave. Propodeal carinae irregular. Petiole flattened dorsally, the flattened area leathery in texture and bordered by blunt carinae. Postpetiole ruguloso-punctate, apically smoother, without a distinct median field. Second tergite rugose medially and punctate on either side. Gastrocoeli moderately deeply impressed, carinate. Thyridium about as wide as the space between them. Third tergite closely finely punctate. The following tergites progressively weakly punctate to mat. Abdomen amblypygous. Female subgenital plate long and V-shaped.

Black. Fore and middle femora, tibiae and tarsi and hind femur orange-red. Hind tibia and tarsus brownish-black, with tarsal segments apically brown. Wings clear hyaline, stigma pale brown, veins darker, brownish-black.

Length: 14 mm. Kolubayiv (1937) mentioned the length as 15-16 mm.

Distribution: Europe.

Specimen: One female seen in the Townes Collection. No reared material seen.

IV. SUBFAMILY MESOSTENINAE

Members of the subfamily Mesosteninae are characterized by having a depressed abdomen, with a moderately long first tergite, with spiracles usually behind the middle and the postpetiole widened posteriorly. Clypeus separate from the face, often convex and its apical margin usually without teeth. Notauli and sternaulus distinct and strongly impressed. Apical truncation of scape strongly oblique. Dorsal rim of metanotum without a small sublateral angular projection, but sometimes with such a projection *just below* the hind margin, and often with a submedian pair of such projections, one opposite each side of the scutellum. Propodeum with one or two trans-verse carinae, usually the basal transverse carina alone present or more prominent. Propodeum without longitudinal carinae except rarely. Epiplura of second tergite narrow, often vestigeal. Tarsal claws simple. Second recurrent vein with a single bulla, not sloping outwards, meeting subdiscoidal vein at a right angle. Areolet usually present. Ovipositor long, without any subapical notch, with ridges on tip of lower valve.

The subfamily Mesosteninae is the same as tribe Mesostenini in Townes (1970). Gupta (1973) raised it to the subfamily level.

The genera that have been associated with the gypsy moth are believed to be parasitic on the host prepupae or larvae within the cocoons. The egg is laid externally on the body of the host larva or prepupa within the cocoon. Emergence is from the pupa, or from the parasite's cocoon within the cocoon of the gypsy moth.

Five species of Mesosteninae have been recorded as parasites of the gypsy moth, *viz.*, *Cryptus cyanator* Gravenhorst, *C. amoenus* Gravenhorst, *C. liparidis* Rnd., *Ischnus assertorius* Gravenhorst and "*Ischnus flavus* Rnd.*"* (Stadler, 1933). Of these, *Cryptus liparidis* and *Ischnus flavus* are *nomina nuda*. They are mentioned by Rudow (not Rondani as subsequent catalogues give), 1918, as "*Cryptus liparidus* Rd." and "*Ichneumon* [not *Ischnus*] *flavus* Rd." = nov. sp., but apparently he never described them.

Reared specimens of the other species were not available. They have not been encountered during the surveys for gypsy moth parasites in Europe and Asia. Their occurrence on the gypsy moth is either accidental, or more likely, doubtful.

The nomenclature of the species is corrected according to the modern treatment of the group. Townes, Momoi and Townes (1965), Townes (1970), and Kasparyan (1981) may be consulted for synonymies and other taxonomic information.

KEY TO THE GENERA AND SPECIES OF MESOSTENINAE ASSOCIATED WITH THE GYPSY MOTH

1. Intercubitii parallel. Median portion of posterior mesosternal carina present between the middle coxae. Sternaulus short, reaching up to 0.6 the length of mesopleurum, not sinuate. Clypeus a little arched, weakly convex. Basal and apical transverse carinae of propodeum equally prominent. Propodeal spiracle short, circular. First tergite with basolateral teeth. . . . . . . 1. Gambrus amoenus (Gravenhorst) Intercubitii convergent towards radius. Median portion of posterior mesosternal carina absent or indistinct. Sternaulus reaching middle coxa and sinuate. Clypeus convex. One of the two transverse carinae of propodeum more prominent than the other. 2

2. Propodeal spiracle elongate, elliptic. Apical transverse carina of pro-podeum strong and strongly arched medially (more prominent than the basal transverse carina). Axillus vein strongly pigmented and diverging from the anal margin of hind wing. Base of first tergite without a lateral tooth. . . . . 2. Meringopus cyanator (Gravenhorst) Propodeal spiracle short, circular. Basal transverse carina of propodeum prominent. Apical carina weak. Axillus vein weakly pigmented and very close and parallel to the anal margin of hind wing. Base of first tergite

with a lateral tooth. . . . . 3. <u>Ischnus inquisitorius</u>
<u>inquisitorius</u> (Mueller)

1. Genus GAMBRUS   (Fig. 74)

Gambrus Foerster, 1869. Verh. Naturh. Ver. Rheinlande, 25: 188.

Medium sized insects. Clypeus subconvex, its apical margin convex and
with a blunt median prominence. Mesoscutum mat and with scattered punctures. Apical carina of propodeum complete, its median portion more or
less bowed forwards, the lateral crests weak. Propodeal spiracle short,
circular. Areolet sub-quadrate, the intercubiti parallel. Base of first tergite
with a lateral tooth on either side. Ovipositor sheath 1.0-2.0 as long as hind
tibia. Hind tibia without a basal whitish band.
  Gambrus is a Holarctic genus with species extending into the northern
parts of the Oriental Region. It is related to *Agrothereutes* and *Aritranis* and
the characters of the three genera merge in some taxa. The above description will identify the genus and further details can be found in Townes (1970)
treatment of the genera of *Agrothereutina*.
  One species, *Gambrus amoenus*, was listed by Howard and Fiske (1911)
as a parasite of the gypsy moth in Europe.

1. GAMBRUS AMOENUS (Gravenhorst)   (Fig. 76)

Cryptus amoenus Gravenhorst, 1829. Ichneumonologia Europaea, 2: 623.
(Cryptus) Aritranis amoenus: Howard and Fiske, 1911. U. S. Dept.
    Agr. Bur. Ent. Bull., 91: 85. Host: *Lymantria dispar*.

This species has been referred to as *Cryptus amoenus*, *Aritranis amoenus*
or *Spilocryptus amoenus* in the literature on the gypsy moth parasites.
Howard and Fiske (1911) first reported it as a parasite of *Lymantria dispar*,
based upon information in their card file. Thompson (1946: 449) cites
Stadler (1933) and Schedl (1936) as his source of information about its being
parasitic upon the gypsy moth.
  The American species *Gambrus nuncius* (Say) is a junior synonym of
Gambrus amoenus (n. syn.), which has been reared from *Callosamia
angulifera*, *Callosamia promethi*, *C. angulifera*, *Samia cynthia*, *Antheraea
polyphemus*, *Acronicta* sp., and *Battus philenor*. According to Townes and
Townes (1962) the normal hosts are *Callosamia* species. The parasite overwinters as a cocoon in the host cocoon and the adult emerges in the following
spring. Its occurrence on the gypsy moth is doubtful or accidental.
  Face closely finely punctate on a granular surface. Clypeus subconvex
and shiny, with scattered punctures. Malar space mat, about equal to the
basal width of clypeus. Occipital carina coming to the base of mandible and
meeting the hypostomal carina near it, or a little erased at that place. Frons
centrally rugoso-striate. Vertex granulose. Interocellar distance 0.85 as
long as ocellocular distance. Upper part of temple mat, as wide as the width
of eye in profile. Lower temple polished. Mesoscutum mat, with scattered
shallow punctures. Scutellum subconvex, with scattered punctures on a
smooth surface. Side of thorax rugoso-punctate, with metapleurum becoming
finely reticulate. Propodeum convex, smoother basally, longitudinally striate

medially between the basal and apical transverse carinae, both of which are prominent, and rugose apicad of the apical transverse carina. First tergite smooth, second mat with scattered punctures, third finely mat and rather closely shallowly punctate and the following tergites progressively subpolished. Ovipositor about 0.66 to 0.70 the length of abdomen. Upper valve of ovipositor tip compressed, in profile view convex.

Black. Mandible medially red. Clypeus medially and scape may be red marked. Antenna brownish-black with flagellar segments 5-9 white, and segments 4 and 10 partly white. Tegula brownish-yellow in European specimen and black in American series. Legs yellowish-brown with fore coxa partly brownish and apices of hind femur and tibia blackish. Hind tibia often laterally also fuscous. Hind tarsus white. Basal three abdominal tergites orange-brown. Fourth and the following tergites black with apices of sixth and seventh tergites and often also the fifth, white.

Male: Propodeum with a faint indication of an areola. Tyloids on seven flagellar segments. Tyloids linear with the median tyloids a little widened and rounded above. Hind coxa black. Abdominal tergites 2, 3 and part of 4 orange-brown, rest black. Seventh tergite with a white triangular mark. Rest as in the female.

Length 9-11 mm. Fore wing 5-9 mm. Ovipositor 3-4 mm.

Specimens from Europe (Germany) and U.S.A. examined, but no reared specimens seen. It may be that all European specimens have been from cocoons imported from North America, and that the species is native only to North America.

Townes and Townes (1962) provided a key to the North American species. Schaffner and Griswold (1934) give biological information on this species.

2. Genus MERINGOPUS (Fig. 77)

Meringopus Foerster, 1869. Verh. Naturh. Ver. Rheinlande, 25: 186.

For the taxonomy of the genus, refer to Van Rossem (1969).

Medium sized insects. Clypeus strongly convex, its apical margin truncate. Malar space 0.8 to 1.2 the basal width of mandible. Mandibular teeth equal. Flagellum normal, not flattened or enlarged. Mesoscutum with smooth lateral areas, punctures sparser than on other parts. Notauli moderately deep, reaching beyond center of mesoscutum. Epomia, sternaulus and prepectal carina complete and distinct. Propodeal spiracle 2.8 to 4.0 as long as wide. Apical transverse carina of propodeum complete, forming weak lateral crests. Areolet pentagonal, large, the intercubiti convergent anteriorly. Axillary vein of hind wing strong and diverging from the anal margin. First tergite stout, without basolateral teeth. Postpetiole flat and wide, its dorsal and lateral carinae distinct. Second tergite mat, with weak sparse punctures to subpolished. Abdomen mat. Ovipositor tip short to elongate, with a nodus.

This genus is separated from other ischnine genera by the presence of clearly visible tentorial pits on the frons. *Buathra* also has conspicuous tentorial pits on the frons, but in *Buathra* the axillus vein in the hind wing is converging towards the anal margin and is usually weakly pigmented.

There are several species in Eurasia and North America. The Eurasian species were revised by Van Rossem (1969).

One species, *Meringopus cyanator* was listed by Howard and Fiske (1911) as a parasite of the gypsy moth. No earlier reference could be traced.

2. MERINGOPUS CYANATOR (Gravenhorst) (Figs. 78, 79)

Cryptus cyanator Gravenhorst, 1829. Ichneumonologia Europaea, 2: 442.
Cryptus cyanator: Howard and Fiske, 1911. U. S. Dept. Agr. Bur. Ent.
     Bull., 91: 85. Host: *Lymantria dispar.*
Meringopus cyanator: Van Rossem, 1969. Tijdschr. v. Ent., 112: 188.
Trachysphyrus cyanator: Griffiths, 1976. Parasites and predators of the
     gypsy moth: A Review, p. 42.

Body with long brownish pilosity. Face convex medially, with a prominance below antennal sockets, closely punctate, punctures superimposed on a granular surface. Malar space granulose, 1.1 as long as the basal width of mandible. Clypeus convex, with scattered punctures on a polished surface. Frons rugose, antennal scrobes smooth, concave with tentorial pits usually well developed (sometimes weak). Vertex punctate. Temple in profile as wide as eye, punctate, the punctures sparser below, the interspaces shiny. Interocellar distance 0.87 as great as ocellocular distance. Occipital carina abruptly erased at a distance equal to the basal width of mandible and not joining hypostomal carina. Thorax strongly rugose. Mesoscutum with polished areas in between irregularly placed punctures. Epomia complete to upper margin of pronotum. Propodeum with apical transverse carina forming weak lateral crests and with an irregularly and weakly formed areola. Central area of propodeum a little convex and narrowed, followed by a longer and steep slope to the apex. Axillary vein in hind wing strongly pigmented and diverging from inner hind margin of the wing. Legs with femora and tibiae normal, not dilated or flattened respectively. Abdomen granulose. Ovipositor tip normal, teeth on lower valve not strong ventrally, nor flanged out or projecting when seen in profile.
     Black. Palpi brown to black. Mandible reddish medially. Antennae brownish-black. Outer orbits with faint reddish lines. A reddish spot on vertex close to eye. Thorax black. Subtegular ridge narrowly reddish. Wings infumated. All coxae and trochanters black, femora reddish-brown, tibiae reddish brown basally and fuscous apically and dorsolaterally, tarsi largely fuscous. Abdomen blackish-brown with a faint violet iridescence. Apical margins of basal 1 or 2 tergites show faint brownish coloration.
     This species is rather close to *Meringopus nigerrimus murorum* (Tschek) from Scandinavia and the Alps, but the latter species has well developed ventral ridges on the ovipositor tip.
     Length: 12-8-15.7 mm. Fore wing 8.8-11.2 mm. Ovipositor 4.5-5.5 mm.
     Distribution: Europe: Germany, Netherlands, Poland, Denmark, Russia.
     Hosts: Lymantria dispar, Panolis flammea, Malacosoma neustria, Diloba coeruleocephala, and *Phragmatobia fuliginosa.*

3. Genus ISCHNUS (Fig. 75)

Ischnus Gravenhorst, 1829. Ichneumonologia Europaea, 1: 638.

Body usually slender. Clypeus small, convex and often pyramidal in profile, its apical margin convex. Mesoscutum mat and with minute punctures. Notaulus extending beyond center of mesoscutum. Sternaulus long and sinuate. Propodeal spiracle circular. Apical propodeal carina variable (in the species discussed here weak and broadly interrupted medially, almost indistinct). Areolet pentagonal, intercubiti convergent towards the radial vein. Axillus vein weakly pigmented and very close to the anal margin, its tip turned towards the margin. Base of first tergite with a lateral tooth on either side. First tergite slender, subpolished. Abdomen finely mat. Ovipositor about 0.5 the length of abdomen.

Ischnus assertorius Gravenhorst was apparently first listed by Rudow (1917) as a parasite of the gypsy moth. The present day name of it is *Ischnus inquisitorius inquisitorius* (Mueller). Townes and Townes (1962) recognized six subspecies of *inquisitorius:* two from Europe, one from Northern Japan, and three from North America. They have also provided a key to separate them.

3. ISCHNUS INQUISITORIUS INQUISITORIUS (Mueller) (Fig. 75)

Ichneumon inquisitorius Mueller, 1776. Zool. Danicae Prodromus,
    p. 151.
Ichneumon assertorius Fabricius, 1793. Entomologia Systematica, 2: 140.
Ischnus assertorius: Rudow, 1917. Ent. Ztschr. Frankfurt, 31: 67.
    Hosts: *Lymantria dispar, Xestia triangulum.*

Face granuloso-punctate. Frons rugose. Clypeus smoother, convex. Vertex granulose, Malar space granulose, about equal to basal width of mandible. Introcellar distance a little less than ocellocular distance (8: 10). Temple and vertex behind lateral ocelli receding, the vertex rather abruptly so. Mesoscutum mat, with fine punctures. Thorax generally otherwise closely punctate, tending to be rugose at places. Propodeum rugulose, its apical transverse carina weak and absent in the central area, laterally forming weak crests. First tergite slender and subpolished, rest of the tergites mat. Ovipositor about half the length of abdomen, its tip finely tapered.

Black, with abdomen reddish-brown. Marks along inner orbits extending on vertical orbits, a mark between and behind lateral ocelli, a mark on the center of clypeus, anterior and upper margins of pronotum dorsally, its hind corner, subtegular ridge, scutellum apically and metascutellum, yellow. Tegula brownish-black. All coxae and trochanters, black. Legs otherwise reddish-brown, with apices of hind femur and tibia fuscous and hind tarsus a little dark-brown. Tip of abdomen black. Male darker, particularly the abdomen brown. Yellow on head more extensive. Male sometimes also has yellow marks on side of thorax and on propodeum.

Length: 8-10 mm. Fore wing 5-7 mm. Ovipositor 3-4 mm.

Distribution: Eurasia. Other subspecies occur in Japan and North America.

Hosts: Lymantria dispar, Pandemis cerasana, Archips rosana (Kasparyan, 1981). *Xestia triangulum* (Rudow, 1917).

According to Townes and Townes (1962) this species is related to the Japanese *Ischnus assimilis* Uchida and *Ischnus yezoensis* Uchida.

V. SUBFAMILY TRYPHONINAE

Members of the tribe Tryphoninae are small to large sized ichneumon-flies that are ectoparasitic on lepidopterous caterpillars and on sawfly larvae. They are characterized by having a medium-sized to large transverse clypeus with its apical margin usually broad and beset with a fringe of long parallel hairs. Male flagellum devoid of tyloids. Sternaulus absent or short. Propodeum usually partly to completely areolated. Sometimes (as in *Netelia*) propodeal carinae reduced or absent. Tarsal claws usually pectinate. Areolet usually present, pointed or petiolate above. Second recurrent vein usually with two bullae. Nervellus intercepted variously. First tergite with a glymma, its spiracles before the middle. Abdomen depressed or in *Netelia* compressed and slender. Ovipositor usually short, sometimes long, its tip without a subapical dorsal notch and usually without conspicuous teeth.

This subfamily contains several tribes, but only one genus *(Netelia)* of the tribe Phytodietini is associated with the gypsy moth. The tribe Phytodietini contains two genera *Phytodietus* and *Netelia*, which are external parasites of lepidopterous larvae. They are characterized by having two spurs on hind tibia, propodeum without carinae or with only sublateral crests and with transverse striations, prepectal carina present, and nervellus vertical or reclivous. The genus *Netelia* has twisted mandibles, long slender brownish colored body, and a compressed and long abdomen.

1. Genus NETELIA (Figs. 68, 133)

Paniscus of authors, not of Schrank.
Netelia Gray, 1860. Ann. Mag. Nat. Hist., (3) 5: 341.

Body long, slender, with fore wing 6 to 23 mm. long. Mandibles twisted, the lower tooth much smaller than the upper. Eye and ocelli enlarged. Eyes emarginate opposite antennal sockets. Basal 0.65± of propodeum with transverse striations, usually with a pair of transverse crests. Nervellus intercepted above the middle. Ovipositor 1.0 to 2.0 as long as the apical depth of abdomen, its tip sharp and slender.

General body coloration brownish or ferrugineous, sometimes with a few blackish marks.

Members of the genus *Netelia* are external parasites of exposed lepidopterous larvae. Several subgenera are recognized (Townes, 1969: 149), but the two species reported from the gypsy moth belong to the subgenus *Netelia*, which has normally pectinate hind tarsal claws, occipital carina complete, nervulus distad of basal vein, lateral carina of scutellum reaching to its apex, and thorax usually without definte yellow markings.

Two species of *Netelia*—as "*Paniscus cephalotes* Holmgren" and "*P. testaceus* Gravenhorst" were reported by Rühl (1914) as parasites of *Lymantria dispar* in Europe. They have since been mentioned by Wolf and Kraube (1922), Meyer (1931) and Thompson (1946) in the gypsy moth literature. I have not come across any rearing records, nor have I seen any reared specimens of them. Both these species are only occasional parasites of the gypsy moth, if they attack it at all. These two species have often been misidentified in the past; so their true identities are uncertain.

The correct name of "*Paniscus cephalotes*" is *Netelia (Netelia) vinulae*

(Scopoli). The identity of *"Paniscus testaceus"* cannot be established and is referred here as *Netelia* sp.

1. NETELIA (NETELIA) VINULAE (Scopoli) Figs. 69-73, 133)

Ichneumon vinulae Scopoli, 1763. Entomologia carniolica..., p. 286.
Paniscus cephalotes Holmgren, 1858, Svenska. Vetensk Akad. Handl.
(N.F.), 2 (8): 31.
Taxonomy: Rühl, 1914: 26. Townes, Momoi and Townes, 1965: 95.
Delrio, 1975: 64. Kaur and Jonathan, 1979: 137.

Female: Head wider than long. Face convex medially and moderately closely punctate. Malar space 0.2 as long as the basal width of mandible. Frons transrugulose. Temple strongly convex, as wide as eye. Occipital carina rather weak and ending shortly before the oral carina. Pronotum, mesoscutum and scutellum finely closely punctate. Mesopleurum moderately punctate and subpolished. Metapleurum and propodeum finely closely striated. Lateral crests of propodeum prominent. Hind tarsal claw strongly bent apically and with about 14 rather fine pectines and 2 ungual bristles. Tibial bristles moderate. Nervulus distad of basal vein by 0.3-0.4 its length and distinctly bent in its upper 0.4. Areolet rather small, oblique, sub-sessile. First tergite stout and swollen, about 3.5 as long as wide at apex.
Male genitalia: Gonoforceps large and exserted, without a apico-dorsal spine. Brace broad, moderately long and supporting a large U-shaped pad apically. Penis valve normally shaped.
Color: Brown. Ocellar triangle lightly infuscate. Stigma yellow.
Length: ♀: 17-19 mm. Fore wing 13.5 to 15 mm. ♂: 13-22 mm.
Distribution: Europe, China, India, Japan, Russia.
Hosts: It is apparently a polyphagous parasite of various common lepidopterous pests of forest trees, including *Lymantria dispar* and *L. monacha* (Kaun and Jonathan, 1979). As no reared specimens could be seen, its occurrence on the gypsy moth could not be confirmed.
The above description is adapted from Kaur and Jonathan (1979), who examined the lectotype of *N. cephalotes*. The type of *N. vinulae* no longer exists.

2. NETELIA (NETELIA) sp.

Paniscus testaceus Gravenhorst: Rühl, 1914. Soc. Ent., 29 (5): 26.
Host: *Lymantria dispar*. Europe.

Rühl (1914) reported *Paniscus testaceus* Gravenhorst as a parasite of *Lymantria dispar*, which record has been repeated in literature. However, no reared specimens are at hand to confirm the identity of the species involved. Nor have any subsequent rearing records been discovered in the literature.
The identity of *N. testaceus* is very doubtful and the type is also lost. This species has often been misidentified in the past. Delrio (1974) who revised the western Palaearctic species of *Netelia*, listed it as *"species incerte sedis"*, mentioning that it is a composite species. Meyer (1936) apparently confused it with *Netelia melanurus* (Thomson) when he reported *melanurus* as a parasite of the gypsy moth in 1936, but not in his taxonomic

treatment of *melanurus* in 1935.

Netelia melanurus is a rather distinct species with a black abdominal tip. Since what Rühl referred to as *testaceus* is uncertain, the record from the gypsy moth is referred here as *Netelia (Netelia)* sp.

Townes, Momoi and Townes (1965) and Delrio (1974) provide further information on Palaearctic *Netelia* species and their hosts.

VI. SUBFAMILY BANCHINAE

The subfamily Banchinae includes moderately long, stout or slender insects with fore wing 1.8 to 16 mm. long. Clypeus small, wide or narrow, usually separated from face by an epistomal groove, its apical margin round, notched or truncate, without an apical fringe of hairs. Mandibular teeth equal to subequal, the upper sometimes broad to obliquely truncate. Male flagellum without tyloids. Sternaulus absent or short. Postpectal carina absent. Propodeum with apical transverse carina evenly arched and usually strong. Tarsal claws usually pectinate. Areolet present or absent. Nervellus intercepted variously. First abdominal segment stout to slender, its spiracles before the middle, or at middle, its dorsolateral carinae usually absent. First sternite not fused with its tergite. Glymma nearly always present. Abdomen depressed, its apex often compressed. Female subgenital plate large, prominent, its apex with a median notch. Ovipositor with a subapical notch, without a node, the lower valve without teeth. Ovipositor and its sheaths long to short.

Members of the subfamily Banchinae are usually internal parasites of lepidopterous larvae. Oviposition is usually within the young larvae. Emergence of the adult parasite is from a cocoon spun after the parasite larva has killed and left its host.

Only one genus *Banchus* has been associated with the gypsy moth. Keys to separate it from other genera and tribes of Banchinae are given by Townes (1970).

1. Genus BANCHUS Fabricius   (Figs. 80, 137, 138)

> *Banchus* Fabricius, 1798. Supplementum Entomologiae Systematicae, p. 209, 233.

For synonymical references refer to Townes (1970). Townes and Townes (1978) treated the Nearctic species. Chandra and Gupta (1977) treated the Oriental species. Aubert (1978) gives a catalog and host records of the Palaearctic species.

Body moderately stout. Face broad above, weakly narrowed toward lower side. Apical margin of clypeus with a median notch. Upper tooth of mandible wider and longer than the lower tooth and its apex with a weak concavity. Malar space 0.5 to 0.8 the basal width of mandible. Eye emarginate near antennal socket. Fourth segment of maxillary palpus nearly always expanded apically, the expansion faint to strong and more expanded in the male. Occipital carina entire or interrupted medially above, joining hypostomal carina above base of mandible. Prepectal carina absent. Apex of scutellum usually with a median point or spine. Propodeum short, its apical transverse carina complete or erased medially. Propodeal spiracle long, elliptic. Areolet usually large, receiving second recurrent vein near the middle. Nervulus distad of basal vein by 0.2 to 0.6 its length. Nervellus intercepted far above the middle. Abdomen apically compressed. First tergite without dorsal or dorsolateral carinae. Epipleura of tergites 2 and 3 about 0.7 as wide as long. Ovipositor very short, its sheaths about 0.13 as long as hind tibia.

Two species of *Banchus* are reported in literature as parasites of the gypsy moth, viz., *Banchus femoralis* Thomson = *hastator* (Fabricius) and *B. falcator* Fabricius = *falcatorius* (Fabricius). The record of the latter species was apparently based upon guess work. None of these species have turned up in subsequent rearings and no voucher specimens are available to confirm their association with the gypsy moth.

KEY TO THE SPECIES OF BANCHUS ASSOCIATED WITH THE GYPSY MOTH

Apex of scutellum with a median spine. Abdominal tergites black, sub-polished. Inner orbits yellow (broadly so in male). Hind femur blackish. Europe. . . . .   1. hastator (Fabricius)
Apex of scutellum with a median point, without a conspicuous spine. Abdominal tergite 2 with a broad reddish-brown band (sometimes apex of tergite and base of tergite 3 also reddish-brown). In males tergites 1-3 largely yellowish-brown. Inner orbits in female black and in male largely yellow. Hind femur yellowish-brown or yellow in male. Europe. Doubtful, almost certainly erroneous record.

2. falcatorius (Fabricius)

1. BANCHUS HASTATOR (Fabricius)(Fig. 137)

Ichneumon hastator Fabricius, 1793. Entomologia Systematica, 2: 167. Europe.
Banchus femoralis Thomson, 1897. Opuscula Entomologica, 22: 2411. Sweden. Syn. by Fitton, 1978.
Banchus femoralis: Kolubayiv, 1934. Acta Ent. Soc. Ent. Csl., 31: 114. Czechoslovakia. Host: *Lymantria dispar.*

Kolubajiv (1934) apparently reared it from the gypsy moth for the first time in Czechoslovakia. No subsequent rearing record has been seen.
Female: Face punctate. Clypeus, inner orbits, frons and vertex, mat with frons a little rough and dull. Frons excavated and with circular striations just above antennal sockets. Malar space 0.7 to 0.8 as long as the basal width of mandible. Interocellar distance 1.2 as long as the ocellocular distance. Tip of flagellum tapered, narrowed and cylindrical. Mesoscutum, mesopleurum and metapleurum punctate, punctures on mesoscutum closer than those on mesopleurum and metapleurum. Scutellum shiny, with minute scattered punctures, its apex with a pointed spine, about 0.25 as long as the length of scutellum. Scutellum with erect hairs surrounding the spine. Propodeum leathery, ruguloso-punctate in basal half and rugose to a little rugose striate in apical half. Its apical transverse carina weak and irregularly represented. Propodeal spiracle long, linear. Abdomen polished, with shallow scattered and minute punctures on basal three tergites. First tergite flat dorsally, without dorsal carinae, a little depressed apicad of the projecting spiracles. Ovipositor short, its sheaths hardly exerted, more or less parallel-sided,

wider and truncate apically.

Black. Face broadly along inner orbits and narrowly along outer orbits, clypeus, scape ventrally, tegula, fore femur on the anterior aspect, fore tibia largely, middle femur dorsally, and middle tibia except apically, and all second trochanters, yellow. Tegula often and apex of scutellum surrounding the spine, brownish to brownish-yellow. Fore and middle femora with black marks. Fore and middle tibiae and tarsi brownish at places. Hind leg largely black or blackish-brown, with apex of femur, basal 0.6 of tibia and base of basitarsus, yellowish-brown. All coxae and first trochanters black. Metapleurum often with a yellow oval spot. Abdomen with brownish or yellowish brown apical margins on tergites 1-3. Basal 2-3 sternites often apically broadly yellow.

Male: Similar to the female in sculpture and coloration, except for the following sexual differences:

Tip of flagellum somewhat attenuate and flattened. Flag setae on flagellar segments not visible (may be present, but small). Fourth segment of maxillary palpus flattened and widened apically. Face largely yellow with only a broad central black stripe. Scutellar spine short. Oval yellow spot on metapleurum absent. Abdominal tergites black. Male claspers with a concave lower margin, tapered apically (appear beak-like).

Length: 11-13 mm. Fore wing 8.5 to 10 mm. Ovipositor 0.5 mm.

Specimens: Several males and females from Europe examined in Townes Collection. No reared specimen seen.

Distribution: Europe.

Hosts: Lymantria dispar, Panolis flammea, Zeiphera diniana (= *griseana.* Hb.) and *Diprion* sp. (after Thompson, 1957).

2. BANCHUS FALCATORIUS (Fabricius)

Ichneumon falcatorius Fabricius, 1775. Systema Entomologiae, p. 332.
       Denmark.
Banchus falcatorius: Fabricius, 1798. Suppl. Ent. Syst., p. 234.
Banchus falcator Fabricius: Morley, 1915. Revision of Ichneumonidae
       in British Museum, 4: 138. Germany. Syn., hosts. [Doubtful
       association with the gypsy moth.].

This species was first reported as a parasite of *Lymantria dispar* by Morley (1915), which has been cited by Aubert (1978). Morley's record was based on a male specimen from Germany, bearing the following data, "Between Wilmannsdorf and Schwangendorf, 6 PM, under willows, also flying. Larva of *L. dispar* (gypsy moth) common." It is clear that the host association was simply by guess-work.

Since there is no definite evidence of the association of *B. falcatorius* with the gypsy moth, and since it has not been reared from that host subsequently, this species is to be removed from the list of ichneumonid parasites associated with the gypsy moth.

VII. SUBFAMILY CREMASTINAE

Members of the subfamily Cremastinae are long slender insects with fore wings 2.5 to 14 mm. long. Abdomen strongly compressed. Eyes bare. Ocelli of male sometimes enlarged. Clypeus small to large, separated from face by an epistomal groove, its apical margin convex. Mandible short, bidentate. Sternaulus absent. Prepectal carina present. Postpectal carina complete. Propodeum usually completely areolated, with longitudinal as well as transverse carinae. Spurs of all tibiae set in a membranous area that is separated from the membranous area of tarsal insertion by a sclerotized bridge (not seen in other ichneumonids). Tarsal claws partly to completely pectinate. Epipleurum of third tergite not separated or only partly separated from its tergite. Female subgenital plate unspecialized, usually not visible. Ovipositor long, with a subapical dorsal notch.

Members of the subfamily Cremastinae are internal parasites of lepidopterous larvae living in concealed situations. Some of them also attack coleopterous larvae, or larvae in exposed situations, like the gypsy moth. Emergence of the parasite is from the mature larva about to pupate.

Only one genus and species, *Pristomerus vulnerator* (Panzer) has been associated with the gypsy moth in Eurasia and North Africa.

1. Genus PRISTOMERUS Curtis (Fig. 81)

Pristomerus Curtis, 1836. British Entomology, 13: 624.

For full synonymical references and relationships with other cremastine genera, refer to Townes (1971). This genus is easily distinguished from all other genera treated in this paper by having a tooth on the lower side of hind femur and ovipositor tip sinuate.

Body moderately slender. Abdomen moderately to strongly compressed apically. Occipital carina usually complete. Lateral carina of scutellum absent, or present only basally. Hind femur often swollen and with a ventral tooth followed by a series of minute teeth, especially in male. Stigma wide. Radial cell short. Areolet absent. Nervellus intercepted near its lower 0.35. First tergite moderately slender, usually longitudinally carinate. Lower edges of first sternite nearly parallel-sided, not touching each other. Thyridium on second tergite transverse or subcircular, near base of the tergite. Epipleurum of second tergite narrow, separated by a crease, and turned under. Apex of male clasper rounded. Ovipositor long, only a little shorter than fore wing, its tip sinuate.

1. PRISTOMERUS VULNERATOR (Panzer) (Fig. 81)

Ichneumon vulnerator Panzer, 1799. Faune Insectorum Germanicae,
　　　　Heft 72, pl. 5. Europe.
Pristomerus vulnerator: Curtis, 1836. British Entomology, 3: 624.
　　　　England.
Pristomerus vulnerator: Barsacq, 1913. Rev. Phytopath. Appl. Paris,
　　　　1(5): 70-73. France. Host: *Lymantria dispar*.

Pristomerus vulnerator: Mokrzecki, 1913. Reports [of the Chief Entomologist to the Zemstov] on Injurious Insects and Diseases of Plants in Govt. of Taurida during the year 1912. Simferopol, 1913, p. 1-23. Host: *Lymantria dispar*.

For fuller synonymical references, hosts and distributional records, refer to Townes, Momoi and Townes (1965).

Barsacq (1913) and Mokrzecki (1913) first reported it parasitic in young caterpillars of gypsy moth in France and Taurida (Russia) respectively. Meyer (1927) reported having reared it in Russia. Obrtel (1949) and Sedivy (1970) gave an extensive host range for this species.

Female: Face punctate, granuloso-punctate along inner orbits. Clypeus smooth. Width of face almost equal to the length of face and clypeus. Malar space about 0.7 the basal width of mandible, punctate. Temple and vertex posteriorly strongly receding from the eye. Temple 0.25 the width of eye. Vertex concave above. Occipital carina close to the lateral ocelli, distant from the ocelli by about the ocellar diameter or slightly more than that. Vertex and frons (excluding the antennal scrobes), closely finely granulose. Interocellar distance equal to the ocellocular distance. Mesoscutum granuloso-punctate, punctures set on a granular surface. Scutellum flat, punctate, subpolished in between punctures. Mesopleurum punctate, interspaces equal to at least the diameter of the punctures and shiny. Metapleurum somewhat coarsely punctate. Pleural area punctate, with interspaces equal to the diameter of the punctures and shiny. Propodeal spiracles situated at or a little below the middle of the spiracular carina (pleural part of basal transverse carina). Propodeum convex, shallowly to moderately punctate. Areola smoother and petiolar area usually transrugose. Propodeal areolation complete. Areola broad, 1.5 times as long as its maximum width at costula. Costula arising at basal 0.33 to 0.4 of the length of areola. Apical closing carina of areola often weak in the middle. Hind coxa strongly granulose. Hind femur with one strong tooth in apical 0.33, followed by a series of minute teeth between it and apex. First and second tergites longitudinally striate. The following tergites subpolished and compressed. Ovipositor long, its sheath 0.66 as long as the fore wing, its tip sinuate.

Male: Eyes strongly diverging ventrally. Vertex constricted. Ocelli large and raised. Ocellocular space almost obliterated. Propodeal carinae strong. Areola a little longer and narrowed apically and basally, widest at costula. Costula arising from basal 0.6 of areola. Pleural area of propodeum and metapleurum somewhat sparsely punctate and subpolished, sometimes a little strongly and irregularly so. Otherwise similar to the female.

Black. Head devoid of brownish orbital marks. Mandible, tegula, and fore and middle legs except their coxae, yellow. Scape and pedicel black to brownish-black. Coxae black. Fore coxa may be partly brown. Hind leg brownish-yellow with black marks on trochanter, femur, and apex of tibia and tarsus except basally. The extent of black on hind leg variable. Abdomen black either wholly or apices of tergites brownish-yellow, or apical tergites largely yellowish-brown.

In males third tergite usually partly to wholly yellowish-brown. Sometimes faint brownish lines seen along inner orbits.

Length: 6-7 mm. Fore wing 4 to 4.5 mm. Ovipositor 4 mm.

Distribution: Eurasia, China, Japan, Korea.

Hosts: *Lymantria dispar* and many other common lepidopterous pests [see Meyer (1927), Obrtel (1949) and Šedivý (1970)].

Pristomerus vulnerator appears to be a variable species occurring in Eurasia. It is closely related to *Pristomerus orbitalis* Holmgren and the two have often been mixed up in European literature. *P. orbitalis* has a wider face, wider than high, wider temples, ocellocular distance greater than inter-ocellar distance, vertex not strongly concave behind, occipital carina considerably away from lateral ocelli, in males lateral ocelli distinctly separated from eye margin, mesoscutum punctate, without granulations, propodeum smoother, propodeal areola narrow and elongate, about 4.0 as long as wide, and propodeal spiracle usually above the middle on the spiracular carina. The hind coxa is minutely punctate, not densely granulose as in *P. vulnerator*. The head is with yellowish-brown orbital stripes encircling the eyes.

VIII. SUBFAMILY ANOMALINAE

The subfamily Anomalinae is characterized by having a coarsely reticulate propodeum, and a long, slender and strongly compressed abdomen. Clypeus often not separated from face by a groove, its apex often with a median point. Occipital carina often at the outer margin of head. Head quadrate. Mandible bidentate. Epomia usually long. Sternaulus absent. Propodeum reticulate wrinkled, without carinae, its apex often produced beyond bases of coxae. Areolet absent. Legs long, slender. First tergite slender, long, without a glymma, its tergite and sternite fused, and spiracle behind the middle. Abdomen strongly compressed. Ovipositor and its sheaths short. Ovipositor with a subapical dorsal notch.

Members of the subfamily Anomalinae are internal parasites of the larvae of Lepidoptera, except that the genus *Anomalon* is reported from Coleoptera. The emergence is from the pupa. The parasite larva spins a flimsy cocoon within the pupae of Lepidoptera. The larval morphology is similar to that of the Metopiinae.

Rudow (1911) reported two species of *"Anomalon"* from *Lymantria dispar* as well as *L. monacha*, *viz.*, *Anomalon flaveolatum* Gravenhorst and *A. pallidium* Gravenhorst. The former species now belongs to *Agrypon* while the latter to *Barylypa*. Why no subsequent cataloger has listed these species as gypsy moth parasites, is unknown to me. Kirchner (1856) reported *Anomalon (Trichomma) enecator* Rossi as a parasite of *Lymantria dispar*. Morley (1915) stated that Gaulle (1908) bred it from *Lymantria dispar* in France. Kovacevic (1925) reported *Barylypa perspicillator* Gravenhorst as a parasite of the gypsy moth in Yugoslavia. This name is a junior synonym of *B. delictor* (Thunberg). None of these parasites have been confirmed as gypsy moth parasites by subsequent rearings, nor have I seen any reared specimens of them.

KEY TO THE GENERA OF ANOMALINAE ASSOCIATED WITH THE GYPSY MOTH

1. Eye surface hairy. Inner margins of eyes strongly convergent toward mandible. Ovipositor sheath 2.0 to 3.8 as long as the apical depth of abdomen. Discoidella and often also the brachiella veins absent.
                                        1. <u>Trichomma</u> (Trichomella)
    Eye surface bare. Inner margins of eyes weakly convergent, or parallel.
        Ovipositor sheath less than 2.0 the apical depth of abdomen. 2

2. Discoidella present. Frons with a median vertical carina. Postnervulus
   intercepted near upper 0.25. . . . .   2. Barylypa
   Discoidella absent. Frons without a median vertical carina. Postnervulus
   intercepted near upper 0.38. . . . .   3. Agrypon

1. Genus TRICHOMMA (Fig. 84)

   Trichomma Wesmael, 1849. Bull. Acad. Roy. Sci. Lett. Beaux-Arts
      Belgique, 16(2): 119, 139.
   Trichomella Szépligeti, 1910. Notes Leyden Mus., 32: 91.

   Inner eye margins moderately to very strongly convergent ventrad. Eyes
with long dense hairs. Frons without a median carina or tooth, often with
transverse wrinkles. Apex of clypeus convex or subtriangular, with a median
tooth that varies from very small to large. Temple narrow. Occipital carina
complete, close to ocelli and reaching base of mandible, separate from hypostomal carina. Lower tooth of mandible a little shorter than the upper. Flagellum moderately long, lower front corner of pronotum truncate. Epomia present. Scutellum flat to convex, carinate laterally. Prepectal carina extending
to 0.5 the height of mesopleurum. Sternaulus absent. Posterior mesosternal
carina interrupted in front of each middle coxa. Propodeum and metapleurum
reticulate. Apex of propodeum extending up to 0.5 the length of hind coxa.
Tarsal claws pectinate on basal half or more. Middle tibia with two spurs.
Nervulus a little distad of basal vein. Intercubitus basad of second recurrent
vein. Discoidella and brachiella veins present of absent. Postnervulus meeting discocubital cell before its middle. Second tergite much longer than the
third. Epipleurum of third tergite not separated by a crease. Ovipositor 2.0
to 3.8 as long as the apical depth of abdomen.
   Species of *Trichomma* are generally parasites of the larvae of Microlepidoptera, particularly those which are concealed in rolled leaves, mining in
soft plant tissues, etc. Oviposition is into the host larva and the adult emerges
from the host pupa by biting off the entire anterior end of the puparium. *Trichomma enecator* (Rossi) was, however, reported as a parasite of *Lymantria
dispar* by Kirchner (1856).

1. TRICHOMMA (TRICHOMELLA) ENECATOR (Rossi)   (Fig. 85, 86)

   Ichneumon Enecator Rossi, 1790. Fauna Etrusca, 2: 48. Italy.
   Anomalon (Trichomma) enecator: Kirchner, 1856. Lotos, 6: 150.
      Czechoslovakia. Host: *Lymantria dispar.*
   Trichomma (Trichomella) enecator: Townes, Momoi and Townes, 1965.
      Mem. Amer. Ent. Inst., 5: 362.

   This species was first reported as a parasite of *Lymantria dispar* (= *Bom-
bax dispar*) by Kirchner (1856) from Czechoslovakia. According to Morley
(1915) it is a parasite of a number of tortricid hosts in England, but was also
bred from the pupae of *Lymantria dispar* by de Gaulle in France. It is listed
as a parasite of *Lymantria dispar* in Europe by several subsequent cataloguers,
like Wolf and Kraube (1922), Meyer (1931), Stadler (1933), Schedl (1936), and
Thompson (1946), etc.
   Female: Face a little convex, minutely punctate as well as polished.

Clypeus polished, pointed apically. Frons rugoso-striate. Malar space 0.2 the basal width of mandible. Interocellar distance 1.1 the ocellocular distance. Mesoscutum rugoso-punctate. Scutellum shallowly rugose, flat dorsally and strongly carinate laterally. Upper 0.3 of pronotum polished and a little swollen medially with widely spaced parallel and strong carinae. Mesopleurum trans-carinate, carinae somewhat weaker centrally and interspaces punctate. Meta-pleurum and propodeum reticulate, with honey-comb like cells. Hind wing with distal abscissa of cubitella, discoidella and brachiella virtually absent. Petiole slender, circular in cross-section. Postpetiole a little swollen. Tergite 2, 1.75 as long as the third. Third and the following tergites strongly compressed laterally and their epipleurae not separated from the tergites by a crease. Ovipositor long, slender, about 3.2 to 3.5 as long as the apical depth of abdomen, and 0.75 as long as the fore wing.

Black. Face, clypeus, mandible, palpi, outer orbital border narrowly, upper margin of pronotum (narrowly to broadly), scutellum centrally, often fore and middle legs largely, and hind trochanters yellow. Postpetiole, hind femur, basal half of hind tibia and whole of tarsi, and ventrolateral aspects of abdominal tergites, yellowish-brown. Hind coxa, trochanter, base of femur, and apical 0.3 to 0.6 of tibia often brown (extent of brown variable). Base of hind tibia often brown.

Length 10-11 mm. Fore wing 5.5 to 6.0 mm. Ovipositor 4.0-4.5 mm.

Specimens: Several females seen in the Townes Collections from Europe, but no specimens reared from the gypsy moth are at hand.

Hosts: Lymantria dispar. Several other hosts are listed by Gauld and Mitchell (1977) and Kasparyan (1981), but not the gypsy moth.

Distribution: Eurasia. Japan.

2. Genus BARYLYPA Foerster (Fig. 87)

Barylypa Foerster, 1869. Verh. Naturh. Ver. Rheinlande, 25: 156.

Inner eye margins weakly convergent ventrad. Eye surface hairless. Frons with a median vertical carina. Apex of clypeus pointed medially. Occipital carina complete, close to posterior ocelli, joining hypostomal carina at or immediately before mandibular base. Lower mandibular tooth about 0.6 as long on the upper. Antenna moderately long, that of female without a white band. Clypeus usually with a median apical tooth. Scutellum short, weakly convex and usually with lateral carina. Lower anterior margin of pronotum without a tooth, lower corner truncate. Epomia running along pronotal collar, not deflected above. Middle tibia with two spurs. Tarsal claws pectinate up to middle. Intercubitus basad of second recurrent vein. Postnervulus intercepted near upper 0.25. Discoidella present. Tergite 2 much longer than tergite 3, its epipleurum separated by a crease. Abdomen apically strongly compressed, with epipleura of tergite 3 onwards not separated by a crease. Ovipositor about 2.0 or long as apical depth of abdomen.

Two species of *Barylypa* are reported in literature as parasites of the gypsy moth, viz., *B. delictor* (Thunberg) and *B. pallida* (Gravenhorst).

1. BARYLYPA DELICTOR (Thunberg) (Fig. 88, 89)

> *Ichneumon delictor* Thunberg, 1822. Mém. Acad. Imp. Sci. St. Péters-
> bourg, 8: 265; 1824, Ibid, 9: 319. Sweden.
> *Barylypa delictor:* Roman, 1912. Zool. Bidr. Från Uppsala, 1: 249.
> *Barylypa perspicillator* Gravenhorst: Kovacevic, 1925. Sumar List.,
> 49(1): 1-5. Yugoslavia. Host: *Lymantria dispar.*

Barylypa perspicillator (Gravenhorst) was first reported from the gypsy
moth by Kovacevic (1925). Meyer (1935) synonymized *perspicillator* under
delictor (Thunberg).
This species is unknown to me. No specimens could be examined. Kas-
paryan (1981) gives a key to the Eurasian species.
Distribution: Eurasia.
Hosts: Several hosts are mentioned by Meyer (1931-36), Šedivý (1957) and
Kasparyan (1981), including *Lymantria dispar.*

2. BARYLYPA PALLIDA (Gravenhorst)

> *Anomalon pallidum* Gravenhorst, 1829. Ichneumonologia Europaea,
> 3: 675. Rudow, 1911. Internat. Ent. Ztschr., 5: 99. Host:
> *Lymantria dispar.*
> *Barylypa pallida:* Schmiedeknecht, 1908. Opuscula Ichneumonologica,
> 19: 1507.

This species was first reported from the gypsy moth by Rudow (1911) but
has not been listed subsequently by Schedl or Thompson. Kasparyan (1981)
provides a key to distinguish it from other Eurasian species. He also lists
B. humeralis Brauns as a synonym of it. It is unknown to me.

3. Genus AGRYPON Foerster (Fig. 90)

> *Agrypon* Foerster, 1860. Verh. Naturh. Ver. Rheinlande, 17: 151.

This genus is rather similar to *Barylypa* and the main differences appear to
be the absence of discoidella vein in the hind wing, frons without a vertical
carina, and the epomia with a different course. The postnervulus is intercepted near upper 0.38, and not very high as in *Barylypa*. Other characters
are mostly similar *(cf.* Townes, 1971, pp 144 and 139). Townes also provides
fuller synonymical references.

4. AGRYPON FLAVEOLATUM (Gravenhorst) (Fig. 90)

> *Ophion flaveolatum* Gravenhorst, 1807. Vergl. Uebers. Zool. Syst.,
> p. 268.
> *Anomalon flaveolatum:* Rudow, 1911. Internat. Ent. Ztschr., 5: 99.
> Europe. Hosts: *Lymantria dispar, L. monacha.*
> *Agrypon flaveolatum:* Townes, Momoi and Townes, 1965. Mem. Amer.
> Ent. Inst., 5: 373.

This has not been reported subsequently from the gypsy moth. It occurs in Europe, China, Japan and Korea. Kasparyan (1981) gives some diagnostic characters and a long list of hosts, but not the gypsy moth. Schmiedeknecht (1908) gave a description of it.

IX. SUBFAMILY OPHIONINAE

Members of the subfamily Ophioninae are characterized by having a slender, long, and compressed abdomen with the first tergite long, tubular and its spiracles placed far behind the middle. Ocelli very large. Clypeus separated from the face by a groove. Tarsal claws pectinate. Epomia absent. Second brachial cell with a long spurious vein, that parallels its hind margin. Areolet absent, the intercubitus far distad of the second recurrent vein. Ovipositor short and with a subapical dorsal notch, without any ridges on the lower valve.

They are medium to large sized species with longer wings. The body color is usually pale brown. Adults are generally crepuscular or nocturnal.

Members of the subfamily Ophioninae are world wide in distribution. They are endoparasites of medium to large sized lepidopterous larvae.

Only two genera, *Ophion* and *Enicospilus* have been associated with the gypsy moth, by one species of each genus. Rearing records are scanty. There are no recent records and no reared material was available for study. Even the identities of the species of the genera *Ophion* and *Enicospilus* are subject to doubt (Gauld, 1976, 1978) and Gauld and Mitchell (1978, 1981).

KEY TO THE GENERA OF OPHIONINAE ASSOCIATED WITH THE GYPSY MOTH

1. Posterior transverse carina of mesosternum broadly interrupted in front of each middle coxa. Mandibles normal, teeth in the same plane and equal. Fore wing without a glabrous area (fenestra). . . .   1. Ophion
    Posterior transverse carina of mesosternum complete or rarely incomplete. Mandibles strongly narrowed and twisted apically, teeth subequal. Fore wing with a fenestra, often with one or more sclerites present.

                                                              2. Enicospilus

1. Genus OPHION   (Fig. 82)

Ophion Fabricius, 1798. Suppl. Ent. Syst., p. 210, 235.

Townes (1971) provides synonymical references and a key to the world genera of Ophioninae. Gauld (1976, 1978) gives information on the British and some European species.

Body long and slender. Color reddish-brown, sometimes with pale stripes on mesoscutum. Fore wing 8 to 21 mm. long. Ocelli large, the lateral ones almost touching the eye. Antenna long, slender, longer than the length of the fore wing. Mandible not narrowed, nor twisted apically, teeth equal or nearly equal. Notauli generally sharply impressed in anterior 0.3 to 0.4. Scutellum moderately convex, usually without lateral or with short lateral carinae. Posterior transverse carina of mesosternum present only as lateral rudiments.

Propodeum partly to completely areolated, or rarely without carinae. Fore wing with pterostigma rather stout. Areolet absent. Discocubital cell without a fenestra, but with a small hairless area below the base of stigma. Fore tibial spur with a membranous flange behind the macrotrichial comb.

Only one species, *Ophion luteus* (Linnaeus) is mentioned in literature as a parasite of the gypsy moth.

1.   "OPHION LUTEUS (Linnaeus)"

Ichneumon luteus Linnaeus, 1758. Systema Naturae, (Ed. 10), 1: 566.
Ophion luteus: Townes, Momoi and Townes, 1965. Mem. Amer. Ent.
     Inst., 5: 317. Synonymical references and distribution. Holarctic.

Kolomiyetz (1958) reported having reared this species from gypsy moth pupae in Siberia in July 1954. It was considered as a rare parasite. This is apparently the first record of an *Ophion* species from *Lymantria dispar*. It has, however, been previously reported several times as a parasite of *Lymantria monacha*.

Gauld (1976, 1978) mentioned that *Ophion luteus* (Linnaeus) has been a commonly misidentified species. According to him *Ophion luteus* of authors is *Ophion slaviceki* (Kriechbaumer). He (1978) provided a key to distinguish the two and the others occurring in Britain.

Which one of the above mentioned two species was actually reared by Kolomiyetz cannot be ascertained without the specimens. The following table will distinguish the two:

Ophion luteus	*Ophion slavickei*
Outer tibial spur of middle leg less than 0.8 as long as the fourth tarsal segment. Smaller species, fore wing 13-15 mm. long.	Outer tibial spur of middle leg 0.8 or more as long as the fourth tarsal segment. Larger species, fore wing 15-18 mm. long.

Characters in common of these two species are: Propodeum with transverse carinae. Malar space less than 0.5 the basal width of mandible. First subdiscoidal cell unevenly hairy. Ocelli a little away from the eye. Antenna with less than 64 segments. Middle tibial spurs unequal in length.

Brown. Interocellar area and propodeum entirely orange-brown to yellowish. Wings slightly yellow-tinged. Thorax uniformly brown, never pale marked except rarely on mesepimeron.

Distribution: Europe.

2.   Genus ENICOSPILUS (Figs. 83, 139)

Enicospilus Stephens, 1835. Illustrations of British Entomology,
     Mandibulata, 7: 126.

Some of the recent studies on the genus are of Townes (1971), and Gauld and Mitchell (1978, 1981).

Body slender. Abdomen long and strongly compressed. Antenna long. Eyes and ocelli large. Malar space short. Mandible wide at base and strongly narrowed and somewhat twisted before the middle, its upper tooth longer than the lower. Notauli indistinct. Scutellum usually long, its lateral carina reaching to apex or to near apex. Posterior mesosternal carina usually complete. Propodeum with basal carina present or rarely obsolete, without other carinae except that oblique or longitudinal wrinkles may be present apically. Pterostigma rather narrow. Areolet absent. Discocubital cell with a small to large glabrous area (fenestra), usually containing 1, 2 or several corneous scleromes. Fore tibial spur without a membranous flange, with only an antennal brush of closely spaced hairs.

General body color pale brown to reddish-brown with abdominal tip often black marked.

Only one species, *Enicospilus merdarius* (Gravenhorst) has been mentioned in literature as parasitic on the gypsy moth. The earlist record could be traced back only to Rühl (1914). There is no recent rearing record confirming its association with the gypsy moth.

Gauld and Mitchell (1981) mention *Enicospilus transversus* Chiu as a parasite of *Lymantria* sp. from Bangalore, India.

2. ENICOSPILUS MERDARIUS (Gravenhorst) (Fig. 83)

Ophion merdarius Gravenhorst, 1829. Ichneumonologia Europaea, 3: 698.
Enicospilus merdarius: Stephens, 1835. Illustrations of British Entomology, Mandibulata, 7: 311.

This species occurs in Europe. It has often been misidentified in literature and does not occur in the Orient (Gauld and Mitchell, 1981). In the absence of reared specimens it is difficult to ascertain if it is really a parasite of the gypsy moth.

Face finely punctate. Clypeus convex, smoother, its apical margin almost smooth and impressed. Upper tooth of mandible about 2.0 as long as the lower. Frons, vertex and temple smoother, subpolished. Interocellar distance about 2.0 the ocellocular distance. Occipital carina complete. Mesoscutum very minutely punctate and subpolished. Scutellum elongate, slightly convex and minutely punctate, its lateral carina strong and reaching apex of scutellum. Mesopleurum finely punctate, its central area often striato-punctate. Prepectal carina arched toward anterior margin of mesopleurum but not touching it. Metapleurum punctate to rugoso-punctate. Propodeum smooth basad of basal transverse carina, circularly reticulo-striate in the depressed area covering petiolar region. Disco-cubital cell with two scleromes in the fenestra, the distal sclerome not distinct, the proximal sclerome pear-shaped, and the central sclerite rounded-oval but its inner side toward quadra irregular and unpigmented. Wings moderately densely hairy. Abdomen strongly compressed, with first tergite tubular. Second and the following tergites with dense short hairs. Thyridium elongate, separated from base of second tergite by about 2.5 to 3.0 its length. Ovipositor short, not longer than the apical depth of abdomen, pointed apically.

Reddish-brown, with orbits, face, frons and clypeus, largely yellowish. Tip of abdomen not black. Sometimes apical segments ventrally a little darker.

Length: 20-24 mm. Fore wing 14-16 mm. Ovipositor 3 mm.

Specimens from various localities in Europe examined in the Townes Collection. No reared material was available.

Hosts: Lymantria dispar (*vide* Rühl, 1914). Kasparyan (1981) lists only

Panolis flammea as its host. Wolff and Krausse (1922), and Meyer (1935) mentioned it from the gypsy moth in Germany and Russia respectively. Stadler, Schedl and Thompson also listed it as such.

This species is very similar to, if not the same as the North American *Enicospilus purgatus* (Say).

X. SUBFAMILY XORIDINAE

Members of the subfamily Xoridinae are parasitic upon wood boring Coleoptera and therefore all records from *Lymantria* are erroneous.

Rudow (1911) listed two species from the gypsy moth, *viz.*, *Xylonomus irrigator* Fabr., and *Xorides praecatorius* Fabr. Both belong to *Xorides*. They are not parasitic upon the gypsy moth. They are therefore removed from the list of gypsy moth parasites.

Another species, *Odontomerus dentipes* Gmelin was recorded by Morley (1908: 11) from *Lymantria monacha*. Morley's record was from an erroneous reporting of Ratzeburg (1844).

XI. SUBFAMILY SCOLOBATINAE

1. OPHELTES GLAUCOPTERUS (Linnaeus)

Members of this genus are parasitic upon *Cimbex* (saw flies) and not on Lepidoptera. The record of its occurrence on *Lymantria dispar,* originating from Rühl (1914) is erroneous. This species is therefore to be removed from the list of parasites of the gypsy moth.

Györfi (1963) mentioned having reared it from the gypsy moth in Hungary. Evidently there was some mix up either in the hosts reared, or in the determinations.

REFERENCES

Anonymous. Exotic parasites of Lymantria dispar established in Pennsylvania and unestablished exotic parasites. Pennsylvania Dept. Environmental Resources Bureau of Forest Pest Management, Report, 3 p.

Aubert, J.F. 1969. Les Ichneumonides ouest-paléarctiques et leurs hotês. 1. Pimplinae, Xoridinae, Acaenitinae. Ouvrage Publée avec le Concours du CNRS. 304 p.

Aubert, J.F. 1975. Ichneumonides pétiolées inédities avec un genre nouveau. Bull. Soc. Ent. Mulhouse. Oct.-Dec. 1974: 53-60.

Aubert, J.F. 1978. Les Ichneumonides ouest-paléarctiques et leurs hôtes 2. Banchinae et suppl. aux Pimplinae. OPIDA, 318 p.

Barsacq, J. 1913. Les Bombyx dissemblabe ou spongieuse Lymantria (Ocneria) dispar L. Rev. Phytopath. Appliqué, Paris 1 (5): 70-73.

Beeson, C.F.C. and Chatterjee, S.N. 1935. On the biology of the Ichneumonidae (Hymenoptera). Indian Forest Res. (N.S.) Ent. 1 (8): 151-168.

Berthoumieu, G.V. 1894-96. Ichneumonides d'Europe et des pays limitrophes. Ann. Ent. Soc. France, 63: 241-274, 505-664 (1894); 64: 213-296, 553-564 (1895); 65: 285-418. Suppl. 393-399 (1896).

Bouché, P. F. 1833. Naturgeschiste der Schadlichen aus nutzlichen Garten-Insekten und die bewarhtesten Mittel zur Verteilung der ersten. Berlin, Nicolai, 176 p.

Britton, W. E. 1935. The gypsy moth. Conn. Agri. Exptl. Sta. Bull. 375: 623-647.

Burgess, A. F. and Crossman, S. S. 1929. Imported insect enemies of the gypsy moth and brown tail moth. U. S. Dept. Agri. Tech. Bull. 86: 1-148.

Campbell, R. W. 1963. Some ichneumonid sarcophagid interactions in the gypsy moth (Porthetria dispar L.) (Lepidoptera: Lymantriidae). Canad. Ent. 95:337-343.

Carlson, R. W. 1979. Family Ichneumonidae. In Krombein et al.: Catalog of Hymenoptera in America north of Mexico, 1: 315-740.

Cecconi, G. 1924. Manuale di Entomologia Forestale. Padua.

Chandra, G. and Gupta, V. K. 1977. Ichneumonologia Orientalis, Part VII. The tribes Lissonotini and Banchini (Hymenoptera: Ichneumonidae: Banchinae). Oriental Insects Monographs 7: 1-291.

Dalla Torre, C. G. de 1901-1902. Catalogus Hymenopterorum hucusque descriptorum systematicus et synonymicus, Vol. 3, 1141 p.

Delrio, G. 1975. Revision des especies ouest-paléarctiques du Genere Netelia Gray (Hym., Ichneumonidae). Studi Sassar. Sez. III-Annalia Della Facolta di Agraria dell'Università di Sassari, 23: 1-126.

Doane, C. C. and McManus, M. (Editors) 1981. The Gypsy Moth: Research toward integrated pest management. U. S. Dept. Agriculture Forest Service, Tech. Bull. No. 1584: 757 p.

Drea, J. J. 1978. A resumé of recent studies made by the European Parasite Laboratory with Lymantria dispar L. and its natural enemies in Europe, Iran, and Japan. [Plant Protection] Zastita Bilja, Beograd 29: 119-125.

Drea, J. J. and Fuester, R. W. 1979. Larval and pupal parasites of Lymantria dispar and notes on parasites of other Lymantriidae (Lep.) in Poland, 1975. Entomophaga 24: 319-327.

Dysart, R. J. 1982. Personal communication.

Fernald, C. H. 1892. The gypsy moth. Mass. Hatch Exp. Sta. Bull. 19: 109 p.

Fitton, M. G. 1978. In Kloet and Hincks. A check-list of British Insects. 2nd Edition. Handbooks of identification of British Insects. Vol. 11 (4, Hymenoptera), 1-159 (Ichneumonidae p. 12-45). Roy. Ent. Soc. London.

Forbush, E. H. and Fernald, C. H. 1896. The gypsy moth. Wright and Porter Printing Co., State Printers, Boston. 495 p.

Fukaya, S. 1936. On the Hymenopterous parasites of Lymantria dispar (L.) (In Japanese). Oyo Dobutsugaku Zasshi 8: 232-335.

Fukaya, S. 1950. (In Yasumatsu and Watanabe, 1964). Not seen.

Fusco, R. A. and Simons, E. E. 1973. (Revised, 1977). A review of some common gypsy moth adult parasites and predators in Pennsylvania. Part I. Native and established species. Dept. of Environmental Resources, Bureau of Forestry, Div. of Forest Pest Management, Pennsylvania, 13 p.

Gauld, I. D. 1976. Notes on British Ophioninae (Hym., Ichneumonidae). Part 3. Ent. Gaz. 27: 113-117.

Gauld, I.D. 1978. Notes on British Ophioninae (Hymenoptera, Ichneumonidae).
    Part 4. Ent. Gaz. 29: 145-149.

Gauld, I. D. and Mitchell, P.A. 1978. The taxonomy, distribution, and host pre-
    ferences of African parasitic wasps of the subfamily Ophioninae. 287 p.
    Commonwealth Inst. Entomology. London.

Gauld, I.D. and Mitchell, P.A. 1981. The taxonomy, distribution and host pre-
    ferences of Indo-Papuan parasitic wasps of the subfamily Ophioninae (Hymen-
    optera: Ichneumonidae). 611 p. Commonwealth Inst. Entomology, London.

Gaulle, J.de 1908. Catalogue systematique et biologique des hyménoptères de
    France. Feuille Jeun. Natural. 38: 64-66, 77-82, 102-104, 120-122, 140-141,
    183-184, 209-210. 234-235, 252-257.

Gravenhorst, J.L.C. 1829. Ichneumonologia Europaea. Vratislaviae, Vol. 1, 830 p.
    Vol. 2, 989 p. Vol. 3, 1097 p.

Griffiths, K. J. 1976. The parasites and predators of the gypsy moth: A review of
    the world literature with special application to Canada. Dept. Environ. Canad.
    Forest Service, Ontario, Report 0-X-243. 92 p.

Griffiths, K.J. 1980. A bibliography of gypsy moth literature. Vols. 1 and 2.
    Canadian Forestry Service, Sault Ste. Marie. Report 0-X-312.

Gupta, V.K. 1962. Taxonomy, zoogeography, and evolution of Indo-Australian
    Theronia (Hymenoptera: Ichneumonidae). Pacific Ins. Monogr. 4: 1-142.

Gupta, V.K. 1975. Ichneumonological Explorations in India, Delhi. 126 p.

Gupta, V.K. and Maheshwary, S. 1977. Ichneumonologia Orientalis, Part IV.
    The tribe Porizontini (Campoplegini) (Hymenoptera: Ichneumonidae). Oriental
    Insects Monogr. 5: 1-267.

Györfi, J. 1963. A Lymantria dispar L. parazitai. Kulonl. Allattani Kozlem
    1963: 50-53.

Hedlund, R.C. and Mihalache, G. 1980. Parasites recovered from pupae of
    Lymantria dispar (Lep.: Lymantriidae) in Romania, 1978. Entomophaga
    25: 55-59.

Heinrich, G.H. 1937. A list and some notes on the synonymy of the types of the
    subfamily Ichneumoninae Ashmead (Hymenoptera) in the collections of the
    British Museum and Hope Dept. of the Oxford University Museum. Ann. Mag.
    Nat. Hist. (10) 20: 257-279.

Heinrich, G.H. 1960-62. Synopsis of Nearctic Ichneumonidae Stenopneusticae with
    particular reference to the northeastern region (Hymenoptera). Parts I-VII.
    Canad. Ent. 92 (Suppl. 15): 1-87; 92 (Suppl. 18): 91-205; 93 (Suppl. 21):
    209-368; (Suppl. 23): 371-505; (Suppl. 26): 511-671; (Suppl. 27): 27: 677-802;
    (Suppl. 29): 29: 807-886.

Heinrich, G.H. 1968. Burmesische Ichneumoninae 4 & 5. Ent. Tidskr. 89: 77-106;
    197-228.

Heinrich, G.H. 1978. Eastern Palearctic Ichneumoninae (In Russian). Acad. Sci.
    USSR 80 p.

Herard, F. and Fraval, A. 1980. La repartition de les enemies naturelles de
    Lymantria dispar (L.) (Lep.:Lymantriidae) au Maroc, 1973-1975. Acta
    Oecologia 1 (1): 35-48.

Herard, F., Mercadier, G., and Abai, M. 1979. Situation de Lymantria dispar (Lep.: Lymantriidae) et de son complexe parasitaire en Iran en 1976. Entomophaga 24: 371-384.

Hinz, R. 1964. Uber einige Typen der Holmgrenschen Gattung Limneria (Hym. Ich. Ophioninae). Entomophaga 9: 67-73.

Horstmann, K. 1969 (1968). Bemerkungen uber die Typusarten von vier Gattungen der Ichneumonidae (Hymenoptera. Opusc. Zool. 102: 1-4.

Horstmann, K. 1974 (1973). Ubersicht uber die europaeischen Arten der Gattung Venturia Schrottky (Hymenoptera, Ichneumonidae). Mitt. Deutsch. Ent. Ges. 32: 7-12.

Hoy, M.A. 1976. Establishment of gypsy moth parasitoids in North America: An evaluation of possible reasons for establishment or non-establishment. In Perspectives in Forest Entomology. Academic Press. p. 215-232.

Howard, L.O. and Fiske, W.F. 1911. The importation into the United States of the parasites of the gypsy moth and the brown tail moth. U. S. Dept. Agr. Bur. Ent. Bull. 91: 1-344.

Kamiya, K. 1934. Studies on the morphology, bionomics, and hymenopterous parasites of the pine caterpillar (Dendrolimus spectabilis Butl.) (In Japanese). Bull. Forest Expt. Sta. Govt. -Gen. Chosen, Korea 18: 1-110.

Kamijo, K. 1962. Natural enemies parasitic on moths attacking poplar. (In Japanese). Rept. Koshunai Forest Tree Breeding Station 1: 83-90.

Kovacevic, Z. 1925. Parasites of Malacosoma neustria L. and Porthetria dispar L. (In Czech). Sumar List 49: 1-5.

Kasparyan, D.R. 1973. A review of the Palearctic ichneumonids of the tribe Pimplinae (Hymenoptera, Ichneumonidae). The genera Itoplectis Forst. and Apechthis Forst. Ent. Rev. 52: 444-455.

Kasparyan, D.R. 1974. Review of the Palearctic species of the tribe Pimplini (Hymenoptera, Ichneumonidae). The genus Pimpla Fabricius. Ent. Review 53 (2): 102-117.

Kasparyan, D.R. 1981. Fauna Hymenoptera USSR. Ichneumonidae (In Russian). Vol. 3, 687 p.

Kaur, R. and Jonathan, K.J. 1979. Ichneumonologia Orientalis, Part VIII. The Tribe Phytodietini from India (Hymenoptera: Ichneumonidae). Oriental Insects Monograph 9: 1-276.

Kirchner, L. 1856. Die von mir erzogenen ichneumonen der Umgegend von Kaptilz. Lotos 6: 107-153.

Kolomietz, N.G. 1958. Parasites of harmful insects in Siberia (English translation). Ent. Rev. 37: 522-534.

Kolubajiv, S. 1934. The results of breeding parasites from their hosts in the State Experiment Institute in Prague in 1929-33. (In Czech). Acta Soc. Ent. Csl. 31: 59-68, 113-120, 155-163.

Kolubajiv, S. 1937. Parasitenverzeichnis und Bestimmungstabelle. In Komarek: Kritisches vort ueber die Beduntung der Insekten parasiten der Nonne. Z. Angew. Ent. 24: 95-117.

Kolubajiv, S. 1937. Notes on the biology of the nun moth and its chief insect para-
    sites (In Czech). Lesn. Pr. 16: 169-199.

Krombein, K. V. et al. 1979. Catalog of Hymenoptera in America north of Mexico.
    Vols. 1-3. Smithsonian Institution Press, Washington.

Leraut, Patrice 1980. Liste systématique et synonymique des Lépidoptères de
    France, Belgique, et Corse. Suppl. à Alexanor et au Bull. Soc. Ent. France
    334 p.

Marsh, P.M. 1979. The braconid (Hymenoptera) parasites of the gypsy moth,
    Lymantria dispar (Lepidoptera: Lymantriidae). Ann. Ent. Soc. America
    72: 794-810.

Matsumura, S. 1931. 6000 illustrated insects of Japan Empire [p. 61].

Meyer, N.F. 1927 and 1929. Schlupfwespen die in Russland in den Jahren 1881-1926
    aus Schaedlingen gezogen sind. Izv. Otdj. Prikladn. Ent. Gossund. Inst.
    Agron. 3: 79-91; 4: 231-248.

Meyer, N.F. 1931. Parasitic insects from harmful rural pests of gardening and
    forestry. (In Russian). p. 98-110.

Meyer, N.F. 1933-1936. Parasititscheskije perepontschatokrylyje sem.
    Ichneumonidae SSSR i sopredetnych stran. Tables systematiques des
    hyménoptères parasites (fam. Ichneumonidae) de l'URSS et des pays limi-
    trophes. Inst. Zool. Acad. Sci. URSS 1933, 1: 1-458; 1933, 2-325; 1934,
    3: 1-271; 1935, 4: 1-535; 1936, 5: 1-340; 1936, 6: 1-356.

Mocsáry, A. (S.) 1885. Data ad cognitionem Ichneumonidum Hungariae I.
    Ichneumones Wesm. [1844]. Magy. Tud. Acad. Math. es Termeszettud
    Kozlem. 20: 51-144.

Mocsáry, A. (S.) 1878. Data ad faunam Hymenopterologicum Sibiricae. Tijdschr.
    v. Ent. 21: 198-200.

Mokrzecki, S.A. 1913. Reports [of the Chief Entomologist to the Zemstov] on
    injurious insects and diseases of plants in Govt. of Tauridia during the year
    1912 (In Russian). Simferopol 1913, 231 p.

Momoi, S. 1961. On some host known Ichneumon flies from Japan, with descrip-
    tions of a new species (Hymenoptera: Ichneumonidae). Kontyu 29: 271-272.

Momoi, S. 1963. New host records of Ichneumonidae of Japan and new homonymy.
    Insecta Matsumurana 26: 54.

Momoi, S. 1970. Ichneumonidae (Hymenoptera) of the Ryukyu Archipelago. Pacific
    Insects 12: 327-399.

Morley, C. 1903-1915. Ichneumonologia Britanica. The Ichneumons of Great
    Britain, Parts 1-5. London.

Morley, C. and Rait-Smith, W. 1933. The hymenopterous parasites of the British
    Lepidoptera. Trans. Roy. Ent. Soc. London, 81: 133-183.

Mosher, F.H. 1915. Food plants of the gypsy moth in America. U. S. Dept. Agri.
    Bull. 250, 39 p.

Muesebeck, C.F.W. and Dohanian, S.M. 1927. A study in hyperparasitism with
    particular reference to the parasites of Apanteles melanoscelus (Ratzeburg).
    U. S. Dept.Agri. Bull. 1487, 35 p.

Muesebeck, C. F. W. and Parker, D. L. 1933. Hyposoter disparis Viereck, an introduced ichneumon parasite of the gypsy moth. J. Agri. Res. 46: 335-347.

Oehlke, J. 1967. Westpaläarktischen Ichneumonidae I. Ephialtinae. In Junk: Hymenopterorum Catalogus (nova editio), Part 2: 1-48.

Oehlke, J. and Townes, Henry K. 1969. Schmiedeknechts Ichneumonidentypen aus der Kollektion des Museums Rudolstadt (Hymenoptera: Ichneumonidae). Beitr. z. Ent. 19: (3-4): 395-412.

Orbtel, R. 1949. K. otrzce druhu rodu Pristomerus Curtis [Notes on species of the genus Pristomerus Curtis] (Hym.: Ich.). Ent. Listy (Folia Ent.) 12: 102-106.

Perkins, J. F. 1953. Notes on British Ichneumoninae with descriptions of new species. Bull. Brit. Mus. (Nat. Hist.) Ent. 3: 105-176.

Perkins, J. F. 1959-60. Handbooks for the identification of British insects. Hymenoptera, Ichneumonoidea, Ichneumoninae, I and II. Vol. 7, part 2 (ai and aii). Roy. Ent. Soc. London. 116 and 213 p.

Picard, F. 1921. Le Bombyx disparate ou spongieuse (Lymantria dispar). Progres Agri. Vitic. 76 (33): 160-165.

Pschorn-Walcher, H. 1974. Gypsy moth (Porthetria dispar): Work in Europe in 1974. Annual Project Statement. Commonw. Inst. Biol. Control. Delemont, Switzerland (Unpublished report).

Rao, V. P. 1967 (1966). Survey for natural enemies of gypsy moth. July 25, 1961 to July 24, 1966. Final Technical Report. Commonw. Inst. Biol. Control Indian Station. 50 p.

Rao, V. P. 1972. Evaluation of hymenopterous parasites of the gypsy moth and study of the behavior of the promising species. Final Technical Report. March 1, 1967 to August 31, 1972. Commonw. Inst. Biol. Control Indian Station 25 p.

Rasnitsyn, A. P. 1981. Gravenhorst's and Berthoumieu's types of Ichneumonidae Stenopneusticae preserved in Wrocław and Cracow, Poland (Hymenoptera, Ichneumonidae). Polskie Pismo Ent. 51: 101-145.

Ratzeburg, J. T. C. 1844. Die Ichneumonen der Forstinsecten in forstlicher und entomologischer Bezeichung. Ein Anhang zur Abbildung und Beschreibung der Forstinsecten. Vol. 1, 224 p.

Ratzeburg, J. T. C. 1852. Ibidem. Vol. 3, 272 p.

Riley, C. V. and Howard, L. O. eds. 1894. Work on the gypsy moth in 1893. Insect Life. U. S. Dept. Agri. Div. Ent. Period. Bull. 6: 338-339.

Romanyk, N. 1965. The study of parasites, predators, and diseases of the gypsy moth (Lymantria dispar) and the possibility of their application in the biological control. Final Technical Report. Buln. Serv. Plagas. For. 65 p.

Rondani, C. 1871-1872. Degli insetti parassiti e delle loro vittime. Bull. Soc. Ent. Ital. 1871, 3: 121-143, 217-243; 1872, 4: 41-78, 229-259, 321-342.

Rondani, C. 1873. Degli insetti novici e dei loro parassitti. Bull. Soc. Ent. Ital. 5: 3-30.

Rudow, F. 1911. Dei Schmarotzer der deutschen Spinner. Bombycidae. Internat. Ent. Ztschr. 5: 90-91, 98-99, 118-119.

Rudow, F. 1917a. Die Ichneumonidengattung Amblyteles und ihre Wirte. Ent. Ztschr. Frankfurt 31: 25-26, 31, 33-35.

Rudow, F. 1917b. Ichneumoniden und ihre Wirte. Ent. Ztschr. Frankfurt. 31: 58-59, 61-62, 66, 76, 71-72.

Rudow, F. 1918-1919. Ichneumon. Ent. Ztschr. Frankfurt. 32: 59, 63-64, 71-72, 75 (1918); 79-80, 84, 88 (1919).

Rühl, M. 1911. Liste neuerdings beschriebener oder gezogener Parasiten und ihre Wirte. Soc. Ent. 26: 31-40.

Rühl, M. 1914. Liste neuerdings beschriebener oder gezogener Parasiten und ihre Wirte. Soc. Ent. 29 (5): 26-30.

Schaefer, P.W. (1981-1982). Personal communication.

Schaffner, J.V. and Griswold, C.L. 1934. Macrolepidoptera and their parasites reared from field collections in the northeastern part of the United States. Misc. Publ. U. S. Dept. Agri. No. 188, 160 p.

Schedl, K.E. 1936. Der Schwammspinner (Porthetria dispar L.) in Euroasien, Afrika, und Neuengland. Monogr. Angew. Ent. 12: 1-242.

Schmiedeknecht, O. 1908-1911. Opuscula ichneumonologica, vol. 4, Ophioninae. pp. 1407-2271.

Šedivý, J. 1957 (1956). A contribution to the knowledge of the ichneumon flies of the tribes Hellwigiini, Anomalonini, and Therionini in Czechoslovakia (Hym: Ichneumonidae) (In Czech). Acta Fauna Ent. Mus. Natl. Praha 1: 127-139.

Šedivý, J. 1963. Faunistische und taxonomische Bemerkungen zu den Ichneumoniden der Tschechoslowakei, Pimplinae, II (Hymenoptera: Ichneumonidae, Pimplinae). Acta Fauna Ent. Mus. Natl. Praha 9: 155-177.

Šedivý, J. 1970. Westpalaearktische Arten der Gattungen Dimophora, Pristomerus, Eucremastus, and Cremastus (Hym., Ichneumonidae). Acta Sci. Nat. Brno 4 (11): 1-38.

Seyrig, A. 1927. Etudes sur les Ichneumonides, II (Hymen.). Eos 3: 201-242.

Shapiro, V.A 1956. The principal parasites of Porthetria dispar L. and the prospects of using them.(In Russian). Zool. Zh. 35: 251-265.

Short, J.R.T. 1978. The final larval instars of the Ichneumonidae. Mem. Amer. Ent. Inst. 25: 1-508.

Simons, E.E., Reardon, R.C., and Ticehurst, M. 1979. Selected parasites and hyperparasites of the gypsy moth, with keys to adults and immatures. U. S. Dept. Agri. Agri Handb. No. 540: 1-59.

Stadler, H. 1933. Ein neuer Ichneumonide aus Schwammspinnerraupen (Lymantria dispar L.). Ent. Anz. 13: 27-30, 43-45, 58-60.

Stafanov, D. and Keremidchiev, M. 1961. The possibility of using some predators and parasitic insects (entomophagous insects) in the biological control of the gypsy moth (Lymantria dispar L.) in Bulgaria.(In Bulgarian). Nauch. Trud. Vis. Lesotech. Inst. 9: 157-168.

Timberlake, P.H. 1912. Technical results from the gypsy moth parasite laboratory. V. Experimental parasitism: A study of the biology of Limnerium validum (Cresson). U. S. Dept. Agri. Bull. Ent. Tech. Ser. 19: 71-92.

Thompson, W.R. 1946. A catalogue of the parasites and predators of insect pests. Section I. Parasite host catalogue. Part 8. Parasites of the Lepidoptera (N-P).

Thompson, W. R. 1957. A catalogue of the parasites and predators of insect pests. Section 2. Host parasite catalogue. Part 4. Hosts of the Hymenoptera (Ichneumonidae). CIBC, Ottawa. 561 p.

Townes, H. K. 1940. A revision of the Pimplini of eastern North America (Hymenoptera, Ichneumonidae). Ann.Ent. Soc. Amer. 33: 283-323.

Townes, Henry K. 1944-45. A catalogue and reclassification of the Nearctic Ichneumonidae (Hymenoptera). Mem. Amer. Ent. Soc. 11: 1-925.

Townes, Henry 1969-71. The genera of Ichneumonidae, parts 1-4. Mem. Amer. Ent. Inst. 11, 12, 13, and 17.

Townes, Henry and Marjorie, 1960. Ichneumon flies of America north of Mexico: 2. Subfamilies Ephialtinae, Xoridinae, Acaenitinae. U. S. Natl. Mus. Bull. 216 (2): 1-676.

Townes, Henry and Marjorie, 1962. Ichneumon-flies of America north of Mexico: 3. Subfamily Gelinae, tribe Mesostenini. U. S. Natl. Mus. Bull. 216 (3): 1-602.

Townes, Henry and Marjorie, 1978. Ichneumon-flies of America north of Mexico: 7. Subfamily Banchinae, tribes Lissonotini and Banchini. Mem. Amer. Ent. Inst. 26: 614 p.

Townes, H., Momoi, S., and Townes, M. 1965. A catalogue and reclassification of the eastern Palearctic Ichneumonidae. Mem. Amer. Ent. Inst. 5: 671 p.

Townes, H., Townes, M., and Gupta, V.K. 1961. A catalogue and reclassification of Indo-Australian Ichneumonidae. Mem. Amer. Ent. Inst. 1: 502 p.

Tragardh, I. 1920. Investigations on the occurrence of the nun moth near Gulov in 1915-1917. (In Swedish). Medd. Stat. Skogsforsoksanst. Stockholm 17 (4): 301-328.

Uchida, T. 1930. Vierter und fuenfter Beitrage zur Ichneumoniden fauna Japans. J. Fac. Agri. Hokkaido Imp. Univ. 25: 242-347.

Uchida, T. 1930. Beitrag zur Kenntnis der Ichneumoniden Fauna der Insel Izu Oshima. Trans. Sapporo Nat. Hist. Soc. 11 (2): 78-87.

Uchida, T. 1941. Die Kriechbaumerschen Typen der Japanischen Ichneumoniden. Trans. Sapporo Nat. Hist. Soc. 16: 227-230.

van Rossem, G. 1969. A study of the genus Meringopus Foerster in Europe and some related species from Asia (Hymenoptera, Ichneumonidae, Cryptinae). Tijdschr. v. Ent. 112: 165-196.

Vasic, K. 1958. Parasitic Hymenoptera of gypsy moth (In Russian). Zast. bilja 41-42: 17-21.

Victorov, G.A. 1957. Species of the genus Enicospilus Stephens (Hymenoptera, Ichneumonidae) in USSR (In Russian). Ent. Obozr. 36: 179-210.

Victorov, G.A. 1958. Material on the taxonomy of the ichneumon-flies of the genus Encospilus Stephens (Hymenoptera, Ichneumonidae) (In Russian). Zool. Zh. 36: 215-221.

Viereck, H. L. 1911. Descriptions of one new genus and eight new species of ichneumon-flies. U. S. Natl. Mus. Proc. 40: 455-480.

Walley, G.S. 1947. The genus Casinaria Holmgren in America north of Mexico (Hymen., Ichneumonidae). Scient. Agri. 27: 364-395.

Wolff, M. and Krausse A. 1922. Die forstlichen Lepidopteren. Jena. 337 p.

Yafaeva, 1959. Ukrainskaya Akad. Selsk. Nauk p. 227. (not seen).

Yasumatsu, K. and Watanabe, C. 1964. A tentative catalogue of insect natural
    enemies of injurious insects in Japan. Part I. Parasite-Predator Host
    Catalogue. Ent. Lab. Kyushu Univ., Fukuoka, Japan. 166 p.

Due acknowledgment is made here for the use of certain figures from published
sources, as indicated below:

Townes (1969-1971) (Figs. 1-6, 43-48, 68, 74-77, 80-84, 87, 90, 114-121).

Finlayson and Hagen, 1979. Final instar larvae of parasitic Hymenoptera.
    Pest Management Papers No. 10, SFU.   (Figs. 122, 123, 138-141).

Howard and Fiske, 1911.   (Fig. 61).

Kaur and Jonathan, 1979.   (Figs. 69-73).

Kasparyan, 1981. (Figs. 85, 86, 88, 89).

Perkins, 1960.   (Figs. 91-113).

Short, 1978. (Figs. 124-137).

van Rossem, 1969. (Figs. 78, 79).

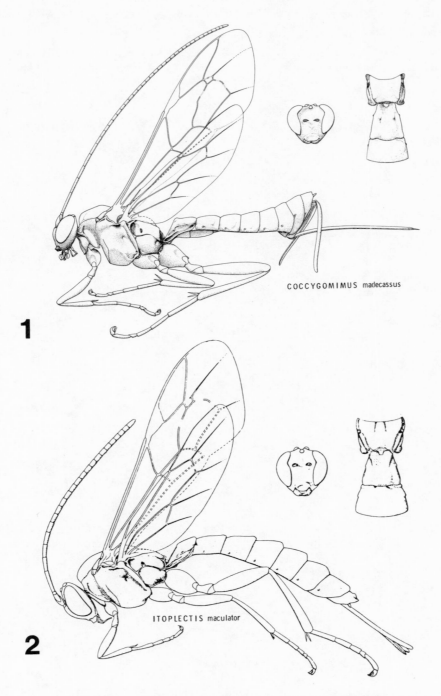

COCCYGOMIMUS madecassus

1

ITOPLECTIS maculator

2

Figs. 1-2. Generic diagrams of: 1, <u>Coccygomimus</u>.
2, <u>Itoplectis</u>.

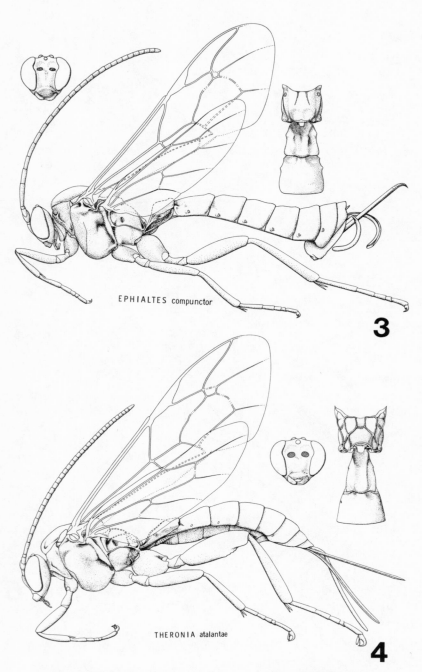

EPHIALTES compunctor

3

THERONIA atalantae

4

Figs. 3-4. Generic diagrams of: 3, Ephialtes.
4, Theronia.

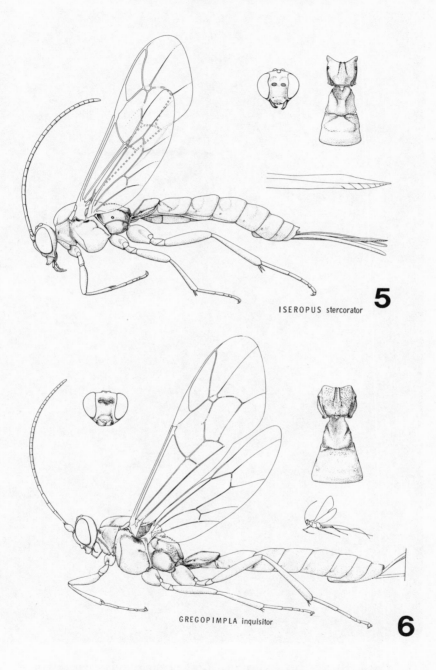

ISEROPUS stercorator

5

GREGOPIMPLA inquisitor

6

Figs. 5-6. Generic diagrams of: 5, Iseropus. 6, Gregopimpla.

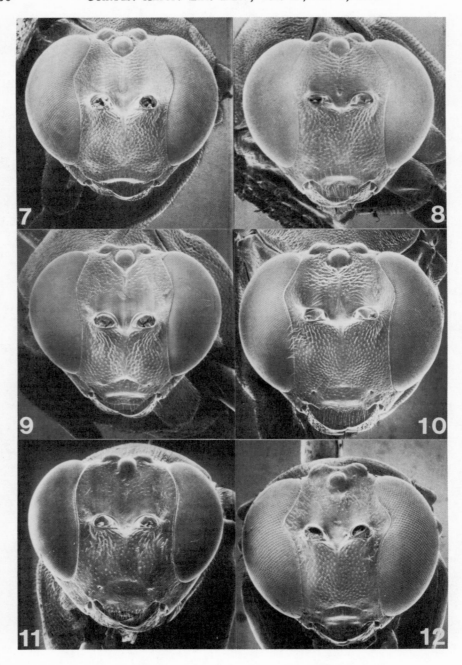

Figs. 7-12. Head, front view of: 7, Coccygomimus turionellae.
8, C. moraguesi. 9, C. instigator. 10. C. disparis. 11, C. pedalis.
12, Itoplectis alternans.

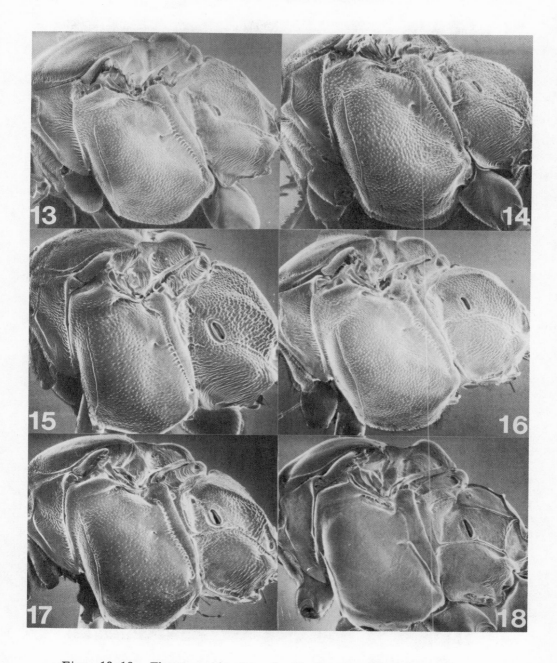

Figs. 13-18. Thorax, side view of: 13, Coccygomimus turionellae.
14, C. moraguesi. 15, C. instigator. 16, C. disparis. 17, C. pedalis.
18, Theronia atalantae fulvescens.

Figs. 19-24. Propodeum of: 19, Coccygomimus turionellae.
20, C. moraguesi. 21, C. instigator. 22, C. disparis. 23, C. pedalis.
24, Theronia atalantae.

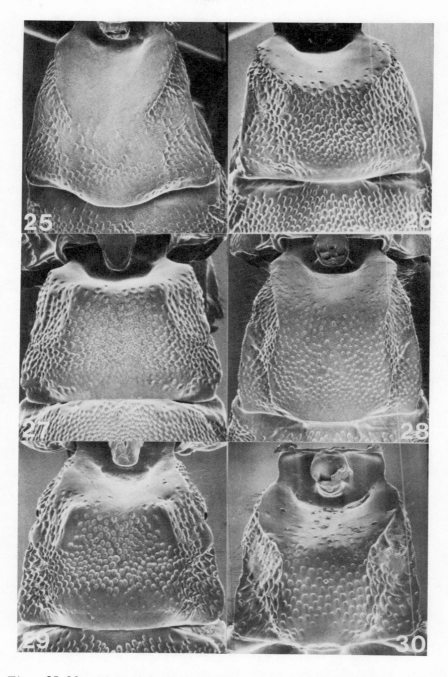

Figs. 25-30. First abdominal tergite of: 25, Coccygomimus turionellae. 26, C. moraguesi. 27, C. instigator. 28, C. disparis. 29, C. pedalis. 30, Itoplectis alternans.

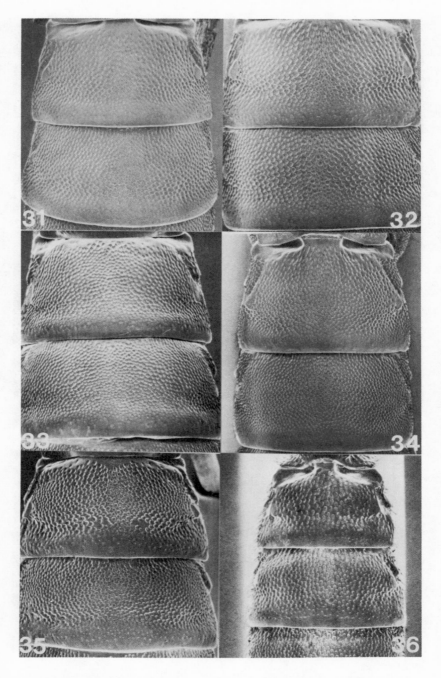

Figs. 31-36. Tergites 2 and 3 of: 31, Coccygomimus turionellae.
32, C. moraguesi. 33, C. instigator. 34, C. disparis. 35, C. pedalis.
36, Itoplectis alternans.

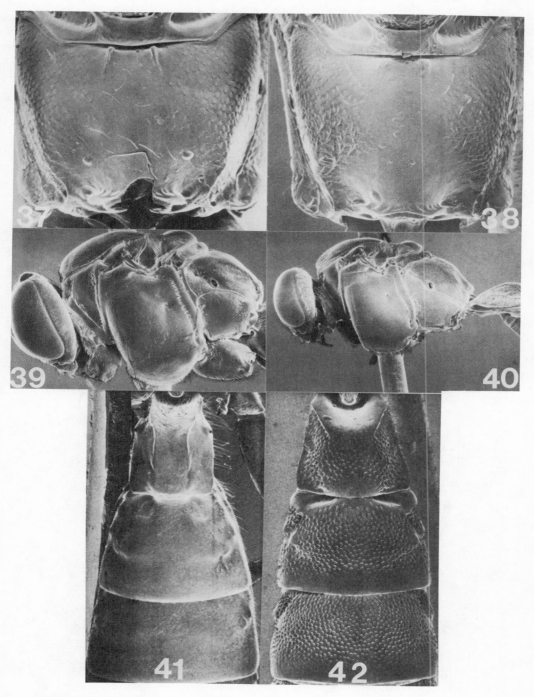

Figs. 37-42. Propedeum of: 37, Itoplectis alternans. 38, I. conquisitor.
Side of thorax of: 39, I. alternans. 40, I. conquisitor. Tergites 1 to 3 of:
41, Theronia atalantae. 42, I. conquisitor.

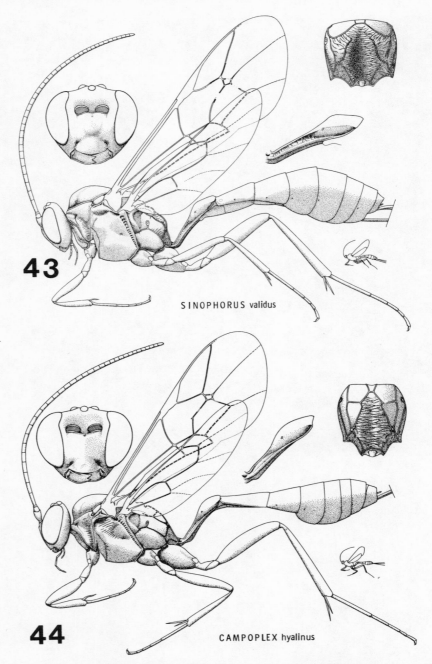

43

SINOPHORUS validus

44

CAMPOPLEX hyalinus

Figs. 43-44. Generic diagrams of: 43, Sinophorus. 44, Campoplex.

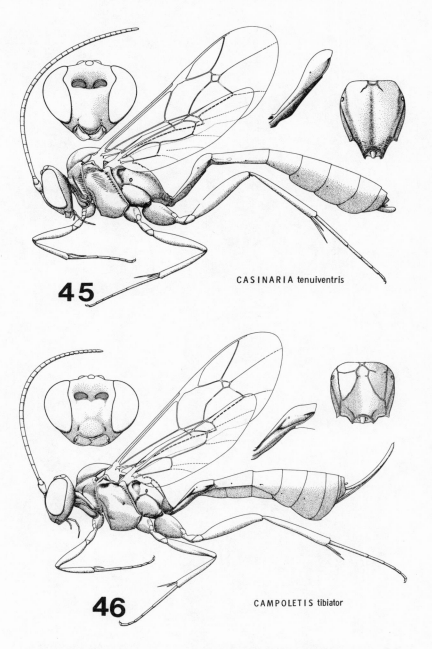

45   CASINARIA tenuiventris

46   CAMPOLETIS tibiator

Figs. 45-46. Generic diagrams of: 45, <u>Casinaria</u>.
46. <u>Campoletis</u>.

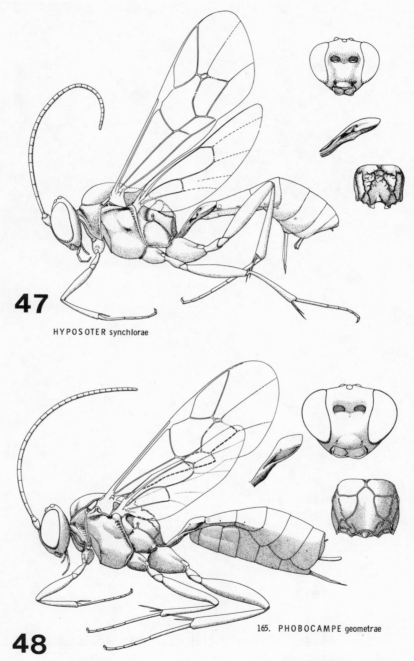

47

HYPOSOTER synchlorae

165. PHOBOCAMPE geometrae

48

Figs. 47-48. Generic diagrams of: 47, Hyposoter.
48, Phobocampe.

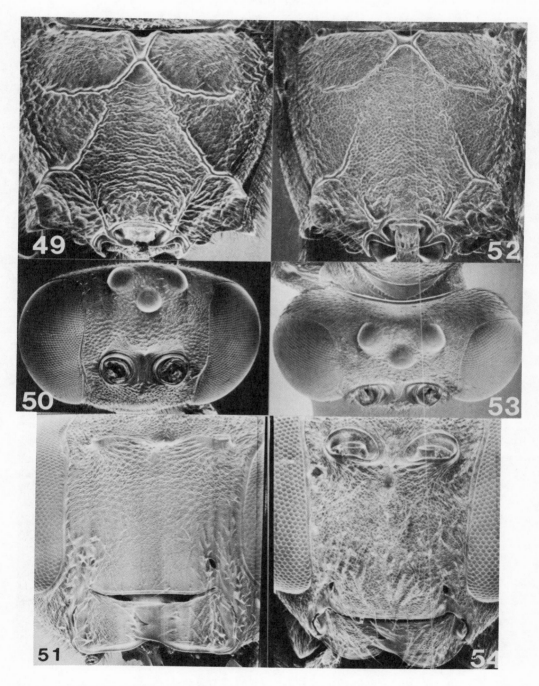

Figs. 49-54. Phobocampe unicincta: 49, Propodeum. 50, Vertex. 51, Face. P. lymantriae: 52, Propodeum. 53, Vertex. 54, Face.

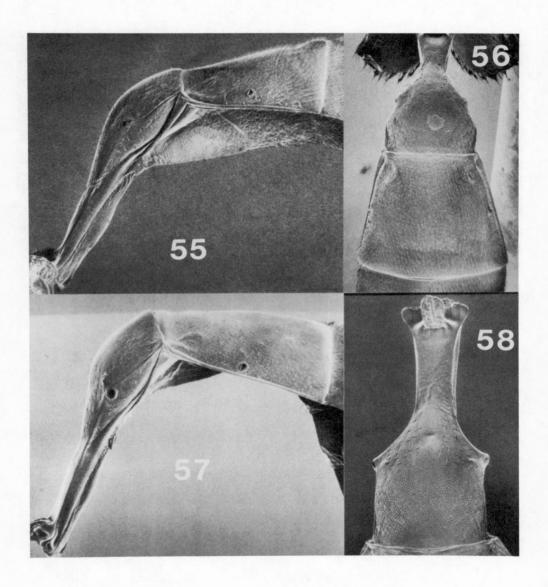

Figs. 55-58. Phobocampe lymantriae, tergites 1-2: 55, side view.
56, Dorsal view. P. unicincta, tergites 1-2: 57, side view. 58, Dorsal
view.

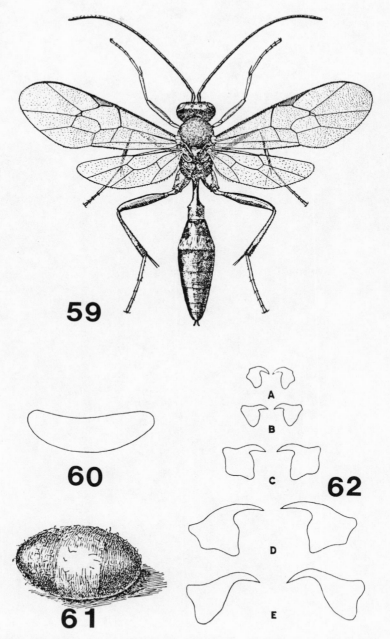

Phobocampe unicincta: 59, Adult. 60, Egg. 61, Cocoon. 62, Larval mandibles, A-E. First to fifth instars.

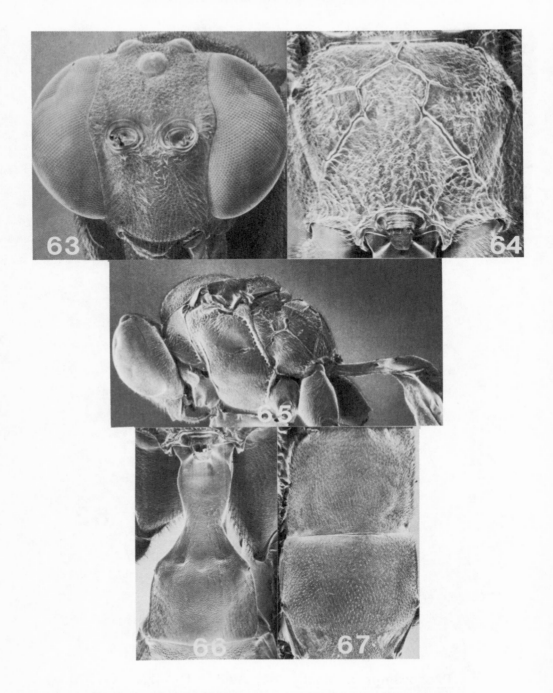

Figs. 63-67. _Hyposoter_ tricoloripes: 63, Face. 64, Propodeum.
65, Thorax. 66, Tergite 1. 67, Tergites 2-3.

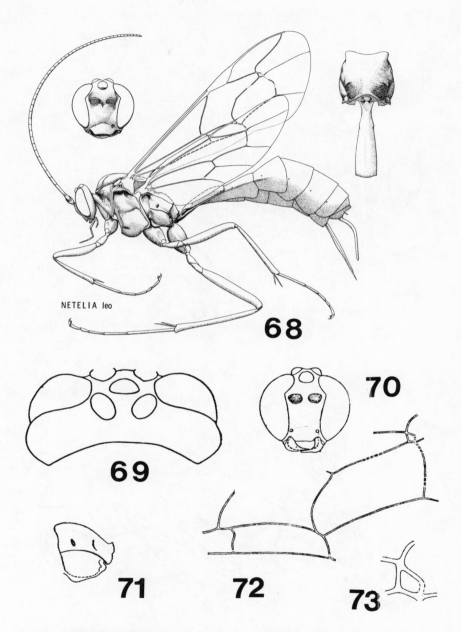

NETELIA leo

68

69

70

71   **72**   **73**

Figs. 68-73. 68, Generic diagram of <u>Netelia</u>.
<u>N. vinulae</u>: 69, Vertex. 70, Head, front view.
71, Propodeum. 72, Part of front wing. 73, Areolet.

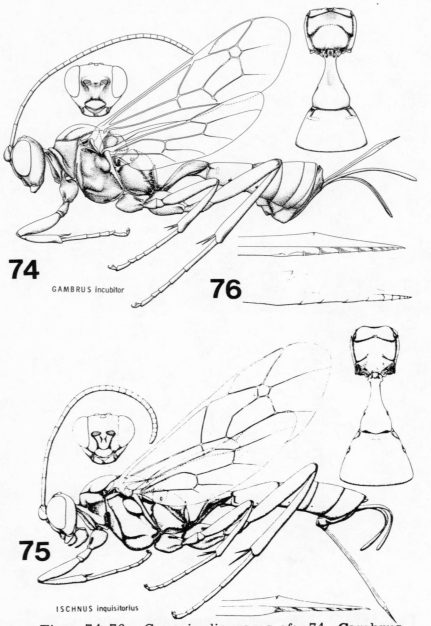

74

GAMBRUS incubitor

76

75

ISCHNUS inquisitorius

Figs. 74-76. Generic diagrams of: 74, Gambrus.
75, Ischnus. 76, Ovipositor tip of G. amoenus.

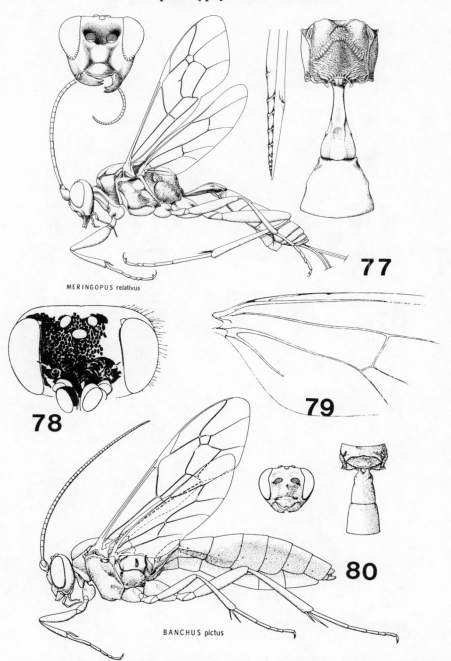

MERINGOPUS relativus

77

78

79

BANCHUS pictus

80

Figs. 77-80. 77, Generic diagram of <u>Meringopus</u>. 78, <u>M</u>. <u>cyanator</u>, head. 79, Hind wing of same. 80, Generic diagram of <u>Banchus</u>.

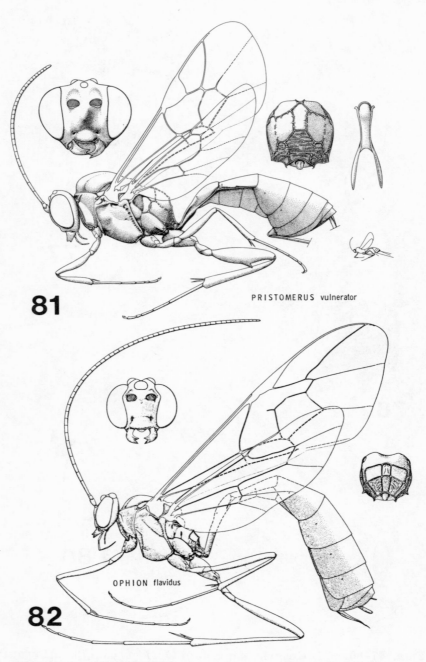

81   PRISTOMERUS vulnerator

82   OPHION flavidus

Figs. 81-82. Generic diagrams of: 81, Pristomerus. 82, Ophion.

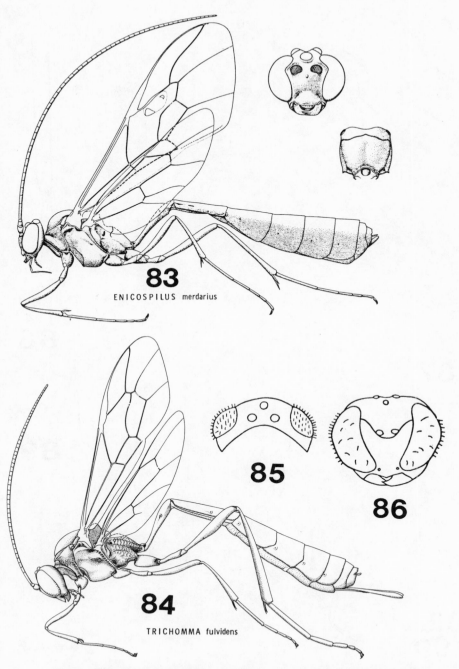

83

ENICOSPILUS merdarius

85

86

84

TRICHOMMA fulvidens

Figs. 83-86. Generic diagrams of: 83, Enicospilus.
84, Trichomma. T. enecator: 85, Vertex. 86, Face.

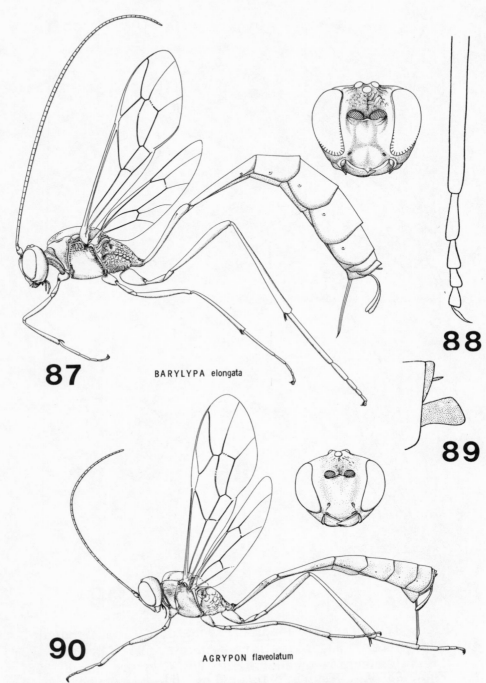

87   BARYLYPA elongata

88

89

90   AGRYPON flaveolatum

Figs. 87-90. Generic diagram: 87, <u>Barylypa</u>.
88, Hind tarsus <u>B</u>. delictor. 89, Male clasper,
<u>B</u>. delictor. Generic diagram: 90, <u>Agrypon</u>.

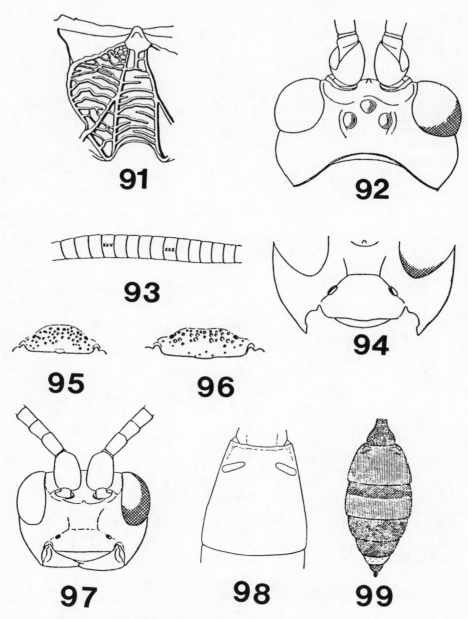

Figs. 91 - 99. Callajoppa cirrogaster: 91, Propodeum.
92, Vertex. 93, Flagellar segments. 94, Clypeus.
Melanichneumon leucocheilus: 95, Clypeus, male. 96,
Clypeus, female. 97, Head. 98, Tergite 2, 99, Pterocormus sarcitorius, abdomen color.

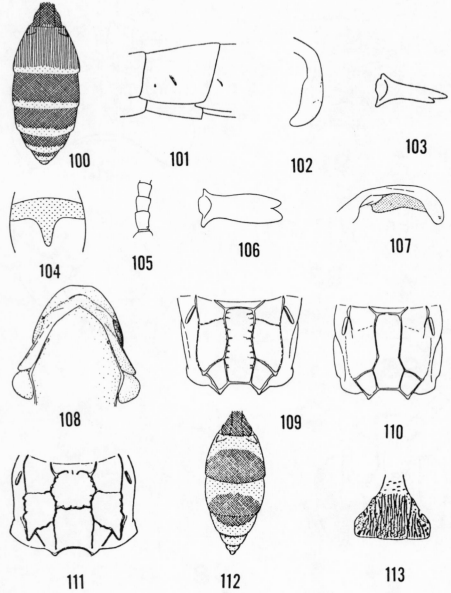

Figs. 100-113. Triptognathus amatorius: 100, Abdomen color.
101, Tergites 2-3. 102, Male penis valve. 103, Mandible.
Spilichneumon occisor: 104, Hypopygium. 105, Basal flagellar
segments. 106, Mandible. 107. Penis valve. 108, Pronotum.
109, Male propodeum. 110, Female propodeum. Amblyteles
armatorius: 111, Propodeum. 112, Abdomen. Pterocormus
sarcitorius: 113, Postpetiole.

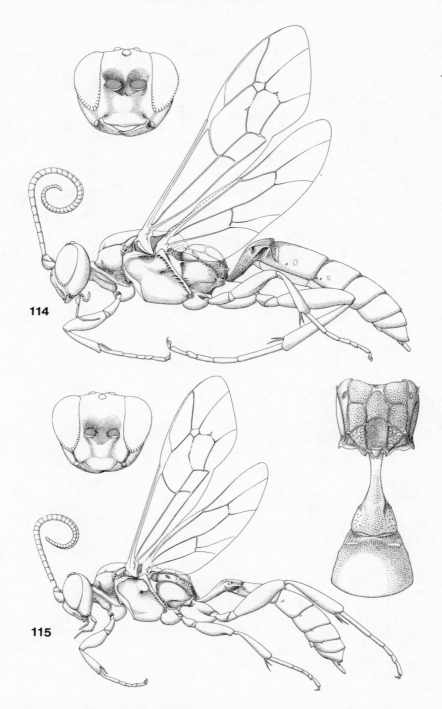

Figs. 114-115. Generic diagrams of: 114, _Ichneumon_. 115, _Melanichneumon_.

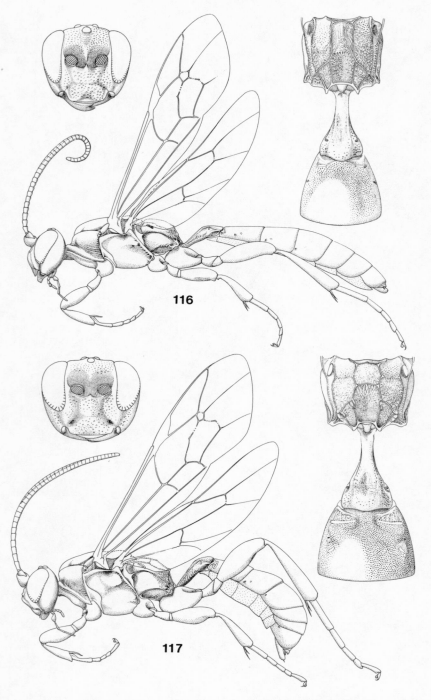

Figs. 116-117. Generic diagrams of: 116, Spilichneumon. 117, Cratichneumon.

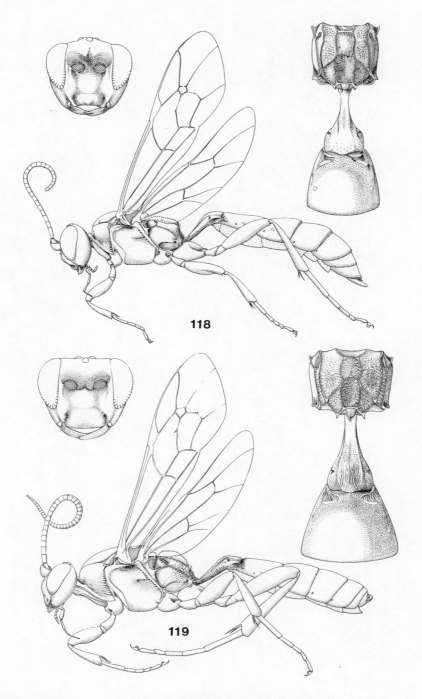

Figs. 118-119.   Generic diagrams of: 118, <u>Chasmias</u>.   119, <u>Triptognathus</u>.

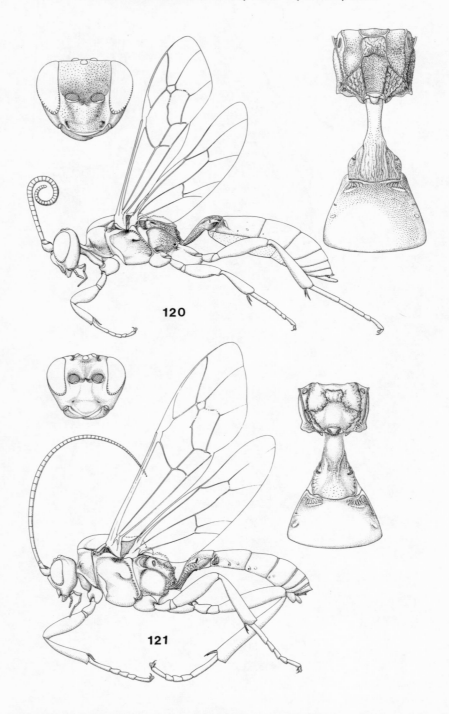

Figs. 120-121. Generic diagrams of: 120, _Pterocormus_. 121, _Cotiheresiarches_.

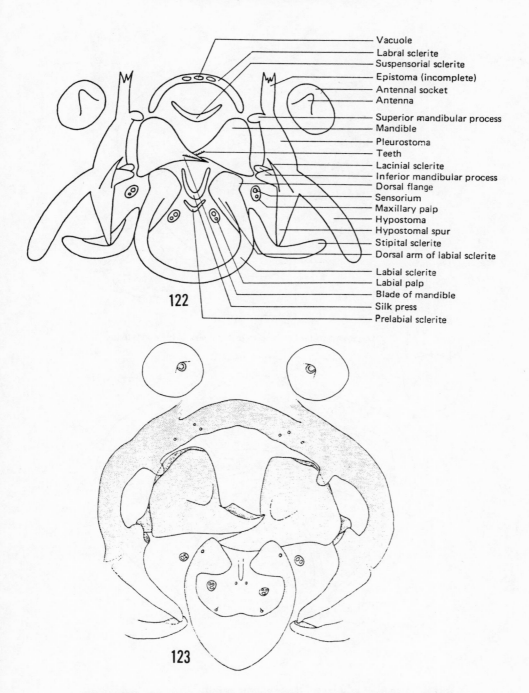

Vacuole
Labral sclerite
Suspensorial sclerite
Epistoma (incomplete)
Antennal socket
Antenna
Superior mandibular process
Mandible
Pleurostoma
Teeth
Lacinial sclerite
Inferior mandibular process
Dorsal flange
Sensorium
Maxillary palp
Hypostoma
Hypostomal spur
Stipital sclerite
Dorsal arm of labial sclerite
Labial sclerite
Labial palp
Blade of mandible
Silk press
Prelabial sclerite

Figs. 122-123. 122, Nomenclature of cephalic sclerites of the ichneumonid larval head. 123, Larval head of Coccygomimus pedalis.

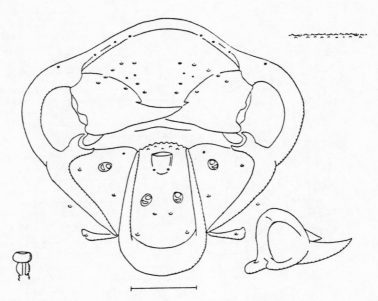

124 Coccygomimus turionellae with posterior view of left mandible

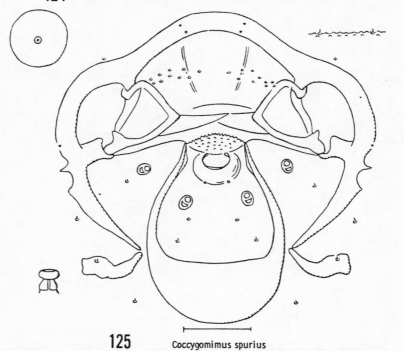

125 Coccygomimus spurius

Figs. 124-125. Larval head of : 124: <u>Coccygomimus turionellae</u>. 125, <u>C</u>. <u>spurius</u>.

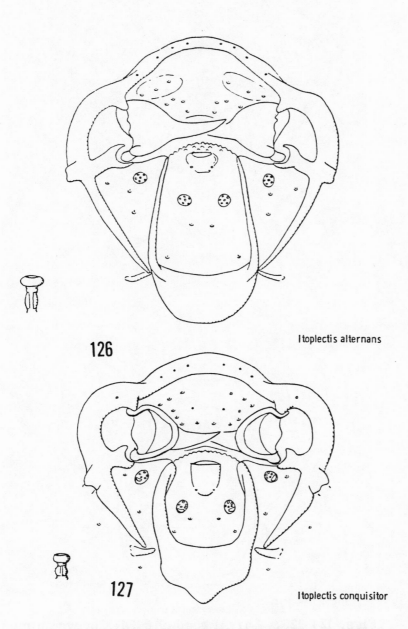

126

Itoplectis alternans

127

Itoplectis conquisitor

Figs. 126-127.   Head sclerites of larvae.

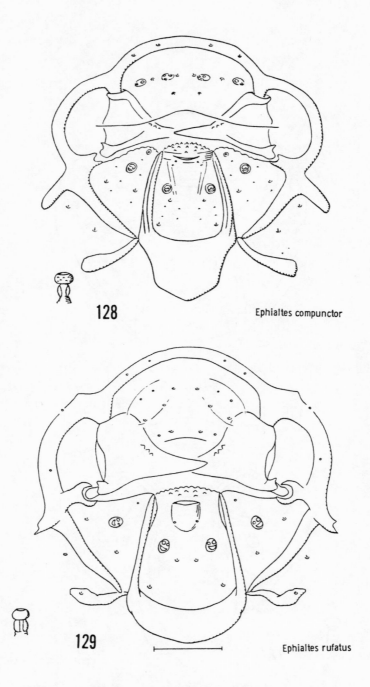

128   Ephialtes compunctor

129   Ephialtes rufatus

Figs. 128-129. Head sclerites of larvae.

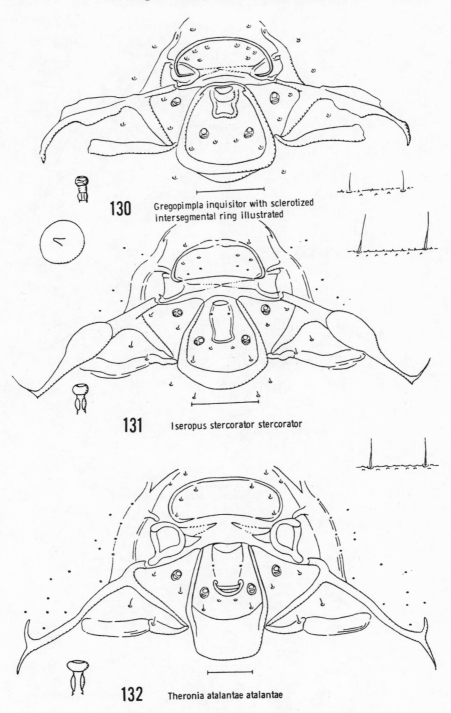

130 Gregopimpla inquisitor with sclerotized intersegmental ring illustrated

131 Iseropus stercorator stercorator

132 Theronia atalantae atalantae

Figs. 130-132. Head sclerites of larvae.

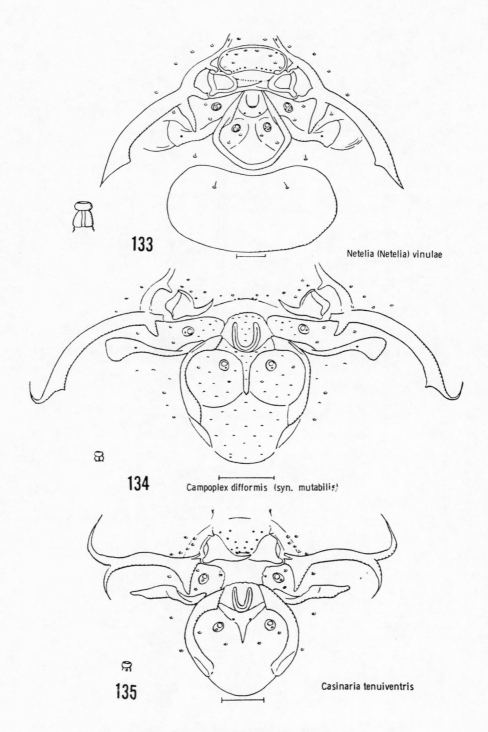

133   Netelia (Netelia) vinulae

134   Campoplex difformis (syn. mutabilis)

135   Casinaria tenuiventris

Figs. 133-135. Head sclerites of larvae.

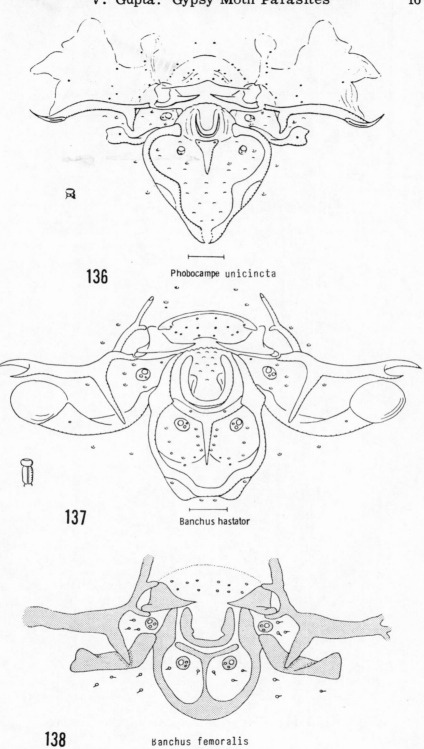

136   Phobocampe unicincta

137   Banchus hastator

138   Banchus femoralis

Figs. 136-138. Head sclerites of larvae.

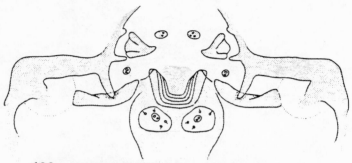

139   <u>Enicospilus macrurus</u>

140   <u>Habronyx</u> <u>nigricornis</u>

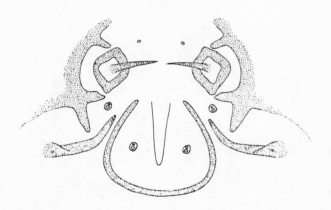

141   <u>Mesochorus fulgurans</u>

Figs. 139-141.   Head sclerites of the larvae.

Errata to: Gupta, Virendra, 1983. The Ichneumonid Parasites Associated with the Gypsy Moth (**Lymantria dispar**) . Contrib. Amer. Ent. Inst. 19 (7): 1-168.

References inadvertently left out . To be inserted on pages 124 to 132.

Gauld, I. D. and Mitchell, P. A. 1977. The Orthopelmatinae and Anomalolinae. Handbooks for the Identification of British Insects. 7, 2b: 1-32.

Gupta, V. K. 1970. Ichneumon Hunting in India. A report of the work done under P.L. 480 research project ... on Ichneumonidae in India. Delhi. Pp. 1-109 + appendix 1-80.

Gupta, V.K. and Tikar, D.T. 1976. Ichneumonologia Orientalis, Part I. The tribe Pimplinae (Hymenoptera: Ichneumonidae). Oriental Insects Monographs 1: 1-312.

Fahringer, J. 1922. Contributions to a knowledge of the habits of some parasitic Hymenoptera with special regard to their importance in the biological control of injurious insects. Z. Angew. Ent. 8 (2): 325-388. (in German).

Nagaraja, H., Dharmadhikari, P.R., and Rao, V. P. 1969. A comparative study of the external morphology of Lymantria obfuscata Wlk. in India and L. dispar (L.) in the U.S.A. Bull. Ent. Res. 59: 105-112 (1968).

Rudow, F. 1888. Einige neue Ichneumoniden. Ent. Nachr. 14: 83-92, 120-124, 129-136.

Uchida, T. 1926. Erster Beitrage zur Ichneumoniden Japans. J. Coll. Agri. Hokkaido Imp. Univ. 18: 43-173.

Uchida, T. 1958. Shin Konchu 8(5): 8.

Walkley, L. M. 1958. Family Ichneumonidae. In: Krombein, K.V. (Ed.) Hymenoptera of America North of Mexico Synoptic Catalog First Supplement to Agri. Monogr. 2. U. S. Dept. Agri. Washington. Pp. 36-62.

P. 132. Under acknowledgment for illustrations add Muesbeck and Parker, 1933 for figures 59, 60 and 62.

Corrections

Page 4	Line 12 from bottom	For Lerut (1980)	Read Leraut (1980)
11	Under species 10	Add (1933) after Morley and Rait-Smith	
11	,, ,, 11	For page 77	Read page 75
12	,, ,, 23	For Rudow (1933)	Read Rudow (1918)
14	Line 15 from bottom	,, Nagaraja et al (1968)	Read Nagaraja et al. (1969)
104	Line 11 and 36	For Kolubayiv	Read Kolubajiv
105	Line 9	For Gupta (1973)	Read Gupta (1970)
110	Line 7 from bottom	For Kraube	Read Krausse
112	Line 5	For Delrio (1974)	Read Delrio (1975)
118	Line 3 from bottom	For Kraube	Read Krausse
124	Line 4 from bottom	For Res.	Read Rec.
127	Reference to Kovacevic Z. should come after Kolabujiv on page 128		

Please let me know if you discover any other omission or correction.

Virendra Gupta

A Synopsis of the World Species of
Desmometopa Loew (Diptera, Milichiidae).

Curtis W. Sabrosky [1]
Systematic Entomology Laboratory,
Agricultural Research Service, U.S. Department of Agriculture

ABSTRACT

The genus Desmometopa Loew is revised, with the number of recognized species in the world raised from 10 to 51, divided for the first time into two subgenera, Desmometopa Loew and Platophrymyia Williston. The classification, identification, morphological characteristics, and biology of adults and larvae are discussed, with numerous new rearing records. A key is provided to the 51 species, 41 of them new, of which 8 are left unnamed because of inadequate material. Two new synonyms are proposed, a queried synonymy is confirmed, and 4 lectotypes are designated.

CONTENTS

[1] Mailing address: Systematic Entomology Lab., USDA, c/o U.S. National Museum, Washington, D.C. 20560.

INTRODUCTION

The small black flies of the dipterous genus Desmometopa Loew are commonly encountered in all faunal regions, and they form a delightfully recognizable group of species because of a black M-shaped frontal vitta delineated by gray frontal triangle, fronto-orbital plates, and two straplike interfrontal plates (figs. 1,2,10). The genus is easily placed in available keys to the genera of Milichiidae (e.g., Hennig 1937). However, some confusion in identification of the species has long indicated that a synopsis and usable key were badly needed.

Classification

For some years, the genus Desmometopa included species now referred elsewhere, notably to Leptometopa Becker, Stomosis Melander, and Neophyllomyza Melander, none of which has the inter-frontal plates although they may have rows of fine setae in the same positions. As recognized for over six decades, however, the genus has been confined to those species having the two straplike or bandlike interfrontal plates, with 17 proposed specific names well scattered from 1820 to 1965. In the present study these are reduced to 10 species and 7 synonyms, only 4 of the recognized species being common. The present study recognizes a total of 51 species, including 41 new species, chiefly Neotropical, well over half of these in the "tarsalis complex". Eight of the 41 are left unnamed at this time, however, because of the inadequacy of available material, but they have been included in the key. I am inclined to believe that even more species will be discovered, as has often been the case in other genera of neglected or overlooked small black flies.

The contrast with recent literature is striking. Modern catalogues for the Nearctic, Neotropical, Oriental, and Afrotropical Regions (Sabrosky, 1965b, 1973, 1977, 1980) listed, respectively, 3, 6, 6, and 8 species, and Duda (1935) recognized 4 Palearctic species. Eliminating duplications, a total of 12 species was recognized, and 3 of these are dropped as synonyms in the present classification.

The discovery of several polished black Neotropical species that might belong to the genus inevitably raised questions about their relationship to Desmometopa and to Litometopa Sabrosky, the latter proposed for a single species from Tanzania. Neither Litometopa nor the polished Neotropical species have interfrontal stripes, but the Neotropical species have 2 rows of interfrontal setae in contrast to their complete absence in Litometopa. Possibly these are degrees of reduction from the interfrontal plates of Desmometopa, but I prefer here to retain the distinctness of the plates as uniquely characteristic of Desmometopa. The polished species may represent a new genus. I may add that Litometopa has other distinct features: only one upper and one lower fronto-orbital bristle on each side; no presutural bristle, and mesonotum almost bare of hairs except for the median acrostical and dorsocentral rows and a few intra-alar hairs, whereas the polished black Neotropical species resemble Desmometopa in all these features.

The genus may be divided into two subgenera on the basis of the structure of the head:

Subgenus Platophrymyia Williston (type species P. nigra Williston, which is a synonym of D. tarsalis Loew, a commonly used--and all too frequently misused--name): Vibrissal angle distinctly produced anteriorly to about a 45° angle (fig. 18), the angle emphasized by shining black lateroventral corner of facial plate, immediately mesad of vibrissal angle and usually warped forward and upward beyond it. Lower margin of head comparatively long and face deeply concave as seen in profile. Epistomal margin, the lower margin of face, more or less strongly warped upward also, and especially so on the midline so as to shorten the face between epistomal margin and apex of the lunule between antennal bases, this warping and shortening, together with a distinct medial facial carina, all combining to accentuate the concave antennal grooves or foveae.

Subgenus _Desmometopa_ s. str. (type species _Agromyza_ _m-atrum_ Meigen, a synonym of _D_. _sordida_ (Fallén): Vibrissal angle not produced anteriorly, only an 80° to 90° angle (e.g., figs. 3,5,6–8), face only slightly concave as seen in profile, the latero-ventral corner of facial plate dull gray like rest of face, and not warped forward. Epistomal margin at most only slightly warped forward and upward, usually not markedly so at midline, the face not materially shortened on midline, facial carina weak, and antennal foveae not accentuated.

Most of the species fall easily into one or the other of the two groups. The characteristics of a few species seem to be somewhat intermediate, or could be misinterpreted by someone using the key without adequate reference material or experience. Accordingly the subgeneric division is not used initially in the key, which is intentionally artificial to bypass some problem species and give priority to accuracy of identification.

The species can be assigned to subgenera as follows, with geographic distribution indicated by faunal regions (Oceanic = the Pacific islands):

Subgenus _Desmometopa_ Loew (22 species)

Widely distributed:
> inaurata (Afrotropical, Neotropical, Australian, Oceanic), microps (Afrotropical, Palearctic, Oriental, Oceanic), m-nigrum (Holarctic, Neotropical, Afrotropical, Oriental, Australian), singaporensis (Oriental, Oceanic, Afrotropical, Palearctic, Neotropical), varipalpis (Australian, Oceanic, Oriental, Afrotropical, Palearctic, Nearctic, Neotropical).

Holarctic: sordida.
Palearctic: sp. H.
Afrotropical: aldabrae, interfrontalis, leptometopoides, magnicornis,
> nudigena, pleuralis, postorbitalis.

Oriental: philippinensis, propeciliata, srilankae, spp. L,M,N.
Australian: ciliata.
Oceanic: terminalis.

Subgenus _Platophrymyia_ Williston (29 spp.)

Nearctic: floridensis, latigena, melanderi, nearctica, parafacialis, saguaro,
> sp. O.

Neotropical: aczeli, argentinica, atypica, blantoni, evanescens, flavicoxa,
> glaucanota, indistincta, lucidifrons, meridionalis, nigrohalteralis,
> obscurifrons, stilbopleura, woldai, spp. I,J,K.

Neotropical to Oceanic: tarsalis.
Oceanic: flavipalpis, gressitti.
Oriental: kandyensis.
Afrotropical: nigeriae.

Five species of the typical subgenus are widely distributed, apparently having been spread in commerce, and _D_. _tarsalis_ in _Platophrymyia_ has moved out into the Pacific islands from its Neotropical homeland. Otherwise, except for a few unexplained species in _Platophrymyia_, the two subgenera make a fairly good separation between Old World (_Desmometopa_) and New World (_Platophrymyia_) species.

There are a number of records of adult Desmometopa being collected on ships and planes (e.g., see varipalpis), and it has been suggested that this indicates how some species were distributed widely in commerce. Another possibility is indicated by the rearing of varipalpis at New York from larvae in potatoes in a ship from Argentina, and of singaporensis in Fiji from onions imported from Australia.

Desmometopa is clearly a name of feminine gender, not neuter as it has often been used in such combinations as D. sordidum. Loew's first species included in the genus was D. tarsalis, which could indicate either masculine or feminine gender, but certainly not neuter. In his accompanying note he used the feminine ablative form "cui Desmometopae." The distinguished European dipterists Hendel and Becker, among others, in writing on milichiids around the beginning of the century used D. sordidum, undoubtedly influenced by the specific names m-atrum and m-nigrum; however, in these the adjectival ending -um (neuter) depends not on the generic name but on the "m" which it modifies, and individual letters of the alphabet are treated as neuter. Konow (1907) early pointed out that Desmometopa was feminine. Grensted (1956) likewise argued that Desmometopa was feminine, compounded with the stem of the Green metopon (a forehead), which is neuter, but with an irregular termination -a that made the name feminine. He followed a ruling of the International Commission on Zoological Nomenclature (1958, Declaration 39), which was incorporated into the International Code (1961, 1964) as Article 29c.

For the record, and to assist in the proper placement of species listed in the literature as "Desmometopa," species either described in or referred to that genus at some period, but now referred elsewhere, are listed at the end of this paper together with a discussion of the special case of Agromyza albipennis Meigen.

Identification

Misidentifications have been rife in Desmometopa, probably in part because of the great similarity in appearance of many of the species. The two interfrontal plates combine with the fronto-orbital plates and frontal triangle to divide the frontal vitta into an M-shaped black area (M as viewed from in front), and this has led to the frequent misuse of the name D. m-nigrum. An early error in using the name D. tarsalis Loew for an Old World species resulted in many misidentifications, in addition to which the name tarsalis has been widely misapplied in the New World because of failure to realize the large number of similar species in what might be called the "tarsalis complex". In all faunal regions, taxonomists have failed to appreciate the large number of species that actually exist in the genus, aided and abetted by the belief--partly true--that several common scavenger species had been widely distributed in commerce. Minor sources of error have been the inability to distinguish females of singaporensis and varipalpis, compared with their very distinct males, and lack of recognition of the sexual dimorphism in the form of the head in microps.

A few examples will suffice. In one of the great New World museums, I found five species identified as D. m-nigrum: true m-nigrum (most of the specimens, fortunately), a few of sordida, and one each of tarsalis, varipalpis, and Leptometopa latipes (Meigen), the latter probably only a curatorial lapsus. Likewise in one of the great European museums there were five species under m-nigrum: true m-nigrum (again, most of the specimens), sordida, varipalpis, inaurata, and a new species near microps. "D. tarsalis" of Malloch's (1914) report on Sauter's Formosa-Ausbeute proved to be a mixture of singaporensis and microps.

The confused usage of names is vividly illustrated in Hawaii, where four introduced species of Desmometopa are currently known to occur. The first record of the genus from the islands, as far as I know, is that by Illingworth at the Feb. 5, 1925 meeting of the Hawaiian Entomological Society (note published 1926), who published D. m-nigrum as "recently" identified by J. M. Aldrich from specimens reared in 1916 "in abundance from macerated hen manure." True m-nigrum has never been found in Hawaii, however, and Aldrich himself corrected his identification to D. tarsalis (see Illingworth 1929). Hardy and Delfinado (1980) in the "Insects of Hawaii" record Illingworth's 1916 rearing in two places, once under their "singaporensis" (based on an earlier identification by me), which is varipalpis, and once under tarsalis, the original usage of Illingworth (on authority of Aldrich). Both may be in error. The specimens in the U.S. National Museum of Natural History reared by Illingworth and identified by Aldrich are singaporensis as I recognize it, not "singaporensis" of Hardy and Delfinado (see discussion under singaporensis). To complicate the picture still further, both true tarsalis and true singaporensis are now known in Hawaii and one or both might also have been present in Illingworth's material. For what it may indicate, however, true singaporensis was also reared from poultry manure in Hawaii years later by Yoshinori Tanada, and I have no records of varipalpis from manure anywhere, and only one record of tarsalis from manure, from cow manure in Guam.

As a result of such situations, published records with commonly used names must be ignored unless the original specimens can be reexamined, at least outside of the Holarctic Region in which the few species are well known. Ordinarily, for the common species, enough specimens have been available to me that I can give an adequate picture of the distribution without verifying or correcting individual published records. If voucher specimens have been checked, however, I record the fact.

The condition of specimens affects the usefulness of many characters. One must ever be alert for immature (teneral) specimens in which proportions and color are unnatural. In particular, the proportions of the head can be greatly affected, especially the breadth of the cheek and the angle of the vibrissal angle. If the face is collapsed, it becomes more concave as seen in profile, and the vibrissal angle may thus appear to be produced and acute. Identification of such examples should be done with caution and attention to other characteristics.

Morphological Characteristics

All species have a similar habitus and community of structure and chaetotaxy, and a full description under the genus permits rather brief descriptions for the species, with concentration on characters differing most among the species. A few comments on these are in order.

Microtomentose: The dorsum of the thorax and parts of the pleuron are densely gray to brownish gray, over the black ground color. This has usually been referred to as pollinose or pruinose or "dusted", but analysis shows that the areas are actually covered with microscopic outgrowths of the cuticle, sometimes curled or curved like minute microtrichia. The terms pollinose and pruinose seem fundamentally inappropriate, and in recent papers I have used the term "tomentose". However, this may imply to some readers woolly or matted hair, and I suggest, and use here, the term microtomentose.

The frons almost always appears slightly longer than broad at vertex, especially in males. Measurements confirm this, although the quotient for longer than broad is usually not as great as anticipated, often only up to 1.2 times. I have mentioned it only for a few extreme cases. Aside from lengths and widths of interfrontal plates, fronto-orbital plates, and frontal triangle, which need no explanation, the appearance of the M-shaped frontal vitta is a most useful character. In two species, the entire frons is uniformly dull brownish or brownish gray, obscuring all plates and the frontal triangle and making these species unusual and aberrant in the genus. Most other species have the frontal vitta velvet black, sometimes slightly subshining, against which background the other parts stand out distinctly. In a number of species (tarsalis and relatives) most of the frontal vitta is gray microtomentose, although not densely so, from at least certain angles.

The frontal triangle is not a definite sclerotized plate, as in most Chloropidae, but nevertheless the triangular microtomentose area is as regular and consistent as the interfrontal plates. It is a continuation of the microtomentum regularly present on the ocellar tubercle. Occasionally the tomentum barely extends beyond the median ocellus, or it may extend far forward between the interfrontal plates to midway of the frons (fig. 10).

The term cheek (rather than gena) is used here for the combined gena plus subgena below the eye. If gena is correct for the upper part of the cheek, it should not be used for the entire area, even though in these tiny flies the subgena is linear and gena s.str. is almost coextensive with cheek. The breadth (height) of the cheek compared to the breadth of a 3rd antennal segment and to the vertical height of an eye are useful characteristics, as well as the development of a polished area along the lower margin of the eye that I propose to call the subocular crescent (cf. figs. 7 and 14 for extremes). Without actual measurement, one can more easily perceive the proportion of cheek to 3rd antennal segment than that of cheek to the much higher and very convex eye, where optical illusion can deceive. However, the antennae are easily knocked off and are often missing in available specimens that are otherwise in good condition, but the cheek:eye relationship can always be determined.

In the row of subgenal setae along the lower margin of the head, the first or second behind the vibrissa is developed in a few species as a strong upcurved subgenal bristle, e.g., in D. ciliata (fig. 11). In most species, the setae form an even row, gradually becoming longer and stronger towards the vibrissa.

The postorbital (dorsal) and postgenal (ventral) areas are usually narrow but in a few species one or both are broader than usual, and in males of one species (microps, fig. 17) they are convex and appear bulging.

The produced vibrissal angle (fig. 18) is explained under the division into subgenera. The area immediately mesad of the vibrissal angle has no special name and I have referred to it, hopefully accurately albeit long and somewhat awkwardly, as the lateroventral corner of the facial plate. In the subgenus Platophrymyia (fig. 18), it is shining black although not smooth and polished, the shine interrupted by fine lines. The warping forward and upward of this area and the whole epistomal margin is distinct and even exaggerated in large specimens, but less distinct and unimpressive in small specimens.

In most species, each 3rd antennal segment is only a little broader than the 2nd segment, and it is referred to in the descriptions as "small." In three new species, however, the 3rd segment of the male is conspicuously enlarged (fig. 21)(see the supplemental "key to males with unusual features"). Other species presently known only from females may also show this, especially the small species with blackish halteres that seem closely related to the three just mentioned.

The palpi are usually gently clavate (e.g., fig. 18), gradually broadening from base to apex, but in a few species they are broad and flat (fig. 4), unusually long, with the development more striking in large specimens. In a few species, the males have unusually long and distinctive palpi (see special key for such males, and cf. figs. 3 and 5). The color of the palpi is often useful, but one can also be deceived. Species with palpi entirely or almost entirely yellow in both sexes, or black in both sexes, are easy to separate. Between those extremes are species with yellow or predominantly yellow palpi in males but with palpi half or more infuscated in females. Occasional specimens of the latter have the palpi entirely infuscated, and while this apparently involves only a small proportion of the specimens, it does diminish the usefulness of the color as a character for the key. Except for one possible exception (sp. H), the broadly flattened palpus (cf. fig. 4) is found in some species of subgenus <u>Platophrymyia</u>.

The geniculate <u>proboscis</u> appears slender in side view, but in dorsal or ventral view the <u>haustellum</u> is often broadened, especially toward the base. It is usually slightly longer than the lower margin of the head and is mentioned only if unusual. The <u>labella</u> are almost always a little shorter than the haustellum, but because they are soft and their condition of expansion differs greatly among individual specimens it is useless to mention comparative lengths. They almost always appear slender, but in reality they are like a furled sail and occasionally specimens will have them expanded nearly to the width of the oral cavity. This should not be interpreted as a specific distinction.

The <u>propleuron</u> is rounded anteriorly in most <u>Desmometopa</u>, but some show part of a ridge dorsally, just below the humerus. In some, there appears to be a ridge on a line between a gray, microtomentose part and a polished anterior declivity, but this line may have appeared more like a ridge because of drying of the specimen. A strong propleural carina is characteristic of the family Chloropidae, and the ridge in these species of <u>Desmometopa</u> might confuse the unwary. It does suggest relationship between the two families, as indicated by recent authors.

The <u>polished areas on the pleuron</u> (figs. 23–27) did not impress me at first, and indeed such a careful observer as Hennig made no mention of them in his monograph of the European species (1937). I have found these polished areas surprisingly uniform within each species throughout the genus, even in such an apparently insignificant and easily overlooked place as the small concave post-spiracular area. The spots may have been misinterpreted as rubbed areas and therefore of no importance. The polished spot on the pleuron posterodorsad of the fore coxa is sometimes difficult to see because the femur at rest is over it and close to the body. In most species the area is large enough that one can usually glimpse it between femur and pleuron, but in species with a small spot, such as <u>interfrontalis</u>, <u>singaporensis</u>, and <u>varipalpis</u>, one could easily miss it and key to the few species that truly have an entirely microtomentose pleuron. Luckily, <u>interfrontalis</u> has a distinctive pattern on the frons, and the common <u>singaporensis</u> and <u>varipalpis</u> have yellowish cheeks and in the males uniquely distinct palpi, and the chance of misidentification because of missing the pleural spot is almost nil. The figures are somewhat stylized and semi–diagrammatic, but they illustrate the major different types of polished areas.

The <u>color of the tarsi</u> needs particular attention. Tarsi that appear yellowish in ventral aspect may actually be infuscated dorsally. Truly yellow tarsi are yellow viewed from any angle. Specimens that have been mounted out of alcohol will almost always be paler. Even though the sharpness of the character is not always all that could be wished, nevertheless it is often useful to distinguish between entirely infuscated tarsi, chiefly yellow tarsi (except for distal tarsomere or two), and species with fore tarsus infuscated and mid and hind tarsi chiefly yellowish.

Yellow <u>fore coxae</u> set off the <u>flavicoxa</u> group, all Neotropical species. Species with black fore coxae are regularly and unquestionably so, but in very teneral specimens the coxae may not be fully colored and one might think they were yellowish. It is a good general rule to be cautious in dealing with teneral individuals.

In most species, and in almost all females, the fore coxa is short and convex, and the fore femur is not significantly longer than mid and hind femora. In a few species (e.g., saguaro and melanderi), the fore coxa and fore femur are strikingly elongate and raptorial or mantislike in appearance (cf. fig. 22). In others, they are slightly to moderately elongate. Within any given species that shows elongation, large specimens show it more distinctly than small ones.

The halteres usually have the stalk brown, but the knob may be lemon yellow or whitish yellow, or it may be brown to blackish. The color of the knob is actually an excellent specific character, especially for mature and clean specimens. Unfortunately, many specimens, especially those from tropical countries, are cleared from fluid, or are teneral, and the halteres are paler than normal and can be misinterpreted.

The length is variable, and precision is difficult at best, but the approximate length is given in order to distinguish in a rough way between the relatively large and stocky species and the tiny species.

Male genitalia: Males of 32 species were dissected, including 13 of the subgenus Desmometopa and 19 of the subgenus Platophrymyia. The genitalia of all are strikingly similar, including those of such widely different species as varipalpis and singaporensis on the one hand and tarsalis on the other, and they have not been described for each species. The postabdominal and genitalic characters agree with the characterization of Griffiths (1972) in most particulars. The postabdomen and genitalia are completely symmetrical. There is no full pregenital sclerite between the 5th segment and the hypopygium, rather only a lateral band of sclerotization along each side of the epandrium. Spiracles 6 and 7 lie in the membrane alongside each of these bands. The bristly cerci are unusually large, convergent ventrally in dissected specimens, as noted by Griffiths, but in dried specimens the mesal margins are parallel and adjoining so that under ordinary magnification they appear as a single shining slightly convex line. This is quite different in appearance from the female abdomen so that unless the abdomen is damaged or collapsed one can almost always be sure of the sex of a specimen without dissection, even though the genitalia are tiny. The epandrium bears on each side a single surstylus, partly fused with the epandrium. The hypandrium is relatively small and slender, incomplete dorsally, the dorsal ends of the arms of the hypandrium bifid. The aedeagus has a short basiphallus (phallophore of Griffiths) and a long and weakly sclerotized to membranous distal section, the two folding back at rest against the long and slender aedeagal apodeme.

Sternites: Dissection of the male abdomens for genitalia revealed more differences in the shape and setation of the sternites, especially of the 5th, than in the genitalia. In most species examined, the 5th sternite was nearly square or broader than long, but in lucidifrons and kandyensis it was decidedly longer than broad. The 5th sternites in some species showed numerous discal setae, up to 70-80 in 8 or 9 very irregular rows, e.g., in ciliata, while others showed few and sparse discal setae, even as few as 4 or 6 (indistincta and parafacialis). Because of the similarity of the male genitalia, relatively few specimens were dissected and the range of variation in shape and setation cannot be stated positively. I would expect variation in the number of discal setae, but consistency in the general pattern of few vs. many setae.

Biology of the Adults

Adult _Desmometopa_ are recorded as visiting various flowers, sometimes in numbers, and they are also taken occasionally in light traps, including black light, and in Medfly traps, Steiner traps, and "fruitfly traps." Numerous collections of adults of _D_. varipalpis show their attraction to odors: "hospital laboratory" (Reading, Penn.), "adults entering sterile operating and surgery area" (Peru, Ind.: John Sillings), "in hospital operating room" (Ogden, Utah: J.B. Marsh), in a Dairy Cheese room (Clovis, N. Mex.: B. Dictson), "in urinal" (Austin, Tex.: M.R. Wheeler), "over outdoor latrine" (Mona I., West Indies: W.F. Pippin), "in septic tank" (Khartoum, Sudan), "ex latrine" (Saipan), "on mud at edges of sewer effluent beds" (Phoenix, Ariz.), "privy trap" (Savannah, Ga.: H.R. Dodge), "abundant in butcher shop" (Austin, Minn.), and a huge number "collected dead in plastic about trunk containing _Cannabis_ _sativa_ (San Francisco, Calif.: Terry Coddington, J.F. Williams, P.H. Arnaud, Jr.). _D_. singaporensis was commonly "collected on decaying giant African snails" (Palau Is., Koror: C.W. Sabrosky), and this species, as "tarsalis", was collected on Guam on several occasions as "feeding adults from human excrement" (Bohart and Gressitt 1951). In South Africa (Transvaal), _D_. m-nigrum was collected off an Impala carcass (L. Braack), and in Ohio they have occurred in great numbers in poultry houses (C.A. Triplehorn). One record that may be open to some doubt: _D_. varipalpis was reported by employees of a filling station in Riverside, Calif., to be "very annoying and hovering around the faces...and occasionally getting into the eyes." This habit is that of _Hippelates_ flies (eye gnats), and perhaps the wrong flies were collected and charged with being the culprits. On the other hand, a female of _D_. singaporensis was collected in Manila, Philippine Is., October 1928, by R.C. McGregor, who pinned this note to the specimen: "The small fly kept pestering me--tried to get into my eye." Perhaps the records do suggest annoyance at times or under some circumstances.

Another interesting habit of the adults is the phoretic relationships that have been observed with predacious insects and spiders, in which adults of _Desmometopa_--as well as some other milichiids--feed on the juices of the prey. Knab (1915) reviewed a number of these observations, as did Peyerimhoff (1917), who added two observations of small flies on the bodies of the prey of asilids. In one case the small flies covering the body of the prey, a bee, were identified as _D_. m-nigrum. Subsequently, Rabaud (1924) recorded an interesting case of _D_. sordida riding not on the body of a dead bee that was prey, but upon the pollen "pâtée" on the hind tibia of a live bee, and apparently feeding at the pollen paste. Richards (1953) observed _D_. sordida on a dead honey bee being sucked by a reduviid, and he later captured specimens of the same species sitting on the same species of bug that was without prey. Most interesting of all, McMillan (1975) made detailed notes on an unidentified species of _Desmometopa_ closely associated with spiders in Western Australia, and called by him "cleaning flies" because of their habits. They not only congregated in numbers on the spiders' prey (bees and cicadas) but also on the spiders themselves. These had become "wet and sticky around their chelicerae and mouths" from feeding on the bees, and the flies were observed "actually feeding ... actively all over the bases, fangs and mouth." McMillan noted that none of the spiders observed attempted to capture or rid themselves of the flies and in fact seemed "to actively cooperate with them in making the cleaning easier by opening their chelicerae."

There may be other instances in the literature, but I have not attempted an exhaustive search for present purposes. I can add seven records from material that I have identified:

_D_. floridensis n. sp.: Lake Worth, Fla., "on asilid prey" (S.W. Bromley).

D. m-nigrum: San Diego, Calif., "feeding at Apis wrapped up by Metargiope spider" (F.X. Williams); Maadi, near Cairo, Egypt, numerous (ca. 75) on a honey bee captured by a spider, Thomisus sp., "hovering in a cluster very closely over bee thorax and sometimes crawling or resting briefly on it, as if to obtain a fluid" (Harry Hoogstraal).

D. sordida: France, "Sur Abeille captura par Harpactor" (Hemiptera, Reduviidae) (H. Manwal); England, with two honeybees labeled "sucked by Desmometopa (L. Parmenter).

D. tarsalis: Jamaica, "flies which attack a large spider Nephila clavipes (H.G. Hubbard); Panama, Canal Zone, "pentatomid" (Michael Robinson).

Biology of the Larvae

Rearing records of Desmometopa show feeding on a wide variety of spoiled, decaying, or rotten plant material, with rare exceptions. Specimens have been personally identified or verified except as noted.

From manure, dung, and sewage:

D. inaurata: poultry: Hawaii (Y. Tanada), Samoa (P.A. Buxton, G. H. Hopkins), Nyasaland (i.e., Malawi) (W.A. Lamborn).
  horse: Samoa (Buxton and Hopkins).
  "manure": Guyana (F.A. Squire).
D. m-nigrum: chicken: Auburn, Ala., and Montgomery Co., Va. (G. Breeden).
D. singaporensis: cattle: Pakistan (L.S. Sohi), Guam (Bohart and Gressitt 1951, as "tarsalis," "in moist cattle excrement, both fresh droppings and when piled as manure.").
  poultry: Hawaii (J.F. Illingworth, Y. Tanada), Samoa (Buxton and Hopkins).
D. sordida: cow: Dallas, Tex. (F.C. Pratt).
D. tarsalis: cow: Guam (J.L. Gressitt).
D. varipalpis: in sewage water: Dade Co., Fla. (J. Porter); in septic tank: Khartoum, Sudan (H.W. Bedford); "breeding in traps of sinks": Brookings, S. Dak. (H. C. Severin); "millions...breeding on the bio-filters" in "trickle sewage filter": Lafayette, Ind. (G.L. Walker); "eri dung" (i.e., feces of Attacus sp., probably A. ricini, a wild silkworm): Coimbatore, South India (Y. Rao).
D. sp.: from stable manure and from toilet pools: Sendai, Japan (Kato and Hori 1952; unverified).

From plant material:

D. ciliata (unverified): from African boxthorn berries (Nikitin 1965).
D. gressitti: "ex papaya log": Truk (R.W.L. Potts).
D. inaurata: "from pods of Inga ingoides infested with olethreutid and cosmopterygid larvae": Dominica (J.F.G. Clarke); "from larvae feeding on rotten cow pea seed": Fiji (W. Greenwood); "from over-ripe coffee cherries": Kenya (T.J. Anderson); "ex maize cob": Sierra Leone (E. Hargreaves); "ex avocado pear fruit": Sierra Leone (E. Hargreaves): "bred from decaying banana skins": Uganda (E.G. Gibbins).
D. interfrontalis: "palm log": Liberia (C.C. Blickenstaff); "reared from rotting lettuce": N. Nigeria (J.C. Deeming); "bred from decaying banana skins": Uganda (E.G. Gibbins).
D. magnicornis: "reared cacao pods": Ibadan, Nigeria (R.W. Williams).
D. melanderi: reared from Opuntia cacti: San Dimas Canyon, Los Angeles Co., Calif. (C.P. Christianson, J.P. Fonseca).

D. meridionalis: "ex rotting Jack fruit": Bahia, Brazil (J.A. Winder).

D. nearctica: "reared from grass": Coachella, Calif. (D.G. Hall, Sr.).

D. saguaro: "ex rotting Saguaro": Pima Co., Ariz. (F.J. Santana); reared from Opuntia cacti: San Dimas Canyon, Los Angeles Co., Calif. (R.E. Ryckman).

D. singaporensis: "from rotten onions" and "from rotten pawpaw stem": Darwin, N.Terr., Australia (G.F. Hill); "from onions imported from Australia": Fiji (H.W. Simmonds); "decaying stump of a papaya tree": Guam, as "tarsalis" (Bohart and Gressitt 1951); "ex Pomalo fruit": Malaya (G.H. Corbett); "larvae feeding on decaying inflorescence of Areca catechu": Malaya (G.H. Corbett); "ex decaying leaves of Brassica oleracea": Malaya; "ex rotten Solanum tuberosum": Malaya; "bred from decaying banana skins": Uganda (E.G. Gibbins).

D. sordida: "reared from grass silage": East Lansing, Mich.

D. tarsalis: "reared from decaying Cereus gigantea": Wickenburg, Ariz. (R.E. Ryckman and C.T. Ames)(Ryckman and Ames 1953); "reared in pond weed": Guam (G.E. Bohart and J.L. Gressitt); "emerged from decaying barrel cactus": Jamaica (E.F. Legner).

D. varipalpis: "reared ex rotting Saguaro": Pima Co., Ariz. (F.J. Santana); "reared from decaying head lettuce": Coachella, Calif. (D.G. Hall, Sr.); "larvae in potatoes": New York City, N.Y. (in ship from Argentina); "ex potatoes": Algiers, Algeria; "fr[om] blue figs": Jerusalem, Israel (J.H. Brair); "reared ex damaged sugar beet roots": Khorassan, Iran (Mir Salavatian); "from decaying melons": Khartoum, Sudan (R. Cottam); "ex rotting potato": Kinshasa, Zaire (M. Wanson); "from rotting mustard stem": Coimbatore, South India; "from rotting pomegranate": Coimbatore, South India; "from larvae on rotting pumpkin": Coimbatore, South India (Fletcher).

D. sp. (woldai?): "reared from Pachycereus pringlei": Baja Calif., Mexico (R.E. Ryckman et al.).

D. sp.: "ex rotting Jack fruit": Bahia, Brazil (J.A. Winder).

Miscellaneous food media

D. inaurata: "ex locust eggs": Zimbabwe (A. Cuthbertson); rotting snails and Drosophila pupa: Hawaii (Hardy and Delfinado 1980).

D. leptometopoides: "reared from mud and debris collected from pools": Accra, Ghana (J.W. Scott Macfie).

D. m-nigrum (not checked; probably tarsalis): "reared from water contained in the axils of the large bracts of decaying Heliconia blossoms": Hawaii (Swezey 1952).

D. singaporensis: "from dead cat": Samoa (P.A. Buxton and G.H. Hopkins).

D. tarsalis (unverified): "breeding in material, possibly bone meal with molasses added, set out for cattle to lick": Oahu, Hawaii (J.L. Gressitt)(Gressitt 1956).

D. varipalpis: "in cadelle culture": Montreal, Canada; 'found living around and depositing their eggs on a fungus growing on formerly preserved and dried sheep hearts': Commerce, Texas (E.C. Hancock); 'found breeding in enormous numbers in a vermiculite-alfalfa meal-brewers' yeast mixture used as a breeding medium for eye gnats, Hippelates flies': Riverside, Calif. (Mulla and Barnes 1957).

D. sp. H: "des galeries du Cossus": Algeria (P. Lesne).

Sources of Material

For convenience of reference, museums and collections are referred to by the name of the city, enclosed in brackets. I am indebted for specimens and assistance to the individuals named.

Amsterdam: Zöologisch Museum (G. Kruseman, Jr.).

Austin (Texas): M.R. Wheeler personal collection.

Berkeley: University of California, Dept. of Entomology (the late Paul Hurd).

Berlin: Zoologisches Museum, Museum für Naturkunde der Humboldt-Universität (H. Schumann).

Budapest: Zoological Section, Hungarian Natural History Museum (F. Mihályi, L. Papp, A. Soós).

Cambridge (Mass.): Museum of Comparative Zoology, Harvard University (N.E. Woodley).

Colombo: Dept. of National Museums, Sri Lanka (through K.V. Krombein).

East Lansing (Mich.): Dept. of Entomology, Michigan State University (repository of R.R. Dreisbach Collection)(the late R.R. Dreisbach).

Eberswalde: Institut für Pflanzenschutzforschung, Bereich Eberswalde, German Democratic Republic (formerly Deutsches Entomologisches Institut, Berlin-Dahlem)(the late W. Hennig, G. Morge).

Gainesville (Fla.): Florida State Collection of Arthropods) (H.V. Weems, Jr.).

Helsinki: Zoological Museum, University of Helsinki, Finland (B. Lindeberg, W. Hackman).

Honolulu: Bernice P. Bishop Museum (the late J.L. Gressitt, N. Evenhuis).

Lansing (Mich.): W.L. Downes personal collection.

Logan (Utah): Dept. of Entomology, Utah State University (W.J. Hanson).

Loma Linda (Calif.): College of Medical Evangelists (R.E. Ryckman).

London: Dept. of Entomology, British Museum (Nat. Hist.) (B. Cogan), including material from the Commonwealth Institute of Entomology (the late F. van Emden).

Ludwigsburg: Staatl. Museum für Naturkunde in Stuttgart, Zweigstelle Ludwigsburg, West Germany (B. Herting).

Lund: Museum of Zoology, University of Lund, Sweden (H. Andersson, R. Danielsson).

Ottawa: Canadian National Collection (J.F. McAlpine).

Paris: Museum National d'Histoire Naturelle, Entomologie (L. Matile, L. Tsacas).

San Francisco: California Academy of Sciences (P.H. Arnaud, Jr.).

Stockholm: Naturhistoriska Riksmuseum, Entomologiska Avdelningen (R. Malaise).

Sydney: School of Public Health and Tropical Medicine, University of Sydney, Australia.

Tucson: Dept. of Entomology, University of Arizona (F.J. Santana).

Tucumán: Instituto Miguel Lillo, Tucumán, Argentina (the late M. Aczél).

Vienna: Naturhistorisches Museum Wien (Ruth Contreras-Lichtenberg).

Washington (D.C.): U.S. National Museum of Natural History (including material from Henk Wolda, Smithsonian Tropical Research Institute, Balboa, Panama).

Acknowledgments

In the preceding list of "Sources of Material", I have indicated my indebtedness to numerous museums, curators, and individuals for the loan of material or the study of specimens in their collections. Beyond this, however, special thanks are due to individuals who loaned holotypes or entire type series, in some cases waiting for years more or less patiently while my understanding of some difficult complexes was evolving. Having all these types before me simultaneously was of enormous advantage. I acknowledge with deep gratitude the cooperation of B. Herting (Ludwigsburg), G. Kruseman, Jr. (Amsterdam), F. Mihályi (Budapest), G. Morge (Eberswalde), H. Schumann (Berlin), A. Soós (Budapest), and N.E. Woodley (Cambridge). Figures 1–3, 5–11, and 14 were drawn by Kathryn M. Conway, and the author's figures were completed and the plates arranged by Linda Lawrence.

Desmometopa Loew

Desmometopa Loew, 1866, Berlin.Ent.Ztschr. (1865) 9: 184 (Cent. 6, no. 96).
     Two species. Type species, _Agromyza_ _m-atrum_ Meigen, 1830, by designation of Hendel, 1903, Wien. Ent. Ztg. 22: 251, = _D_. _sordida_ (Fallén).
Platophrymyia Williston, 1896, Trans. Entomol. Soc. London 1896: 426. Type
     species, _P_. _nigra_ Williston, 1896, by monotypy. (Synonymy by Sabrosky, 1973).
Desmetopa (error) Hendel, 1902, Wien. Ent. Ztg. 21: 262–4.
Liodesmometopa and _Liodesma_ Duda, 1935, Natuurhist. Maanblad 24: 24, 25.
     Subgenus of _Desmometopa_. Type species, _D_. _atra_ Duda, 1935, by original designation, = _D_. _sordida_ (Fallén)(Synonymy queried by Hennig 1937, confirmed here).

There is great uniformity of color, structure, chaetotaxy, and general habitus among the species of this genus, and the characters given here are usually not mentioned further in the individual descriptions.
     Small flies (1–2.5 mm), black in ground color, sometimes entirely so, some species with yellow to orange-yellow color on antennae, palpi, propleura, halter knobs, fore coxae, and tarsi, especially mid and hind tarsi.
     Frons with black M-shaped frontal vitta delineated by the frontal triangle and the fronto-orbital and interfrontal plates (e.g., figs. 1,2,10); each fronto-orbital plate with 2 lateroclinate upper orbital and 2 mesoclinate lower orbital bristles; inner and outer vertical, ocellar, and postocellar bristles strong; a row of 4 to 6 short setae on each interfrontal plate. Eye large. Cheek usually narrow, 1/3–1/2 the breadth of a 3rd antennal segment, but broad in a few species like _D_. _m-nigrum_. Postorbital and postgenal areas usually narrow. Distinct lunule projecting between bases of antennae. Face with more or less distinct median carina, and oral margin warped forward and upward, leaving subcircular and concave antennal grooves (foveae). Vibrissal angle either strongly produced forward to about a 45° angle (fig. 18), with face decidedly concave in profile (subgenus _Platophrymyia_), or not so produced, the angle about 80° to 90° and face not or weakly concave in profile (figs. 6–8) (subgenus _Desmometopa_). Palpus usually moderately large, clavate, occasionally broad and flat (fig. 4), and even capitate or elongate fusiform in males of species showing sexual dimorphism (figs. 3,5). Proboscis commonly slender, geniculate, haustellum often as long or longer than lower margin of head, labella usually as long as haustellum or nearly so. Antenna usually small, 3rd segment only a little larger than 2nd segment; arista slender, micropubescent.
     Thorax black, subshining dorsally, densely microtomentose, usually gray to brownish gray, brighter gray on pleuron, the pleuron occasionally entirely gray microtomentose (four species) but in most species with some bare and polished areas (figs. 23–27), in a few species almost entirely polished. Chaetotaxy: 1 humeral, 1 + 1 notopleural, 1 presutural, 1 postalar, 1 dorsocentral, 2 scutellar, 1 small propleural, and 1 sternopleural pairs of strong bristles; a 2nd short

dorsocentral anterior to but close to the strong dorsocentral and usually not conspicuous; prescutellar, supra-alar, and intra-alar bristles are much weaker and usually not noticeable but may sometimes be developed. Mesonotum densely haired but scutellum bare; pleuron predominantly bare, only the sternopleuron with some hairs.

Legs usually slender and without striking characteristics except for a few species with fore coxa elongate in males, and sometimes fore femur as well, raptorial in appearance (e.g., fig. 22).

Wing venation differing little throughout the genus (cf. Curran, 1934, p. 337, fig. 16; Hennig, 1937, fig. 36), the 2nd vein and 2nd sector of costa very long, 2nd, 3rd, and 4th veins ending near apex of wing, and usually the 3rd and 4th veins slightly convergent toward apex of wing; costa extending to 4th vein (M), with both humeral and subcostal breaks, the costa between the breaks with up to 16 erect to semi-erect dorsal setae, usually fine but in a few species (ciliata, etc.) coarse and well spaced.

KEY TO THE WORLD SPECIES OF Desmometopa

(For cautionary remarks on interpreting some characters, see the introductory section on "Morphological Characteristics." Bracketed characters may be useful, in addition to regular parts of a couplet. This key is followed by a "Key to males with unusual characteristics.")

1.   Pleuron of thorax entirely gray microtomentose [Note: If thinly so, as in
         lucidifrons, concave areas such as anterior slope of sternopleuron will be
         shining and the microtomentum difficult to see] . . . . 2.
--   Pleuron with polished black spot posterodorsad to fore coxa, entirely or
         chiefly on anterior slope of sternopleuron, in a few species pleuron
         extensively polished . . . . 5.
2.   Cheek with broad polished subocular crescent that is 2/5 height of cheek,
         extending as broad band along entire lower margin of eye (fig. 15); [all tarsi
         yellowish except distally] (Gambia, Nigeria). . . . . 1. D. pleuralis, n. sp.
--   Not so, cheek with narrower subocular crescent that is either linear or
         widened anteriorly (figs. 14, 20) . . . . .3.
3.   Entire frons dull, heavily gray to brownish gray microtomentose, viewed from
         most angles the interfrontal and fronto-orbital plates only weakly
         demarcated; lunule yellow, large and long, extending to or beyond apices of
         2nd antennal segments (Nigeria) . . . . .2. D. nigeriae, n. sp.
--   Not so, frontal vitta entirely or chiefly subshining velvet black, at least
         frontal triangle and fronto-orbital plates sharply demarcated; lunule
         blackish, small, shorter than preceding . . . . 4.
4.   Mesonotum bright gray microtomentose with yellowish to golden cast; all
         tarsi infuscated; subocular crescent linear (fig. 14); frons with M-shaped
         frontal vitta subshining velvet black, interfrontal plates strong and distinct
         (widespread) . . . . 3. D. inaurata Lamb
--   Mesonotum dark gray to brownish gray microtomentose; mid and hind tarsi
         yellow except distal tarsomere or two; subocular crescent widened
         anteriorly (fig. 20); frons with anterior 2/5 glistening, interfrontal plates
         weak and obscure, evanescent (Trinidad). . . . . 4. D. lucidifrons, n. sp.

5.* Cheek exceptionally broad for _Desmometopa_, appearing equal to breadth of
      3rd antennal segment or nearly so and often with broad and triangular
      polished subocular crescent (fig. 7), or postgenal area broad (fig. 17), or
      both, postorbital area also broad in males of two species . . . . 6.
      *[Note: A possible new species from Algeria, sp. H, tentatively associated
      with the broad-cheeked species, is represented by two teneral males and
      may not belong here. It has a narrow subocular crescent and somewhat
      broad and flat palpus (cf. fig. 4). If it should prove to have the cheek
      obviously narrower than 3rd antennal segment, it would pass to couplet 43,
      but it agrees with neither of the species there.]
--   Cheek not so, much narrower and obviously less than breadth of 3rd antennal
      segment, any shining subocular crescent usually linear (figs. 3,5,11), broad
      in only a few species; postgenal and postorbital areas always narrow . . .10.
6.   Thoracic pleuron chiefly polished, including entire propleuron and area almost
      surrounding anterior spiracle (fig. 28) (Texas, Calif.) . . . .
      . . . . 5. _D._ _latigena_, n. sp.
--   Thoracic pleuron predominantly dull, gray microtomentose, including entire
      propleuron and area surrounding anterior spiracle . . . . .7.
7.   Palpus yellow on basal third to half or more, slender clavate in both sexes,
      gradually broadening from base to apex . . . . 8.
--   Palpus black, somewhat broad and flat distally in male (not as extreme as fig.
      4); [halter brown; polished spot on pleuron bilobed] (Algeria, 2 males). . . . .
      . . . . .6. _D._ sp. _H_
8.   Polished spot on pleuron relatively large, bilobed, the dorsal lobe an
      elongate-oval anteroventral area of mesopleuron (fig. 23); knob of halter
      yellow . . . . .9.
--   Polished spot on pleuron relatively small, not bilobed, not with adjoining
      polished area on mesopleuron; knob of halter brownish; [postgenal area
      wide in both sexes, and in male the postorbital area broad to vertex, both
      areas convex and bulging (fig. 17)] (Afrotropical and Oriental Regions, to
      Guam) . . . . 7. _D._ _microps_ Lamb
9.   Postorbital and postgenal areas in male exceptionally broad, shining black,
      the former fairly broad up to vertical bristles, the latter continued forward
      nearly to vibrissa as a band nearly half as broad as cheek (fig. 13), in
      female the band distinct but postorbital and postgenal areas narrower
      (Kenya, Uganda) . . . . .8. _D._ _postorbitalis_, n. sp.
--   Postgenal area only slightly broadened and not continued forward as a broad
      band, the subocular shining area triangular, the postorbit narrowed dorsally
      (fig. 7) (widespread, especially Holarctic Region). . . . .
      . . . . 9. _D._ _m-nigrum_ (Zett.)
10.   Each fronto-orbital and interfrontal plate wide, and frontal triangle long,
      hence the black M of frontal vitta with exceptionally narrow sections, at
      least inner ones little over half as wide as an interfrontal plate (fig. 10)
      (Afrotropical). . . . . 10. _D._ _interfrontalis_ Sabr.
      Each fronto-orbital and interfrontal plate narrower and frontal triangle
      shorter, the black M of frontal vitta more conspicuous (except in
      obscurifrons), each section almost always equal to or wider than a plate . .
      . . . . 11.
11.   Fore coxa yellow or predominantly so, contrasting with black thorax, often
      elongate in male but not or only slightly elongate in female; [vibrissal
      angle strongly produced, to a 45° angle] (_flavicoxa_ group of subgenus
      Platophrymyia)(Neotropical spp.) . . . . 12.
--   Fore coxa entirely black or brown-black. . . . . .23.

12. Pleuron chiefly bright gray microtomentose with large polished black spot posterodorsad to fore coxa and including the adjoining anteroventral area of mesopleuron (as in fig. 24); propleuron and areas surrounding anterior spiracle entirely and heavily gray microtomentose, and mesopleuron predominantly so; propleuron black in ground color . . . . 13.
-- Pleuron chiefly, sometimes almost entirely, polished black (cf. fig. 27), including propleuron (at least ventrally), or areas surrounding anterior spiracle, or both, and mesopleuron entirely or chiefly; propleuron orange-yellow in some species . . . . .16.
13. Interfrontal plates notably short, narrow and evanescent, each appearing to consist of 2 to several small, separate microtomentose spots surrounding bases of interfrontal setae (Panama). . . . . .11. D. evanescens, n. sp.
-- Not so, plates distinct, if short the frons approximately square . . . . .14.
14. Frons clearly longer than broad (about 1.25x), entirely black . . . . 15.
-- Frons relatively broad, approximately square, at least in female (male unknown), narrowly yellow along anterior margin (s. Brazil), 1 female). . . .
. . . . 12. D. sp. I
15. Male: Palpus entirely black (Neotropical) . . . . .13. D. woldai, n. sp.
[See woldai for possible species that key here]
-- Male: Palpus orange-yellow on more than basal half (Argentina) . . . .
. . . . .14. D. flavicoxa Hendel
[Note: The holotype of flavicoxa, a female, has black palpi, but the male could be either black or orange-yellow. I have assumed the latter, from an available male from Argentina. Thus far I cannot separate or satisfactorily associate females of woldai and flavicoxa.]
16. Frontal vitta shining or subshining velvet-black, viewed at any angle . . . . .17.
-- Frontal vitta changeable, velvet-black at some angles, but at others the anterior 2/5 dull brownish gray like frontal triangle; [pleuron almost entirely polished anterior to pleural suture; propleuron orange-yellow; halter knob yellow; all tarsi black] (Panama; 1 male) . . . . .15. D. sp. J
17. Knob of halter yellow or whitish yellow . . . . .18.
-- Knob of halter brown-black, concolorous with stalk; [pleuron chiefly polished, mesopleuron entirely so] (Argentina) . . . . 16. D. nigrohalteralis, n. sp.
18. Small area behind anterior spiracle gray, microtomentose, sometimes continuous with broad or narrow band of microtomentum along dorsal and posterior margins of mesopleuron (fig. 27). . . . . 19.
-- Postspiracular area polished, as is mesopleuron almost entirely. . . . . 21.
19. Palpus yellow on basal half or more, gently clavate in both sexes; propleuron black or chiefly so. . . . . .20.
-- Palpus entirely black in male (female unknown), broad and flat (cf. fig. 4); propleuron orange-yellow (s. Brazil, 1 male). . . . . 17. D. sp. K
20. Mesopleuron chiefly polished black, sometimes appearing entirelyso, but postspiracular depression always with small patch of gray microtomentum and usually a narrow band of same along posterior margin of mesopleuron (fig. 27), occasionally the two narrowly connected along dorsal margin (Belize, Mexico, Panama). . . . . .18. D. glaucanota, n. sp.
-- Mesopleuron with a broad band of gray microtomentum dorsally and posteriorly that extends across anterior spiracle and propleuron (cf. fig. 26) (Florida, Georgia) . . . . .19. D. floridensis, n. sp.

21. Frons shining black; interfrontal plates without microtomentum, shining, only weakly distinguished from the shining frons; antennal fovea thinly microtomentose and rather shining; pteropleuron chiefly polished, at least on lower half; [male with palpus broad and flat (cf. fig. 4)] (3 males, Peru and Costa Rica, have orange-yellow propleuron; 2 females, Colombia and Ecuador, have propleuron black). . . . . 20. D. indistincta, n. sp.
-- Frontal vitta subshining, velvet black; interfrontal plates microtomentose, well marked against velvet black vitta; antennal fovea densely bright gray microtomentose; pteropleuron entirely gray microtomentose. . . . . 22.
22. Male with propleuron strikingly orange-yellow and palpus broad and flat (cf. fig. 4); female apparently distinguishable only by geographic distribution and association with male (Argentina, Brazil, Uruguay, Bolivia, Peru) . . . .
. . . . .21. D. meridionalis, n. sp.
-- Male with propleuron black and palpus clavate; female as noted in preceding (Mexico to Panama) . . . . 22. D. blantoni, n. sp.
23. Abdomen entirely black. . . . . 24.
-- In male, margin of 4th tergum, all of 5th, and all terminalia orange-yellow; female unknown but assumed to show some color at apex of abdomen (Palau Is.). . . . . 23. D. terminalis, n. sp.
24. Thoracic pleuron predominantly dull or subshining, gray microtomentose, including entire propleuron and area surrounding anterior spiracle (cf. fig. 23) . . . . .25.
-- Propleuron chiefly polished black (fig. 26), rest of thoracic pleuron anterior to pleural suture usually predominantly so . . . . 47.
25. Frontal vitta dull and gray to brownish gray, viewed at most angles, interfrontal plates only weakly contrasting . . . . 26.
-- Frontal vitta subshining, velvet black, the gray interfrontal plates sharply distinct, or frons shining on anterior half . . . . 28.
26. Lunule, antenna, palpus chiefly, and proboscis black . . . . 27.
-- Lunule, antenna in part, palpus chiefly, and proboscis yellow (Neotropical)
. . . . .24. D. obscurifrons, n. sp.
27. Epistomal margin and lateroventral corner of facial plate strongly warped forward well in advance of and accentuating the vibrissal angle (cf. fig. 18); [male with fore coxa and fore femur elongate] (Texas, Mexico) . . . . . .
. . . . .25. D. parafacialis, n. sp.
-- Epistomal margin and lateroventral corner of facial plate only weakly warped forward, the vibrissal angle about a 70° angle (fig. 12); parafacial midway not or only linearly visible in profile; cheek with comparatively broad polished subocular crescent, half as broad as cheek; [male unknown but probably fore coxa and fore femur not elongate] (Panama, Ecuador, Peru; 2 females, s. Brazil and Trinidad, key here and may be conspecific) . . . .
. . . . .26. D. atypica, n. sp.
28. Vibrissal angle not produced anteriorly, the angle 80° to 90° (figs. 3,5,11), face only weakly concave in profile; lateroventral corner of facial plate dull gray like rest of face, and not warped forward . . . . 29.
-- Vibrissal angle produced anteriorly to about a 45° angle (cf. fig. 18), face deeply concave in profile, emphasized by shining black lateroventral corner of facial plate, mesad of vibrissal angle, which is warped forward often beyond vibrissal angle . . . . 46.
*29. Knob of halter yellow. . . . . 30.
-- Knob of halter brown to black . . . . 36.

[*Note: Teneral specimens or those mounted out of fluid may be pale, and
caution must be exercised. Species with brown to blackish halteres usually
retain at least a brownish tint, however, even though the paleness suggests
otherwise. The marginal specimens are all in the tiny species, 1.25–1.5
mm, whereas most of the species with yellow halteres are larger, 2.2–2.5
mm, and considerably bulkier.]

30. Polished spot on thoracic pleuron, posterodorsad to fore coxa, relatively
large, including elongate-oval area along anteroventral margin of
mesopleuron (fig. 24); 2nd subgenal seta behind vibrissa developed as a
strong bristle, subequal to vibrissa (fig. 11) (Australia). . . . .
. . . . .27. D. ciliata Hendel

-- Polished spot on thoracic pleuron relatively small, confined to anterior slope
of sternopleuron, the mesopleuron not polished anteroventrally (fig. 25); no
outstanding subgenal bristle subequal to vibrissa, although in one species
(leptometopoides) the subgenal setae quite long and becoming longer
toward vibrissa, and the second may be stronger than the others. . . . . . .31.

31. Males. . . . . .32.
-- Females. . . . . .34.

32. Hind tibia unusually broad and flat, resembling that of Leptometopa (fig. 19);
palpus clavate, gradually enlarged distally (West Africa; ? Cape Province) .
. . . . .28. (male) D. leptometopoides, n. sp.

-- Hind tibia slender; palpus elongate and greatly broadened distally (figs. 3,5) . .
. . . . .33.

33. Palpus fusiform elongate, often considerably so, tapering to acutely angled
apex (fig. 3); cheek wider than in singaporensis (cf. figs. 3,5) (widespread) .
. . . . 29. (male) D. varipalpis Mall.

-- Palpus capitate, abruptly broadened, rounded distally (fig. 5); cheek narrower
than in varipalpis (cf. figs. 3,5) (widespread) . . . .
. . . . 30. (male) D. singaporensis Kert.
. . . .(syns.: D. tristicula Hendel, D. palpalis Meij.)

34. Cheek comparatively narrow, usually barely over 1/10 the height of an eye
(cf. fig. 5 of male); each fronto-orbital plate not appearing broad, lower
section in particular approximately equal to that of an interfrontal plate
(fig. 2). . . . . .35.

-- Cheek obviously broader, about 1/5 height of eye (fig. 3), each fronto-orbital
plate relatively broad, obviously greater than breadth of an interfrontal
plate (fig. 1)(widespread) . . . . .29.(female) D. varipalpis Mall.

35. Cheek black in ground color, heavily gray microtomentose; frontal triangle
large and interfrontal and fronto-orbital plates relatively broad, the
sections of black M of frontal vitta relatively narrow, posterior arms of
interfrontal plates separated from frontal triangle by approximately their
own width (West Africa; ?Cape Province) . . . .
. . . . .28. (female). D. leptometopoides, n. sp.

-- Cheek yellowish in ground color, thinly microtomentose; sections of black
M-shaped frontal vitta relatively broad (fig. 2), interfrontal plates usually
more widely separated from frontal triangle, by twice their own width
(widespread). . . . . .30. (female) D. singaporensis Kert.
. . . .(syns.: D. tristicula Hendel, D. palpalis Meij.)

36. Subocular crescent narrow, linear to sublinear (cf. figs. 5,14);chiefly tiny
species, 1.25–1.5 mm. . . . . .37.

-- Subocular crescent broadened behind vibrissa, subtriangular (fig. 8); relatively
large and bulky species, 2–2.5 mm (Holarctic). . . . . .31. D. sordida (Fallén)

37. Frontal triangle and ocellar tubercle virtually coextensive, the gray
      microtomentum of triangle barely or not at all extending anterior to
      median ocellus . . . . 38.
  -- Frontal triangle longer, extending into median part of frontal vitta. . . . . .42.
38. Palpus black in both sexes . . . . 39.
  -- Palpus partly orange–yellow, at least on basal third to half. . . . . 40.
39. Third antennal segment of male relatively small, little larger than 2nd
      segment; costa between humeral and subcostal breaks with only 5 erect,
      well–spaced setae (Malaya, 1 male). . . . . 32. D. sp. L
  -- Third antennal segment of male exceptionally large (as in fig. 21); costa
      between breaks with 7–8 dorsal setae (West Africa). . . . .
      . . . . .33. D. nudigena, n. sp.
40. At least mid and hind tarsi yellow except distally. . . . . .41.
  -- All tarsi black (Marshall and Palau Is., New Hebrides; ?Philippines) . . . .
      . . . . 34. D. flavipalpis, n. sp.
41. Palpus heavily infuscated distally and below in both sexes; antenna black in
      both sexes; polished pleural spot bilobed anteriorly (cf. fig. 23)(Sri Lanka;
      ?Philippines). . . . . 35. D. srilankae, n. sp.
  -- Palpus chiefly yellow in male, sometimes infuscated at tip, in female heavily
      infuscated distally; antenna black in female, 3rd segment orange–yellow on
      basoventral half in male; polished pleural spot not bilobed anteriorly
      (Malaya, Java, Thailand). . . . . .36. D. propeciliata, n. sp.
42. At least mid and hind tarsi yellow except distally. . . . . .43.
  -- All tarsi infuscated (Nigeria, Ivory Coast) . . . . .37. D. magnicornis, n. sp.
43. Second subgenal seta behind vibrissa developed as a strong bristle, standing
      out among the shorter setae in subgenal row (cf. fig. 11) . . . . .44.
  -- Subgenal setae even, none developed as an outstanding bristle although setae
      may lengthen gradually toward vibrissa. . . . . .45.
44. Polished subocular crescent distinct, although linear; male unknown (Taiwan,
      1 female). . . . . 38. D. sp. M
  -- Subocular crescent absent or indistinct; 3rd antennal segment of male
      exceptionally large (cf. fig. 21) (Philippine Is.) . . . .
      . . . . .39. D . philippinensis, n. sp.
45. Polished subocular crescent distinct, although linear; palpus black in female
      (male unknown) (Malaya; Philippines). . . . . 40. D. sp. N
  -- Subocular crescent absent or indistinct; palpus yellow in both sexes, slightly
      infuscated at apex (Aldabra). . . . . .41. D. aldabrae, n. sp.
46. Polished pleural spot small, not including any area of mesopleuron (cf. fig.
      25); palpus ordinary, gently clavate; abdominal tergum 5 of male elongate,
      longer than terga 3 and 4 combined (Sri Lanka) . . .42. D. kandyensis, n. sp.
  -- Polished pleural spot relatively large, including anteroventral area of
      mesopleuron (as in fig. 24); palpus of male especially broad and flat (fig. 4);
      abdominal tergum 5 of male not elongate, barely longer than tergum 4
      (South Pacific: Caroline, Gilbert, and Marshall Islands, Guadalcanal, New
      Hebrides) . . . . 43. D. gressitti, n. sp.
47. Mesopleuron microtomentose dorsally and posteriorly, the dorsal band
      extending to anterior spiracle, though sometimes narrowly (as in fig. 26) . .
      . . . . .48.
  -- Mesopleuron extensively polished black, the immediately postspiracular area
      polished (as in fig. 28) . . . . .51.
48. Frons dull gray from most angles of view, the interfrontal plates obscure . . .
      . . . . .49.
  -- Not so, frontal vitta subshining velvet black, interfrontal plates well
      contrasted . . . . .50.

49. Fore leg of male raptorial in appearance, mantislike, the fore coxa and femur elongate, latter obviously longer than other femora and with anteroventral and posteroventral rows of short, even, straight spinelike bristles (fig. 22); fore coxa and femur of female longer than usual, coxa more or less elongate but not as extreme as in male; [wing whitish, veins whitish yellow] (Calif., Ariz.) . . . . 44. D. saguaro, n. sp.
-- Not so, fore coxa and fore femur not elongate (circum-Caribbean, also Hawaii to Solomon Islands) . . . . 45. D. tarsalis Loew
50. Subocular crescent relatively broad, obviously over 1/2 breadth of cheek, broad anteriorly, similar to fig. 20 but usually not triangular; Nearctic (Calif. to Utah and Texas, N.Y. to Kans. and Ga.). . . . .
. . . . 46. D. nearctica, n. sp.
-- Subocular crescent comparatively narrow, obviously less than 1/2 breadth of cheek [Warning: teneral specimens will appear like nearctica]; South Temperate (Argentina; 1 Peru). . . . . 47. D. argentinica, n. sp.
51. Knob of halter lemon yellow; pteropleuron gray microtomentose; frontal vitta usually dull . . . . 52.
-- Knob of halter black; pteropleuron polished black; frontal vitta subshining black, extremely broad, the interfrontal plates narrow and lower fronto-orbital plates reduced to mere lines (Mexico, 1 female). . . . .
. . . . .48. D. sp. O
52. Frontal vitta dull gray-black from above and in front, with velvet black spot along each side of ocellar tubercle . . . . 53.
-- Frontal vitta chiefly subshining velvet black, thinly gray on anterior half between and beside the interfrontal plates (s. Brazil) . . . .
. . . . 49. D. stilbopleura, n. sp.
53. Fore leg of male raptorial in appearance, mantislike, with coxa and femur elongate, fore femur obviously longer and larger than other femora and anteroventrally with row of short, even, stout spines, posteroventrally with row of similar but weaker spines (like fig. 22); fore coxa and femur of female longer than usual, without spine rows, the coxa more or less elongate though not as extreme as in male (Calif. to Texas and Mexico) . . .
. . . . 50. D. melanderi, n. sp.
-- Fore leg not raptorial, fore coxa and femur not elongate, without spines, fore femur of approximately same length as other femora (Argentina). . . . .
. . . . .51. D. aczeli, n. sp.

Key to males with unusual characteristics
(Males, or palpi of males, unknown for atypica, evanescens, nigeriae, stilbopleura, woldai, and spp. I, J, M, N, O)

1. Palpus unusually enlarged and extended (figs. 3,5), chiefly yellow . . . . .2.
-- Palpus not so, either ordinary clavate, or black if broad and flat. . . . . 3.
2. Palpus elongate, fusiform, often considerably so, tapering to rounded but acutely angled apex (fig. 3) (widespread). . . . . 29. D. varipalpis Mall.
-- Palpus capitate, abruptly broadened, rounded distally (fig. 5) (widespread) . . .
. . . . .30. D. singaporensis Kert.
3. Hind tibia greatly broadened, resembling Leptometopa (fig. 19) (West Africa; ? Cape Province) . . . . .28. D. leptometopoides, n. sp.
-- Hind tibia slender. . . . . .4.
4. Postorbital area broadened up to vertex, postgenal area broader than usual (figs. 13,17) . . . . .5.
-- Upper portion of postorbital area narrow, linear, and postgenal area narrow (as in fig. 14) . . . . .6.

5. Postgenal and postorbital areas convex, appearing bulging (fig. 17); polished spot on pleuron relatively small, not bilobed, not with adjoining polished area on mesopleuron (cf. fig. 25); knob of halter brownish (Afrotropical and Oriental Regions, to Guam) . . . . .7. _D. microps_ Lamb

-- Postgenal and postorbital areas flat, not appearing bulging; polished spot on pleuron large, bilobed, the dorsal lobe an elongate-oval anteroventral area of mesopleuron (cf.fig. 23); knob of halter yellow (Kenya, Uganda) (cf. fig. 23). . . . . 8. _D. postorbitalis_, n. sp.

6. Fore leg raptorial in appearance, mantislike, with fore coxa and fore femur elongate, latter obviously longer than other femora and often more or less incrassate (fig. 22). . . . . .50. _D. melanderi_, n. sp.
21. _D. meridionalis_, n. sp.
44. _D. saguaro_, n. sp.

-- Not so, fore leg not raptorial, fore coxa and fore femur not or only slightly elongate . . . . .7.

7. Third antennal segment large (fig. 21); palpus gently clavate . . . .
. . . . .37. _D. magnicornis_, n. sp.
33. _D. nudigena_, n. sp.
39. _D. philippinensis_, n. sp.

-- Third antennal segment small; palpus broad and flat (fig. 4). . . . .
. . . . .43. _gressitti_,
20. _indistincta_, 50. _melanderi_, 21. _meridionalis_, 25. _parafacialis_, 44. _saguaro_, 17. sp. K)

1. _Desmometopa pleuralis_, n. sp.
(Fig. 15)

Pleuron entirely dull gray microtomentose; cheek relatively broad (fig. 15).

Male, female. Chiefly black; palpus entirely yellow in both sexes; knob of halter yellowish; all tarsi yellowish; distal 1 or 2 tarsomeres infuscated on mid and hind tarsi, and distal 3 on fore tarsus.

Frons with M-shaped frontal vitta subshining velvet black, the gray interfrontal and fronto-orbital plates and rather long frontal triangle sharply distinct; cheek relatively broad, approximately equal to or a trifle greater than breadth of 3rd antennal segment and 3/10 the height of an eye, with relatively broad, shining black subocular crescent that is 2/5 breadth of cheek and continues from slightly broader postgenal area (fig. 15); face only weakly concave in profile, vibrissal angle rounded, about a 90° angle, approximately opposite anterior margin of eye; a strong subgenal bristle developed, immediately below and behind vibrissa; 3rd antennal segment small in both sexes; palpus short, gently clavate.

Thorax entirely gray microtomentose, including entire pleuron. Fore coxa not elongate. Length, 1.5 mm.

Holotype male, GAMBIA: Bakau at Tropic Bungalow, "swept in meadow rich in flowers, at the beach," Nov. 7, 1977 [Lund]. Allotype female, NIGERIA: Ibadan, Aug. 25, 1962 (D.C. Eidt; Malaise trap)[Ottawa].

This species differs from most other _Desmometopa_ in having the pleuron entirely dull, without a polished spot. Other species with dull pleuron have narrower cheek and narrower subocular crescent.

The specific name is an adjective derived from the Greek _pleura_, side.

2. _Desmometopa nigeriae_, n. sp.

Frons uniformly dull, heavily brownish gray microtomentose, interfrontal and fronto-orbital plates not strongly contrasting with frontal vitta; pleuron entirely gray, without polished black spot.

Female (male unknown). Chiefly black, with yellow to orange-yellow lunule, 1st and 2nd antennal segments, palpus chiefly, and proboscis; all tarsi yellow except distal segment or 2; knob of halter yellow.

Frons uniformly dull brownish gray from most angles of view, the interfrontal and fronto-orbital plates and frontal triangle scarcely evident, not strongly contrasting with frontal vitta and thus not demarcating the usual black "M" of most species of Desmometopa; interfrontal plates long, posterior ends at level of foremost upper orbital bristles; cheek narrow, 1/3 breadth of 3rd antennal segment and about 1/10-1/11 the height of eye, with linear subocular crescent; face concave in profile, vibrissal angle produced to a 45° angle, accentuated by shining black lateroventral corner of facial plate, which projects even beyond vibrissal angle; 3rd antennal segment small, little larger than 2nd segment; palpus large, clavate; proboscis long and slender.

Thorax entirely dull gray to brownish gray microtomentose, including entire pleuron, with no trace of the usual shining black spot posterodorsad to fore coxa. Fore coxa large but not especially elongate. Length, 2.5 mm.

Holotype female, NIGERIA: Olokemeji, 1914 (J.C. Bridwell) [Washington]. Paratype female, NIGERIA: Ibadan, July 4, 1962 (D.C. Eidt, Malaise trap)[Ottawa].

This species is obviously closely related to the Neotropical obscurifrons, but it differs strikingly by having the pleuron entirely gray microtomentose and thus the two are placed some distance apart in the key. D. nigeriae also differs in having the 3rd antennal segment entirely black, but antennal color is probably not significant and may be variable. The holotype has the 2nd antennal segment all orange-yellow, but in the paratype it is blackened dorsally.

The specific name is a noun in the genitive case, from the name of the country, Nigeria.

3. Desmometopa inaurata Lamb
(Fig. 14)

Desmometopa inauratum Lamb, 1914, Trans. Linn. Soc. London, ser. 2 (Zool.), 16: 363 (Seychelles Islands)[London].
D. ciliata sensu Malloch, 1924, Proc. Linn. Soc. N.S. Wales 49: 336 (New South Wales); 1934, Insects of Samoa, Pt. VI (Diptera), Fasc. 8: 327 (Samoa).
D. inauratum; Lamb in Bezzi and Lamb, 1926, Trans. Entomol. Soc. London 1925: 563 (Rodriguez).
D. M-nigrum; Bezzi, 1928, Diptera Brachycera and Athericera of the Fiji Islands [British Mus. (Nat. Hist.)], p. 163.
D. semiaurata Sabrosky, 1958, Stuttg.Beitr. Naturk. 4: 4 (Tanzania) [Ludwigsburg]. N. syn.
D. inaurata; Sabrosky, 1958. loc. cit.: 4 (Tanzania).
D. inaurata; Hardy, 1972, Proc. Hawaiian Entomol. Soc. 21: 160 (Hawaii).
D. inaurata; Hardy and Delfinado, 1980, Insects of Hawaii, 13 (Diptera: Cyclorrhapha III): 354-5 (Hawaii; figs. of head and male genitalia).

Pleuron entirely dull gray microtomentose; mesothorax with the bright gray microtomentun with slightly yellowish to decidedly golden tint.

Male, female. Almost entirely black, yellowish only on antenna and palpus (more so in male than in female), and knob of halter; antenna usually all black in female, occasionally orange-yellow to base of 3rd antennal segment, in male base of 3rd segment largely orange-yellow; palpus chiefly orange-yellow in both sexes, infuscated apicoventrally, more infuscated in female than in male.

Frons with M-shaped frontal vitta subshining velvet black, the fronto-orbital and interfrontal plates, and large frontal triangle gray microtomentose and sharply distinct; upper orbital plate much broader than lower; apex of frontal triangle nearly midway of frons, at level of foremost upper orbital bristles; cheek 1/2 to nearly 3/5 breadth of 3rd antennal segment and 1/8-1/5 the height of an eye, chiefly gray with linear and sometimes scarcely visible subocular crescent (fig. 14); vibrissal angle not produced, the angle about 80° to 90° , and mesad of it the facial plate flat and dull gray, 3rd antennal segment small in both sexes; palpus clavate in both sexes.

Thorax entirely dull gray microtomentose, including entire pleuron; mesonotum bright gray microtomentose, usually with distinctive yellowish to golden tint. Fore coxa and all tarsi infuscated, the fore coxa not elongate. Length, 2.5 mm.

The yellowish to golden tint of the mesonotal microtomentum will distinguish _inaurata_ at once from the other species with all gray pleuron and indeed from all other known species of _Desmometopa_. Dirty or greasy specimens cannot of course be fairly judged. The subocular crescent is narrower than in _pleuralis_ and _lucidifrons_ (cf. figs. 14,15,20). Teneral females might be confused with females of _singaporensis_, but the latter has the polished black area posterodorsad to the fore coxa, as usual in most _Desmometopa_.

After seeing much more material from a wide range of localities, I now believe that _D. semiaurata_ Sabrosky was based on less brightly colored variants of _inaurata_. Moreover, in smaller specimens, called "_semiaurata_", the cheek measures proportionately wider than in larger specimens. The subocular crescent is normally narrow, but in some specimens, probably as a result of a slightly teneral condition, the cheek shows a triangular folding below the polished area, giving an appearance somewhat like that of _D. m-nigrum_.

Distribution: Widespread, suggesting its dissemination in commerce, but it seems curious that I know no records from the Oriental Region. It is widespread in the Afrotropical Region from West Africa (Gambia) to northeast Africa (Sudan) and south to South Africa (Cape Province), plus Rodriguez, Mauritius, and the Seychelles. In the Neotropical Region, I have records from Brazil, Guyana, and El Salvador, and from the West Indies (Bahamas, Puerto Rico, Virgin Islands, Dominica, the Grenadines, and St. Vincent). In Oceanica, I know it from Hawaii, Marquesas Islands, Samoa, Ellice Islands, and Fiji, and from Australia (Queensland, New South Wales, Victoria, and South Australia).

4. _Desmometopa_ lucidifrons, n. sp.
(Fig. 20)

Frons with short, weak, evanescent interfrontal plates, and anterior 2/5 of frons shining.

Male, female. Black, gray to brownish gray microtomentose; knob of halter lemon yellow; mid and hind tarsi yellow except for distal tarsomere or two.

Frons longer than broad (1.2x), with anterior 2/5 glistening, not a smooth and polished appearance but with a peculiar sheen that helps to obscure the interfrontal plates, the posterior 3/5 the usual subshining velvet black, fronto-orbital plates and frontal triangle distinct but interfrontal plates obscure and weak, evanescent, appearing to consist of 3 or 4 small, separate, microtomentose areas surrounding bases of interfrontal setae, the posterior ends of the plates about at level of foremost upper orbital bristles; frontal triangle long and narrow, much longer than broad at base, apex about at level of posterior ends of the plates; cheek narrow, barely over 1/3 breadth of 3rd antennal segment and 1/9 the eye height, with distinct polished subocular crescent that is broadly triangular anteriorly (fig. 20); face deeply concave in profile, the vibrissal angle produced anteriorly nearly to a 45° angle, lateroventral corner of facial plate shining and warped forward to exaggerate the angle; 3rd antennal segment small in both sexes; palpus gently clavate in both sexes, that of male not enlarged; proboscis slender in side view, but from above or below haustellum broadened toward base, about 0.70x the distance between the vibrissae.

Thoracic pleuron gray microtomentose, including entire propleuron and areas surrounding anterior spiracle; anterior slope of sternopleuron, posterodorsad to fore coxa, rather shining but entirely thinly microtomentose, without the customary polished spot of most species of Desmometopa. Fore coxa and fore femur not elongate. Length, 2 mm.

Holotype male, allotype, and 2 paratypes (male, female), TRINIDAD: Simla, Arima Valley, Feb. 6-12, and 20-26 (allotype), 1966 (S.S. and W.D. Duckworth)[Washington].

The peculiar sheen of the anterior part of the frons, which helps to obscure the already short and weak, evanescent interfrontal plates, is a distinctive feature of this species. Also, the species is apparently the only one of the subgenus Platophrymyia in which the pleuron is entirely gray microtomentose, although thinly so and hence somewhat shining on the anterior slope of the sternopleuron. A Panamanian species, evanescens, has a frons suggestive of lucidifrons, but evanescens has a large polished pleural spot.

The specific name is a noun in apposition derived from the Latin verb luceo, to shine, combined with frons.

5. Desmometopa latigena, n. sp.

Cheek as in m-nigrum (fig. 28), broad and subequal to breadth of 3rd antennal segment, with large subtriangular subocular crescent; pleuron anteriorly chiefly polished.

Male, female. Chiefly black or brownish black; antenna and palpus black in both sexes; knob of halter yellow; fore coxa brownish to black; proximal tarsomere or two of mid and hind tarsi yellowish.

Frons with M-shaped frontal vitta dull, rather heavily bluish gray microtomentose, the gray fronto-orbital and interfrontal plates and frontal triangle only moderately distinct, a narrow velvet black spot along each side of frontal triangle, which is long and reaches middle of frons; cheek subequal to breadth of 3rd antennal segment and about 1/4 the eye height, with large subocular crescent 1/2 or more as broad as cheek, approximately as figured for m-nigrum (cf. fig. 7); vibrissal angle acute, nearly a 45° angle, and lateroventral corners of facial plate shining black; 3rd antennal segment small in both sexes; palpus gently clavate in both sexes.

Dorsum of thorax dark gray microtomentose; anterior 1/2 of pleuron chiefly polished, including propleuron, most of mesopleuron, and anterior slope of sternopleuron, the anterior spiracle almost surrounded by polished areas (fig. 28); mesopleuron gray posteriorly. Fore coxa not elongate. Length, 1.7 mm.

Holotype male, allotype, and 13 paratypes (12 males, 1 female), TEXAS: Big Bend National Park, Dagger Flats, 3500 ft., May 11, 1959 (W.R.M. Mason; "Ex Yucca torrei"[Ottawa]; 3 female paratypes, CALIFORNIA: 2, Riverside Co., Cottonwood Spring, Apr. 12, 1950 (P.D. Hurd)[Berkeley], and 1, San Bernardino Co., San Bernardino Mts., Hidden Valley, May 5, 1928 (A.L. Melander) [Washington].

The broad cheek will readily distinguish this species from all other Desmometopa except m-nigrum, sp. H, and postorbitalis, and from these by the predominantly polished pleuron. The California females are far removed from the main series but seem to be conspecific. The fore coxa is definitely black, compared with brownish yellow in the Texas specimens. However, most of the latter appear to be slightly immature and not fully colored.

The specific name is a noun in apposition derived from the Latin latus, broad, combined with gena, cheek.

6. Desmometopa sp. H

Two males from ALGERIA: Edough, August 1907 (P. Lesne; "des galeries du Cossus de Moskat")[Paris], are teneral, and the species is left unnamed until the characters can be properly described. The cheek appears to be broad, which would place it near m-nigrum, but the subocular crescent is narrow and bandlike, unlike m-nigrum and similar species. The 3rd antennal segment is small. The black palpus is almost capitate rather than clavate, broad and flat although not as extreme as figured for gressitti (fig. 4). The black fore coxa is somewhat elongate, and the fore femur distinctly elongate compared to the other femora. As in m-nigrum and postorbitalis, the polished black spot posterodorsad to fore coxa is large, including an anteroventral area of mesopleuron, and bilobed (cf. fig. 23).

7. Desmometopa microps Lamb
(Figs. 16,17)

Desmometopa microps Lamb, 1914, Trans. Linn. Soc. London, ser. 2 (Zool.), 16: 364 (Seychelles Is.)[London].
D. microps; Sabrosky, 1977, in Delfinado and Hardy (eds.), A Catalog of the Diptera of the Oriental Region 3: 271.
D. microps; Sabrosky, 1980, in Crosskey (ed.), Catalogue of the Diptera of the Afrotropical Region, p. 686.
D. tristicula part; Hennig, 1941, Ent. Beihefte aus Berlin-Dahlem 8: 177 [Pilam and Chipun, Formosa, specimens].

Near sordida but with broad postgenal and postorbital areas.
Male, female. Black, only the basal 1/3 of palpus orange-yellow in both sexes.
Frons with distinct interfrontal plates against velvet black frontal vitta; frontal triangle large, equilateral, apex well in advance of median ocellus and anterior to level of posterior ends of interfrontal plates; cheek relatively broad (figs. 16,17), 7/10 the breadth of 3rd antennal segment and 1/4-1/3 the eye height, with broadly rounded (female) to broadly triangular (male) polished black subocular crescent; in male, postorbital area broad and convex up to vertex, with a bulging appearance (fig. 17), as is the postgenal area, but in female the postorbital area narrow but the subshining postgenal area, while relatively narrow compared to that of male, is broader than in other females of the genus (fig. 16); 3rd antennal segment small in both sexes; face weakly concave, vibrissal angle not produced anteriorly, only an 80°-90° angle; palpus moderately long, clavate in both sexes.
Thorax predominantly bright gray microtomentose, brighter gray on sides, including entire propleuron and area around anterior spiracle, with comparatively small polished black spot, posterodorsad to fore coxa, that is not bilobed and does not include an anteroventral area of mesopleuron (as in fig. 25). Fore coxa not elongate. Length, 2-2.5 mm.
This is one of the most distinctive species of the genus in the male sex, but the female is more easily confused except by the experienced eye. In the male,

the postorbital area, which extends broadly to the vertex (fig. 17), is convex and together with the similar postgenal area can best be described as bulging. Females of microps are easily confused with those of sordida, and older records of "sordida" from the Oriental Region seem likely to be microps. The cheeks of the two are similar in breadth and in the appearance of the polished subocular crescent. The postorbital area, so broad in the male, is not obviously different in females from that of sordida, but it is still clearly broader in microps. If the specimen is teneral and the head collapsed, which is all too often in available material, this feature can easily be misinterpreted. The difference in the color of the palpi is a consistent difference, but if the palpi are withdrawn into the oral cavity and the proboscis folded back, the color of the basal halves of the palpi is not at all evident and those of microps might easily be misinterpreted as entirely black.

Although microps is placed in the key near m-nigrum and other broad-cheeked species, it is apparently not closely related. D. latigena has an extensively polished pleuron, and m-nigrum, postorbitalis, and sp. H all have a large, bilobed polished spot that includes an anteroventral area of the mesopleuron (fig. 23).

Distribution: Afrotropical Region (Seychelles, Tanzania, West Cameroun), widespread in Oriental Region and bordering areas of Palearctic Region, from Afghanistan and West Pakistan to Nepal, Manchuria and Japan, south to the Ryukyu Islands (Okinawa), Taiwan, Java, Malaya, Thailand, and Sri Lanka; also Guam.

8. Desmometopa postorbitalis, n. sp.
(Fig. 13)

Cheek broad, subequal to breadth of 3rd antennal segment, but postorbital area broad below and rather broad dorsally up to vertical bristles, and continued forward below eye as a broad subocular band (fig. 13).

Male, female. Chiefly black or brown-black; 1st and 2nd antennal segments and part of 3rd reddish yellow; palpus yellow on basal 1/2 in female, chiefly yellow in male; knob of halter yellow; fore coxa and all tarsi black.

Frons with upper orbital plates obviously broader than lower plates, the latter and the interfrontal plates narrow and the sections of the M-shaped, subshining, velvet black frontal vitta relatively broad; frontal triangle short, apex not quite opposite upper ends of interfrontal plates; cheek broad, subequal to breadth of 3rd antennal segment and over 1/4 the eye height; postorbital area in male broad up to vertical bristles, widening below, postgenal area shining black and continued forward beneath the eye as a broad band 1/2 or more as broad as cheek (fig. 13), in female the postorbital and postgenal areas not as broad as in male; vibrissal angle approximately a 90° angle, the lateroventral corner of facial plate flat and dull gray; 3rd antennal segment small in both sexes; palpus gently clavate, not large.

Thorax dark gray microtomentose; pleuron, posterodorsad to fore coxa, with bilobed polished black spot, including adjoining broad anteroventral marginal area of mesopleuron (as in fig. 23). Fore coxa somewhat elongate, but fore femur only moderately so. Length, 2.25 mm.

Holotype male, allotype, and seven paratypes (3 males, 4 females), UGANDA: Kigezi Province, Mabungo, 6000 ft., Nov. 1934 (J. Ford)[London, paratypes Washington]. Other paratypes: male, female, UGANDA: Kigezi District, Mabungo Camp, 6000 ft., Nov. 18, 1934 (J. Ford)[London, Washington]; male, female, Kigezi District, Mt. Muhavura, Sept. 29, 1934 (F.W. Edwards)[London]; KENYA: male, Molo, Mau Escarpment, 2420 m, Dec. 1911 (Alluaud and Jeannel) [Paris].

This species is superficially similar to _m-nigrum_ and has been confused with it in the past because of the broad cheek. The key and figures will adequately distinguish the two species, especially the males because of the broad upper part of the postorbital area and broader postgenal area in _postorbitalis_. Females alone will be more difficult: the postgenal area is narrower than in the male but there is still the shining band continuing forward below the eye.

The specific name is an adjective referring to the postorbital area.

9. _Desmometopa m-nigrum_ (Zetterstedt)
(Figs. 7,23)

Agromyza M nigrum Zetterstedt, 1848, Dipt. Scand. 7: 2743 (Sweden).

Desmametopa [sic] _niloticum_ Becker, 1903, Mitt. Zool. Mus. Berlin 2: 188 (Nile Valley, Egypt).

Desmometopa m-nigrum; Malloch, 1924, Proc. Linn. Soc. N.S. Wales 49: 336 (Australia).

D. M. nigra; Duda, 1935, Natuurhist. Maandblad 24: 25.

D. m-nigrum of most European and American authors (e.g., Hennig, 1937; Melander, 1913).

With unusually broad cheek and large triangular polished subocular crescent.

Male, female. Predominantly black, including all tarsi, only palpus in part and knob of halter yellow; 1st and 2nd antennal segments and base of 3rd sometimes dark reddish; body chiefly gray to brownish gray microtomentose.

Frons with M-shaped frontal vitta subshining velvet black and especially large and distinct because of narrow gray interfrontal and fronto-orbital plates, the latter narrower on lower orbital plate than on upper; frontal triangle usually not extending quite to middle of frons and ending opposite or only slightly in advance of posterior ends of interfrontal plates; cheek especially broad, appearing equal to breadth of 3rd antennal segment or nearly so, and over 1/4 the height of an eye, with large triangular polished subocular crescent (fig. 7); postgenal and postorbital areas relatively narrow; vibrissal angle an 80° - 90° angle, the latero-ventral corner of facial plate dull gray; face gently concave in profile; 3rd antennal segment small in both sexes; palpus gently clavate, not large.

Thorax dark gray microtomentose, pleuron chiefly so but with large bilobed polished black spot posterodorsad to fore coxa, the spot including an adjoining broad anteroventral marginal area of the mesopleuron (fig. 23). Fore coxa and fore femur not elongate in either sex. Length 2-3 mm.

This well known species is distinctive because of the broad cheek with triangular polished subocular crescent (fig. 7), as figured by Hennig (1937, fig. 34). Most specimens in collections are correctly identified, yet some confusion is possible, as noted in the introduction. Johnson (1913) recorded _m-nigrum_ from Biscayne Bay and Lake Worth, Florida (Mrs. Slosson), as determined by Coquillett. I have not found these specimens, but they may have been _sordida_, judging from a Dallas, Tex. specimen of _sordida_ [Washington] that was identified by Coquillett as _m-nigrum_. Aldrich also at one time misidentified Hawaiian material as _m-nigrum_, but the species is not known to occur there (see discussion under "Identification" in the Introduction).

This species, along with _postorbitalis_ and sp. H, not only has a characteristically broad cheek but also a characteristically bilobed polished spot on the pleuron (fig. 23). _D. latigena_, the fourth member of the group of broad-cheeked species--_microps_ is not really one of these, is quite different in having the pleuron chiefly polished (fig. 28).

D. m-nigrum is virtually cosmopolitan, probably spread in commerce. It occurs widely in the Palearctic Region, but most records are from southern Europe, especially the Mediterranean Subregion. I have also seen material from the Azores and Israel, and from northern Africa (Egypt, Algeria, Morocco). I have seen numerous specimens from the Afrotropical Region, ranging from sub-Saharan countries south to Cape Province, and from the Cape Verde Islands and St. Helena to the Seychelles and Madagascar. In North America it is recorded from Iowa and Michigan to New York and New Hampshire, south through the Atlantic and Gulf Coast States to Texas, and from Arizona and California, as well as Bermuda. Neotropical records are scattered but significant and suggestive of transport in commerce: Mexico, Cuba, Dominica, Barbados, Ecuador, and Chile. In the Oriental Region I have seen it only from India (Assam), Sri Lanka, and Pakistan. I have seen it from Australia, from several localities in New South Wales, as correctly recorded by Malloch (1924). I have not seen it from any of the Pacific Islands, and published records from Fiji by Bezzi (1928) and from Hawaii by Illingworth (1926) and others are misidentifications.

10. Desmometopa interfrontalis Sabrosky
(Figs. 9,10)

Desmometopa interfrontalis Sabrosky, 1965, Stuttg. Beitr. Naturk. 138: 3
　　(Tanzania) [Ludwigsburg].

With broad interfrontal and fronto-orbital plates, and correspondingly reduced sections of the M-shaped frontal vitta.

Male, female. Chiefly black, palpus chiefly yellow, knob of halter yellow, fore coxa yellowish to brown, and all tarsi chiefly yellow.

Frons with interfrontal and fronto-orbital plates broad and frontal triangle long, 1/2 to 2/3 length of frons, so that the sections of M-shaped frontal vitta are greatly narrowed, the inner arms of the M little over half as wide as an interfrontal plate; cheek over 1/2 as broad as 3rd antennal segment and 1/6 the height of an eye, entirely gray, without trace of subocular crescent; vibrissal angle not produced, about a 90° angle; 3rd antennal segment small in both sexes; palpus clavate.

Thorax heavily gray microtomentose, brownish gray on dorsum, brighter gray on pleuron, with small polished black spot posterodorsad to fore coxa (cf. fig. 25), the mesopleuron with anteroventral polished area; fore coxa not elongate. Length, 1.5 mm.

Distribution: Cameroun, Ivory Coast, Liberia, Namibia, Nigeria, Tanzania, Uganda, and Zaire. An exceptionally long series (151 specimens) was "bred from decaying banana skins" at Mulago, Kampala, Uganda, Sept. 30, 1936 (E.G. Gibbins)[London].

The broad interfrontal stripes and parafrontals and the narrowed "M" of the frontal vitta make this species unique in the genus.

11. Desmometopa evanescens, n. sp.

Frontal vitta shining velvet black, the interfrontal plates short, obsolescent.

Female (male unknown). Chiefly black, only the knob of halter and the mid and hind tarsi, except for distal segment or two, yellow.

Frontal vitta shining velvet black from all angles of view, gray frontal triangle and dark gray fronto-orbital plates distinctly delineated, triangle ending at or near middle of frons and at or almost at level of the foremost upper orbital bristles, the plates narrow throughout their length; interfrontal plates short and

weak, posterior ends well anterior to apex of frontal triangle, each plate linear
and usually interrupted, consisting of 2 to several small spots of microtomentum
surrounding bases of interfrontal setae; cheek narrow, slightly over 1/4 the
breadth of 3rd antennal segment and nearly 1/10 the height of an eye, with narrow
polished black subocular crescent that is linear posteriorly but widens anteriorly
(similar to fig. 18, but wider anteriorly); vibrissal angle produced to a 45° angle,
accentuated by shining black lateroventral corner of facial plate warped forward
even beyond the angle; face strongly concave in profile, parafacial not visible; 3rd
antennal segment not large; palpus gently clavate.

Thorax chiefly and heavily dark gray microtomentose; pleuron chiefly so but
with large polished black spot posterodorsad to fore coxa and including adjoining
area of mesopleuron (as in fig. 24), also with a median to posteromedian polished
spot on sternopleuron. Fore coxa not elongate, fore femur slender. Length, 1.5
mm.

Holotype and 8 paratypes, PANAMA: Panamá Province, Las Cumbres,
various dates Jan. 31–Feb. 28, 1981 (holotype Feb. 28)(H. Wolda, at flowers of
Aristolochia)[Washington].

The weak and evanescent interfrontal plates are suggestive of _lucidifrons_ but
that species has entirely gray microtomentose pleuron. Some of the specimens
are slightly teneral, but even the darkest have the fore coxae slightly infuscated
proximally. I believe the species is best associated with the _flavicoxa_ group,
because the fore coxae are at least partly yellowish. Should it be judged
otherwise at couplet 11, _evanescens_ would run easily--at least disregarding the
"sharply distinct" interfrontal plates mentioned in couplet 25, 2nd part--to the
Old World species _kandyensis_ and _gressitti_ in couplet 46, both of which have long
and distinct interfrontal plates.

The specific name is a participle, from the Latin _evanesco_, to vanish or die
away.

12. _Desmometopa_ sp. I

One female, BRAZIL: Sao Pãulo, Nova Teutonia, July 7, 1937 (F. Plaumann)
[Helsinki] apparently represents a distinct species, but I forgo naming it until
material of both sexes is available. The apex of the long frontal triangle is
midway on the frons, approximately on a level with the posterior ends of the short
interfrontal plates and with the foremost upper orbital bristles. The cheek is
narrow, 1/3 the breadth of 3rd antennal segment and 1/9 the height of an eye,
with linear subocular crescent. The face is only moderately concave in profile
because the vibrissal angle is only slightly produced, although almost a 45° angle.
The pleuron is predominantly gray, including the entire propleuron and the areas
surrounding the anterior spiracle; the polished pleural spot is fairly large,
including an elongate-oval anteroventral area of the mesopleuron (as in fig. 24),
and there is a median polished spot on the sternopleuron. The halter knob is
brownish yellow, and I believe it is brown but immature rather than a discolored
yellow. The 3rd antennal segment is small and the fore coxa short, features usual
in females but not necessarily true of the males.

13. _Desmometopa_ _woldai_, n. sp.

Fore coxa yellow, palpus black, and propleuron entirely gray microtomentose.

Female. Chiefly black, with strikingly contrasted yellow fore coxa; knob of
halter yellow; mid and hind tarsi yellow except distally.

Frons with frontal vitta subshining velvet black, the fronto-orbital plates, narrow interfrontal plates, and frontal triangle gray and distinct; frontal triangle moderately long, its apex usually at level of foremost upper orbital bristles; cheek narrow, 1/3 breadth of 3rd antennal segment and 1/10 the height of an eye, with linear polished subocular crescent that slightly widens anteriorly; face deeply concave in profile, vibrissal angle strongly produced to a 45° angle, the lateroventral corner of facial plate shining black and warped forward even beyond the angle; 3rd antennal segment small; palpus clavate.

Thorax bright gray microtomentose; pleuron with large polished black spot posterodorsad to fore coxa, including anteroventral area on mesopleuron (as in fig. 24); sternopleuron with small to large polished spot mesally, just dorsal to apex of mid coxa; propleuron entirely gray microtomentose, and anterior spiracle completely surrounded by gray areas. Fore coxa not elongate, as usual in females. Length, 1.75-2 mm.

Holotype female and 144 paratypes, all females, PANAMA: Las Cumbres, Panamá Province, Nov. 1980-Feb. 1981 (H. Wolda, on flowers of Aristolochia pilosa) [Washington].

The long series of topotypic females permitted a useful evaluation of variation. The characters proved to be very consistent. The polished sternopleural spot was always present, though varying in extent from a small round spot (common) to a large subquadrate spot (uncommon, only a half dozen specimens). One female, not a paratype, had exceptionally short and narrow interfrontal plates, but it is probably only a variant. Some 32 other specimens are more or less damaged and not included in the type series.

Most species of the flavicoxa group have the thoracic pleuron predominantly polished, whereas flavicoxa and woldai have the sternopleuron chiefly gray microtomentose with a mesal polished spot, usually small.

A series of specimens, partly from widely scattered localities without good series, partly specimens in poor condition, will key to woldai but may represent different species. I leave them unidentified for the present. All males have small 3rd antennal segment, ordinary clavate palpus, and slightly elongate fore coxa. The various specimens differ from woldai as follows:

All tarsi yellowish: 3 females, Mexico, Costa Rica, and Colombia.

All tarsi black: 2 males, Panama (Canal Zone).

Sternopleuron entirely dull gray microtomentose: 2 males, Peru, Dominica.

Larger palpus in male: Trinidad, Tobago, Panama (Canal Zone).

Probably woldai: Mexico (Baja Calif., Michoacan), Honduras, Panama (Canal Zone), Colombia, and possibly a male from southern Brazil (Nova Teutonia).

The specific name is in the genitive case. It is dedicated to the collector of the magnificent series, Henk Wolda, who has found many interesting species of Desmometopa and Chloropidae visiting the flowers of Aristolochia in Panama.

14. Desmometopa flavicoxa Hendel

Desmometopa flavicoxa Hendel, 1932, Konowia 11: 143 (n. Argentina) [Ludwigsburg].

Species with narrow cheek, produced vibrissal angle, subshining velvet black frontal vitta, yellow fore coxa, and entirely gray microtomentose propleuron; male (presumed) with partly orange-yellow palpus.

Female (holotype). Chiefly black; antenna and palpus black; knob of halter yellow; legs predominantly black, fore coxa yellow and contrasting strongly with black pleuron; tibiae narrowly yellow basally; mid and hind tarsi chiefly yellow, with distal tarsomere or two infuscated. Length, 2 mm.

Frons with frontal vitta entirely subshining velvet black, with ocellar triangle and fronto-orbital plates gray microtomentose; interfrontal plates distinct but thickly microtomentose, ending posteriorly about midway of frontal vitta and just short of apex of frontal triangle; cheek narrow, not 1/3 breadth of 3rd antennal segment and 1/10 the height of an eye, gray-black with narrow polished subocular crescent that is broader anteriorly; vibrissal angle produced as in tarsalis, the angle about 45°, face decidedly concave in profile, and lateroventral corner of facial plate shining black; 3rd antennal segment small in both sexes.

Thorax heavily dark gray microtomentose except for large polished black spot posterodorsad to fore coxa, including a broad anteroventral area of the mesopleuron (as in fig. 24), and a small polished black spot centrally on the sternopleuron, the propleuron entirely gray and anterior spiracle completely surrounded by gray areas. Fore coxa not elongate.

Through the kindness of Dr. B. Herting, I received for study a specimen that is undoubtedly the holotype, although it was not so marked. The specimen is labeled "Mis. Tacaagle/XI.25. Lindner/D.Chaco-Exped." and "Desmometopa flavicoxa H. [apparently in Hendel's handwriting]/F. Hendel det.", all agreeing with the information published by Hendel, and in the appropriate collection. It agrees with Hendel's description except that it is a female and not a male, which was either a typographical error or an understandable mistake in these small black flies with small genitalia. I have seen only one other specimen that can be associated with it, a male that differs only in having the palpus partly yellow, a not unexpected difference in males, and in even shorter and narrower interfrontal plates.

Male, "Bemgerg" [sic!, probably Puerto Bemberg], Misiones, March 14-30, 1945 (Hayward, Willink, and Golbach)[Tucumán]. As described for female, but interfrontal plates narrower and shorter, not reaching middle of frons and ending well in advance of frontal triangle; palpus orange-yellow on more than basal half, clavate, not broad and flat.

The combination of yellow fore coxa and entirely gray microtomentose propleuron distinguishes this species from all but woldai. The latter has entirely black palpus in the male.

15. _Desmometopa_ sp. J

One male, PANAMA: Almirante, Bocas del Toro Province, Dec. 10, 1952 (F.S. Blanton)[Washington], is in poor condition and will not be named at this time. It appears to represent a distinct species in the flavicoxa group. It is the only one of that group with the frontal vitta dark gray microtomentose on the anterior half. The palpi are missing, unfortunately. The yellow fore coxa is elongate, and the prosternum and propleuron are concolorous with the fore coxa. Other characters that can be noted are the pleuron almost entirely polished, knob of halter yellow, and fore tarsus black. The other tarsi are discolored and may have been reddish or yellowish. Like all others in the flavicoxa group, the vibrissal angle is strongly produced and the face in profile is deeply concave. The 3rd antennal segment is small.

16. _Desmometopa_ nigrohalteralis, n. sp.

Fore coxa yellow, pleuron almost entirely polished, and knob of halter black.

Male, female. Almost entirely black, only fore coxa and mid and hind tarsi, except for distal segment or two, yellow; propleuron pitch black.

Frons with frontal vitta velvet black; frontal triangle large and chiefly polished, with slight microtomentum on ocellar tubercle; interfrontal plates exceptionally short and weak, little more than 1/3 length of frons and not strongly marked; face deeply concave in profile and vibrissal angle produced to a 45° angle; antennal grooves deeply concave and heavily gray microtomentose; cheek narrow, 1/4 breadth of 3rd antennal segment and 1/10 the eye height, with polished subocular crescent that is linear posteriorly but wider anteriorly; 3rd antennal segment small in both sexes; palpus of male somewhat broad and flat but not as extreme as figured for gressitti (cf. fig. 4).

Thoracic pleuron anterior to pleural suture all polished, as is the pteropleuron. Fore coxa of male somewhat elongate but fore femur not so, both slender. Length, 1.75–2 mm.

Holotype male and allotype, ARGENTINA: Villa Padre Monti, Tucumán-Burruyacu, Jan. 17–Feb. 7, 1948 (R. Golbach)[Tucumán].

The black halter will distinguish this species from all others in the flavicoxa group, and indeed from most other species of Desmometopa, in addition to the uncharacteristic polished frontal triangle.

The specific name is an adjective referring to the black halteres.

17. Desmometopa sp. K

This species appears to be distinct, but I leave the lone male unnamed for the present in the hope that additional material will be forthcoming. It is very close to meridionalis in general habitus, with elongate yellow fore coxa, raptorial fore leg, small 3rd antennal segment, and black and broad and flat palpus (cf. fig. 4) but the mesopleuron has a broad gray band of microtomentum dorsally and posteriorly, beginning at the anterior spiracle, and the fore femur is thinly gray microtomentose, whereas meridionalis has polished black mesopleuron, and the posterior surface of the fore femur is shining and chiefly polished.

BRAZIL: male, Rio de Janeiro, Nietheroy, July 20, 1915 (P.G. Russell)[Washington].

18. Desmometopa glaucanota, n. sp.
(Fig. 27)

Fore coxa yellow; pleuron anterior to pleural suture chiefly polished, with isolated postspiracular patch of gray microtomentum.

Male, female. Chiefly black, with palpus yellow on basal 1/2, knob of halter yellow, fore coxa yellow, mid and hind tarsi yellow in female but infuscated dorsally in male.

Frons with M-shaped frontal vitta subshining black and the long interfrontal plates distinctly delineated; frontal triangle long acute, its apex well past the posterior ends of the interfrontal plates; cheek very narrow, 1/4 breadth of 3rd antennal segment and 1/12 height of an eye, with linear subocular crescent that is slightly broader anteriorly; face deeply concave and vibrissal angle strongly produced to a 45° angle, the lateroventral corner of facial plate highly shining black and well warped forward and upward; antennal grooves deeply concave and densely gray microtomentose; 3rd antennal segment small in both sexes; palpus large, clavate.

Thoracic pleuron anterior to pleural suture almost entirely polished, with postspiracular patch of gray microtomentum, and a narrow margin of same gray on mesopleuron posteriorly and occasionally dorsally (fig. 27); propleuron polished on lower 1/2; pteropleuron gray microtomentose. Fore coxa and fore femur in male slender and elongate, not incrassate and raptorial, in female ordinary. Length, 1.75–2 mm.

Holotype male, allotype, and 4 male paratypes, BELIZE: Corozal Town, Aug. 30, 1967 (G. and R. Lacy)[Washington]. Other paratypes [all Washington]: MEXICO: male, Veracruz, Nov. 1963 (N.L.H. Krauss), and female, Territory Quintana Roo, Cancún, July 16, 1974 (D.J. Pletsch). PANAMA: 95 females, Panamá Province, Las Cumbres, various dates 1980-82 (H. Wolda; on flowers of <u>Aristolochia</u>).

In addition to the type series, about 60 other females from Las Cumbres, Panama, not in suitable condition, were identified.

The pleuron is characteristic of this species, with the isolated spot of gray microtomentum in the flat area just behind and a little above the anterior spiracle. This is sometimes only a small patch, and that area often appears shining, especially if the upper margin (notopleural ridge) is projecting and the postspiracular area is thus depressed and concave. One will need to rotate the specimen and view it at different angles to be sure. This spot is usually not connected to any linear strip of microtomentum along the dorsal margin of the mesopleuron.

The pattern of microtomentum on the pleuron is virtually the only difference I find from <u>floridensis</u>. In the long series available, there is almost always a polished break posterior to the postspiracular patch of microtomentum (fig. 27), and rarely is there a continuous narrow band of microtomentum along the dorsal margin of the mesopleuron. In <u>floridensis</u>, on the contrary, both dorsal and posterior bands of microtomentum on the mesopleuron are broad (similar to fig. 26), and the microtomentum on the post-spiracular area does not appear as an isolated patch.

The specific name is a noun in apposition derived from Latin <u>glaucus</u>, gray, plus <u>nota</u>, referring to the spot of gray posterior to the anterior spiracle.

19. <u>Desmometopa floridensis</u>, n. sp.

Fore coxa yellow; vibrissal angle produced; frontal vitta velvet black; pleuron extensively polished but broadly gray microtomentose behind anterior spiracle and along dorsal and posterior borders of mesopleuron.

Male, female. Chiefly black; palpus yellow on approximately basal 1/2; knob of halter and fore coxa yellow; tarsi brown-black in male, but in female mid and hind tarsi yellow except distally.

Frons with M-shaped frontal vitta subshining velvet black; frontal triangle not extending quite to middle of frons, ending approximately opposite posterior ends of the narrow interfrontal plates; cheek narrow, slightly over 1/3 breadth of 3rd antennal segment and 1/9 height of an eye, subocular crescent strongly broadened anteriorly; face deeply concave and vibrissal angle strongly produced to a 45° angle; lateroventral corner of facial plate shining black and oral margin strongly warped forward, especially at midline, shortening face and accentuating the concavity of the antennal grooves; 3rd antennal segment small in both sexes; palpus clavate.

Thorax dark gray microtomentose; pleuron anterior to pleural suture extensively polished black, including part of propleuron, but mesopleuron broadly gray along dorsal and posterior margins, including gray postspiracular area (as in fig. 26). Fore coxa of male slightly elongate. Length, 2 mm.

Holotype male, allotype, and 4 paratypes (2 males, 2 females), FLORIDA: Seminole Co., Sanford, Aug. 14, 1965 (G.W. Desin; Steiner trap)[Washington, paratypes in Gainesville]. Other paratypes [Washington except as noted]: GEORGIA: male, Savannah, Feb. 3, 1954 (H.R. Dodge). FLORIDA: 7 females, Clewiston, June 20, 1953 (M.R. Wheeler)[Austin], 2 females, Vero [Beach], Feb. 25, 1937 (J.R. Malloch); 3 females, Lake Worth, Jan. 16, 1929, "on asilid prey"

("S.W. Bromley Colln."); 3 females, Lee Co., Sanibel Island, May 11, 1973 (W.W. Wirth, Malaise trap); male, Sweetwater, Apr. 15, 1969 (M.J. Kuck, "ragweed"); male, Merritt Island, Mar. 12, 1956 (H.V. Weems, Jr.)[Gainesville].

The pattern of gray microtomentum on the pleuron will distinguish this species from the very similar species glaucanota (see discussion under that species).

The specific name is an adjective derived from the name of the state of Florida, the provenance of most of the specimens in the type series.

20. Desmometopa indistincta, n. sp.

Yellow fore coxa, unusually shining frons, and almost entirely polished pleuron.

Male, female. Chiefly black; propleuron (and sometimes humerus) and fore coxa yellow; mid and hind tarsi yellow except for distal 2 or 3 tarsomeres; knob of halter yellow; palpus sometimes obscurely yellowish toward base.

Frons shining black except for microtomentose ocellar tubercle and parafrontals; interfrontal plates long, shining, without microtomentum and hence only weakly distinguished from frontal vitta, the long frontal triangle likewise shining and not microtomentose anterior to median ocellus; cheek narrow, less than 1/2 breadth of 3rd antennal segment and 1/10 height of an eye, with polished subocular crescent that is broader anteriorly; antennal grooves deeply concave, densely gray microtomentose; face strongly concave in profile, vibrissal angle produced to a 45° angle, and shining black lateroventral corner of facial plate conspicuous and warped forward even beyond vibrissal angle; 3rd antennal segment small in both sexes; palpus strongly clavate, broad and flat distally in male (as in fig. 4), less broad in female.

Thoracic pleuron almost entirely polished, gray microtomentose only on posterior slope and on upper part of pteropleuron, about the wing base. Fore coxa moderately elongate in male but fore leg not raptorial, both ordinary in female. Length, 2-2.25 mm.

Holotype male and male paratype, PERU: Iquitos, March-April 1931 (R.C. Shannon); allotype, COLOMBIA: Rio Raposo, Feb. 1965 (V.H. Lee, light trap). Paratypes: COSTA RICA: male, Turrialba, Nov. 1922 (Pablo Schild)[all Washington]. ECUADOR: female, Pompeya, Napo R., Pastaza, May 14-22, 1965 (L. Peña)[Ottawa].

The shining frons with indistinct interfrontal plates and the almost entirely polished pleuron will distinguish this species from others of the flavicoxa group.

The specific name is an adjective from the Latin indistinctus, indistinct or obscure.

21. Desmometopa meridionalis, n. sp.

With yellow fore coxa, and orange-yellow propleuron in male, sometimes in female; pleuron anterior to pleural suture chiefly polished black.

Male, female. Predominantly black, strikingly marked in male with elongate yellow fore coxa and orange-yellow propleuron; knob of halter yellow; mid and hind tarsi yellow in both sexes except for distal 2 or 3 tarsomeres.

Frons with M-shaped frontal vitta subshining, velvet black, the dark gray interfrontal and fronto-orbital plates and frontal triangle distinct against that background; interfrontal plates extend back only to midlevel of frons and end approximately opposite apex of long frontal triangle; cheek narrow, 1/3 breadth of 3rd antennal segment and 1/9 the height of an eye, with linear subocular crescent that is wider anteriorly; face deeply concave as seen in profile, vibrissal angle produced to a 45° angle; polished black lateroventral corner of facial plate warped forward, and with the strong facial carina leaves deeply concave, heavily gray microtomentose antennal grooves; 3rd antennal segment small in both sexes; palpus of male strongly clavate, broad and flat distally (cf. fig. 4), especially striking in large specimens, palpus of female long but not broadly expanded.

Thoracic pleuron anterior to pleural suture almost completely polished, except for narrow dorsal and posterior margins of mesopleuron and sparse microtomentum on ventral portion of propleuron; postspiracular area polished; pteropleuron gray microtomentose. Fore leg in male raptorial in appearance, fore coxa elongate, 4 times as long as greatest width, with a row of rather strong bristles, and fore femur clearly longer and larger than other femora, with an anteroventral row of short, even, well-spaced bristles; in female, fore femur only slightly elongate. Length, 1.5 - 3 mm (large males).

Holotype male, allotype, and 13 paratypes (11 males, 2 females), BRAZIL: Nova Teutonia, 300-500 m, various dates (holotype, June 1964)(Fritz Plaumann)[Ottawa]. Other paratypes: BRAZIL [all Washington]: 4 males, 3 females, Nova Teutonia, Sept. 1949 (2 females) and Apr. 1950 (F. Plaumann); male, Rio Grande do Sul, Pelotas, Oct. 20, 1956 (C.M. Biezanko); male, Bahia, Itabuna, Apr. 1973 (J.A. Winder); male, 3 females, São Paulo, Maua, May (N.L.H. Krauss). URUGUAY: male, Montevideo, Jan. 25, 1965 (E.F. Legner) [Washington]. PERU: 3 males, Iquitos, Mar.-Apr. 1931 (R.C. Shannon) [Washington]. BOLIVIA: male, Chulumani, Yungas, Dec. 19-25, 1955, 1700 m (L.E. Peña) [Ottawa]. ARGENTINA [Tucumán, except for last 7 specimens]: 8 males, 2 females, Villa Padre Monti, Tucumán-Burruyacu, Jan. 27-Feb. 7, 1948 (R. Golbach); 2 males, 1 female, Santiago del Estero, Montepotrere, Apr. 13, 1952 (A. Willink); female, Mendoza, Vista Flores, Jan. 31, 1950 (M.L. Aczél); 5 males, Tucumán, Villa Padre Monti, Jan. 17-Feb. 7(3) and Mar. 7, 1948 (R. Golbach); male, Tucumán: Alpechiri, Nov. 29, 1946 (R. Golbach); male, Tucumán, San Javier, Nov. 18, 1946 (R. Golbach); male, Tucumán, [locality illegible], Nov. 23-28, 1951 (Aczél & Golbach); 2 males, Salta, San Lorenzo, Jan. 20 (M.L. Aczél); 4 males, Salta, Urundel, Feb. 8-12, 1949 (M.L. Aczél)[Tucumán]; male, Salta, Bella Vista, Embarcación, Apr. 20, 1927 (R.C. Shannon); female, Misiones, Posa, May (N.L.H. Krauss); female, Corrientes, 39 mi. s. Goya, Dec. 13, 1976; male, Jujuy, Zapia, Apr. 10, 1927 (R.C. Shannon)[Washington]; 2 females, Salta, El Carmen, 27 km s. Molinos, 1900 m, Oct. 6, 1968 (L. Peña)[Ottawa]; female, Catamarca, El Arenal, Oct. 3-4, 1968 (L. Peña)[Ottawa].

The males have the propleuron consistently orange-yellow in the fairly long series available, and the palpus broad and flat as in _gressitti_ (cf. fig. 4), and these provide striking differences from _blantoni_ in which the males have black propleuron and ordinary clavate palpus. However, I have found no way to separate females of the two species except by geography. The fore coxa and fore femur of _meridionalis_ are similar to those of _saguaro_ (cf. fig. 22), but narrower.

The specific name is a Latin adjective meaning southern, referring to the geographic distribution of the specimens in the type series.

22. _Desmometopa_ blantoni, n. sp.

Yellow fore coxa, predominantly polished pleuron, and black propleuron.

Male, female. Chiefly black, only fore coxa and knob of halter yellow, and in the female mid and hind tarsi yellowish except for distal tarsomere or two.

Frons with M-shaped frontal vitta subshining, velvet black, the interfrontal and fronto-orbital plates brownish gray and distinct but narrower than usual, the interfrontal plates short, posterior ends about at midlevel of frons and about opposite apex of frontal triangle; triangle mostly shining anterior to median ocellus, the microtomentum usually confined to ocellar tubercle; cheek narrow, 1/4 breadth of 3rd antennal segment and 1/15 height of an eye, with linear subocular crescent that is slightly wider anteriorly; face deeply concave in profile, vibrissal angle produced to a 45° angle (cf. fig. 18), the lateroventral corner of facial plate shining black and warped forward; antennal grooves deeply concave, densely gray microtomentose; 3rd antennal segment small in both sexes; palpus gently clavate in both sexes, only slightly larger in male.

Thoracic pleuron entirely polished anterior to pleural suture, except for sublinear gray margin posteriorly on mesopleuron and sternopleuron; pteropleuron entirely gray microtomentose. Fore coxa moderately elongate in male, nearly 4 times as long as greatest breadth, with short but strong bristles ventrally. Length, 1.5 mm.

Holotype male, allotype, and a female paratype, PANAMA: Canal Zone, Camaron, Ft. Kobbe, June 23, 1952 (F.S. Blanton, light trap) [Washington]. Paratypes [Washington except as noted]: PANAMA : female, Canal Zone, Colón, July 2-14, 1979 (E. Broadhead et al., canopy fogging of Hura crepitans L. in humid forest). COSTA RICA: male, Cartago, Nov. 1965 (N.L.H. Krauss); 2 females, [Farm] La Caja, 8 km w. of San José, 1930 (H. Schmidt)[Eberswalde].
EL SALVADOR: male, San Andrés, Apr. 7, 1952 (P.A. Berry). MEXICO: 3 males, 1 female, Nayarit, 15 km n. of Chapalilla, July 19, 1951 (P.D. Hurd)[Berkeley]; male, Jalisco, Barra de Navidad, Sept. 1965 (N.L.H. Krauss); female, Veracruz, Fortin de Las Flores, June 1964 (F.S. Blanton, light trap).

This species is closest to meridionalis. The males appear to be distinct on the basis of the color of the propleuron and the form of the palpus (see under meridionalis), but I have been unable to separate females of the two species.

The specific name is in the genitive case, and is dedicated to my good friend, F.S. Blanton, who carefully saved much interesting material from his years in the Panama Canal Zone.

23. Desmometopa terminalis, n. sp.

Almost entirely black, with knob of halter yellow and 5th tergum and male terminalia orange-yellow.

Male. Black, dark gray microtomentose, only knob of halter yellow and apex of abdomen orange-yellow, including posterior edge of 4th tergum, all of 5th, and all terminalia.

Frons broader than usual in male, approximately square; frontal vitta dull black, not velvet black, but the gray interfrontal and fronto-orbital plates and frontal triangle are distinct; interfrontal plates relatively long, divergent, widely separated, the interval between them much wider than between one of them and a fronto-orbital plate; frontal triangle relatively short, its apex only slightly in advance of median ocellus; cheek about 1/2 breadth of 3rd antennal segment and 1/5-1/6 height of an eye, with subtriangular subocular crescent that is at its widest 2/5 as broad as cheek; face weakly concave in profile, vibrissal angle not produced and approximately a 90° angle, the lateroventral corners of facial plate dull gray and not developed; 3rd antennal segment relatively large, nearly reaching lower margin of face, but not strikingly enlarged as in magnicornis; palpus small, clavate.

Thoracic pleuron densely bright gray microtomentose, including all propleuron and areas surrounding anterior spiracle, with large polished black area posterodorsad to fore coxa that includes an anteroventral area of mesopleuron but is not bilobed (cf. fig. 24). Fore coxa ordinary, convex and not elongate, only 2/3 length of fore femur. Length, 1.25 mm.

Holotype male, PALAU ISLANDS: Koror Island, Mar. 15, 1953 (J.W. Beardsley)[Honolulu].

The yellowish terminalia are unique in the genus as far as known, and thus I have presumed to describe this single specimen. As in other Old World species, the vibrissal angle is not produced. Possibly it is near ciliata. The subgenal bristles are present on one side only, and the 2nd from the vibrissa is upturned and longer than the others, although still much shorter and weaker than a vibrissa. The setae on the section of costa between the costal breaks are few in number, which would also tend to associate the species with cilata, although there are other species which also show a small number of such setae.

The specific name is an adjective referring to the terminalia and the apex of the abdomen, from the Latin referring to boundaries.

24. Desmometopa obscurifrons, n. sp.

Frons uniformly dull, heavily brownish gray microtomentose, interfrontal and fronto-orbital plates and ocellar triangle not strongly contrasting with frontal vitta; pleuron with large polished black spot posterodorsad to fore coxa.

Male, female. Chiefly black, but with considerable yellow to orange-yellow color on lunule, all antennal segments except narrowly dorsally, palpi chiefly, and proboscis; all tarsi yellow except distal tarsomere or two; knob of halter whitish yellow.

Frons uniformly dull brownish gray from most angles of view, often changeable to greenish brown from certain angles, the interfrontal and fronto-orbital plates and the frontal triangle scarcely evident, not strongly contrasting with frontal vitta and thus not delineating the usual black "M" of most species of Desmometopa; cheek narrow, 1/4 breadth of 3rd antennal segment and 1/12 height of an eye, with linear polished subocular crescent; face concave in profile, the vibrissal angle produced to a 45° angle, and lateroventral corner of facial plate, mesad of the vibrissal angle, at least partly shining black; 3rd antennal segment small in both sexes; palpus clavate; proboscis especially elongate and narrow.

Thorax heavily brownish gray microtomentose; pleuron with large polished black spot posterodorsad to fore coxa, including broad anteroventral area on mesopleuron (cf. fig. 24), this coxa not elongate in male. Length, 2.25 mm.

Holotype male, PANAMA: David, Chiriqui, 2200 ft., July 24, 1964 (A. Broce, light trap); allotype, Panamá Province, Las Cumbres (H. Wolda, on flowers of Aristolochia pilosa)[Washington]. Paratypes [Washington except as noted]: PANAMA: 8 females, Panamá Province, Las Cumbres (Henk Wolda, on flowers of Aristolochia pilosa); 4 males, 5 females, Almirante, Arraijan, and in the Canal Zone, Mojinga Swamp at Ft. Sherman, and Summit Gardens (all, F.S. Blanton). MEXICO: male, female, Vera Cruz, Fortin de Las Flores, June 1964 (F.S. Blanton, light trap). EL SALVADOR: female, Santa Tecla, June 3, 1958 (O.L. Cartwright). COSTA RICA: male, San José, July (H. Schmidt); 2 females, [Farm] La Caja, 8 km w. of San José, 1930 (H. Schmidt)[Eberswalde]. COLOMBIA: 4 males, Rio Raposo, May 1964 and Feb. 1965 (V.H. Lee, light trap). ECUADOR: male, 22 females, Pichilingue, 1976 and April 1978 (E.J. Mendoza); female, Rircay Azuay, Oct. 31, 1954 (R. Levi-Castillo); female, Sta. Domingo, Pichincha, June 19, 1965, 600 m (L. Peña) [Ottawa]. TOBAGO: male, two females, St. John Prov., Charlotteville, Mar. 14-21, 1979 (D. Hardy and W. Rowe).

This characteristic species is distinct from all other Desmometopa except nigeriae by the uniformly dull frons. The interfrontal plates, parafrontals, and frontal triangle are present but indistinct, and one does not see the usual black M-shaped frontal vitta of almost all species of Desmometopa.

A small series from PERU: Huanuco, Tingo Maria, Apr. 19-24, 1969 (P. & P. Spangler) and Iquitos, Mar-Apr. 1931 (R.C. Shannon), and another from TRINIDAD: Simla, Arima Valley, Feb. 6-12 and 13-19, 1966 (S.S. and W.D. Duckworth, black light) [both Washington] are referred here tentatively. The pleuron appears to vary from a large polished spot posterodorsad to fore coxa, through limited polished areas, in some cases only a narrow anteroventral area of mesopleuron, to entirely gray pleuron, but the immaturity of some specimens introduces an element of uncertainty.

The specific name, a noun in apposition, is derived from Latin obscurus, indistinct, plus frons.

25. Desmometopa parafacialis, n. sp.

Frons heavily gray microtomentose, antenna and palpus black, and parafacial visible in profile.

Male, female. Black, densely gray to bright gray microtomentose, only knob of halter yellow; tarsi sometimes yellowish basally, especially in females.

Frons densely gray to brownish gray microtomentose viewed from any angle, fronto-orbital plates distinct but interfrontal plates indistinct or not visible from in front, slightly shining and therefore visible from behind, the plates rather narrow and short, posterior ends at level of foremost upper orbital bristles, which are also about the level of apex of frontal triangle; cheek bright gray, of moderate width, barely over 1/2 breadth of 3rd antennal segment and 1/5 height of an eye, with narrow polished subocular crescent that widens anteriorly and continues as a polished area halfway up a parafacial, which is visible in profile; face deeply concave, vibrissal angle strongly produced anteriorly, at about a 45° angle, even beyond level of anterior margin of frons; lateroventral corner of facial plate shining black and warped forward even beyond vibrissal angle; 3rd antennal segment small in both sexes; palpus gently clavate in female but very broad and flat in male (cf. fig. 4).

Thoracic pleuron densely gray microtomentose, including entire propleuron and areas surrounding anterior spiracle, with polished black spot posterodorsad to fore coxa that includes anteroventral area of mesopleuron (cf. fig. 24). Fore coxa and fore femur in male somewhat elongate, former nearly 3 times as long as broad. Length, 1.5-2 mm.

Holotype male, allotype and 39 paratypes (27 males, 12 females), TEXAS: Austin, Nov. 9 and 13, 1958 (Lynn Throckmorton)[holotype, allotype, and paratypes in Washington, paratypes in Austin]. Other paratypes, TEXAS: 2 males, 10 mi. s. Charlotte, Sept. 13, 1955 (W.L. Downes)[Lansing], 2 females, Llano River, Kimble Co., May 23, 1972 (W.W. Wirth, Malaise trap)[Washington]; 2 males, Big Bend National Park, Oak Spring, 4500 ft., May 1, 1959, and Panther Junction, 3500 ft., May 14, 1959 (both, J.F. McAlpine)[Ottawa]. MEXICO: 3 males, 10 mi. ne. San Luis Potosí, 6200 ft., Aug. 22, 1954 (R.R. Dreisbach) [Lansing], and 2 males, same data (J.G. Chillcott)[Ottawa]; 3 males, Hidalgo, Pachuca, 1700 ft., July 29, 1954 (J.G. Chillcott)[Ottawa]; male, female, Nayarit, Ahuacatlán, July 18-22, 1951 (P.D. Hurd, on flowers of Donnellsmithia Hintonii) [Berkeley]; male, Durango, Nombre de Dios, Aug. 5, 1951 (P.D. Hurd, [flowers of?] Keysenhardtia polystachya) [Berkeley]; male, Puebla, Tehuacán, June 23, 1951 (P.D. Hurd) [Berkeley].

The parafacial visible in profile will separate this species from most of those in the genus, and certainly from those in the subgenus Platophrymyia with dull and densely microtomentose frons.

The specific name is an adjective referring to the parafacials.

26. Desmometopa atypica, n. sp.
(Fig. 12)

Frons heavily gray microtomentose, antenna and palpus black, and parafacial not visible in profile.

Female. Black, heavily gray microtomentose; knob of halter, and mid and hind tarsi except for distal segment or two, yellow; palpus obscurely yellowish dorsally toward base.

Frons as described for parafacialis, the short and narrow interfrontal plates sometimes visible as slightly shining lines, from other angles not evident; cheek narrow, 1/2 breadth of 3rd antennal segment and 1/7-1/8 eye height, with polished subocular crescent almost 1/2 breadth of cheek (fig. 12); parafacial midway not visible in profile; face moderately concave in profile, vibrissal angle somewhat produced anteriorly but not as much as in typical members of the subgenus, about a 70° angle; lateroventral corner of facial plate shining black and slightly warped forward; 3rd antennal segment small, palpus gently clavate.

Thoracic pleuron densely gray microtomentose, including entire propleuron and areas surrounding anterior spiracle, with rather large polished spot postero-dorsad to fore coxa, including anteroventral area of mesopleuron (cf. fig. 24). Fore coxa not elongate, and with no suggestion that it might be elongate in males, but this is not certain. Length, 1.5 mm.

Holotype female and 12 paratypes, all females, PANAMA: Panamá Province, Las Cumbres, Nov. 19 (holotype), 21, and 22, 1980, Jan. 24, Feb. 22, Nov. 16, and Dec. 1, 1981, and 4 undated (H. Wolda)[Washington]; female, ECUADOR: Sto. Domingo, Pichincha, June 19, 1965, 600 m (L. Peña)[Ottawa]; female, PERU: Iquitos, March-April 1931 (R.C. Shannon)[Washington].

This species is obviously very close to parafacialis, but the narrow parafacial and the less distinctly produced vibrissal angle will distinguish it. In side view the heads are quite different, that of parafacialis with head somewhat elongate and long axis of eye diagonal, that of atypica with head not elongate and long axis of eye vertical.

I have isolated females from far distant places that may indicate a wider Neotropical distribution, but unrecognized species may be involved, and they are not included in the type series: BRAZIL: São Paulo, Nova Teutonia, Nov. 1958 (F. Plaumann)[Ottawa]; TRINIDAD: Simla, Arima Valley, Feb. 20-26, 1966 (S.S. and W.D. Duckworth)[Washington].

The specific name is an adjective referring to the atypical appearance of the head compared with typical members of the subgenus Platophrymyia.

27. Desmometopa ciliata Hendel
(Figs. 11, 24)

Desmometopa ciliata Hendel, 1919, Ent. Mitt. 8: 200 (Sydney, New South Wales) [Budapest].

Dark gray, resembling a small sordida but with knob of halter yellow; strong subgenal bristle.

Male, female. Chiefly black or black-brown; knob of halter yellow.

Frons slightly broader than long, with velvet black M-shaped frontal vitta delineated by strong gray interfrontal and fronto-orbital plates and frontal triangle, the interfrontal plates moderately long and strong, extending posteriorly to level of hindmost upper orbital bristles, the fronto-orbital plates broader than usual; frontal triangle short and approximately equilateral, apex not midway of frons; cheek 2/3 breadth of 3rd antennal segment and 1/5-1/4 height of an eye, with relatively broad polished subocular crescent, and 2nd subgenal seta behind the vibrissa developed as a strong, upcurved bristle (fig. 11); face only weakly concave in profile, and vibrissal angle not produced, a broadly rounded 80°-90° angle; 3rd antennal segment small in both sexes, little larger than 2nd segment; palpus clavate.

Thorax dark leaden gray microtomentose; pleuron chiefly gray, including entire propleuron and areas surrounding anterior spiracle, with large polished black spot posterodorsad to fore coxa that includes an anteroventral area of mesopleuron, the anterior margin of spot more or less straight (fig. 24), not bilobed as in m-nigrum (cf. fig. 23). Fore coxa of male not elongate. Section of costa between humeral crossvein and subcostal break with 8-10 coarse and well-spaced setae. Length, 2 mm.

Hendel's original series consisted of five specimens. Through the kindness of Dr. F. Mihályi of the Hungarian National Museum at Budapest, I was loaned a male and 3 females of the original series, each specimen labeled "Typus". The male, which is in good condition and bears Hendel's identification label, is hereby designated lectotype and has been so labeled. I have also seen the fifth syntype, now a paralectotype, in the Museum in Vienna.

I have seen barely a dozen additional specimens of the species, from several localities in New South Wales, and from South Australia and the Australian Capital Territory. It may be an endemic Australian species.

Hendel's brief characterization gave few details and emphasized the bristling of costa between the costal breaks, 8-10 well-spaced bristles in ciliata but 14-16 in other species. The type series reveals that an even better character is the development of a strong subgenal bristle, often subequal to a vibrissa. The association of this character and the smaller number of coarse setae on the costa before the subcostal break is found in ciliata and a few Oriental species and may be said to link these as a "ciliata group".

Malloch's D. ciliata was a misidentification of D. inaurata Lamb, both in 1924 (p. 336) from New South Wales and in 1936 (p. 327) from Samoa, as revealed by Malloch-labeled specimens in Washington.

28. Desmometopa leptometopoides, n. sp.
(Fig. 19)

Of the sordida group but with yellow halter knob, small polished spot on pleuron, and Leptometopa-like broad and flat hind tibia.

Male, female.   Chiefly black, densely gray microtomentose; palpus orange-yellow, infuscated distally and below, more so in female than in male; knob of halter yellow.

Frons with interfrontal and fronto-orbital plates and frontal triangle distinctly well developed, the sections of M-shaped frontal vitta narrower than usual; frontal triangle long, its apex at middle of frons, posterior portions of interfrontal plates separated from triangle by approximately the width of one plate; cheek of moderate width, 1/2 breadth of 3rd antennal segment and 1/8 height of an eye, with narrow but distinct polished subocular crescent; face weakly concave in profile, vibrissal angle 80° to 90°, not produced anteriorly in profile; 3rd antennal segment small in both sexes, only a little larger than 2nd segment; palpus clavate.

Thoracic pleuron densely gray microtomentose, including all propleuron and areas surrounding anterior spiracle; polished spot on pleuron posterodorsad to fore coxa rather small and confined to anterior slope of sternopleuron, no polished area anteroventrally on mesopleuron (cf. fig. 25). Fore coxa not elongate; hind tibia of male broad and flat (fig. 19) as in males of Leptometopa. Length, 1.5 - 2 mm.

Holotype male, allotype, and one female paratype, LIBERIA: Suakoko, July 1, 1952 (C.C. Blickenstaff)[Washington]. Paratypes: GHANA: 5 females, Accra, Aug. 16, 1945 (M.A. Locke)[Washington]; female, Accra, Dec. 1921 (J.W. Scott Macfie, "reared from mud and debris collected from pools" [London]; female, Legon, Apr. 6, 1969 (O.W. Richards, at light)[London]. CAMEROUN: 8 females, Victoria, Dec. 22, 1920 (L.H. Booth)[London]. N. NIGERIA: male, 3 females, Zaria, May 23, 1966 (J.M. Lyall, "Tenebrio/Trib culture window") [London]. TUNISIA: male, Bou Hedma, Apr. 11, 1976 (M. Olsson) [Lund].

One female, not a paratype, is not in good condition but appears to be this species. If correct, it would represent a considerable extension of the known range: CAPE PROVINCE, Mossel Bay, Dec. 15, 1928 (R.E. Turner)[London].

The striking feature of broad and flat hind tibia in the male, resembling that of Leptometopa, is unique in Desmometopa. Females are much less distinct, however. The occasional lengthening of the subgenal setae might suggest ciliata, but the small polished pleural spot of leptometopoides separates it from ciliata and associates it with varipalpis and singaporensis. The elongated palpi of these two species easily distinguish the males from leptometopoides, but females are less distinctive. Females of varipalpis have a definitely broader cheek, but females of singaporensis are much closer. Teneral females of leptometopoides have yellowish cheeks and could on that basis alone have been confused with singaporensis. Associated males give the best basis for identification.

The specific name is an adjective derived from the generic name Leptometopa plus oides, like.

29. Desmometopa varipalpis Malloch
(Figs. 1,3,6,25)

Desmometopa varipalpis Malloch, 1927, Proc. Linn. Soc. N.S. Wales 52: 7 (New South Wales) [Sydney].
D. tarsalis Loew; Hennig, 1937, Fam. 60a. Milichiidae et Carnidae, p. 44, in Lindner (ed.), Fliegen Palaeark. Region, Lfg. 115.
D. singaporensis (tarsalis of European records); Hennig, 1939, Arb. Morph. Taxon. Ent. Berlin-Dahlem 6: 87-88.
D. M-nigrum (Zetterstedt); Wolcott, 1951, Jour. Agr. Univ. Puerto Rico 32(3): 529 (Puerto Rico, at least in part: the San Juan specimen, now in Washington, is varipalpis).
D. varipalpis; Lee, Crust, and Sabrosky, 1956, Proc. Linn. Soc. N.S. Wales 80: 339 (footnote on presumed holotype).
D. singaporensis; Hennig, 1965, Stuttg.Beitr. Naturk. 139: 2, fig. 1 (Iran; figs. of male and female palpi).
D. singaporensis; Sabrosky, 1973, Family 75, p. 2, in A Catalogue of the Diptera of the Americas South of the United States.
D. varipalpis (singaporensis, authors, in part); Sabrosky, [1977], p. 271, in Delfinado and Hardy (eds.), A Catalog of the Diptera of the Oriental Region, p. 271.
D. singaporensis; Hardy and Delfinado, 1980 (June 4), Insects of Hawaii 13: 355-6 (Hawaii; figs. of head and male genitalia).
D. varipalpis (singaporensis, authors, in part); Sabrosky, 1980 (July 10, Family Milichiidae, p. 687, in Crosskey (ed.), Catalogue of the Diptera of the Afrotropical Region.

Polished pleural spot small, not including an area of mesopleuron; fronto-orbital plates relatively broad (fig. 25); palpus of male strikingly elongate, fusiform (fig. 3).

Male, female. Black, heavily gray microtomentose; cheek yellowish in ground color; 1st and 2nd antennal segments almost always reddish, contrasting with black 3rd segment; palpus partly yellow, extensively so in the enlarged palpus of male, yellow on proximal 1/2 in female; knob of halter yellow; mid and hind tarsi yellowish except distal 2, rarely 3, tarsomeres.

Frons (fig. 1) with velvet black frontal vitta, the interfrontal and fronto-orbital plates and frontal triangle gray and distinct; frontal triangle equilateral or slightly longer, reaching about to middle of frons; fronto-orbital plates especially broad, each almost twice width of an interfrontal plate, and without or almost without break between upper and lower orbital plates, unlike singaporensis (cf. figs. 1,2); cheek (fig. 3) over 1/2 breadth of 3rd antennal segment and about 1/5 height of an eye, with narrow polished subocular crescent; parafacial narrowly visible throughout in profile; face weakly concave in profile, vibrissal angle about 80° - 90°, not strongly produced, the lateroventral corner of facial plate dull and not warped forward; 3rd antennal segment small in both sexes; palpus gently clavate in female, but elongate fusiform in male (fig. 3), narrowed apically, usually large and long, occasionally in small specimens short and not so narrowed apically.

Thoracic pleuron heavily and extensively gray microtomentose, with small polished spot posterodorsad to fore coxa, mesopleuron not polished anteroventrally (fig. 25). Fore coxa and femur not elongate. Section of costa between the humeral and subcostal breaks with many (12-14) short, semierect setae, each only a little longer than diameter of costa. Length, 2.5 mm.

A specimen found in Malloch's collection years ago, still before me but to be returned to Australia, is undoubtedly the holotype of varipalpis. It agrees perfectly with the specimen data and the description but is labeled "Desmometopa varicornis Type." No doubt Malloch changed this in publication upon realizing that he really meant the name to refer to the palpus and not the antenna. It is a male, not a female as stated by Malloch. He clearly described the palpi as "large" and "lanceolate," an attribute not then recognized as characteristic of males only.

Males of this species are very distinctive because of the elongate fusiform palpi, unique in the genus and comparable only with singaporensis which also has a large palpus but a capitate one (cf. figs. 3,5). However, females are much less distinctive and may easily be confused with those of singaporensis and perhaps with leptometopoides, and even with some other species (e.g., m-nigrum) if the specimens of the last named are teneral with collapsed cheeks. Two characteristics distinguish varipalpis from singaporensis in both sexes, and these are useful for females: (1) in varipalpis (fig. 1) the fronto-orbital plates are broad throughout, without an obvious break between the upper and lower orbital plates, whereas in singaporensis (fig. 2) the fronto-orbital plates are narrower, especially the lower orbital plate, and there is a distinct break and narrowing from upper to lower sections; (2) the cheek is wider in varipalpis than in singaporensis (cf. figs. 3,5), a consistent difference but one that can be tricky because in teneral individuals the cheek tends to fold longitudinally toward the eye, thus narrowing the cheek.

The antennae are usually entirely black in singaporensis, but with reddish 1st and 2nd segments in varipalpis. However, enough specimens of singaporensis also have these segments more or less reddish that the character cannot be relied upon, although the bright segments in typical varipalpis will be a supporting character.

Males of _varipalpis_ have the additional characteristic of elongate fusiform palpi, but in a few cases, almost always small individuals, the palpi are relatively short and small and in such cases they appear less acute apically. In some long series that are available from the same place and time, there are usually a few small specimens with small palpi, which encourages me to believe that the condition of small palpi represents only an occasional variant. In small specimens of _singaporensis_, small palpi also occur, and such individuals are difficult to identify with assurance except for the reliable character of the narrower parafrontals. In males, the larger the specimen, the more elongate and conspicuous appear the palpi, whereas in small specimens the palpi are shorter and apically they are less acutely angled.

There has been almost no confusion in the use of the name _varipalpis_, except for Hennig's synonymizing of it with _palpalis_, undoubtedly misled by Malloch's incorrect statement of the sex of the holotype. The real confusion has been in the use of the name _singaporensis_, to which I myself, regretfully, have contributed. See the discussion under that species for the effect of the lectotype designation by which _singaporensis_ must be applied to _palpalis_ for the species with elongate and capitate palpi in the males, leaving _varipalpis_ the valid name for the species with elongate and fusiform palpi. Hardy and Delfinado (1980) figured the quite different palpi of the two species, with _singaporensis_ under the synonymous name _tristicula_ (their fig. 143c) and _varipalpis_ under the name _singaporensis_ (their fig. 143a). For further discussion of the confused usage, see the introductory section on "Identification". If it will alleviate the pain of name changing, I can note that even if _singaporensis_ had been restricted to the species here called _varipalpis_, the frequently used name _palpalis_ would have had to be changed to the rarely used _tristicula_, which has priority.

Distribution: Widespread, occurring in all faunal regions, summarized as follows to show the wide range (records from planes and ships not included):

Nearctic: Canada: Quebec (Montreal); U.S.A.: records from 23 states, Wash.-Mich.-N.Y., south to Calif.-Texas-Fla.

Neotropical: Bolivia, Brazil (Santos, São Paulo), El Salvador, Guatemala, Mexico (Veracruz), Panama, and islands Antigua, Clipperton Island, Cuba, Galapagos, Puerto Rico, St. Vincent, Virgin Islands.

Palearctic: Algeria, Egypt, Iran, Iraq, Israel, Saudi Arabia. [The record from Hyères, southern France, published by Séguy (1934) as _D_. _albipennis_ (Meigen) with _singaporensis_ in synonymy, and mentioned by Hennig (1937) as _tarsalis_, was actually based on _D_. _m-nigrum_, from the female specimen in Paris].

Afrotropical: Djibouti [as French Somaliland], Ghana, Kenya, South Yemen (Aden), Sudan, Uganda, Yemen, Zaire, also Ascension Island.

Oriental: India (West Bengal, Tamil Nadu), Singapore, Thailand.

Australian: New South Wales.

Oceanica: New Guinea, Bismarcks, Bonins (Chichi Jima), Carolines (Lukunor Atoll, Truk), Guam, Hawaii, Johnston Island, Marianas (Saipan), Marshalls (Eniwetok Atoll, Jaluit Atoll, Kwajalein, Ujelang Atoll), New Hebrides (Espiritu Santo), Volcanos (Iwo Jima), Wake Island.

A number of records suggest the probable importance of commerce in the distribution of _varipalpis_: Guam, from planes (China Clipper and Honolulu Clipper); Honolulu, Hawaii, plane from New Zealand and New Caledonia; Liverpool, England, "from ship in Liverpool docks ex Canada"; Port Adelaide, South Australia, "taken on ship ex Kuwait"; Valparaiso, Chile, "on board Santa Inez" [undoubtedly the specimen called _m-nigrum_ by Malloch, 1934b]; New Orleans, La. from Central America; New York, N.Y., "larvae in potatoes in ship from Argentina", and "plane from Buenos Aires"; Norfolk, Va., "specimens in soil with potatoes from Colombia".

Perhaps it is only chance, but I have seen only one specimen of singaporensis bearing any similar information, although the rearing records noted under "Biology" would indicate the similar possibility of transport in commerce or movement of people.

Interesting historical records are furnished by a male labeled "Cuba/Poey" and "Loew Coll." and a female with similar labels [both Cambridge], and a male, "Havana, Cuba" "27.1.'69" [Washington, very old handwritten label, undoubtedly 1869], showing that the species has been in the New World for well over a century.

30. Desmometopa singaporensis Kertész
(Figs. 2, 5)

Desmometopa singaporensis Kertész, 1899, Termész. Füzetek 22: 194
    (Singapore)[Budapest].
D. tarsalis Loew (syn., singaporensis); Hendel, 1907, Wien. Ent. Ztg. 26: 242
    [but at least Egypt and Aden records refer to varipalpis].
D. tarsalis; Malloch, 1914, Ann. Mus. Nat. Hungar. 12: 309 (male, female)
    [Budapest].
D. tristicula Hendel, 1914 (Jan. 27), Suppl. Ent. 3: 96 (Formosa) [Eberswalde].
    N. syn.
D. palpalis de Meijere, 1914 (Oct. 15), Tijd. Ent. 57: 251 (Java, Sumatra)
    [Amsterdam].
D. tarsalis; de Meijere, 1914 (Oct. 15), Tijd. Ent. 57: 251 (Java, Sumatra;
    females).
D. m-nigrum (Zetterstedt) Illingworth, 1926, Proc. Hawaiian Entomol. Soc.
    6(2): 224 (Hawaii, det. Aldrich).
D. tarsalis (female) and D. palpalis (male); Bezzi, 1928, Diptera Brachycera
    and Athericera of the Fiji Islands, p. 162-3 (Fiji; also Fiji, "from onions
    imported from Australia").
D. tarsalis; Illingworth, 1929, Proc. Hawaiian Ent. Soc. 7(2): 233
    ("Correction" from Aldrich of his earlier identification of m-nigrum).
D. palpalis; Malloch, 1934, Insects of Samoa, Pt. VI (Diptera), Fasc. 8: 327-8
    (Samoa; fig. of male head and palpus; first to note sexual dimorphism in the
    palpi of this species).
D. palpalis; Hennig, 1939, Arb. Morph. Taxon. Ent. Berlin-Dahlem 6: 88-89,
    fig. 8 (head of male, side view).
D. tarsalis; Hennig, 1941, Ent. Beihefte aus Berlin-Dahlem 8: 177 (Formosa, at
    least in part; I have seen the Hokuto specimens).
D. palpalis; Bohart and Gressitt, 1951, Bull. Bishop Mus. 204: 46 (Guam).
D. palpalis; Hardy, 1952, Proc. Hawaiian Ent. Soc. 14(3): 474 (Hawaii).
D. singaporensis; [Hardy?], 1972, Proc. Hawaiian Ent. Soc. 21(2): 160 (Correct
    name for species known in Hawaiian literature as palpalis, teste Sabrosky).
D. palpalis; Sabrosky, 1973, Family 75, p. 2, in A Catalogue of the Diptera of
    the Americas South of the United States (Brazil, Puerto Rico).
D. singaporensis (syn., palpalis) and D. tristicula; Sabrosky, [1977], Family
    Milichiidae, p. 271, in Delfinado and Hardy (eds.), A Catalog of the Diptera of
    the Oriental Region.
D. tristicula; Hardy and Delfinado, 1980 (June 4), Insects of Hawaii 13: 356,
    358 (Hawaii, figs. of head and male genitalia).
D. singaporensis (syn., palpalis); Sabrosky, 1980 (July 10) Family
    Milichiidae, p. 687, in Crosskey (ed.), Catalogue of the Diptera of the
    Afrotropical Region.

Polished pleural spot small, not including an area of mesopleuron; fronto-orbital plates relatively narrow (fig. 2); palpus of male broadly expanded, capitate (fig. 5).

Male, female. Black, heavily gray microtomentose; cheek yellowish in ground color; antenna usually black but sometimes 1st and 2nd segments reddish; palpus partly yellow, extensively so in male but only proximal 1/2 in female; knob of halter yellow; mid and hind tarsi sometimes yellowish on proximal 2 to 3 tarsomeres.

Frons with velvet black frontal vitta, the interfrontal and fronto-orbital plates and frontal triangle gray and distinct; each fronto-orbital plate moderately narrow, especially the lower orbital plate, and almost always with a more or less distinct break between upper and lower sections, just anterior to foremost upper orbital bristle (fig. 2); frontal triangle longer than broad at base, its apex about midway on frons; cheek about 1/2 breadth of 3rd antennal segment and 1/8-1/9 height of an eye, with narrow polished subocular crescent (fig. 5); parafacial usually not visible in profile; face weakly concave, vibrissal angle not produced, about an 80° – 90° angle, lateroventral corner of facial plate dull gray and not warped forward; 3rd antennal segment small in both sexes; palpus gently clavate in female, but in male elongate and broad, capitate (fig. 5).

Thoracic pleuron densely and extensively gray microtomentose, with small polished spot on pleuron posterodorsad to fore coxa, no adjoining polished area anteroventrally on mesopleuron (cf. fig. 25). Fore coxa and femur not elongate. Costal setae as in _varipalpis_. Length, 2.5 mm.

Lectotype, female, "Singapore/Biró 1898," "M-nigrum" [Biró label], "Desmometopa/singaporensis/typus [in red ink] Kert./ det. Kertész," "typus" [printed in red on large red-bordered label], "tarsalis Lw./ det. Hendel." Paralectotypes: 6 females, "Singapore/ Biró 1898" [Budapest]; 3 females, same data [Vienna]. Of the paralectotypes, 5 in Budapest are the narrow-cheeked species, conspecific with the lectotype; the other four are _varipalpis_.

Kertész did not state the number of specimens in the type series, nor did he designate a holotype. The stated range of length (2.3-2.5 mm) and the use of the plural "die Exemplare" show that he had at least more than one specimen, and thus the single example now labeled "typus" is technically not a holotype. It seems reasonable to conclude that the seven specimens in the Museum at Budapest bearing identical labels "Singapore/Biró 1898" as published by Kertész, and standing in the collection under the name label "tarsalis Lw. (singaporensis Kert.) as revised by Hendel, are all syntypes. All are females, and these are now before me. Hendel (1907) also referred to "Typen" in the Museum in Vienna, and I have seen those three, also females with identical data to those in Budapest, and here also considered syntypes. Six of the Budapest specimens, including the one labeled "typus," are a narrow-cheeked species agreeing with _palpalis_ de Meijere, and one is a broader-cheeked form agreeing with _varipalpis_ Malloch. The three in Vienna agree with the latter. I have designated and labeled as lectotype the specimen in Budapest bearing the label "typus." Another female from Singapore, collected by Biró, is labeled 1895 and thus cannot be considered a syntype, although it is possible that it was before Kertész and he overlooked the difference in date. It is _varipalpis_.

There was a strong temptation to designate the one wide-cheeked example in Budapest as lectotype, which would have saved the name singaporensis for the species often (e.g., by Hennig 1937)--but not always--called that. However, such a designation would have been inconsistent with the labeling of the single specimen as "typus," it would have disagreed with the original description, and it would have been contrary to the majority of the specimens. The labeling of "typus" is not a binding consideration under the International Code of Zoological

Nomenclature, because it was not published, but some taxonomists may consider that such labeling should fix the status of the specimen and I consider it desirable to avoid that possible argument, as well as to respect the author's choice. In view of the mixed series, perhaps that "choice" was unintentional, but on the other hand it may have indicated the choice of an individual that seemed to Hendel most typical in the type series. Most of the description is generalized and applies in most particulars to both species, but the characters of 'cheeks very narrow' and 'antennae black' apply to the narrow-cheeked species and not to the other, varipalpis, which has a broader cheek-although not exceedingly so--and partly reddish antennae, differences that are clearly evident in the type series.

The narrow-cheeked form is that described by de Meijere as palpalis, from Salatiga, Java (D. van Leeuwen) and Deli, Sumatra (de Bussy). The palpi were described as "sehr gross, löffelförmig." I have before me, loaned from the Museum at Amsterdam by Dr. G. Kruseman, two males of the cited data. The one from Java was labeled "Type" by de Meijere, although not so published, and I here designate it as lectotype, the other being paralectotype. Three females of the same species are also present in that collection, labeled tarsalis by Becker (Pasuruon, Java) and by de Meijere (Semarang, Java, and Deli, Sumatra). The last two specimens were published by de Meijere as tarsalis at the same time that he described the males as palpalis.

D. tristicula Hendel was briefly and inadequately described from "♂♀", kindly loaned me for study from Eberswalde by Dr. G. Morge. One specimen is obviously a female, with genitalia clearly evident. The other, presumably the male referred to, bears a red label "Typus" but this was not so published and the type series really consists of two syntypes. The "male" is unrecognizable; it is now headless, the mid and hind legs are missing, and the abdomen, which has had the distal half sliced off, appears filled with eggs! Hennig (1941) recorded "1 Typus plus 1" in his list of the insects of Formosa, but this was a curatorial list and cannot be considered a definitive nomenclatural act. In other cases, all of Hendel's series have been marked "Typus", and the lack of such a label on the second specimen is probably a preparator's error. I hereby designate as lectotype the second specimen, the female, which is in excellent condition. Dr. Morge also sent 17 other specimens, mostly from H. Sauter's collecting in Formosa, identified as tristicula, most of which had been recorded by Hennig (1941). Four of these, from H. Sauter's collecting at Tainan, Nov. 1909 (1 female) and Hokuto, Dec. 1912 (2 males, 1 female) are tristicula, but most of the others are microps. Unfortunately, the abdomen is missing from each of the available males of tristicula, but the capitate palpus is characteristic of that sex.

The species that I recognize as singaporensis is nearest varipalpis, and although males of the two are easily distinguished by the form of the strongly developed palpi, identification of females is much more difficult, as discussed in detail under varipalpis. Females of singaporensis have a slightly narrower cheek and narrower fronto-orbital plates with a distinct break between upper and lower orbital plates. It must be kept in mind, as noted under varipalpis, that small males have less strongly developed and less conspicuous palpi, and this can sometimes be quite deceiving.

Hennig (1939) unfortunately misquoted Malloch as saying that only the females of palpalis have the enlarged and brightly marked palpi, whereas Malloch (1934a) had positively stated that such palpi were "characteristic of the male only" and that females had small palpi as in m-nigrum and ciliata. Probably Hennig was influenced by his natural acceptance of Malloch's earlier (1927) statement (erroneous!) that varipalpis was based on a female specimen.

As far as usage is concerned, there has been a great deal of confusion, to which I too have contributed. The name singaporensis has been applied to both broad- and narrow-cheeked species, especially in the female sex in which the two species are difficult to separate. Because of the confusion and the relatively few publications involved, there is no overwhelming amount of usage that needs special consideration. Either choice would have resulted in synonymy. Moreover, if singaporensis had been restricted to the broad-cheeked form (varipalpis), the name most commonly used for the narrow-cheeked form, i.e., palpalis, would have had to be changed anyway as a junior synonym of tristicula. For further discussion of the confusion, see the introductory paragraphs under "Identification".

Distribution: Chiefly Old World, and chiefly Oriental and the Pacific Islands. I can record specimens from the following:

Neotropical: Brazil (Santos, S.P.), Puerto Rico.

Afrotropical: Ivory Coast, Seychelles, South Africa (Transvaal), Uganda.

Oriental (including some chiefly Palearctic countries with Oriental sections): Afghanistan, Cambodia, China (Szechuan), India (Assam, Bengal), Java, Korea, Malaya, Pakistan, Philippine Islands, Ryukyu Islands (Okinawa), Singapore, South Viet Nam, Sri Lanka, Sumatra, Taiwan, Thailand.

Pacific Islands: Bismarck Archipelago (New Britain), Caroline Islands (Merir Island, Palau Islands, Truk), Fiji, Gilbert Islands (Tarawa Atoll), Guam, Hawaii, Mariana Islands (Saipan), Marshall Islands (Kwajalein), New Hebrides, Solomon Islands (Guadalcanal, Russell Group), Tahiti, Tonga, Yap.

31. Desmometopa sordida (Fallén)
(Fig. 8)

Madiza sordida Fallén, 1820, Oscinides Sveciae, p. 10 (Sweden) [Stockholm].
Agromyza M atrum Meigen, 1830, Syst. Beschr. 6: 170 (usually cited m-atrum).
Desmometopa (Liodesma) atra Duda, 1935, Natuurhist. Maanblad 24: 25, 38
    (Saarland, West Germany) [Berlin]. (Synonymy confirmed).
D. sordida (Fallén)(?Liodesma atra Duda) Hennig, 1937, Milichiidae et
    Carnidae, in Lindner, Fliegen Palaeark. Region, Fam. 60a: 43.

Entirely black or dark brown, including knob of halter; cheek with distinct subocular crescent.

Male, female. Entirely black or dark brown, including palpi, halteres, and all tarsi.

Frons subshining black, the dark gray interfrontal and fronto-orbital plates and frontal triangle distinct; interfrontal plates moderately long, widely divergent, posterior ends barely overlapping apex of frontal triangle; cheek over 3/5 breadth of 3rd antennal segment and nearly 1/6 height of an eye, with large rounded subocular crescent (fig. 8); face weakly concave in profile, vibrissal angle about 90°, not produced; 3rd antennal segment small in both sexes; palpus clavate.

Thoracic pleuron densely brownish microtomentose, including all propleuron and areas surrounding anterior spiracle, a polished black spot posterodorsad to fore coxa that only narrowly or not at all encroaches on anteroventral area of mesopleuron. Fore coxa not elongate, convex, about 3/5 length of fore femur. Length, 2-2.5 mm.

D. sordida, whose junior synonym m-atrum is type of the genus, is a dark, nondescript species without particularly distinctive characters. In the Holarctic Region its black halteres distinguish it. It has apparently not been distributed widely in commerce as has D. m-nigrum, and published records in the Oriental and Afrotropical Regions are suspect.

It is widespread in North America and the Palearctic Region. I have seen numerous specimens from England, Sweden, and Russia (Leningrad Oblast) south to Spain and Israel, also Manchuria and Japan.

In the "Catalog of the Diptera of the Oriental Region" (Sabrosky [1977], p. 271), I listed sordida from "India, Indonesia, Philippines" but indicated possible misidentifications. Up to the present time, I have not seen true sordida from the Oriental Region. Two specimens from Semarang, Java, August 1905 (Jacobson), borrowed from the Museum in Amsterdam as sordida, proved to be D. microps Lamb, and this is a likely species to be confused with sordida.

Likewise, in the "Catalogue of the Diptera of the Afrotropical Region" (Sabrosky 1980), I listed sordida from "Cameroun, ?Kenya, Tanzania", but these records too may now be doubted. The specimen on which the Cameroun record was based is before me, and it is a female of microps as I now recognize. The Tanzania record was also based on a female, recorded earlier by me (Sabrosky 1965a), and I suspect that this was also microps, the female being easily confused with that of sordida. The Kenya record was published by Séguy (1938), and should be checked for this same possibility.

Through the friendly cooperation of Dr. H. Schumann of the Museum für Naturkunde, Humboldt-Universität zu Berlin, I received for study the two specimens on which Duda founded his Desmometopa atra. One is teneral, with head and thorax collapsed. The other, which I here designate as lectotype, bears the following labels: "8 8 19, St. Wendel/Rheinl. Duda [printed], Piomadiza n. gen., atra Duda♀, Typus [printed, colored label]." The name labels are in Duda's handwriting. Duda apparently changed his mind on the generic name and its status, before publishing. The other specimen, which now becomes a paralectotype, is labeled "6 8 20, 5a, atra Duda♂ [actually a female and so published], Typus [printed, colored label]." Its locality was published as "Habelschwerdt (Schlesien)."

Duda described the face of atra, in contrast to Desmometopa, as "poliert glänzend" and the frons as "glänzend und unbereift," hence the name Liodesma. However, the shining appearance is an artifact. The entire head has been wetted with a dark, syrupy substance. The specimens are simply the common D. sordida, and I can confirm the synonymy suggested by Hennig (1937) from his reading of the description.

32. Desmometopa sp. L

A lone male, MALAYA: Pahang, Tahan River, George V National Park, Nov. 5, 1959 (H.E. McClure, light trap)[Washington], apparently represents a distinct species, with characters as given in the key, but it will not be described and named until further material is available. It is tiny (1 mm), with narrow interfrontal plates and short frontal triangle that together result in an unusually large M-shaped frontal vitta. The palpus is short clavate and black, and the 3rd antennal segment is small, only a little larger than the 2nd segment, both structures undoubtedly the same in females. The fore coxa is not elongate. The polished black spot posterodorsad to fore coxa appears to include, as in sordida, a very narrow anteroventral area on the mesopleuron.

33. Desmometopa nudigena, n. sp.

Tiny dark species of the sordida habitus, but with all tarsi partly yellow, and cheek lacking a polished subocular crescent.

Male, female. Black, dark gray microtomentose; all tarsi yellow except for distal segment or two.

Frons with velvet black frontal vitta, the gray interfrontal and fronto-orbital plates and frontal triangle distinct; interfrontal plates very narrow, practically linear, hence the sections of the M-shaped frontal vitta unusually broad, the plates long, with posterior ends opposite apex of the short frontal triangle which is virtually coextensive with ocellar tubercle; cheek narrow, 1/4 breadth of 3rd antennal segment and 1/12 eye height, gray, without polished subocular crescent; face weakly concave, vibrissal angle an 80° - 90° angle and not produced anteriorly; lateroventral corner of facial plate flat, gray like rest of plate; 3rd antennal segment small in female but large in the now headless male (cf. fig. 21); palpus clavate.

Thorax densely dark gray microtomentose, including entire propleuron and areas surrounding anterior spiracle; a large polished black spot posterodorsad to fore coxa and including an elongate-oval anteroventral area of the mesopleuron (cf. fig. 24). Fore coxa and fore femur ordinary, not elongate. Section of costa between costal breaks with 8 erect, coarse, black, well-spaced setae. Length, 1.25 - 1.5 mm.

Holotype female and female paratype, GAMBIA: Bakau, Botanical Garden, Nov. 21, 1977 (Cederholm et al.)[Lund]. Paratypes: GAMBIA: (all, Cederholm et al.): female, 6 km n. Kartung, Nov. 20, 1977, "swept in very dense forest with glades" [Lund]; male, female, at road junction to Situ Sinjang, about 2.5 km se. Kafuta, Mar. 1, 1977 [Lund, Washington]. IVORY COAST: female, Savane à Imperata, May 5, 1971 (D. Lachaise, "inflorescence de _Cussonia_") [Paris]. SIERRA LEONE: Female, Taninahur, Feb. 14, 1925 (E. Hargreaves) [London]. NIGERIA: female, Ibadan, Dec. 4, 1962 (D.C. Eidt) [Ottawa].

The head of the male was accidentally and irretrievably lost, but not before it was noted that the 3rd antennal segment was unusually large, in which feature it resembles _magnicornis_ and _philippinensis_ (cf. fig. 21).

This species is one of a small group characterized by black-brown halteres and, in the male, by extra large 3rd antennal segment. It is differentiated in the key from the other species with these features, _magnicornis_ and _philippinensis_ by its longer frontal triangle, but in these small species the difference does not appear great. Should there be confusion, however, different combinations of other characters served to mark these as distinct species, in addition to the length of the triangle:

nudigena: subgenal setae fine and even, without outstanding bristle; all tarsi yellowish except for distal segment or two.

philippinensis: 2nd subgenal seta behind the vibrissa bristlelike, longer than the others; all tarsi yellowish except distally.

magnicornis: subgenal setae fine and even; all tarsi black.

Possibly _D_. _aldabrae_ belongs in this group also, although the 3rd antennal segment is only moderately enlarged in the male. Should it be involved in the possible confusion, it is easily distinguished by almost entirely yellow palpi in both sexes. The subgenal setae are even and the tarsi are yellowish except distally.

The specific name is a noun in apposition compounded from the Latin _nudus_, bare, plus _gena_, cheek.

34. _Desmometopa flavipalpis_, n. sp.

Small dark species of the _sordida_ habitus, but males with yellow palpus, and both sexes with short frontal triangle.

Male, female. Black, except for yellow palpus in male, slightly infuscated at tips, and basal 1/3 to 1/2 of palpus orange-yellow in females.

Frons subshining, velvet black, interfrontal plates distinct but of moderate width, the sections of M-shaped frontal vitta broad; frontal triangle short, apex barely anterior to median ocellus; face weakly concave, the vibrissal angle not produced forward, about an 80° - 90° angle; cheek narrow, barely over 1/3 breadth of 3rd antennal segment and 1/11 eye height, with linear polished subocular crescent that is slightly wider anteriorly; subgenal setae rather long and lengthening toward vibrissa; 3rd antennal segment small in both sexes; palpus short, clavate.

Thorax dark gray microtomentose, including entire propleuron and areas surrounding anterior spiracle, with large polished black spot posterodorsad to fore coxa that includes an anteroventral area of the mesopleuron (cf. fig. 24). Fore coxa not elongate. Length 1.75–2 mm.

Holotype male, allotype, and 4 paratypes (3 males, 1 female), MARSHALL ISLANDS: Jaluit Atoll, Majurirek Island, Apr. 26, 1958 (J.L. Gressitt, "Hernandia flowers")[Honolulu]. Other paratypes: MARSHALL ISLANDS: 10 males, Jaluit Atoll, Jabor Island, Apr. 27 and May 1, 1958 (J.L. Gressitt, two of Apr. 27 labeled "Crotalaria"). CAROLINE ISLANDS: 2 males, 1 female, Ulithi Atoll, Falalop Island, April 30, 1952 (J.W. Beardsley); female, Lamotrek Atoll, Lamotrek Island, Feb. 5, 1953. PALAU ISLANDS: female, Angaur Island, May 1, 1954 (J. W. Beardsley). NEW HEBRIDES: male, Efate, Vila, 0–100 m, Feb. 1969 (N.L.H. Krauss). [Most paratypes in Honolulu, paratypes in Washington; New Hebrides paratype in Washington].

This species is similar to others of the sordida group in the Oriental and Pacific areas, but it can be distinguished by the combination of characters used in the key. Two noteworthy features are the very short frontal triangle and the conspicuous yellow palpi of the males. Females are much less distinctive because the palpi are heavily infuscated distally and apparently sometimes entirely black.

The specific name is an adjective compounded from the Latin flavus, yellow, and palpus, feeler.

Two females from the Philippines have not been included in the type series, but they appear to be this species: Samar, Osmeña, May 23, 1945 (K. L. Knight, at light), and Calicoan Island, July 27, 1945 (F.F. Bibby, "from dead land crab") [Washington].

35. Desmometopa srilankae, n. sp.

Tiny species near ciliata, but with bilobed polished black pleural spot.

Male, female. Chiefly black; palpus yellowish on basal half; all tarsi yellow except for several distal segments; knob of halter black.

Frons with velvet black M-shaped frontal vitta, the arms of the M broad, interfrontal plates long and slender, posterior ends at or posterior to level of hindmost orbital bristles; frontal triangle short, apex barely before median ocellus; cheek 1/3 breadth of 3rd antennal segment and 1/9 height of an eye, gray, with distinct but narrow subocular crescent, 2nd subgenal seta behind vibrissa more or less well developed; face weakly concave in profile, vibrissal angle not produced, about an 80° angle, and lateroventral corner of facial plate flat and dull, not polished; 3rd antennal segment small, only little larger than 2nd; palpus clavate.

Thorax dark gray microtomentose; pleuron chiefly gray, including entire propleuron and areas surrounding anterior spiracle, with large, bilobed (cf. fig. 23), polished black spot posterodorsad to fore coxa, including an anteroventral area of mesopleuron. Fore coxa and fore femur not elongate in male. Section of costa between costal breaks with few coarse setae, 6–7 in number. Length, 1.5 mm.

Holotype male, allotype, and 3 female paratypes, SRI LANKA: Kandy District, Udawattakele, 1800 ft., Nov. 19, 1976 (G.F. Hevel, R.E. Dietz, S. Karunaratne, D.W. Balasooriya)[Washington, paratype in Colombo]. Other paratypes: SRI LANKA: 2 females, Northwest Province, Bangadeniya, 4 mi. nne. Chilaw (Brinck & Cederholm, on flowers)[Lund, from the Lund University Ceylon Expedition 1962].

One male, PHILIPPINES: Luzon, La Trinidad, May 1914 [Helsinki] appears to belong here but it is in poor condition and decision on its identity must await better material from those islands.

The bristlelike 2nd subgenal seta, the bilobed pleural spot, and the unusually sparse setae on the costa between the costal breaks will spot this species as near _ciliata_.

In the holotype, both interfrontal plates are interrupted midway, but they are continuous in the other specimens and the interruption is undoubtedly an aberrant condition.

The specific name is a noun in the genitive case, from the name of the country Sri Lanka.

36. _Desmometopa_ propeciliata, n. sp.

Tiny species near _ciliata_ and _srilankae_, differing from the latter in a combination of characters as shown in the key, and from _ciliata_ in having brown–black halter and narrower cheek with linear subocular crescent.

Male, female. Chiefly black; 3rd antennal segment orange–yellow on basoventral half in male; palpus chiefly yellow in male, broadly infuscated distally in female; all tarsi yellowish toward base.

Frons velvet black, frontal vitta delineated by distinct but slender interfrontal and fronto–orbital plates, the frontal triangle short, apex not exceeding median ocellus; cheek of moderate width, 1/3 breadth of 3rd antennal segment and 1/12 height of an eye; 2nd subgenal seta behind vibrissa a rather strongly developed bristle, suggestive of _ciliata_ (cf. fig. 11); face weakly concave in profile with vibrissal angle not produced, about an 80° angle, and latero–ventral corner of facial plate dull gray and flat; 3rd antennal segment small, little longer than 2nd segment; palpus clavate.

Thorax dark brownish gray microtomentose; pleuron chiefly gray, including entire propleuron and areas surrounding anterior spiracle, with large polished black spot posterodorsad to fore coxa, and including narrow anteroventral area of mesopleuron, the polished spot not bilobed (cf. fig. 24). Fore coxa not elongate. Section of costa between costal breaks with 7 coarse and well spaced setae, each much longer than diameter of costa. Length, 1.5 mm.

Holotype male, allotype, and 3 paratypes (male, 2 females), MALAYA: Pahang, Tahan River, George V National Park, Nov. 5, 1959 (H.E. McClure, light trap)[Washington]. Other paratypes: THAILAND: female, Nonthaburi, Dec. 20, 1958 (Manop)[Washington]. JAVA: 3 males, 1 female, Bogor, Apr.–May 1954 (A.H.G. Alston)[London].

This species is closest to _srilankae_ and the combination of characters for each is shown in the key. The developed subgenal bristle and long, coarse, well–spaced setae on costa between the humeral and subcostal breaks clearly relate the species to _ciliata_.

The specific name is an adjective from the Latin _prope_, near, plus _ciliata_, the name of a similar species.

37. _Desmometopa_ magnicornis, n. sp.
(Fig. 21)

Entirely black, narrow cheek lacking a subocular crescent, and male with exceptionally large 3rd antennal segment.

Male, female. Entirely black to brown–black, including palpi, halteres, and tarsi.

Frons with velvet black frontal vitta, the gray interfrontal and fronto-orbital plates and frontal triangle distinct; inter-frontal plates long, 2/3 length of frons and extending posteriorly to level of apex of the approximately equilateral frontal triangle, the plates widely divergent and interval between them wider than that between one of them and adjacent fronto-orbital plate; cheek narrow, 1/3 breadth of 3rd antennal segment and 1/10 height of an eye, entirely gray without visible subocular crescent; face weakly concave, vibrissal angle about an 80° angle, not produced, and lateroventral corner of facial plate not developed; 3rd antennal segment of male exceptionally large, 2 to 3 times the length and breadth of 2nd segment and extending to lower margin of face (fig. 21); palpus small, clavate.

Thoracic pleuron densely dark gray microtomentose, including entire propleuron and areas surrounding anterior spiracle; a large polished black spot posterodorsad to fore coxa and including anteroventral area on mesopleuron (cf. fig. 24). Fore coxa convex, short. Length, 1.5 mm.

Holotype male, allotype, and 6 paratypes (1 male, 5 females), NIGERIA: Ibadan, Gambani Forest, Feb. 1965 (R.W. Williams, "reared cacao pods") [Washington]. Other paratypes: NIGERIA: female, Ibadan, Jan. 7, 1963 (D.C. Eidt, Malaise trap)[Ottawa]. IVORY COAST: female, 'Lamto', Frange, Afrayomum, Mar. 20, 1971 (D. Lachaise) [Paris].

This entirely black species is easily distinguished by the absence of a subocular crescent on the cheek, and by the exceptionally large 3rd antennal segment of the male. The reared series was obviously mounted from fluid and is paler in color than one would expect. The paratype from a Malaise trap is fully colored and is the basis of the color description. See discussion at end of nudigena for separation of the three species with black-brown halteres and extra large 3rd antennal segments in the male, nudigena, philippinensis, and magnicornis.

The specific name is an adjective referring to the large antennal segment.

38. Desmometopa sp. M

A single female from TAIWAN (as Formosa): Chipun, July 1912 (H. Sauter)[Eberswalde], identified in the collection as D. tristicula, appears to represent a new species near sordida but with narrow cheeks and sublinear polished subocular crescent. The interfrontal and fronto-orbital plates are fairly broad although nowhere near the pattern in interfrontalis. A relatively strong subgenal bristle is present, although not as strongly developed as in ciliata (cf. fig. 11). At least the first two tarsomeres are yellowish in all tarsi. The 3rd antennal segment is small, which is normal for females. The fore coxa is a little longer than usual, and it is possible that the male will be found to have elongate fore coxa, although not necessarily so. The polished black spot posterodorsad to the fore coxa is bilobed and includes an anteroventral area on the mesopleuron (cf. fig. 23).

39. Desmometopa philippinensis, n. sp.

Tiny species near ciliata, but with large 3rd antennal segment in both sexes.

Male, female. Chiefly black or black-brown; all tarsi yellowish except distally; knob of halter brown.

Frons with broad velvet black frontal vitta distinctly delineated by interfrontal and fronto-orbital plates and short frontal triangle, apex of latter not reaching middle of frons; cheek less than 1/2 breadth of 3rd antennal segment and 1/8 height of an eye, uniformly gray, without polished subocular crescent; 2nd subgenal seta developed bristlelike (cf. fig. 11); face only weakly concave, the vibrissal angle not produced, an 80° - 90° angle, and lateroventral corner of facial plate flat and dull; 3rd antennal segment in both sexes larger than usual, and larger in male than in female, in male similar to but not as extreme as in fig. 21; palpus clavate.

Thoracic pleuron brownish gray microtomentose, including entire propleuron and areas surrounding anterior spiracle, with large polished black spot posterodorsad to fore coxa that includes an anteroventral area on mesopleuron, anterior margin of spot approximately straight, not bilobed (cf. 24). Fore coxa of male not elongate. Section of costa between the costal breaks with few (8-9) coarse and well-spaced setae. Length, 1.5 mm.

Holotype male, allotype, and a female paratype, PHILIPPINES: Manila (Robert Brown)[Washington].

The large antennae will distinguish this species in the ciliata group. See discussion at end of nudigena. Unlike the other species with black-brown halteres and extra large 3rd antennal segments in the male, nudigena and magnicornis, this species has a developed subgenal bristle, second behind the vibrissa, and outstanding in the subgenal row.

The specific name is an adjective referring to the Philippine Islands.

40. Desmometopa sp. N

Two females, and possibly a third, appear to represent a new species but it will not be named at this time. It is one of several tiny species "near" sordida in the sense of generally dark appearance, black palpi and black halteres. The large polished black pleural spot posterodorsad to the fore coxa includes an anteroventral area of the mesopleuron (cf. fig. 24). The 3rd antennal segment is small and the fore coxa short, but these features are usual in females and do not necessarily indicate the appearance of the males.

MALAYA: Two females, Pahang, Tahan River, George V National Park, Nov. 5, 1959 (H.E. McClure, light trap), and Perak, Pulau Panghor, Apr. 1, 1959 (R. Traub, light)[both, Washington]. One female, PHILIPPINES: Port Bauge, Jan. 1915 [Helsinki] is not in good condition but is tentatively associated here.

41. Desmometopa aldabrae, n. sp.

Small species with palpus almost entirely yellow in both sexes.

Male, female. Black, with predominantly yellow palpus, slightly infuscated at extreme apex; knob of halter brownish; tarsi yellow except for distal tarsomere or two.

Frons relatively broad, nearly square, the interfrontal plates, broad fronto-orbital plates, and frontal triangle gray and distinct, the sections of the M-shaped frontal vitta relatively narrow; frontal triangle moderately long, its apex nearly midway on the frons and well anterior to the posterior ends of interfrontal plates; cheek narrow, no more than 1/3 breadth of 3rd antennal segment and 1/8 the height of an eye, subocular crescent absent or indistinct; face weakly concave, vibrissal angle an 80° - 90° angle, not produced anteriorly; 3rd antennal segment small in female, moderately large in male; palpus clavate, approximately same in both sexes.

Thorax heavily gray microtomentose, propleuron and area surrounding anterior spiracle entirely so, a small polished black spot posterodorsad to fore coxa that barely or not at all encroaches on the mesopleuron (cf. fig. 25). Fore coxa and femur of male ordinary, not elongate. Costa between costal breaks with 8-9 fine and erect dorsal setae. Length, 1.25-1.5 mm.

Holotype male, ALDABRA: South Island, Flamingo Pool, Jan. 21-22, 1968 (B. Cogan, A. Hutson), and allotype, Dune Jean-Louis, Mar. 13-20, 1968 (Cogan and

Hutson, at light)[London, collected on the Aldabra Atoll Royal Society Expedition 1967–68].

This is a distinctive little species, dark but lightened by the predominantly yellow palpi in both sexes.

The specific name is a noun in the genitive case, from Aldabra.

42. Desmometopa kandyensis, n. sp.

Subshining velvet black frontal vitta, short and equilateral frontal triangle, and small polished black pleural spot.

Male, female. Black, heavily gray microtomentose; knob of halter yellow; tarsi somewhat yellowish basally in female.

Frons with M–shaped frontal vitta subshining velvet black viewed from any angle, the interfrontal and fronto–orbital plates and frontal triangle gray microtomentose and sharply distinct; interfrontal plates long, reaching level of uppermost fronto–orbital bristles and opposite apex of very short frontal triangle which is virtually coextensive with ocellar tubercle; cheek narrow, little over 1/3 breadth of 3rd antennal segment and 1/9 eye height, with narrow polished subocular crescent; face moderately concave in profile, vibrissal angle produced to a 45° angle and lateroventral corner of facial plate shining black and warped forward; 3rd antennal segment small in both sexes; palpus gently clavate in both sexes.

Thoracic pleuron heavily gray microtomentose, including entire propleuron and areas surrounding anterior spiracle, only a small polished black spot postero-dorsad to fore coxa, not at all encroaching on mesopleuron (cf. fig. 25). Fore coxa slightly elongate in male, but fore femur not so. Abdominal tergum 5 unusually long, longer than 3 and 4 combined; sternum 5 likewise elongate, much longer than broad and longer than sternum 4, with numerous discal setae (40–50).

Length, 1.5 mm.

Holotype male, allotype, and 2 male paratypes, SRI LANKA: Kandy District, Udawattakele, 1800 ft., Nov. 19, 1976 (G.F. Hevel, R.E. Dietz, S. Karunaratne, D.W. Balasooriya)[Washington, one paratype in Colombo].

The combination of velvet black frontal vitta and short frontal triangle will separate this species from all but gressitti, from which it is easily distinguished by the several characters noted in the key. Few species in the genus have the pleural spot so restricted, not encroaching on or including an anteroventral area on the mesopleuron.

The specific name is an adjective based on the name Kandy District.

43. Desmometopa gressitti, n. sp.
(Fig. 4)

Frontal triangle short, polished pleural spot large, and both halter knob and tarsi infuscated.

Male, female. Black, heavily gray to brown–gray microtomentose, halter knob and all tarsi infuscated.

Frons with frontal vitta subshining velvet black, with long and distinct interfrontal plates, their posterior ends at or posterior to level of uppermost orbital bristles and opposite apex of short frontal triangle, which is barely if at all in advance of median ocellus; cheek narrow, 2/3 breadth of 3rd antennal segment and less than 1/8 height of an eye, with narrow polished subocular crescent; face deeply concave in profile, the vibrissal angle produced anteriorly to a 45° angle, and lateroventral corner of facial plate shining and warped forward beyond vibrissal angle; 3rd antennal segment small in both sexes; palpus clavate, small in female, but conspicuously broad and flat in male (fig. 4).

Thoracic pleuron heavily gray microtomentose, including entire propleuron and areas surrounding anterior spiracle, with large polished black spot postero-dorsad to fore coxa that includes an elongate-oval anteroventral area on mesopleuron (cf. fig. 24). Fore coxa and fore femur slightly elongate in male, the former nearly 3 times as long as broad. Abdominal tergum 5 of male not elongate, barely longer than tergum 4; sternum 5 large, approximately square, with numerous discal setae (40-45). Length, 2-2.25 mm; occasional males as small as 1.5 mm.

Holotype male, allotype and 16 paratypes (10 males, 6 females), MARSHALL ISLANDS: Jaluit Atoll, Jabor Island, Apr. 25 (allotype), 26, 27, and May 1 (holotype), 1958 (J.L. Gressitt)[Honolulu]. Other paratypes [Honolulu except as noted]: MARSHALL ISLANDS: 21 males, 5 females, Jaluit Atoll, Majurirek Island, Apr. 26, 1958 (J.L. Gressitt; 4 labeled "_Hernandia_ flowers"); 2 males, 1 female, Jaluit Atoll, Pinlep Island, Apr. 25, 1958 (Gressitt). CAROLINE ISLANDS: male, Kusaie Island, Matanluk (Yepan), 16 m, Jan. 23, 1953 (Gressitt, light trap); female, Ponape Island, s. of Nanponmal, Jan. 17, 1953 (J.F.G. Clarke); 2 females, Truk, S. Valley Mt. Tonaachau, Moen, Apr. 2, 1949 (R.W.L. Potts; one labeled "ex papaya log"; 2 males, 1 female, Ulithi Atoll, Falalop Island, Apr. 30, 1952 (J.W. Beardsley). GILBERT ISLANDS: Butaritari Atoll, Butaritari Island, Dec. 1957 (N.L.H. Krauss). SOLOMON ISLANDS: 4 males, Guadalcanal, 1944 (C.O. Berg)[Washington]. NEW HEBRIDES: 6 males, 3 females, Efate Island, Vila, 0-100 m, Feb. 1969 (N.L.H. Krauss)[Washington]; 6 males, same locality, Feb. 1970 (N.L.H. Krauss)[London].

I have also seen a number of other specimens that duplicate the above records, but their poor condition prevents their inclusion in the type series. One that does add slightly to the known distribution is a male from the New Hebrides, Espiritu Santo, Sept. 1944 (K.L. Knight)[Washington]. One male from Australia, Northern Territory, Darwin, Sept. 1908 (Lichtwardt)[Eberswalde] is possibly this species, but the halter knob is quite yellowish, possibly the result of the teneral condtion of the specimen.

The nearest relative appears to be _flavipalpis_, which occurs on some of the same islands. The two species share the same combination of short frontal triangle and infuscated tarsi and knob of halter, but they differ in the color and shape of the palpus. The difference in color of the palpus, all black in _gressitti_ and partly (female) to chiefly (male) yellow in _flavipalpis_, might be denigrated as possibly mere color variation, but the palpal shape is certainly more significant. In males of _gressitti_, most noticeable in average to large specimens, the palpi are very broad and flat (fig. 4), whereas in _flavipalpis_ they are gently clavate, as in the female.

I dedicate this species to the memory of the collector, J. Linsley Gressitt, my friend of many years, entomologist extraordinary in the Pacific area, who perished in a plane crash in China.

44. _Desmometopa_ _saguaro_, n. sp.
(Fig. 22)

Chiefly polished pleuron, with narrow stripe of gray microtomentum dorsally on mesopleuron behind anterior spiracle; in male, fore coxa and femur elongate, raptorial in appearance.

Male, female. Black; knob of halter yellow; basal tarsomere at least partly yellow on all legs.

Frons gray microtomentose from most angles of view, obscuring the interfrontal plates which are discontinuous, a series of shining spots about bases of interfrontal setae; frontal triangle only slightly extended anterior to median ocellus; cheek about 2/3 breadth of 3rd antennal segment and 1/6 eye height, with polished subocular crescent of moderate width; parafacial visible in profile; face deeply concave in profile, the vibrissal angle produced to a 45° angle, lateroventral corner of facial plate shining black and warped forward even beyond vibrissal angle; 3rd antennal segment small in both sexes; palpus clavate, in male very broad and flat, as in fig. 4 but longer, and at rest projecting even beyond the antennae.

Thoracic pleuron chiefly polished, including entire propleuron and most of meso- and sternopleuron, the mesopleuron gray microtomentose dorsally and posteriorly, a narrow dorsal stripe reaching to anterior spiracle. Fore leg raptorial in appearance in male, fore coxa and fore femur greatly elongate (fig. 22), the former about 4 times as long as broad and its apex approximately opposite base of wing, fore femur 1.6 times as long as mid femur, somewhat incrassate, with anteroventral and posteroventral rows of short, even, straight spines or spinelike bristles, the postero-ventral weak; fore coxa and fore femur of ordinary size in female, neither elongated nor enlarged. Length, 1.5-2.5 mm.

Holotype male, allotype, and 13 paratypes (6 males, 7 females), ARIZONA: Pima Co., Saguaro National Monument (F.J. Santana, "ex rotting Saguaro"), collected Mar. 10-June 25, 1960, emerged in laboratory at Tucson, Mar. 18-July 8[Washington, paratypes at Tucson]. Other paratypes: ARIZONA: female, Tucson, Aug. 8, 1937 (O. Bryant)[San Francisco]. CALIFORNIA: male, Andreas Canyon, Palm Springs, Mar. 11, 1955 (W.R.M. Mason)[Ottawa]; male, Morongo Valley, Oct. 5, 1934 (A.L. Melander)[Washington].

In addition, but not part of the type series because of teneral condition, I have seen a long series from CALIFORNIA: Los Angeles Co., San Dimas Canyon, Apr. 16, 1958, reared June 11, 1958 (R.E. Ryckman, ex Opuntia)[Loma Linda and Washington], and one male, ARIZONA: Maricopa Co., Wickenburg, Aug. 1950 (H.K. Gloyd, light)[Washington].

The raptorial fore legs are unlike most other Desmometopa. The nearest species is melanderi, and I am a little uncertain about their relationship. Both have been collected in San Dimas Canyon. Aside from the gray postspiracular stripe, however, there are a few tangible differences: the palpi in saguaro are definitely longer and broader, the fore coxa and fore femur are longer, but the spine rows on the femur are weaker than in melanderi, and the tarsi are yellow.

The specific name is a noun in apposition from the common name of the giant cactus from which the holotype and topotypic specimens were reared.

45. Desmometopa tarsalis Loew
(Figs. 18, 26)

Desmometopa tarsalis Loew, 1866, Berl. Entomol. Ztschr. (1865) 9: 184 (Cent. 6, no. 96) (Cuba)[Cambridge].
Platophrymyia nigra Williston, 1896, Trans. Entomol. Soc. London 1896: 426 (St. Vincent)[London](Synonymy by Sabrosky 1973).
D. tarsalis; Bohart and Gressitt, 1951, Bull. Bishop Mus. 204: 99(Guam).
Desmometopa sp.; Hardy, 1952, Proc. Hawaiian Entomol. Soc. 14: 474(Hawaii).
D. tarsalis; Hardy and Delfinado, 1980, Insects of Hawaii 13 (Diptera Cyclorrhapha III): 357-8, figs.

Chiefly polished pleuron with gray stripe to anterior spiracle, gray frons, and normal (not elongate) fore coxa and fore femur.

Male, female. Black; knob of halter yellow; mid and hind tarsi except distal tarsomere or two, and usually basal tarsomere of fore leg, yellow.

Frons with interfrontal and fronto-orbital plates and frontal triangle distinct but most of the M-shaped frontal vitta dull, gray microtomentose viewed from most angles, only narrow, long-oval velvet black spots flanking the frontal triangle; cheek narrow, 2/5 breadth of 3rd antennal segment and 1/9 eye height, with moderately narrow polished subocular crescent (fig. 18); face deeply concave in profile, vibrissal angle produced anteriorly to a 45° angle, lateroventral corner of facial plate shining black and warped forward even beyond vibrissal angle; 3rd antennal segment small in both sexes; palpus gently clavate, not enlarged in male.

Thoracic pleuron chiefly polished, especially propleuron ventrally and mesopleuron chiefly, the latter gray microtomentose posteriorly and dorsally, the gray extending anteriorly to anterior spiracle (fig. 26); typically the sternopleuron chiefly gray with large polished spot on middle so that the pattern of gray microtomentum is that of a thick U, open posteriorly. Fore coxa and fore femur not elongate, in both sexes short and not raptorial. Length, 1.5-2 mm.

The characters given in the key will serve to distinguish the species. It keys near saguaro, but the two are not necessarily related: the development of raptorial front legs in that species makes it obviously distinct from tarsalis.

Loew described tarsalis from 'male and female', without recording the number of specimens. Through the kind cooperation of Norman E. Woodley, I received for study from the Museum of Comparative Zoology at Harvard University four specimens that I accept as the type series, glued on two card points, each labeled with a small silver square [meaning Cuba, collected by Gundlach], an old printed label "Loew Coll.", and a red MCZ label "Type 13443". One of the points, with two females, also has a label, "tarsalis m.", apparently in Loew's handwriting. The other point has a male and a female, and I have labeled and here designate the male as the lectotype. The holotype of P. nigra was studied at the British Museum (Nat. Hist.) some years ago.

This is a widely distributed, yet also widely misidentified, Neotropical species that has been transported to Hawaii, and to Wake Island and the South Pacific. Most of the South Pacific records are from the end of World War II or later, and could have been associated with the movements of American military forces to and among the islands. However, at least some introductions may have occurred earlier; Bohart and Gressitt (1951) found it to be "one of the commonest flies on Guam" in 1945.

Distribution (confirmed records, after revision):

Nearctic: Arizona, Texas.

Neotropical: Mexico (13 states from Baja California and Tamaulipas south to Yucatán), Guatemala, Belize, El Salvador, Nicaragua, Honduras, Costa Rica, Panama, Colombia, Ecuador, Venezuela, Guyana, Tobago, Grenada, St. Vincent, Barbados, St. Lucia, Dominica, Montserrat, Virgin Islands, Puerto Rico, Bahamas, Dominican Republic, Cuba, Jamaica.

Pacific: Galapagos, Hawaii, Wake Island, Marianas (Guam, Saipan, Tinian), Fiji, Solomons (Guadalcanal).

Misidentifications: tarsalis of European authors, at least of records from Europe, is usually varipalpis, of the material I have seen.

Sabrosky's (1965b) tarsalis in the Nearctic Catalog refers chiefly to new species described in this paper, except for Arizona and Texas records in part.

Johnson's (1913) record of tarsalis from Biscayne Bay, Fla. (Mrs. Slosson) actually refers to Milichiella sp. near cinerea (Coquillett).

Bezzi's (1928) record of tarsalis from Fiji was based on females of singaporensis. [I have, however, seen other material of true tarsalis from Fiji].

Malloch's (1914) tarsalis from Formosa is singaporensis, from the specimens in the Museum at Budapest. Another female, Takao, Formosa, April 17, 1907, not published by Malloch but apparently identified by him, is microps.

46. Desmometopa nearctica, n. sp.

Velvet black frontal vitta, chiefly polished propleuron, stripe of gray microtomentum across dorsal edge of mesopleuron to anterior spiracle.

Male, female. Black; knob of halter yellow; mid and hind tarsi with 2 to 3 proximal tarsomeres yellow, basal tarsomere on fore tarsus sometimes yellowish, at least toward base.

Frons with M-shaped frontal vitta subshining velvet black and the gray interfrontal and fronto-orbital plates and frontal triangle distinct; interfrontal plates strong and moderately long, extending posteriorly to level of foremost upper orbital bristles; frontal triangle moderately long, apex at or slightly anterior to level of posterior ends of interfrontal plates; cheek narrow, 1/2 breadth of 3rd antennal segment and 1/7 the eye height, with polished subocular crescent that is wider anteriorly and continuous with a polished, sometimes narrowly visible parafacial; face deeply concave, vibrissal angle produced anteriorly to a 45° angle, the latero-ventral corner of facial plate shining black and warped forward so as to exaggerate the vibrissal angle; 3rd antennal segment small in both sexes; palpus gently clavate, not enlarged in male.

Thoracic pleuron predominantly polished, including propleuron (except narrow dorsal strip) and much of mesopleuron, latter with gray stripe of microtomentum along posterior and dorsal margins up to anterior spiracle (cf. fig. 26); sternopleuron gray microtomentose above and below, broadly polished centrally, often with vertical gray stripe. Fore coxa and fore femur short, not elongate in male. Length, 1.5 mm.

Distribution: California to Georgia, north to Kansas and New York.

Holotype male, allotype, and 2 male paratypes, CALIFORNIA: Joshua Tree National Monument, Quail Springs, Oct. 5, 1934 (A.L. Melander)[Washington]. Other paratypes [Washington except as noted]: CALIFORNIA (all A.L. Melander except Coachella specimen): female, Riverside Co., Whitewater, near Palm Springs, Oct. 27, 1934; 3 females, Mill Creek, San Bernardino Mts., Aug. 17, 1952; female, Seven Oaks, sw. San Bernardino Co., July 28, 1953; female, Joshua Tree National Monument, May 18, 1946; male, Mountain Home Canyon, w. side San Bernardino Mts., Aug. 9, 1948; male, San Diego Co., Borrego Desert, Tubb Canyon, w. edge of Anza-Borrego State Park, Nov. 9, 1945; female, Coachella, Nov. 20, 1930 (D.G. Hall, reared from grass). ARIZONA: 3 females, Maricopa Co., Buckeye, July 15, 1960 (Ed Schulz, Steiner lure); female, Portal, June 5-9, 1972 (W.W. Wirth, Malaise trap); 2 males, 1 female, Douglas, Aug. 8, 1955 (R.R. Dreisbach)[East Lansing]. UTAH: male, Uintah Co., Bonanza, July 11, 1974 (G.E. Bohart, Tamarix). TEXAS: 2 males, 9 females, Big Bend National Park, various localities, May 1-22, 1959 (J.F. McAlpine, W.R.M. Mason)[Ottawa]; 3 females, Big Bend National Park, Boquillas Canyon, June 20, 1953 (W.W. Wirth). MEXICO: male, female, Nuevo León, Vallecillo, June 2-5, 1951 (P.D. Hurd)[Berkeley]. GEORGIA: 3 females, Tifton, Sept. 24, Oct., and Oct. 16, 1896. KANSAS: 2 females, Manhattan, Apr. 9, 1934 (C.W. Sabrosky) and Aug. 1945 (N.L.H. Krauss); male, Douglas Co., Oct. 4, 1937 (H.M. Smith); 2 males, Atwood, July 23, 1954 (W.L. Downes)[Lansing]. IOWA: male, female, Des Moines, May 17, 1951 (A.H. Sturtevant). NEW YORK: male, female, Cold Spring Harbor, Long Island, August. DISTRICT OF COLUMBIA: Washington, Aug. 23, 1907 (W.L. McAtee).

A few specimens from Colesville, Md. (W.W. Wirth) and Chittenango, N.Y. (D.J. Peckham)[Washington] are puzzling. The frontal vitta is slightly gray microtomentose and thus suggestive of tarsalis, but the localities are far removed from the known range of that species. All specimens available from the two localities are females, so male genitalia cannot be checked. A different species may be involved, but for the present the specimens are considered here as odd variants of nearctica.

The specific name is an adjective referring to the Nearctic Region.

47. Desmometopa argentinica, n. sp.

Agreeing with nearctica in all particulars except proportion of width of subocular crescent to width of cheek (see key), the crescent more evenly rounded throughout, and cheek slightly wider.

Holotype male, allotype, and 5 paratypes (3 males, 2 females), ARGENTINA: Salta, Urundel, Feb. 8–12, 1949 (M. Aczél)[Tucumán]. Other paratypes: ARGENTINA [all Tucumán]: male, Jujuy, Sierra Zaple, Jan. 30, 1949 (M. Aczél); 8 males, 4 females, Santiago del Estero, Monte Potrero, Apr. 13, 1952 (A. Willink). PERU: female, Iquitos, Mar.–Apr. 1931 (R.C. Shannon)[Washington].

In addition to these I have a female from Urundel and 4 males, 10 females from Monte Potrero that are too teneral to include in the type series.

This species is extremely close to nearctica, and like that species it is also very similar to tarsalis, differing in having subshining velvet black frontal vitta. The proportion of width of subocular crescent to width of cheek is obvious in fully mature specimens. Unfortunately, in teneral specimens the collapse of the cheek affects the lower microtomentose portion and narrows it so that the subocular crescent appears over 1/2 the width of the cheek and thus agreeing with nearctica. Most available specimens of argentinica are somewhat teneral. The same tendency in nearctica does no harm because it merely exaggerates the characteristic proportion of crescent to cheek in that species.

The specific name is an adjective derived from the name of the country of origin of the type series.

48. Desmometopa sp. O

A single female undoubtedly represents a distinct species, but in the absence of additional material and males it is left unnamed. The extensively polished pleuron and black halter will distinguish it from other species. The pleuron is microtomentose only on the posterior slope, behind the sterno- and pteropleuron. The vibrissal angle is not produced anteriorly. The lateroventral corner of the facial plate is shining black as in the subgenus Platophrymyia, but not warped forward. The head structure and black halter suggest sordida, but of course the polished pleuron is quite unlike that heavily microtomentose species. The 3rd antennal segment is small, as usual in females. The fore coxa is slightly longer than usual in females, and this may indicate an elongate fore coxa in the male of the species.

Female, MEXICO: Cuernavaca, July 1965 (N.L.H. Krauss) [Washington].

49. Desmometopa stilbopleura, n. sp.

Predominantly polished black pleuron, lemon-yellow knob of halter, and mid and hind tarsi with proximal 2 to 3 tarsomeres yellow.

Female. Black; knob of halter lemon yellow; mid and hind tarsi with proximal 2 to 3 tarsomeres yellow.

Frons with frontal vitta gray microtomentose anteriorly and centrally, broadly velvet black on upper 2/5, on each side of frontal triangle, and anteriorly mesad of the fronto-orbital plates; interfrontal and fronto-orbital plates and frontal triangle strong and distinct; frontal triangle long, apex midway on frons; cheek moderately narrow, 1/2 breadth of 3rd antennal segment and about 1/7 eye height, with polished subocular crescent that broadens anteriorly and continues as narrow polished parafacial visible in profile; face deeply concave, the vibrissal angle strongly produced to a 45° angle, accentuated by shining black lateroventral corner of facial plate which is warped forward beyond vibrissal angle; 3rd antennal segment small; palpus clavate, large, nearly as broad as 3rd antennal segment.

Thoracic pleuron predominantly polished anterior to pleural suture, including propleuron and all but narrow margins on mesopleuron and sternopleuron. Fore coxa slightly longer than usual for females, and this may indicate an elongate fore coxa in the male of the species. Length, 2.5-3(holotype) mm.

Holotype and 2 paratypes, all females, BRAZIL: São Paulo, Nova Teutonia, 300-500 m (F. Plaumann), one paratype Sept. 1965, the others Nov. 1962 [Ottawa, paratype in Washington].

This is such a distinctive species that I have named it even in the absence of males. The microtomentum on the frons is more limited than usual, but it is so consistent in these specimens that I judge it to be characteristic of the species. Typically, in species with dull frons, the gray microtomentum is heavier and more extensive than in this species, completely covering the frons except for a usually elongate-oval velvet black area along each side of the ocellar tubecle. The breadth of the palpus suggests that in the male the palpus will be broadly expanded and flattened (as in fig. 4).

The specific name is a noun in apposition compounded from the Greek stilbo, shine, plus pleura, side.

50. Desmometopa melanderi, n. sp.

Pleuron polished anterior to pleural suture; frons chiefly dull; fore coxa and fore femur of male elongate, raptorial, at least in large specimens.

Male, female. Black; knob of halter yellow; mid and hind tarsi sometimes partly yellow from certain angles, but usually at least infuscated dorsally.

Frons chiefly gray microtomentose except for velvet black areas flanking ocellar tubercle; interfrontal and fronto-orbital plates and frontal triangle subshining and distinct; frontal triangle long, apex nearly midway on frons; cheek narrow, 1/2 or barely over 1/2 breadth of 3rd antennal segment and 1/7 eye height, with relatively broad polished subocular crescent 1/2 as broad as cheek, anteriorly becoming a polished parafacial narrowly visible in profile; face deeply concave in profile, vibrissal angle well produced anteriorly to a 45° angle, the lateroventral corner of facial plate shining black and warped forward even beyond the vibrissal angle, accentuating the angle; 3rd antennal segment small in both sexes; palpus clavate, broad and flat in male (especially striking in large males), but not as broad distally as in fig. 4.

Thoracic pleuron anterior to pleural suture entirely polished or virtually so, including entire propleuron and area surrounding anterior spiracle, gray microtomentose posterior to pleural suture, including entire pteropleuron. Male with fore leg raptorial in appearance, fore coxa and fore femur elongate (similar to fig. 22), especially evident in large specimens, the coxa over 3 times as long as broad, and femur incrassate and 1.3 times as long as other femora, with a row of short, thick, even spines, and a postero-ventral row of similar but weaker spines; fore coxa with numerous short but strong spines; in female fore coxa and fore femur only slightly if at all elongate, without spines or spinelike bristles. Length, 2-3 mm (large males).

Distribution: California, Arizona, Utah, Texas, Mexico.

Holotype male, allotype, and 23 paratypes (19 males, 4 females), CALIFORNIA: San Bernardino Co., Verdemont, San Gabriel Mts., various dates, including May 1, 1946 (holotype) and June 28, 1945 (allotype)(A.L. Melander)[Washington].

Other paratypes [Washington except as noted]: CALIFORNIA [collector A.L. Melander, except as noted]: 3 males, s. San Bernardino Co., Morongo Valley, Oct. 5, 1954; female, Palm Springs, May 6, 1946; 2 males, 1 female, sw. San Bernardino Co., Upper Santa Ana River, June 18(male) and Sept. 2, 1950; male, Ortega Highway, Mariana River, May 15, 1946; male, San Bernardino Mts., Mill Creek, Aug. 17, 1952; male, San Diego Co., Yaqui Well, w. edge of Anza-Borrego State Park, May 10, 1951; 8 males, 7 females, Los Angeles Co., San Dimas Canyon, Nov. 24, 1957, reared Dec. 16, 1957 - Jan. 3, 1958 (C.P. Christianson, J.P. Fonseca; ex Opuntia), and 1 female, same locality, Feb. 2, 1958, reared Mar. 11 (R.E. Ryckman) [Loma Linda]; 4 males, 3 females, Whittier, 1910, 1911 (P.H. Timberlake). ARIZONA: 2 males, Superior, May 18, 1950 (A.L. Melander, "Datura flower"); male, Globe, Oct. 13, 1948 (F.H. Parker); female, Yarnell Heights, May 31, 1935 (P.W. Oman); male, female, Baboquivari Mts., Apr. 25, 1947 (A.L. Melander); male, Yavapai Co., Cherry, Sept. 1968 (Judson May); male, female, Bowie, Dos Cabezas Mts., Oct. 8, 1916 (E.G. Holt); female, Tucson, Nov. 15, 1936 (O. Bryant)[San Francisco]; female, Portal, Sept. 13, 1960 (H.F. Howden)[Ottawa]; female, Congress, Yavapai County, Apr. 23-26, 1967 (D.M. Wood)[Ottawa]; male, Lower Bear Canyon, Tucson, Apr. 13-15, 1967 (D.M. Wood)[Ottawa]. UTAH: male, Washington Co., Leeds Canyon, 1 mi. nw. Leeds, July 19, 1970 (G.F. Knowlton et al.)[Logan]. TEXAS: 17 males, 14 females, Austin, Nov. 9-23, 1958 (Lynn Throckmorton)[Austin]; male, Austin, July 29, 1950; male, Austin, Oct. 27, 1901 (A.L. Melander); 4 males, 2 females, Big Bend National Park, various localities, May 11-26, 1959 (J.F. McAlpine, W.R.M. Mason)[Ottawa]; 2 females, 10 mi.s. Charlotte, Sept. 13, 1955 (W.L. Downes)[Lansing]; male, female, Brewster Co., 25 mi. s. Marathon, Aug. 31, 1977 (Larry Bezark, "collected on Baccharis glutinosa"). MEXICO: 2 males, Ciudad Victoria, Sept. 1965 (N.L.H. Krauss); 4 males, Morelos, Cuernavaca; March (2) and May 1945, and Apr. 1959 (Krauss); male, Morelos, Hacienda Cocoyotla nr. Cuatlan del Rio, July 31, 1944 (Krauss); female, Michoacán, Morelia, June 1965 (Krauss); male, Durango, Nombre de Dios, Aug. 6, 1951 (P.D. Hurd, "Asclepias")[Berkeley]; 3 males, Durango, 11 mi. w. Durango, June 20, 1964 (J.F. McAlpine)[Ottawa]; male, Hidalgo, Ixmiquilpán, 1700 ft., July 29, 1954 (J.G. Chillcott)[Ottawa]; 8 males, 1 female, Hidalgo, Pachuca, 1700 ft., July 29, 1954 (J.G. Chillcott)[Ottawa]; México, Atlacomulco, 8500 ft., Aug. 18, 1954 (J.G. Chillcott), and male, Teotihuacán, 6900 ft., Aug. 12, 1954 (Chillcott)[Ottawa]; 7 males, 2 females, Nayarit, Ahuacatlán, July 18-22, 1951 (P.D. Hurd, 4 males, 1 female "on fls. of Donnelsmithia Hintonii M & C")[Berkeley]; male, Nayarit, 15 km n. of Chapalilla, July 19, 1951 (P.D. Hurd)[Berkeley]; 16 males, 1 female, San Luis Potosí, 10 mi. ne. San Luis Potosí, Aug. 22, 1954 (R.R. Dreisbach) [East Lansing], and 3 males, 1 female, same locality and date, (J.G. Chillcott)[Ottawa]; 2 males, 1 female, San Luis Potosí, 5 mi. e. Ciudad del Maiz, Aug. 23, 1954 (Dreisbach) [East Lansing].

The raptorial fore legs of the male immediately suggest saguaro, but that species has the dorsal stripe on the mesopleuron extending up to the anterior spiracle, as in tarsalis, and other differences are noted under saguaro. It may be noteworthy that both saguaro and melanderi were collected in San Dimas Canyon, but the former in the spring and the latter in midwinter. The possible consistency or significance of this apparent seasonal difference is not known.

The specific name is a noun in the genitive case, named in honor of my old friend, the late A.L. Melander, ardent collector whose material will enrich entomological studies for years to come.

51. Desmometopa aczeli, n. sp.

Predominantly polished pleuron; knob of halter yellow; all tarsi infuscated; fore leg of male not raptorial.

Male, female. Black; halter knob yellow.

Frons heavily gray microtomentose except for velvet black spot on each side of ocellar tubercle; interfrontal and fronto-orbital plates and frontal triangle distinct because slightly shining; interfrontal plates narrow and short, posterior ends at level of foremost upper orbital bristles; frontal triangle large, apex at middle of frons and opposite posterior ends of the short interfrontal plates; cheek moderately narrow, 1/2 breadth of 3rd antennal segment and 1/6 eye height, with narrow polished subocular crescent that continues as a shining parafacial visible in profile; face weakly concave in profile, vibrissal angle not produced, about 80°, lateroventral corner of facial plate shining black but not strongly warped forward; 3rd antennal segment small in both sexes; palpus gently clavate, not enlarged in male.

Thoracic pleuron virtually entirely polished black anterior to pleural suture, the pteropleuron and posterior slope dull, microtomentose. Fore coxa and fore femur not elongate, without spines, fore femur slightly incrassate but little longer than other femora. Length, 1.75-2 mm.

Holotype male and paratype male, ARGENTINA: Mendoza, Vista Flores, Jan. 31, 1950 (M.L. Aczél); paratype male, Mendoza, Cacheuto, Feb. 5, 1953 (M.L. Aczél)[Tucumán, paratype in Washington].

Even though the various characters lead this species to the final couplet with melanderi, the two are not closely related. That species is large, with raptorial fore legs, strongly produced vibrissal angle, and deeply concave face. In its fundamental characters, aczeli is closer to stilbopleura, for which the chiefly yellow mid and hind tarsi and the slightly microtomentose frons are distinctive.

The specific name is a noun in the genitive case, named in memory of Martin Aczél, enthusiastic entomologist whose untimely death cut short a fruitful career in Argentine entomology and the taxonomy of Diptera.

SPECIES OF "Desmometopa" NOW REFERRED ELSEWHERE

Agromyza albipennis Meigen 1830: Agromyza (Agromyzidae) (See separate
    discussion after this list)
A. annulimana von Roser 1840: synonym of Leptometopa latipes (Meigen)
A. annulitarsis von Roser 1840: ditto
Madiza annulitarsis Zetterstedt 1848: ditto
Desmometopa anuda Curran 1936: Neophyllomyza
D. approximatonervis Lamb 1914: Neophyllomyza
D. fascifrons Becker 1907: synonym of Leptometopa niveipennis (Strobl)
Opomyza flavipes Meigen 1830: synonym of Phyllomyza securicornis Fallén
Madiza griseola Wulp 1871: synonym of Tethina illota Haliday (Tethinidae)
Desmometopa halteralis Coquillett 1900: Leptometopa
Agromyza latipes Meigen 1830: Leptometopa
Desmometopa luteola Coquillett 1902: synonym of Stomosis innominata (Williston)
Agromyza minutissima Wulp 1897: preoccupied name, renamed Desmometopa
    wulpi Hendel, now in Neophyllomyza

Siphonella niveipennis Strobl 1898: Leptometopa
Desmometopa simplicipes Becker 1907: synonym of Leptometopa niveipennis
    (Strobl)
Desmometopa wulpi Hendel 1907: new name for Agromyza minutissima Wulp,
    preoccupied: Neophyllomyza.

Agromyza albipennis Meigen

Agromyza albipennis Meigen, 1830, Syst. Beschr. 6: 171 (Europe).
A. albipennis; Becker, 1902, Ztschr. Hymenop. Dipt. 2: 339 [Two specimens in
    Winthem Collection in Vienna and one in Paris found to belong to Agromyza].
A. albipennis; Hendel, 1931, Agromyzidae, in Lindner, Fliegen Palaeark. Region,
    Fam. 59: 98.
Desmometopa albipennis (Meigen) Séguy, 1934, Faune de France 28: 641
    [Synonyms listed as D. tarsalis Loew of Becker 1907 and D. singaporensis
    Kertész].
A. albipennis Meigen, 1976, Abbildung der europaeischen zweiflügeligen
    Insecten, nach der Natur, Pars III. Beitr. Ent. 26: pl. ccxvi, fig. 7.

The species has long been treated as a valid species in the family
Agromyzidae. However, Séguy (1934) referred it to Desmometopa "sec typ.", with
tarsalis Loew (sensu Becker) and singaporensis Kertész as synonyms. The species
was said by Meigen to be in the Winthem collection, and that collection contains
two examples under the name albipennis, both Agromyza as noted by Becker
(1902) in his review of the Meigen types. Meigen's original description does not
apply to either of the species present in the syntype series of D. singaporensis
(q.v.) in such features as shining black-green body, white halteres, whitish wings,
and rather large black antennae, which are indeed features of Agromyza
albipennis. Further, the wing figured by Meigen himself (1976) is that of
Agromyza, not Desmometopa. I believe, therefore, that the Winthem specimens
must be regarded as the original syntypes, hence in Agromyza, and that the Paris
specimen was a later and erroneous addition, and misidentified. Incidentally, it is
actually D. m-nigrum (Zett.) and not D. singaporensis!

LITERATURE CITED
(other than references in synonymies)

Becker, Th. 1902. Die Meigen'schen Typen der sogen. Muscidae acalypterae (Muscaria holometopa) in Paris und Wien. Ztschr. Hymenop. Dipt. 2: 209-256, 289-320, 337-355.

_____ 1907. Desmometopa. Wien. Entomol. Ztg. 26: 1-5.

Bezzi, M. 1928. Diptera Brachycera and Athericera of the Fiji Islands. British Museum (Nat. Hist.), pp. viii + 220.

Bohart, G.E., and J.L. Gressitt. 1951. Filth-inhabiting flies of Guam. Bernice P. Bishop Mus. Bull. 204: vii + 152.

Curran, C.H. 1934. The families and genera of North American Diptera. 512 pp., New York.

Duda, O. Beitrag zur Kenntnis der Paläarktischen Madizinae (Dipt.). Natuurhist. Maandblad 24: 14-16, 24-26, 37-40.

Grensted, L.W. 1956. On the gender of the generic names Desmometopa and Leptometopa (Dipt., Milichiidae). Entomol. Monthly Mag. 92: 405.

Gressitt, J.L. 1956. Desmometopa tarsalis Loew [Note]. Proc. Hawaiian Entomol. Soc. 16(1): 4.

Griffiths, G.C.D. 1972. The phylogenetic classification of Diptera Cyclorrhapha with special reference to the structure of the male postabdomen. Series Entomologica (Dr. W. Junk N.V.) 8: 341 pp.

Hardy, D.E., and M.D. Delfinado. 1980. Insects of Hawaii, 13 (Diptera: Cyclorrhapha III): 451 pp.

Hennig, W. 1937. Milichiidae et Carnidae. (Fam.) 60a: 91 pp., in Lindner, Die Fliegen der Palaearktischen Region, Lfg. 115.

_____ 1939. Beiträge zur Kenntnis des Kopulationsapparates und der Systematik der Acalyptraten. II. Tethinidae, Milichiidae, Anthomyzidae und Opomyzidae. (Diptera). Arb. Morphol. Taxon. Entomol. Berlin-Dahlem 6: 81-94.

_____ 1941. Verzeichnis der Dipteren von Formosa. Entomol. Beih. Berlin-Dahlem 8: iv + 239.

Illingworth, J.F. 1926. Desmometopa m-nigrum (Zett.)[Note]. Proc. Hawaiian Entomol. Soc. 6(2): 224.

_____ 1929. Desmometopa tarsalis Loew [Note]. Proc. Hawaiian Entomol. Soc. 7(2): 233-4.

International Commission on Zoological Nomenclature. 1958. Declaration 39: Review under Copenhagen Decision 85 of the Rules relating to the gender to be attributed to certain classes of generic names...Opin. Declar. Internat. Commn. Zool. Nomen. 19(4): i-xviii.

_____ 1961, 1964. International Code of Zoological Nomenclature, pp. xviii + 176. (2nd Edition, 1964, pp. xx + 176).

Johnson, C.W. 1913. Insects of Florida. I. Diptera. Bull. Amer. Mus. Nat. Hist. 32: 37-90.

Kato, M., and Katsushige Hori. 1952. Studies on the associative ecology of insects. VI. Larval association of flies during the summer in Sendai and its vicinity, Japan. Tokyo Univ. Sci. Rpts., ser. 4 (Biol.), 19: 238-246.

Knab, F. 1915. Commensalism in Desmometopa (Diptera, Agromyzidae). Proc. Entomol. Soc. Wash. 17: 117-121.

Konow, F.W. 1907. [Footnote to review of Becker 1907 on Desmometopa]. Ztschr. Hymenop. Dipt. 7: 335.

Malloch, J.R. 1914. Formosan Agromyzidae. Ann. Mus. Nat. Hungarici 12: 306-336, pls. 9-10.

_____ 1924. Notes on Australian Diptera. No. iii. Proc. Linn. Soc. N.S. Wales 49: 329-338.

_____ 1927. Notes on Australian Diptera. No. x. Proc. Linn. Soc. N.S. Wales 52: 1-16.

_____ 1934a (June 23). Insects of Samoa, Part VI. Diptera, Fasc. 8: 267-328 [British Mus. (Nat. Hist.)].

_____ 1934b (Nov. 24). Acalyptrata (concluded). pp. 393-489, pl. 8, text-figs. 69-84, in Diptera of Patagonia and South Chile, Part VI, fasc. 5 [British Mus. (Nat. Hist.)].

McMillan, R.P. 1975. Observations on flies of the family Milichiidae cleaning Araneus and Nephila spiders. Western Australian Naturalist 13: 96.

Meigen, J.W. (ed., G. Morge). 1976. Dipteren-Farbtafeln nach dem bisher nicht veröffentlichten Original-Handzeichnungen Meigens. Pars III: Farbtafeln CLXI-CCCV. Beitr. Entomol. 26(2): colored plates as noted.

Melander, A.L. 1913. A synopsis of the dipterous families Agromyzinae, Milichiinae, Ochthiphilinae and Geomyzinae. Jour. N.Y. Entomol. Soc. 21: 219-273, 283-300, pl. 8.

Mulla, M.S., and M.M. Barnes. 1957. On laboratory colonization of the eye gnat, Hippelates collusor Townsend. Jour. Econ. Entomol. 50: 813-816.

Nikitin, M.I. 1965. Insects from boxthorn berries and other non-commercial fruits in New South Wales. Australian Jour. Sci. 27: 264-267.

Peyerimhoff, P. de. 1917. Phorésie et commensalisme chez les Desmometopa. Bull. Soc. Entomol. France 1917: 215-218.

Rabaud, E. 1924. Le commensalisme de Desmometopa sordida Fall. La Feuille des Naturalistes 45(n.s.): 18-19.

Richards, O.W. 1953. [On commensalism of Desmometopa with predacious insects and spiders]. Proc. Roy. Entomol. Soc. London, Ser. C., 18: 55-56.

Ryckman, R.E., and C.T. Ames. 1953. Insects reared from cacti in Arizona. Pan-Pacific Entomol. 29: 163-4.

Sabrosky, C.W. 1965a (Feb. 1). East African Milichiidae and Chloropidae (Diptera). Stuttg. Beitr. Naturk. 138: 1-8.

_____ 1965b (Aug. 23). Family Milichiidae. pp. 728-733, in Stone et al., A catalog of the Diptera of America North of Mexico. Agric. Handbook 276: iv + 1696.

_____ 1973. 75. Family Milichiidae, pp. 1-12, in A catalogue of the Diptera of the Americas south of the United States (Museu de Zool., São Paulo, Brazil).

_____ 1977. Family Milichiidae. pp. 270-274, in Delfinado and Hardy, A catalog of the Diptera of the Oriental Region (University Press of Hawaii, Honolulu).

_____ 1980. 75. Family Milichiidae. pp. 686-689, in Crosskey, Catalogue of the Diptera of the Afrotropical Region (British Museum, Nat. Hist., London).

Séguy, E. 1934. Diptères (Brachycères)(Muscidae Acalypterae et Scatophagidae). Faune de France 28: 832 pp.

_____ 1938. Diptera L. Nematocera et Brachycera. pp. 319-380, in Jeannel (ed.), Mission Scientifique de l'Omo, Tome IV (Zoologie).

Swezey, O.H. 1952. Insects from decaying blossoms. Proc. Hawaiian Entomol. Soc. 14(3): 357.

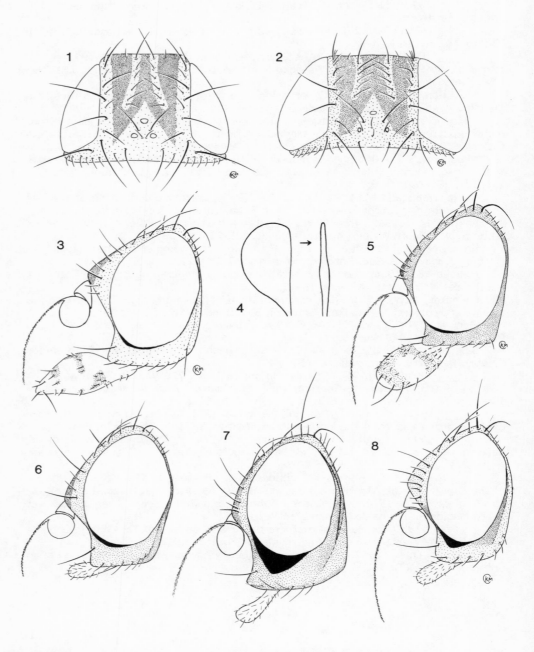

Desmometopa: Dorsal aspect of head, (1) varipalpis, (2) singaporensis; (4) palpus of gressitti; lateral aspect of head, (3) varipalpis male, (5) singaporensis male, (6) varipalpis female, (7) m-nigrum female, (8) sordida.

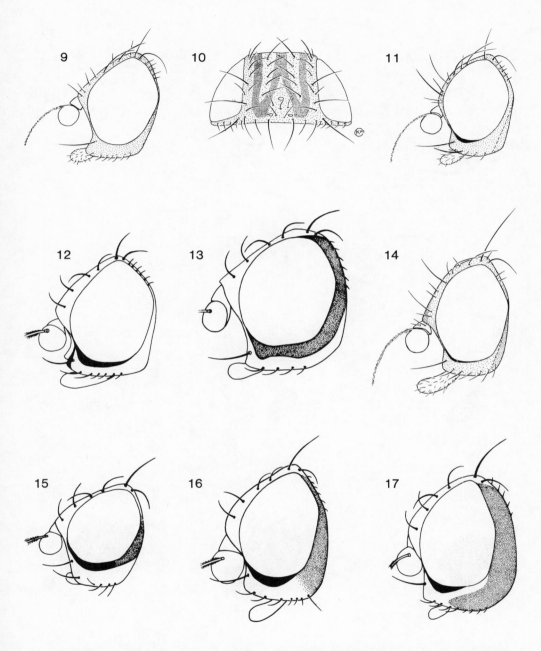

<u>Desmometopa</u>: Dorsal aspect of head, (10) <u>interfrontalis</u>; lateral aspect of head, (9) <u>interfrontalis</u>, (11) <u>ciliata</u>, (12) <u>atypica</u>, (13) <u>postorbitalis</u> male, (14) <u>inaurata</u>, (15) <u>pleuralis</u>, (16) <u>microps</u> female, (17) <u>microps</u> male.

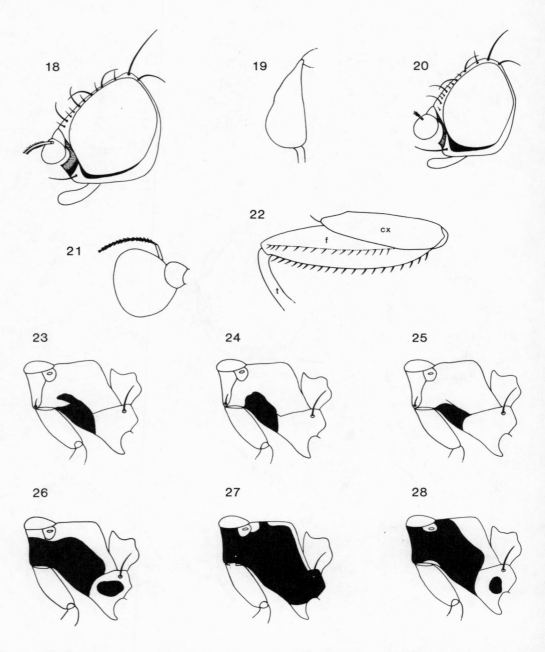

Desmometopa: Lateral aspect of head, (18) tarsalis, (20) lucidifrons; (19) hind tibia, leptometopoides male; (21) antenna, magnicornis male; (22) fore leg, saguaro; left thoracic pleuron, semi-diagrammatic, (23) m-nigrum, (24) ciliata, (25) varipalpis, (26) tarsalis, (27) glaucanota, (28) latigena.

INDEX
(Invalid names underlined)